ADVANCES IN CIVIL ENGINEERING AND BUILDING MATERIALS IV

SELECTED AND PEER REVIEWED PAPERS FROM THE 2014 4TH INTERNATIONAL CONFERENCE ON CIVIL ENGINEERING AND BUILDING MATERIALS (CEBM 2014), 15–16 NOVEMBER 2014, HONG KONG

Advances in Civil Engineering and Building Materials IV

Editors

Shuenn-Yih Chang
Department of Civil Engineering, National Taipei University of Technology, Taiwan

Suad Khalid Al Bahar
Kuwait Institute Scientific Research, Safat, Kuwait

Adel Abdulmajeed M. Husain
Kuwait Institute Scientific Research, Energy and Building Research Center (EBRC), Safat, Kuwait

Jingying Zhao
International Science and Engineering Research Center, Wanchai, Hong Kong

CRC Press
Taylor & Francis Group
Boca Raton London New York Leiden

CRC Press is an imprint of the
Taylor & Francis Group, an **informa** business

A BALKEMA BOOK

CRC Press/Balkema is an imprint of the Taylor & Francis Group, an informa business

© 2015 Taylor & Francis Group, London, UK

Typeset by MPS Limited, Chennai, India
Printed and Bound by CTPS Digiprints, Hong Kong

Published by: CRC Press/Balkema
          P.O. Box 11320, 2301 EH Leiden, The Netherlands
          e-mail: Pub.NL@taylorandfrancis.com
          www.crcpress.com – www.taylorandfrancis.com

ISBN: 978-1-138-00088-9 (Hardback)
ISBN: 978-1-315-69049-0 (Ebook PDF)

Table of contents

Building materials preparation

Coastal engineering

Computer simulation and CAD/CAE

Construction technology

Detection of building materials

Engineering management

Geotechnical engineering

Hydraulic engineering

Others

Road engineering

Structural engineering

Transportation engineering

Urban planning

Advances in Civil Engineering and Building Materials IV – Chang et al (eds)
© 2015 Taylor & Francis Group, London, ISBN: 978-1-138-00088-9

Preface

Following the great progress made in civil engineering and building materials, the 2014 4th International Conference on Civil Engineering and Building Materials (CEBM 2014) aimed at providing a forum for presentation and discussion of state-of-the-art development in Structural Engineering, Road & Bridge Engineering, Geotechnical Engineering, Architecture & Urban Planning, Transportation Engineering, Hydraulic Engineering, Engineering Management, Computational Mechanics, Construction Technology, Building Materials, Environmental Engineering, Computer Simulation & CAD/CAE. Emphasis was given to basic methodologies, scientific development and engineering applications.

This conference is co-sponsored by Asia Civil Engineering Association, the International Association for Scientific and High Technology and International Science and Engineering Research Center. The purpose of CEBM 2014 is to bring together researchers and practitioners from academia, industry, and government to exchange their research ideas and results in the areas of the conference. In addition, the participants of the conference will have a chance to hear from renowned keynote speakers Prof. XIAO-YAN LI from University of Hong Kong.

We would like to thank all the participants and the authors for their contributions. We would also like to gratefully acknowledge the production supervisor Janjaap Blom, Léon Bijnsdorp, Lukas Goosen, who enthusiastically support the conference. In particular, we appreciate the full heart support of all the reviewers and staff members of the conference. We hope that CEBM 2014 will be successful and enjoyable to all participants and look forward to seeing all of you next year at the CEBM 2015.

November, 2014

Prof. Shuenn-Yih Chang
Dr. Suad Khalid Al Bahar
Dr. Adel Abdulmajeed M. Husain
Dr. Jingying Zhao

Advances in Civil Engineering and Building Materials IV – Chang et al (eds)
© 2015 Taylor & Francis Group, London, ISBN: 978-1-138-00088-9

Sponsors and Committees

GENERAL CHAIR

David Packer, *International Science and Engineering Research Center, Hong Kong*

SCIENTIFIC COMMITTEE CHAIR

Konstantinos Giannakos, *Fellow of ASCE, University of Thessaly, Greece*
Suad Al-Bahar, *Kuwait Institute for Scientific Research, Kuwait*

INTERNATIONAL COMMITTEE & SCIENTIFIC COMMITTEE

Ali Rahman, *Jundi-Shapur University Technology, Iran*
Anuchit Uchaipichat, *Vongchavalitkul University, Thailand*
Ata El-kareim Shoeib Soliman, *Faculty of Engineering El-Mataria, Egypt*
Badreddine Sbartai, *Université de Skikda, Algeria*
David S. Hurwitz, *Oregon State University, USA*
Du Jia-Chong, *Tungnan University, Taiwan*
Fu-Jen Wang, *National Chin-Yi University of Technology, Taiwan*
Hosein Rahnema, *Yasuj University, Iran*
Jian Yang, *University of Birmingham, UK*
Jianping Han, *Lanzhou University of Technology, China*
Jianqing Bu, *Shijiazhuang Tiedao University, China*
Jia-Ruey Chang, *MingHsin University of Science & Technology, Taiwan*
Konstantinos Giannakos, *University of Thessaly, Greece*
Lan Wang, *Inner Mongolia University of Technology, China*
Malagavelli Venu, *BITS, Pilani, India*
Manjeet Singh Hora, *Director, MANIT Bhopal, India*
Mehmet Serkan Kirgiz, *Hacettepe University, Turkey*
Moghadas Nejad, *Amirkabir University of Technology, Iran*
Mohammadreza Vafaei, *Imenrah Consulting Engineers Co., Iran*
Mohammadreza Yadollahi, *Universiti Teknologi Malaysia, Malaysia*
Mubiao Su, *Shijiazhuang Tiedao University, China*
Niyazi Ugur Kockal, *Akdeniz University, Turkey*
Ruey Syan Shih, *Tumgnan University, Taiwan*
Shengcai Li, *Huaqiao University, China*
Sina Kazemian, *SCG Consultant Company, Selangor, Malaysia*
TaeSoo Kim, *Hanbat National University, Korea*
Zakiah Ahmad, *Coordinator of Quality Unit, Malaysia*
Zhixin Yan, *Lanzhou University, China*

TECHNICAL COMMITTEE CHAIR

Chang, Shuenn-yih, *National Taipei University of Technology, Taiwan*
Suad Al-Bahar, *Kuwait Institute for Scientific Research, Kuwait*

TECHNICAL COMMITTEE

Deng-Hu Jing, *Southeast University, China*
Dong Yongxiang, *Beijing Institute of Technology, China*

Jiachun Wang, *Xia men University of Technology, China*
Jianrong Yang, *Kunming University of Science and Technology, China*
Jingying Zhao, *International Association for Scientific and High Technology, Hong Kong*
Kuixing Liu, *Tianjin University, China*
Li Liping, *Shandong University, China*
Liangbin Tan, *Kunming University of Science and Technology, China*
Lu Qun, *Tianjin Institute of Urban Construction, China*
Ming Zhang, *Henan Institute of Engineering, China*
Patrick Tiong Liq Yee, *Universiti Teknologi Malaysia, Malaysia*
Tianbo Peng, *Tongji University, China*
Xujie Lu, *Jianghan University, China*

PUBLICATION CHAIRS

Jingying Zhao, *International Association for Scientific and High Technology*

CO-SPONSORED BY

Asia Civil Engineering Association
International Association for Scientific and High Technology
International Science and Engineering Research Center

Architecture

Advances in Civil Engineering and Building Materials IV – Chang et al (eds)
© 2015 Taylor & Francis Group, London, ISBN: 978-1-138-00088-9

Intervention construction techniques in monumental rammed earth architecture in Spain through ministry archives (1980–2013)

L. García-Soriano, C. Mileto & F. Vegas López-Manzanares
Instituto Universitario de Restauración del Patrimonio. Universitat Politècnica de València, Spain

ABSTRACT: This research aims to analyse the intervention techniques proposed in monumental rammed earth architecture restorations in Spain through state-funded projects over the last thirty years. This work was based on files from the Spanish Ministry of Culture and the Ministry of Development, which have been in charge of managing budgets for heritage interventions. Therefore this research aims to carry out a comprehensive analysis of the techniques proposed in the restoration of monumental rammed earth buildings over the last thirty years (1980–present), as well as to extract initial conclusions on how they have fared over time.

Keywords: rammed earth architecture, intervention, ministry archives in Spain

1 INTRODUCTION

This article should be viewed as part of a broader investigation which aims to analyse and study the restoration work carried out on rammed earth architecture in Spain from 1980 until the present, taking into account both the intervention criteria proposed and the construction techniques used in the interventions.

The ministerial structure in charge of the interventions on Spanish heritage was a basic source of information from which numerous case studies could be selected for this research on the restoration of rammed earth architecture in Spain.

It should be noted that since this study focuses on the analysis of interventions with public funding, particularly from the state, the buildings intervened on are monuments, and rammed earth constructions of vernacular architecture are excluded from the analysis.

The analysis of the work presented here focuses basically on the interventions in the construction techniques proposed for the actions.

1.1 *Aim of the study*

The main aim of this article is a global study which systematically and homogeneously covers the study and analysis of intervention techniques used on monumental rammed earth architecture under centralised supervision from the state through the various ministerial structures.

The main goal of this study is to analyze the similarities and contrasts in interventions which have a same building technique.

2 RESEARCH METHODOLOGY AND CASE STUDY SELECTION

As regards state actions relating to interventions on the architectural heritage, a series of case studies were selected from the state archives on heritage interventions. We therefore worked with material from the archives of the Spanish Cultural Heritage Institute, IPCE (Ministry of Culture) and the archives of the Ministry of Development. The selection from these archives of interventions carried out on buildings constructed using the rammed earth technique has provided a series of case studies that are fairly homogeneously distributed throughout Spain if we take into account the parts of the country in which rammed earth architecture is more common (fig. 1).

Figure 1.   Geographical distribution of case studies.

2.1 The IPCE archives

The task of compiling the information from the archives was carried out based on a database list from the General Archives of the IPCE. The list for intervention work carried out in the period 1980–2011, 2,779 files in total, was obtained from this database. A case-by-case search was then carried out in order to select only the buildings intervened which had originally been built using the rammed earth construction technique. This resulted in a shorter list made up of 102 intervention files relating to 78 buildings, which were included in the database.

It should be noted that a global analysis of the evolution over time of the files in these archives shows that of all the files selected, approximately 89% are from the 1980s, 6 are from the 1990s (6%), while 5 files are recorded in the period 2000–2011 (5% of the total sample) (García et al. 2014).

This variable distribution over time clearly shows that most of the files from these archives date back to the 1980s. A more detailed analysis of the distribution over time for this decade shows that it appears in decreasing order, with most of the files dating back to the first half of the 1980s, since with the creation of the Spanish Historical Heritage Law in 1985 powers and competences were gradually ceded to regional administrations and a decentralisation process began.

2.2 The cultural 1% programme through the archive of the Ministry of Development

The cultural 1% programme for conserving and enriching historical heritage began in 1985.

During the first decade of its existence only 200 actions were carried out and at the time it was considered essential to establish criteria for action and priorities in the interventions. Most of the funds for this programme, developed jointly with the Ministry of Culture, are provided by the Ministry of Development (Sánchez Llorente 2010) and an inter-ministry agreement was reached to establish priorities with the creation of a joint committee between the Ministry of Culture and the Ministry of Development. This cooperation between the two ministries began on the 3rd November 1994 and is still in place. In order to carry out an analysis of the interventions on rammed earth architecture funded by this programme we worked with information from the Archive of the Ministry of Development. A total of 77 intervention actions on rammed earth buildings were selected from the data base. This was based on the list of files generated within the cultural 1% programme in the Ministry of Development (where intervention projects from the 2004 joint committee to the present are found), consisting of 627 intervention files in total (García et al. 2015).

Therefore, by researching both archives a database was drawn up with a compilation of 179 actions on rammed earth buildings, carried out throughout Spain over the last three decades (fig. 2).

Figure 2. Images of some of the interventions analysed (Author: L. García-Soriano).

A detailed fiche was created for the analysis and assessment of these case studies for each intervention and generated a database that allows us to carry out the most objective analysis possible of the techniques used in each case.

3 ANALYSIS OF THE CONSTRUCTION TECHNIQUES USED

It can be stated that the intervention construction techniques used in the different cases are closely linked to at least two factors: the original construction technique (generally associated with the type of building) and the author's own intervention criteria.

The interventions proposed in civil and religious architecture are often indirect, that is to say, the intervention is carried out on other elements of the building, such as floors, ceilings, roofs, stairs, etc. These interventions have an indirect effect on the walls while direct actions on rammed earth walls are generally for cleaning or treating the surfaces or structural actions (stitching cracks and fissures). In contrast, in military and defensive architecture, walls are generally the main protagonists and interventions on them are carried out directly.

In the case of this research, given that most of the buildings which form the study sample are military (castles, defensive walls, citadels, towers....) the analysis of the construction techniques used in the interventions should be viewed as relating mostly to this typology.

3.1 Intervention projects in the 1980s

This group is made up of the oldest files from the sample. It should be noted that in these projects documentation us scarce and that in some cases the construction details are not completely defined in the

Figure 3. Current image of the Castle of Tabernas, Almería (Author: L. García-Soriano).

Figure 4. Current image of the Castle of Nogalte (Puerto Lumbreras, Murcia) (Author: L. García-Soriano).

project as the intention was to define them during the work. Nevertheless, if we analyse all the interventions from the 1980s, one fairly common criterion is the search for aesthetic harmony between new and old, trying to respect the principle of distinguishability. In some cases, such as the intervention by Alfredo Vera Botí of the Walls of Aledo (Murcia, 1980) the text specifies that "all the finishes will be as similar as possible to the original, but showing the differentiation line in them so that there is no confusion between the original construction and the new construction and so that the work introduced now may be reversible and identifiable".

In most cases, this attempt to establish a relation with the original building promoted the use of the original construction technique, rammed earth, although there is also a small proportion of projects in which other techniques were used, for example consolidation with masonry or brick, more frequent in interventions in previous decades. Nevertheless, at this point there had already been some major interventions carried out on rammed earth buildings using the original technique, as in the case of the initial restoration phases by Ismael Guarner of the Walls of Niebla, which became a point of reference for numerous other projects. Although the original technique was used, cement was added as a bonding agent in most projects, that is to say, although the technique used was original the materials were not, and at times new elements were even introduced to bind the original walls to the new ones (metal structures and meshes . . .). When the walls were very deteriorated and had lost part of their sections, interventions tended to propose attaching a new rammed earth wall to the original wall using auxiliary elements made mostly out of smooth steel. An example of this is the intervention by Roberto Puig Álvarez on the Castle of Tabernas (Almería, 1983) which stated "as the walls of this castle are built in rammed earth, we will use the same construction system replacing the rammed earth mix with cement-lime mortar with arid soil and tinting the mortar to match the tones of the original wall, adding formwork and putlog holes with planks whose size was deduced from the remains of the wall" (fig. 3).

3.2 Intervention projects in the last decade

The main objective of the actions from the more recent projects, carried out in the first decade of the 21st century, continues to be the general search to establish a harmonious relationship between the new intervention and the existing building, both with chromatic and constructive integration. This is why the original construction technique, rammed earth, is used in the interventions for these cases. As regards materials in these projects, the use of original material is also more frequent, as is the case of the 2004 intervention on the Castle of Nogalte (Puerto Lumbreras, Murcia) which states that "once all the existing structures are consolidated, general filling will be carried out on them. To do so the traditional construction system will use rammed earth with lime mortar for the oldest part and rammed earth reinforced with layers of lime mortar in the extension, just as in the original construction" (file 13-30033-00562-04) (fig. 4).

Although this desire to use the same original materials exists in some cases, albeit not as many, new materials like cement are incorporated, perhaps because authors are more familiar with them. However, these materials are completely foreign to the original constructions and could cause future pathologies, as can be observed from older interventions. Among examples we could mention the 2010 intervention carried out on the Castle of San Juan in Calasparra, where it is stated that "the filling of the rammed earth construction will be carried out following the traditional method, similar to those already executed: paste prepared using natural soil, stabilised with slaked lime and a small amount of white cement and colourant if necessary" (file 13-30013-01908-09).

Just as above, the union between the original and the new material is one of the main problems to be resolved. In these more modern projects, the most frequent proposals are those which suggest improved bonding of the new materials with the existing ones increasing the gripping surface with elements acting as rods or connectors. There are two types of proposals: those which aim for all the elements to be executed with the original materials and those which introduce contemporary materials. The first group is composed of interventions proposing the use of wooden stakes for this union, as wood was the material used originally in the execution of the wall (stays) and there is reasonable certainty that this material is compatible with the earth of the wall itself. Nevertheless, most of the cases analysed fall into the second category and are cases in which the union is planned using current materials, primarily fibreglass.

4 CONCLUSIONS

From this global analysis it is possible to formulate some general conclusions relating to the interventions carried out on monumental rammed earth architecture over the last thirty years.

In the 1980s, despite proposals using different techniques (masonry, brick ...) which had inherited criteria and proposals from previous decades, the original rammed earth construction technique continued to be the most popular option and was therefore the most widely used in intervention projects. As mentioned above, despite restoration architects' interest and desire to use the original construction technique, which may have still been relatively unknown or unexplored at the time, the decision was taken to incorporate modern materials such as cement into the mix, in order to improve the quality of the walls proposed in the intervention. New elements, usually in smooth steel, were also proposed to improve the union between the original and added material.

Time has revealed the successes and failures of these interventions and this has influenced the more modern proposals for intervention, which seek to improve the technical solutions and respect the original elements. This is why in these interventions the current construction technique is still valued as the most appropriate option for these interventions, but the current aim is to also use the original materials, without additions. This change is undoubtedly the result of the increasing study and knowledge of rammed earth technique over these last few decades, which has enabled professionals to use the technique more rigorously, having seen that the use of cement in the mix often caused pathologies in the wall, such as the appearance of salts and efflorescence which in an advanced state may even lead to major material losses. In addition, the materials used in the bonding elements also varied in relation to the initial proposals from the 1980s, as it has been observed that the behaviour of steel in these structures was not optimum, and in many cases material

incompatibilities have caused detachments, uncovering the metal rods, exposed to the elements so that they continue to deteriorate, in turn, damaging the wall.

With all this in mind, it can be stated that past experiences contributed to reinforcing some criteria in the proposals, while rejecting or modifying others. It is important to state that this text does not aim to be a criticism, but rather aims to show that each moment has its positive contributions, both in the more successful proposals and the more unfortunate ones, since these all contribute to the improvement of future interventions. As regards the more modern interventions it is not yet possible to assess the extent of their accuracy, since it is still too soon to know how these materials are going withstand the passing of time. This is why this work should be understood as a small step in learning about the interventions in monumental rammed earth architecture.

NOTE

This study is part of a research project funded by the Spanish Ministry of Science and Innovation "La restauración de la arquitectura de tapia en la Península Ibérica. Criterios, técnicas, resultados y perspectivas" (Ref.: BIA 2010–18921; main researcher: Camilla Mileto) and is part of the research work of "La restauración de la arquitectura de tapia de 1980 a la actualidad a través de los fondos del ministerio de Cultura y del Ministerio de Fomento del Gobierno de España" by PhD student L. García-Soriano.

REFERENCES

Dossiers of the Ministry of Culture, IPCE Archive (1980–2011).
Dossiers of the Ministry of Development Archive (2004–2013 1% Cultural Program).
García Soriano, L., Mileto, C. & Vegas, F., 2014. "La restauración de la tapia en España a través de las financiaciones ministeriales" in La Restauración de la Tapia en la Península Ibérica. Criterios, Técnicas, Resultados y Perspectivas. Argumentum/TC Cuadernos.
García Soriano, L., Mileto, C., Vegas, F. & La Spina, V., 2015. "Restoration of monumental rammed earth buildings in Spain in the last decade according to the 1% cultural program" in Earthen architecture. Past, present and future. Balkema, Taylor & Francis, London.
Sánchez Llorente, A., 2010. El 1% Cultural. Una visión práctica. Revista de Patrimonio Cultural de España 3 (La economía del patrimonio cultural), pp. 129–142.

New methods and development trend of modern Lingnan architectural creation

Changyong Chen

School of Architecture/State Key Laboratory of Subtropical Building Science, South China University of Technology, Guangzhou, Guangdong, China

ABSTRACT: The paper summarizes innovative methods adopted in Lingnan architecture which has experienced shifts from simplicity to exquisiteness, from formal implicitness to space construction, from simple technology to green integration and also from aesthetic monotonousness to harmonious diversity; besides, the paper points out some restrictions to Lingnan architectural thoughts, including overstress on practices and neglect of theories, loss of cultural characters, and disconnection of talent inheritance; finally some solutions are brought up for the future development of Lingnan architectural creation, including predictive protection for Lingnan architectural works and reestablishment of academic traditions and a scientific platform.

Keywords: Lingnan, regional architecture, new methods, restriction, subtropical climate

1 INTRODUCTION

With practical and innovative characters, modern Lingnan architects brought up the architectural concept of integrating modern architectures with regional cultures quite early. Focusing on Lingnan culture, they established the unique Lingnan architectural school and have made enormous achievements. They combine regional climates, cultures, and techniques, develop green techniques for ventilation, thermal insulation and moisture protection, and advocate design concepts like "integrating western and Chinese cultures" as well as "three views and two characteristics". A large number of boutique buildings have been created since the development of Lingnan architecture concept in the 1950s. Representative works include White Swan Hotel in Guangzhou, which attracted wide attention in China when it was built in the early years, some national gold-medal-winning works like the Museum of the Tomb of the Nanyue King, as well as recent influential buildings like the extension project for Nanjing Massacre Memorial Hall, the Chinese Pavilion at the World Expo in Shanghai, etc.[1,2] But what changes will modern Lingnan architecture go through? What's the design trend? The answers to such questions are given in the following, supported by summary and analysis of some creative works.

2 THE SHIFT FROM SIMPLICITY TO EXQUISITENESS

Lingnan architectures built in the early years of the new China were quite simplified due to inadequate materials then. Materials employed at that time are often appreciated more about basic performance than decorative. Decoration is usually about utilization of the structure or the building itself, like pursuing a rich façade effect with horizontal or vertical sun louvers. And a rich and varied space system is usually created by combining blocks in different ways. Various façade effects are achieved by arranging the shutters of the air-conditioner in an intersectional manner. To sum up, architectures at that time display only simplicity due to their loyalty to space and construction. Representative works include Canton Fair Building, SCUT teaching buildings, SUMS teaching buildings, etc. Such buildings were designed in concise combined architectural forms, with their facades formed through combined sun louvers in horizontal and vertical manner to achieve rich shadow changes.

However, advanced technologies, which have experienced rapid development, have been widely used in common buildings in recent years, largely improving their construction precision and transforming them from simple to exquisite. Nowadays the exquisiteness of buildings is achieved by novel technologies and materials which bring fresh aesthetic experiences. For instance, the architectures in recent years are made more artistic and interesting owing to the wide application of green technologies, building skins, parameterized control method, and new materials like titanium, pottery clay and copper plate. Besides, the architects keep digging unknown artistic characteristics of traditional materials to pursue "poetic architectures"[2]. They make innovations in the laying of common bricks by making various combinations to achieve bump effect, leaving regular gaps in brick wall to allow

Picture 1. Culture and sports center of pingyuan county, Guangdong Province.

Picture 2. One view of "Jin Shi Lan Collection" museum.

sunshine into the room and form fantastic light & shadow effect, and combining various construction materials to generate new aesthetical experiences.

The façades of the Culture and Sports Center of Pingyuan County, Guangdong Province, designed by us, define the simplicity of Lingnan architectures to the best. The Center measures 20 m in height and 200 m in length, so it looks quite slim. The façades are subject to a three-segment design. To be specific, the base of this building is an evacuation platform, the middle part is structured with walls and glass windows, and the roof is framed with cambered floating slabs. It looks quite stretchable generally. According to the design of the north and south facades, dry-hang stones are used to make the facades look thick. In terms of its detailing, we employ zigzag bay-windows to protect the Center from direct sunshine. On one side are walls made of dry-hang stones and on the other side are vertical glass windows. We attach great emphasis on detailing and conduct some special treatment for the structural design by using steel structures for large-span space and concrete structure for the rest. Then we employ aluminum alloy plates for decorative finish. The unified finish materials make the building look quite streamlining and dynamic. To sum up, the façade design explains the simplicity of Lingnan architecture in the best way. Refer to Picture 1.

"Jin Shi Lan Collection" Museum is our attempt to make the architecture exquisite with materials. The project is a private museum with simple building functions. It is mainly used to display collected stone carvings. We make the facades exquisite by combining plain concrete with timberwork. The contrast between the kinds of materials highlights characteristics of both. The only decoration of the concrete outer wall is the formwork patterns totally displaying the rough nature of concrete. On the other hand, timbers with warm patterns are used for floor slabs and doors, giving a touch of exquisiteness compared with the rough concrete. The rails are built with section steel which is hard and small in size. And the contrast between the steel and the concrete makes the former more exquisite and the latter rougher. We have also made an attempt in another building by applying old bricks, glass, timber, and section steel to create the effect of exquisiteness. Refer to Picture 2 and Picture 3.[3]

A more detailed design division in domestic architectures is witnessed by the architectural shift from

Picture 3. Another view of "Jin Shi Lan Collection" museum.

simplicity to exquisiteness, which not only marks the change of design concepts, but also shows the constant improvement of construction technologies in Lingnan region. With the development of manufacturing industry in China, the design and construction of buildings are becoming more and more accurate, making Lingnan architectures increasingly exquisite and elegant in terms of cities and buildings.

3 THE SHIFT FROM FORMAL IMPLICITNESS TO SPACE CONSTRUCTION

How to inherit traditional architectural cultural heritages is an issue faced by Lingnan architecture. Architects tend to resonate with architectural elements and symbols only in history. But it is possible to apply traditional elements in modern architectures by different means.[2] For example, pan walls, black bricks and pitched roofs can be applied for new buildings directly or after some industrial transformation by means of designing them into joist steel or making use of skin patterns. Modern architectures are connected to the past by these symbols, achieving its formal implicitness. Influential architectures of this type include the former residence of Mr. Liang Qichao in Jiangmen, the Museum of the Tomb of Nanyue King, White Swan Hotel in Guangzhou, etc.

The formal implicitness has been applied to the design of hotels in Lotus Mountain Villa in Liuzhou. The project, located beside a river, is a five-star hotel

Picture 4. One view of the hotels in Lotus Mountain Villa in Liuzhou.

Picture 5. One view of Nansha Western Cultural centers in Guangzhou.

Picture 6. Another view of Nansha Western Cultural Center in Guangzhou.

for foreign guest reception. It covers 900 mu with about 60,000 m² of floor area and is surrounded by mountains. Arranged in the architecture are the main building of the hotel, annex for entertainment and single inns. The design of this project focuses on the water culture of Lingnan region. Water is introduced into the project area through low-lying areas beside the river so that water becomes a theme of the hotel, which echoes with the fame of Lingnan as the Region of Lakes and Rivers. Built against waters, the hotel enjoys a quite open layout. Surrounded by subtropical plants all around, it almost merges with these plants. The facades of the architecture are of some characteristics shared by local residential houses with the combined flexible hipped roof and double-pitch roof as well as some decorative materials made of timber and local stones. On one side of the river, an open heritage park has been built, around which construction is limited for conservation of this area. The heritage park is also a symbol of the local culture. The timbers applied to this building are of warm color and the roof tiling of gray, consistent with local residential houses in the choice of colors. Abovementioned symbols have adequately reflected characteristics of local culture in Liuzhou by means of implicitness and quotation. Refer to Picture 4.

However, we have realized that overstress on symbols might lead to some negative results. For example, the application of some architectural symbols which are too far away from modern people's life may only lead to the disconnection between the content and the style of the architecture. Moreover, some traditional elements are so incompatible with modern architectural space as to apply them to a modern architecture may make it look like a "fake antique". Therefore, we made new attempts to improve those designs which overstress traditional symbols and express the culture through creation of space atmosphere.

We applied the space creation method to the design of Nansha Western Cultural Center in Guangzhou. The base of the project is located beside a lake in Nansha, Guangzhou with limited scale. Instead of expressing the building attribute in manner of symbolic implicitness, we apply the space creation method to embody the mysterious atmosphere and inherent characters of the church there. Embraced on three sides by water, the building enjoys a quite pleasant view. Inspired by the relationship between space and water, we employ the transparent glass for the architecture and make glass the theme. We use cambered glass curtain walls for the building on the three sides embraced by water, so that the interior space and the exterior waters become integrated and the water and the sky merge in one color. People making their prayers here will have their heart baptized by nature and obtain the peace that is unavailable in cities. The glass curtain walls of the architecture are designed into curved surfaces, which means they become smaller upward. This kind of design creates an upward space trend and a mysterious space atmosphere. A plank road sets off from the platform in the north, which is surrounded on three sides by water. At the end of the road is a simple line-styled statue. The statue, its reflection in water and the platform together form a complete "Cross", which echoes the space characteristic of the church artistically. Symbols have been totally abandoned in this project which concentrates on the significance of space. And we have made a breakthrough in the creation of the atmosphere and have expressed the church culture in a great manner. Refer to Picture 5 and Picture 6.[4]

The essence of architecture is always space. We should not limit our architectural design to collage of symbols. The right thing for us to do is to dig more of traditional Chinese philosophy and space concepts to refine characteristics of traditional space and apply

Picture 8. Another View of Tea Market Planning Project in Fangcun, Guangzhou.

Picture 7. One View of Tea Market Planning Project in Fangcun, Guangzhou.

them to modern architectural design in purpose of expressing our traditional culture appropriately. The shift from formal explicitness to space creation comes from our constantly deepening understanding of the regional culture and represents our new attempt to respond to traditional culture in a more mature way.

4 THE SHIFT FROM SIMPLE TECHNOLOGY TO GREENING INTEGRATION

Greening of buildings in Lingnan region in the early years was achieved through some simple technologies, like space layout, greening applications, sun louvers, etc. However, the greening technologies for buildings have experienced rapid development along with other building technologies. Accordingly, the greening technologies employed in Lingnan architectures shift from simple and energy-conserving to multidimensional & dynamic integration. And these greening technologies are further developed with the application of relevant software for sunlight, ventilation, and thermal insulation. Green buildings become under the spotlight in Lingnan region after the issuing of national building energy-saving standards and green building evaluation standards.[2] Some breakthroughs have been made in green buildings in Lingnan region due to the application of advanced technologies. For example, turbine generators are arranged for the Pearl River Tower project in Guangzhou by making use of the ventilation effect of the refuge storey. Such generators generate electricity under the wind tunnel effect of the high-rise building. The electricity is used to power the building and the extra power, if any, will be sent to the grid, contributing to energy conservation.

A green and comprehensive design characters the tea market planning project in Fangcun, Guangzhou. Advanced modern architectural technologies are employed to show high-tech green aesthetics. Comprehensive green technologies are applied to the project in all aspects. For plant greening, subtropical plants are arranged all over underground plazas, ground

platforms, roof gardens and aerial platforms in the building, making the building merge with the building space and adjusting the micro-climate of the high-rise building. Besides, a rainwater collection system has been set up to increase the usage rate of recycled water. Moreover, green skins and breathing double glazing are applied to curtain walls of the building. In terms of the detailing, construction of skins for high-rise buildings has been repeatedly deliberated to achieve exquisiteness and green aesthetics. In the next design step, we will conduct comprehensive analysis of ventilation, shading, and energy conservation as well as quantitative analysis of green technologies to increase the green efficiency of the whole building. Refer to Picture 7 and Picture 8.

In a word, a systematic analysis of green technologies is necessary for the construction of Lingnan architectures adaptive to the warm wet climate here. We should integrate various technologies through multidimensional and interdisciplinary researches, and put forward technical measures for subtropical green building design after collaborative researches into architectural designs, technologies and energy conservation. These researches will lead to significant development of Lingnan architectures as well as applicable advanced green technologies in this region. To the very end, Lingnan architecture will pursue its development on a really scientific and rational path.

5 THE SHIFT FROM AESTHETIC MONOTONOUSNESS TO HARMONIOUS DIVERSITY

As part of the Chinese civilization, the local architectural aesthetics in Lingnan region is also experiencing great changes, following the pace of the Chinese culture. The fate of Lingnan architectures is closely related to that of our nation. For example, Sun Yat-Sen Memorial Hall built before the foundation of China reflects the "westernized Chinese style" and shows the memorial function of traditional Chinese buildings with the long-span modern technologies. In the early days after foundation of China, the application of national symbols, i.e., buildings designed with large roofs, was quite popular. Facade perpendicular lines were designed in some buildings with a Soviet Union style. And there were still other buildings showing a lively modern architectural style.[2] However, Lingnan architectures began a diversified development after the reform and opening-up when various ideological trends were introduced to this region. In this period, Lingnan architectures were influenced by various styles, modernism, post-modernism, traditional Chinese and western classical continental style, etc. That hodgepodge style defines the consumerism of Lingnan region at such time, reflects its social commerciality and secularity, and embodies its pop culture.

In the culture dominated by consumption, the European continental style prevailed and became the most popular aesthetic idea. The problem is that it is also an architectural style that could not be defined accurately. To be general, the European continental style refers to a classical architectural style across the continent of Europe, in which classic European architectural symbols are widely used. Driven by the development of real estate, this style has been frequently seen in city architectures across Lingnan region. Star River, a residential quarter in Guangzhou, shows the architectural style of "South California" with the red roof and the imitative detailing of western architectures catering to the taste of city upstarts. The Agricultural Bank Tower in the New City of Pearl River, Guangzhou, with its classical western architectural style, the dome, the arch, and various sculptures and decorations absolutely the same as those of European architectures, is totally a reproduction of a classical European architecture and absolutely incompatible with other modern buildings in the city. Of course, there are also some buildings allegedly focusing on traditional Chinese culture, such as the residential quarter named "Yunshan Poetry" in Guangzhou with its Matou walls and black bricks often used in Hui-style architectures. But the traditional architectural symbols are just used as publicity stunts for sales of such architectures, which is quite popular in the commercial society. The design of such architectures with different backgrounds, deeply influenced by fast food cultures, often shows preference for the technique of collage and some philistinism.

From a deeper level, the architectural aesthetics are also influenced by social culture. In those early years, architecture was deeply influenced by political factors. To be specific, the monotonous social value led to monotonous architectural thought. So buildings at that time were only built to fulfill people's basic needs for residence. And the architectural designs showed nothing but monotonousness with no care for the spiritual world of human beings and led only to aesthetical fatigue. Since the reform and opening-up, China has experienced a diversified social development and Chinese cities began to show diversities. That was when the architectural aesthetics began to free itself from monotonousness. However, some architects borrow

from foreign designs without any change or lay too much emphasis on consumption culture, making their designs all commercial, lurid, and absolutely incompatible with the city. And this costs us more social value of the architectural design.[2] This is something we should reflect upon. We are obligated to lead the architectural thought in Lingnan region to a better path, encourage architectural concepts deeply rooted in local community, and finally accomplish a diversified architectural culture based on the local culture.

6 RESTRICTIONS TO LINGNAN ARCHITECTURAL THOUGHTS

Although Lingnan architectural thoughts possess some characters, contribute to series of excellent works, and exert certain influence on domestic architectures, we should never overlook the restrictions to these thoughts behind all the successes. One problem about Lingnan architectural school is its overstress on practices and neglect of theories. Despite the active practices of Lingnan architecture and its deep influence on domestic architecture, it lags behind in the academic respect, so obviously that such lagging-behind can no longer be covered by its superficial prosperity. The sustainable development and prosperity of architectural creation will not be possible without deep reflection upon theories and constant interrogation into fundamental architectural problems. What are the fundamental academic issues and spiritual cores of Lingnan architecture? What are the academic stands? How can we build an academic system without loss in the characteristics of Lingnan? These are key issues for Lingnan architecture scholars to reflect upon.

Another restriction to Lingnan architectural school is caused by its loss of cultural characters. The development of internationalization, the invasion of western cultures and the fading of local cultures have all impacted the development of Lingnan architecture and modern Lingnan architects are confused by their multiple cultural identities. Some scholars even "doubt the existence of Lingnan architectures". Indeed, nowadays there are not so many Lingnan architectural designs that really show the local cultural characters. The so-called Lingnan architectures now refer most of the time to those designed by Lingnan architects while the fact is we can barely see architectural designs that are actually adaptive to the subtropical climate here and show the local cultures in Lingnan region. Lingnan architectures are unconsciously put into a dilemma in which they lose their typical characters.

The disconnection of talent inheritance is the third restriction to Lingnan architectural school. In those early years, we have great Lingnan architects like Xia Changshi and Chen Boqi who brought up the concept of Lingnan architecture, Mo Bozhi and She Junnan who promoted its development, and He Jingtang who carried forward its spirit. However, we now barely see talented architects who are capable of carrying Lingnan architecture forward. The new generations of Lingnan architects have been deeply influenced by western architectural trends and have difficulty resonating with traditional Lingnan culture. That's why they barely think about the future of Lingnan architecture, let alone carrying its spirit forward. As a result, Lingnan architecture is facing the lack of talents. To sum up, the disconnection of talented architects and the defection of education in Lingnan region have truly held up the development of Lingnan architecture.

7 FUTURE DEVELOPMENT OF LINGNAN ARCHITECTURAL CREATION

First of all, we should take some prospective measures to protect important historical Lingnan buildings. A lot of international prizes have been awarded to Lingnan architectures since 1950. And some modern Lingnan architectural designs are recorded in many influential academic magazines. However, some great architectural works are torn down and inappropriately transformed in the process of rapid urbanization, their value ruthlessly marred. And this may be explained by the wrong idea of relevant authorities about historical value in modern architectures. They tend to think that only buildings with a long history are worth protection. In other words, they do not see the necessity to protect modern buildings. What does not occur to them is that great Lingnan architectures will become valuable historical works with time going on. But if they are damaged before that, it will be too late for us to try to protect them. And we will follow the same old disastrous road repeatedly. Therefore, the first thing to do is to list a batch of Lingnan buildings as heritages, conserve them, study more, and dig more of their academic values.

Besides, the academic tradition of Lingnan architecture should be reestablished. During the past decades, many valuable design theories have been put forward by Lingnan architects along with a batch of outstanding architectural works. But these achievements are far from being enough to establish of a systematic architectural theory school. There is still a long way for Lingnan architectural school to go to be widely acknowledged by the theoretical circle like "the Chicago School". It need take researches in more fundamental academic issues about Lingnan academic school, the summary of its theoretical cores, and more intensive explorations by architects and scholars into the academic issues. Only in this way an academic center for Lingnan architecture will be built. This requires not only architectural designs but also more academic stands and opinions to be brought up based on intensive theoretical study. Besides, it is urgent to build an educational system with local characteristics of Lingnan region to nurture more talented architects.

It will take constant effort and exploitation of generations so as to lay a solid theoretical foundation for Lingnan architectural school.

Last but not least, a scientific platform should be built for Lingnan architecture. One of the most striking characteristics of Lingnan architecture is that its building space is adaptive to the subtropical climate here, which could only be done with comprehensive technical support. So far, a national key laboratory for subtropical building science has been built in SCUT, serving as a great technical platform for further development of Lingnan architecture. Besides, we should make use of various new technologies for comprehensive analysis and conduct quantitative analysis through technical integration. By doing this we can rectify deficiencies in empirical traditional Lingnan architectural design, provide scientific basis for Lingnan architecture, and lead it to a brighter future. In a word, we should combine Lingnan architecture's requirements for sunshading, ventilation and thermal insulation with selection of building materials, design of building structures, and application of new construction technologies to quicken the development of its greening technologies, promote its scientific progress, and lay a scientific foundation for Lingnan architecture.

ACKNOWLEDGEMENT

This paper is under the Study on the Creative Thought of Contemporary Regional Architecture in Lingnan, as the Youth Program in 2011 of Guangdong Provincial Philosophy & Social Science Research Program during the 12th Five-Year Plan, Program No.: GD11YYS01).

ABOUT THE PICTURES

Picture 2 and Picture 3 are from Bibliography 3, Picture 5 and Picture 6 from Bibliography 4, and the rest from the project department.

REFERENCES

[1] He Jingtang, Architectural Design Ideology of Lingnan – Review and Prospect of 60 years. Architectaral Journal, 2009 (10).
[2] Chen Changyong, Xiao Dawei. Analyzing the Trend of the Regionalism architecture practicing in view of Lingnan, Architectaral Journal, 2010 (02).
[3] Chen Changyong. Basic discussion on tectonics and regional architecture creation. Advanced Materials Research, 2011-03-22.
[4] Xiao Dawei, Chen Changyong. The Faith of Light, the Feelings of Water – An Essay on The Creation of Western Culture Research Center. South Architecture, 2002 (09).

Advances in Civil Engineering and Building Materials IV – Chang et al (eds)
© 2015 Taylor & Francis Group, London, ISBN: 978-1-138-00088-9

New concepts in modern Lingnan architecture creation

Changyong Chen

School of Architecture/State Key Laboratory of Subtropical Building Science, South China University of Technology, Guangzhou, Guangdong, China

ABSTRACT: This paper summarizes novel concepts in modern Lingnan architecture creation, including the rational and practical functional view that expresses the local characters, the simple and suitable ecological view that echoes the regional climate, the universally shared cultural view that defines the local culture, and the regional artistic view that is adaptive to various changes. The characteristics and connotations of these views are analyzed through some actual project cases.

Keywords: Lingnan, regional architecture, new concepts, creation

1 INTRODUCTION

How to carry forward traditional Chinese culture in modern architectures is a problem every Chinese architect is faced with, including those in Lingnan region. As a matter of fact, this problem has been considered by generations of Lingnan architects. They have summarized some design concepts with local characteristics, achieved series of outstanding architectural works by constant practices, and become increasingly influential in the domestic architectural circle.[1] The simple subtropical idea, which is brought up by famous architects during 1950–1970 and the representative Lingnan works and their design concepts during 1970–2000, has been summarized in many literatures about study on modern Lingnan architecture. Besides, there is the design concept of "two views and three characteristics"[2] deeply influencing the creation of modern Lingnan architectural works as well as the domestic academic circle. Buildings built under its influence have been widely seen across China, known as the "South China Phenomenon". All these design ideas, as the essence of Lingnan architectural thoughts, come down in one continuous line and deserve theoretical analysis in systematic way.

Therefore, in the following, these new concepts in creation of Lingnan architecture will be summarized and demonstrated by some of our project designs. In the process of summarizing these concepts, we find that the inheritance of local cultural characteristics and the response to the special regional climate are quite significant in Lingnan architecture. Its design concepts show certain uniqueness and its innovative ideas cover the rational & practical functional view, the simple & suitable ecological view, the universally shared cultural view and the design view adaptive to various changes.

2 THE RATIONAL AND PRACTICAL FUNCTIONAL VIEW THAT EXPRESSES THE LOCAL CHARACTERS

Modern Lingnan architecture has been deeply influenced by modernism and the functional view runs through the whole design process. Early Lingnan architects tended to attach great importance to the functions of architectures. The core of the functional view is to analyze the suitability of building functions, spaces, technologies, and materials and apply the best technologies. To be specific, the first step they usually took is to analyze the functions and streamline of a building, and design a compact space layout to streamline the building and "save every inch of land". Then they employed decorative frames like the sun louvers and the roof ventilation facilities which should fulfill both functional and decorative ends. Lastly they repeatedly considered the detailing for the frames which should be suitable and reasonable for the best cost performance. Consequently, the forms and functions of modern Lingnan buildings based on such functional view are closely linked, creating a new modern architectural style with internal and external unity. Here in the following is a practical case.

The Culture and Sports Center of Pingyuan County, Guangdong Province, is designed according to the local conditions to express the functional reasonability in the best way. It is a public building for sports activities. Located in the center of Pingyuan County, it borders the Pingyuan Avenue in the east, the Pingyuan Middle School in the north, and the newly planned city axis in the south. Behind the center are ranges of mountains on which the citizens are able to get a birds'eye view of the whole center. It covers $20,000\,m^2$, $12,000\,m^2$ of which is taken by the stadium which is able to accommodate 2,000 people.

The youth activity center takes up another 6,000 m^2 and the exhibition hall 2,000 m^2. The design of the project was started from the year of 2007. In 2011, it was completed and put into use. Lying in the core area of the city, it is supposed to meet needs of the citizens as well as the government. According to the demand of the owner, a youth center and a stadium should be built, and the latter should be used for commerce, exhibition and mass meetings, etc. So we applies the "mixing-function" design strategy to meet these diversified requirements and combine buildings with similar functions into a large one. Besides, we employed the variable space design to obscure the demarcation of building functions and make the building more flexible for various ends. For instance, the measures of large-span space were adopted with reconsideration of space so that the space could be used as a competition venue, as well as exhibition hall and shopping mall. Besides, great emphasis has been laid on the space connectivity and mobility in the design to make the division and integration of the architectural space possible. In this manner, the building can be used more flexibly. To sum up, with the large space in the first floor of the building, it is able to be used as: a competition venue with auxiliary space where various facilities are arranged to meet requirements of dressing, judgment, lighting, strong-current and light-current, etc; a mass meeting venue with special platform and sound system to meet requirements of local governmental meetings; an agricultural trade exhibition hall because the large space of the stadium and the exhibition hall on the east are divided with fire resisting shutters and can be integrated for exhibition of large-scale agricultural trade & vehicles during the annual "Orange Festival".

Lingnan architects give top priorities to functions and try the best to meet the owner's requirements by applying various means to analyze the building space combination, functional streamlining and façade design. They make every effort to pursue the functional diversity of buildings on a reasonable basis. Besides, they lay emphasis on cost control and advocate composite use and investment reduction. Moreover, they apply right technologies and make the best use of practical materials and construction methods to create buildings of a unique style. In brief, the scientific rationality is the most outstanding characteristic of Lingnan architecture, which also reflects the character trait of Lingnan people to take the low profile and be practical.

3 THE SIMPLE AND SUITABLE ECOLOGICAL VIEW THAT RESPONDS TO THE REGIONAL CLIMATE

Lingnan region is dominated by the typical subtropical climate with its high temperature, humidity and miasma. So another important concept that characterizes Lingnan architecture is its response to the regional climate. Lingnan architects attach great importance to sunshading, thermal insulation and ventilation. Buildings in Lingnan region of the early years were largely

limited by backward technical and material conditions and only inferior technologies were adopted to respond to the climate. Such technologies have been constantly summarized by local folks and finally a series of architectural design methods adaptive to the subtropical climate in Lingnan region are concluded, including: cooling group buildings by strengthening ventilation with cold lanes; adopting pile dwelling or void ground floor to meet the ventilation and rain-proof requirements of individual buildings; applying sun louvers to reduce heat; growing various plants to regulate the climate; arranging cross ventilation to drive away extra heat; applying light colors to buildings to reduce heat absorption, etc. These design methods are not achieved with advanced technologies but simple crafts which define the unique and diversified style of Lingnan architectures with subtropical characteristics. To sum up, another significant idea in the creation of Lingnan architecture is to take full consideration of the subtropical local climate and apply simple but practical ecological design methods. Some design cases are given below to give a summary of ventilation, sunshading, waterproofing, and thermal insulation methods.

The void ground floor is widely used in Lingnan buildings for ventilation as well as cost reduction. Due to the subtropical climate here, residences on the ground floor are often subject to humidity, poor ventilation and serious mosquito issue, thus quite unpopular in the market. Since the 1980s when such a problem was noticed, as a coping approach, the void ground floor has been adopted in some residences to avoid the ventilation barrier and improve the ventilation effect of the whole community. Besides, it provides some recreational space for populated residential communities. The void ground floor with multiple benefits has won strong support from the government which has issued the policy that the space of the void ground floor used for outdoor activities of the residents is not included in the plot ratio computation to encourage the application of this design method. Such void ground floor is still widely used for high-rise buildings in Guangzhou till now. More often than not, a multitude of subtropical plants and various activity platforms are arranged in the void ground floor to provide pleasant outdoor activity space for residents. With the increase of city plot ratio and building floors, void space is also arranged at the top floor of some large commercial podium buildings, forming an aerial green platform. In this manner, an open system with subtropical characteristics is formed among the empty space at different elevations and floors, making the Lingnan cities and buildings adaptive to the subtropical climate here.[3,4]

The most special part in the ventilation treatment of Lingnan architecture is the arrangement of the cross ventilation, which is a quite old design method used in ancient buildings and still used now. At present, the southing season wind is made use of in the design of Lingnan residential buildings to achieve the cross ventilation from north to south and take away the extra heat. We have applied this design to gardens

Picture 1. One view of culture and sports centers of Pingyuan county, Guangdong province.

Picture 2. Another view of culture and sports center of Pingyuan county, Guangdong province.

and balconies of our house project in Enping City, Guangdong to achieve the cross ventilation from north to south. This simple method has also been employed in some complicated modern buildings in recent years, like the public buildings in Sanya, Hainan Province, Zhanjiang, Guangdong Province, Beihai, Guangxi Province, or the lobbies and public space in some luxury hotels where the cross ventilation is used instead of air conditioning. A very typical example is the Marriott Hotel in Sanya, where the cross ventilation is arranged instead of air conditioning and it works quite well.

We have also applied the similar method to the Culture and Sports Center of Pingyuan, Guangdong Province. The air conditioning is mainly achieved by the cross ventilation in consideration of the characteristics of the southern climate. Shutters are arranged on top of the venue in over 5% of the total area. Moreover, an appropriate number of openable windows are arranged at the ground floor and the podium level to achieve natural ventilation. A large roof is employed to prevent rainwater from going into the venue through shutters. The roof reaches out for 2.5 m–3 m and works effectively for rain proofing and ventilation. The youth center in the north is of an atrium with enclosures on three sides. The cross ventilation is achieved between the roofless atrium and the building hallway. In this manner, the micro climate within the building is well improved. In a word, full consideration of local climate characteristics has been taken into architectural design of the project with suitable "low-technology" strategy adopted, which largely reduces construction cost as well as the cost for later operation and maintenance. Refer to Picture 1 and Picture 2.

Another example to be given is the Mangrove Resort Hotel in Beihai, in the design of which various means have been employed in response to the local climate. The project is located at the mangrove of Beibu Gulf, in the hilly area in suburb Beihai nearby Beibu Gulf. The slope is gentle at the top of the hilly area but steep at the foot. The seascape can be seen from the base which is surrounded by farmland with superior surroundings. The project is a resort hotel with hot springs. Its main hotel covers about 25,000 m² with about 200 guest rooms as well as the spring reception centers, SPA, large banquet halls, lecture theatres, and courtyard-style spa villas. The natural ventilation method is applied to the hotel lobby instead of the mechanical conditioning with conditioners. An independent layout is adopted for the hotel lobby. On both sides of the lobby are respectively public area and guest room area while its front and back are completely connected so that the guests are able to get a view here. Screen doors are arranged at the front and back of the lobby, which are usually open to achieve the cross ventilation effect. The clerestories at the lobby are openable. When they are open, the stack effect can be achieved, leading to better ventilation effect. Besides, a green courtyard is enclosed by guest rooms in the hotel, which makes the ventilation effect even better. As for water proofing and sunshading, architectural symbols borrowed from the Dong Minorities of Guangxi Province are employed, such as the relatively large slope eave which reaches far out to prevent rainwater from going into the building and enriches the façade effect of the building. The balconies and small slope cornices arranged in the guest rooms work effectively for sunshading and water proofing and enrich the façade effect of the rooms. As for greening, a lot of plants are grown around the hotel and many waterscapes arranged in the central area of the enclosure. In this way, the open lobby, verandah, waterscapes, and tropical plants are combined together to realize the integration of interior and exterior spaces and form a cyclic micro climate which helps in air conditioning. In brief, the simplest technical measures and the scientific space organization work so effectively to fulfill the building's requirements for sunshading, ventilation, water proofing, and ecological energy conservation, and express the simple idea of ecology in Lingnan architecture. Refer to Picture 3.

4 THE UNIVERSALLY SHARED CULTURAL VIEW THAT RESPONDS TO THE REGIONAL CULTURE

The Lingnan culture lies in itself about hedonism and civil life. Its secular social characteristic, which is quite strong, is also reflected in the decoration of Lingnan buildings which is of various forms and tells wonderful folk stories catering to common people. Meanwhile, the integration of modern building space with classical gardens is quite important in Lingnan architecture. Architectural articles are usually arranged within the gardens, a means of adding mass culture into elegant

Picture 3. The Mangrove Resort Hotel in Beihai.

Picture 5. Another View of the Villa Project in Enping, Guangdong Province.

Picture 4. One View of the Villa Project in Enping, Guangdong Province.

culture to make the gardens suit both refined and popular tastes. Furthermore, the aesthetic intention of local people is a significant factor to be considered and expressed in the design of cultural buildings, hotels, residential buildings in Lingnan region. The specific methods include: applying local architectural symbols and decorating modern buildings with traditional forms & façade ornaments; employing local construction materials like the black bricks, timbers, etc.; expressing the characteristics of local landscape architectures and adopting subtropical plants with local characteristics, etc. These design methods respond to modern Lingnan architecture in a great way.

The White Swan Hotel in Guangzhou, designed by Mo Bozhi, expresses the universally shared artistic view in a great way. As a typical modern building, the hotel looks quite concise and lofty in general. It totally defines the "elegance" of modern architectures. On the other hand, some folk cultures have also been integrated into the shared space of the hotel, like the hometown waterscape of a traditional style. Waterscapes are built in the atrium on a large scale. There is a pavilion on the rockery and spring water flows down under the pavilion. In this manner, the hotel shows both the modernity and secularity in response to folk cultures.

The villa project in Enping, Guangdong Province, designed by us, also echoes the local culture in its own way. Located in Enping, Jiangmen, the project is a group of low-rise residential buildings with a limited scale. We kept thinking about how to express traditional residential concepts and folk cultures in modern residence. And we tried to apply some folk architectural symbols to modern architectures in hope of combining modernism with local folk culture. The project is generally of modern style with intersected blocks and concise facades. Based on this, we applied some traditional architectural symbols, like choosing a traditional courtyard layout, laying the outer wall with black bricks in a hollow manner, conducting coping, arranging plants, etc. The popular folk culture has been expressed by the architectural design with the integration of these symbols. Refer to Picture 4 and Picture 5.

However, we also deeply realize that to apply traditional architectural symbols directly to modern architectures, though helping express local culture explicitly, may sometimes make the architectural design seem superficial, stiff and kitsch. It may even lead to disconnection between the form and the content, and some other negative issues.[1]

5 THE ARTISTIC VIEW ADAPTIVE TO CHANGES: NEW REGIONAL FORM

Practicality and reasonability are quite significant in Lingnan architecture. A design overstressing the art form is rarely seen in Lingnan. The regularity of architectural space is quite important and the rich space effects are mainly achieved by intersection, integration, and infiltration. As a result, Lingnan architecture is always characterized with simplicity. However, the overstress on regularity may lead to the stiffness of architecture and monotonousness of architectural form, and deprive architecture of artistic peculiarity. To this end, Lingnan architects make constant efforts to pursue changes based on certain rules which keep them practical. Obviously the position line in the space or facades of Lingnan buildings reveals the fact that Lingnan architects place great emphasis on basic aesthetic composition and precision of scales. But based on this, they also pursue changes in space and composition by adding oblique lines or camber lines on plain network or applying small variable elements to a regulated plane layout. These approaches have broken the monotonousness of Lingnan architecture caused by overstress on regularity and have achieved unexpected artistic effect by making the finishing point.

Picture 6. The Intelligence Valley Project in Zhuzhou.

Picture 7. Tianyin Mansion in Guangzhou.

The Intelligence Valley in Zhuzhou (2011–2012), designed by us, is a persuasive case. As a core complex in Zhuzhou Vocational Education City, the project involves a high-rise teaching building, an intelligence center, a library, a science & technology center, and a community center. It is a large complex around the rectangle central plaza, which is sunken with many movable platforms at different elevations, making the building look like "The Valley of Intelligence". Inspired by the Dwelling in the Fuchun Mountains, we made some attempts to integrate traditional Chinese cultural concepts into modern buildings. We put forward the long-scroll design in which the building contour lines and elevations lie in insulation to demonstrate the integration of the complex. As for design of facades, we sought changes in regularities and tried to make a change to the boundaries by introducing undulant oblique lines which turn out quite artistic. As for planes, we introduced oblique lines or broken lines to the foursquare position lines for varied space. Moreover, we arranged an oval glass atrium as the finishing point to the entire complex. It agrees with the exhibition function of the science & technology center. The introduction of some elements to the regular architectural layout adds great artistic effect to the whole complex. Refer to Picture 6.

Another project in which some similar design concept is used is the Tianyin Mansion in Guangzhou (2003–2010). The project is located in the east of Guangzhou, a corner at the traffic junction of Zhongshan Road, where the peripheral traffic is quite heavy. It occupies over 8,000 m^2 of land, while designed with the scale of 80,000 m^2 floor area and construction height below 100m. The whole building serves as a commercial complex. The first to sixth floors are used for business, and the seventh to twenty-seventh floors for office work or hotels. The layout of the building is of an "L" type and the two main building blocks are arranged in an intersecting manner.[4] In consideration of the cost, a regular layout was adopted for the high-rise building to facilitate the computation of structures. Based on this, we also made some changes to the facades or forms to enrich the façade effect. The architectural design gets rid of monotonousness because of the introduction of flexible curves to the regular outer boundaries. Curtain walls were arranged at one side of the two main building blocks, making the building unique without jeopardizing the regularity of

the building layout. Besides, we built a long and oval glass box at the east of the building, making it a symbol of the city and the finishing point to the whole building. According to the as-built effect drawing, such design with appropriate changes is quite successful in freeing itself from the monotonous shape of commercial buildings and making the building more symbolic.[5] Refer to Picture 7.

6 CONCLUSIONS: THE NEW LINGNAN VIEW THAT EMBRACES EVERYTHING

Lingnan architecture is quite influential as a regional school with several outstanding characteristics. First of all, a marine culture dominates Lingnan region which lies at the intersection of Chinese and western cultures, where the architectural design is exposed to open thoughts and advanced western design concepts; besides, Lingnan culture is quite open to embrace almost everything, which means that it is able to absorb various cultures and integrate characteristics of local, western, and Southeast Asian architectures while making constant improvements and innovations; last but not least, the special climate in Lingnan region makes it necessary to solve the issues of ventilation, thermal insulation and sunshading and leads to a flexible & diversified subtropical architectural style.

To sum up, among all Lingnan architectural concepts, the practical and rational functional view indicates the method of solving different functional problems with a multidimensional analysis and embodies the realistic scientific spirit; the simple and suitable ecological view responds to the regional climate with appropriate technologies, making the architecture adaptive to the climate and showing the emphasis of Lingnan architecture on ecology; the commonly shared cultural view suggests that Lingnan architecture should suite both refined and popular tastes to carry forward folk cultures; the adaptive artistic view represents a design method which advocates enriching regional architectural style by making appropriate changes and innovations in regional building shape. These four views cover fundamental architectural thoughts, aesthetic views, and specific design methods. They supplement each other and together form the essence of creative thoughts in Lingnan architecture.

ACKNOWLEDGEMENT

This paper is under the Study on the Creative Thought of Contemporary Regional Architecture in Lingnan, as the Youth Program in 2011 of Guangdong Provincial Philosophy & Social Science Research Program during the 12th Five-Year Plan, Program No.: GD11YYS01).

ABOUT THE PICTURES

Picture 7 is sourced from Bibliography 5, and the rest from the project teams.

REFERENCES

[1] Chen Changyong, Xiao Dawei. Analyzing the Trend of the Regionalism Architecture Practicing in View of Lingnan, Architectaral Journal, 2010 (02).

[2] He Jingtang. Architectural Creation of Lingnan – Review and Prospect of 60 years. Architectaral Journal, 2009 (10).

[3] Chen Changyong, Xiao Dawei. Building Medium Space System in Lingnan Cities, New Architecture, 2009 (08).

[4] Chen Changyong. Connect The Public Space–One Way to Improve the Resident Space Environment with High Density. Huazhong Architecture, 2006 (12).

[5] Chen Changyong, Liu Xiao. The Design of the City Commercial Complex in the Premise of Respect for the Urban Environment: The Idea of the Creation Methods about the Guangzhou Tianying Commercial Building. Huazhong Architecture, 2011 (03).

Advances in Civil Engineering and Building Materials IV – Chang et al (eds)
© 2015 Taylor & Francis Group, London, ISBN: 978-1-138-00088-9

A study of the inheritance lineage of regional architectural culture in the modern architectural creation of China

Changyong Chen, Dawei Xiao & Jin Tao
School of Architecture/State Key Laboratory of Subtropical Building Science, South China University of Technology, Guangzhou, Guangdong, China

ABSTRACT: This paper studies the works that have won the Grand Prize of the Chinese Architectural Association in Celebration of the 60th Anniversary of the People's Republic of China, through comparison and analysis of statistics, and summarizes the change of inheritance context of traditional regional architectural cultures in modern architectural creation. It also studies the development trends of Chinese regional architectural cultures at various times by pointing out their characteristics of attachment to symbolic significance, abstract forms, dimensional development, etc., and analyses the causes of such characteristics. In addition, the paper demonstrates the significance of traditional regional cultural inheritance for modern Chinese architectural creation. Based on this, it further points out the dilemmas of culture, form, and content faced with the regional cultural inheritance, and also explores the ways to solve these problems.

Keywords: regional culture, inheritance, modern architectural creation, context

1 INTRODUCTION

Modern building technologies and design thoughts have been widely disseminated in China since the last century, improving urban and dwelling environments, satisfying our rapidly growing material needs, and becoming the mainstream of architectural creation in China. However, the inheritance of traditional Chinese architectural cultures has been neglected in modernism, which has resulted in some negative impacts. Modernism emphasizes uniform design methods and overlooks differences of regional cultures. Therefore, buildings built based on modernism are uniform, cold, and indifferent, barely reflecting the unique Chinese architectural cultures of five thousand years. What is worse is that during city development, a large quantity of historical blocks and traditional buildings have been torn down for construction of modern projects, leaving the original city mechanism and landscape sabotaged and causing immeasurable losses. To sum up, there are two sides to modern architectural development, and the inheritance and protection of regional cultures in modern architectural creation must be placed on the agenda of rapid urbanization.[1]

In view of the previously mentioned facts, this paper studies the works that have won the Grand Prize of the Chinese Architectural Association in Celebration of the 60th Anniversary of the People's Republic of China, through classification and statistical analysis of works related to regional architectural cultures, and it sums up the inheritance context of regional cultures in modern architectural development. In addition, the relationship between modern buildings and traditional regional cultures is re-examined to determine the changes of regional cultural concepts in modern architectural design, and the causes of such changes. This is done to recall architects' attention to traditional regional architectural cultures.

2 OBJECTS OF STUDY

A multitude of buildings has been constructed since the foundation of China, making it impossible for us to list all of them here. Therefore, we have selected some excellent works for classified study. The 300 architectural works selected by the Chinese Architectural Association in celebration of the 60th Anniversary of the People's Republic of China are typical enough to represent excellent architectural works of various periods.[2] Of course, these prize-winning works are too small in number to reflect all the characteristics of modern building practices, but at least they are classic enough to represent mainstream concepts in the architectural circle, and thus serve as great samples for analysis.

There exists a subtle relationship between modern buildings and traditional architectural cultures; it can be seen to be the expression of architectural forms or the integration of design concepts. When considering the characteristics of modern buildings and traditional architectural cultures and the relationship between the two, we divide the selected works into three categories respectively with "traditional characteristics," "combined traditional and modern characteristics," and "modern characteristics."

Table 1. Statistics of the various categories of prize-winning works in the four periods (from statistics of [2]).

	Traditional culture	Combined tradition and modernity	Modernism	Traditional western culture	Others
1950–1959	8	14	5	15	
1959–1978	2	4	16	1	
1979–1999	12	21	58	1	
1999–2009	7	22	116	2	1

The first category is of "traditional characteristics," which means works in this category are completely subject to traditional regional architectural concepts and are of certain cultural background. A certain number of such works, mainly including museums, restored historic relics, some special temples, and sacrificial buildings, have been retained in various periods.

The second category is of "combined traditional and modern characteristics," which means works in this category are subject to both traditional and modern architectural concepts. This category falls into two subcategories. The first one includes "traditional Chinese and modern" works mainly designed based on traditional concepts, but built with modern technologies. Such works absorb some techniques of modernism while presenting also "archaizing" characteristics, but basically, they show the style of eclecticism. Buildings in this subcategory were most built in the early years of the new nation, like the "Top Ten Buildings" in Beijing. The second subcategory includes "modern and traditional Chinese" buildings, which are mainly designed based on modern concepts, but also absorb traditional regional architectural concepts and present relatively abstract architectural forms, like the Fragrant Hill Hotel in Beijing, the Museum of the Tomb of the Nanyue King in Guangzhou, the Zhejiang Art Museum in Hangzhou, etc.

The third category refers to "modern" buildings, which are designed based on modern architectural concepts. Such buildings are built with modern construction technologies and are characterized with functional rationality, concise and novel architectural forms, and internationality.

3 TRANSFORMATION OF REGIONAL ARCHITECTURAL CONCEPTS AND CAUSES

3.1 Expansion of modernism

According to the statistical data of the 300 prize-winning works, the ratio of buildings constructed with modernist techniques has been on a constant rise, from 14.71% in the early 1950s to 78.38% currently. Modernism, once the minority, accordingly becomes the mainstream (see Table 1 and Figure 1). The dramatic rise of the number of modern buildings is consistent with reality. The wide dissemination of modernism is related to economic, social, and many other factors. At the design concept level, the horizon of domestic

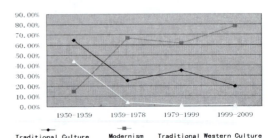

Figure 1. The ratios of the prize-winning works with various characteristics to total works of the four periods (from the analysis of bibliography 2). Remarks: the statistics of "traditional" buildings include "those completely designed based on traditional concepts" and "those combining traditional and modern concepts".

architectural circle has been widely open since the reforms and opening up of China. Some novel western architectural trends have been constantly introduced to China, exercising deep influences on the creation concepts of Chinese architects. Meanwhile, the undergraduate education of domestic architectural schools aims at nurturing professional architects and provides a curriculum system built on modernism. Consequently, a large number of modern architects have been fostered, laying a solid foundation for modernism. In addition, our economy has experienced a rapid growth in the past thirty years. To keep up with such a rapidly growing economy and to satisfy domestic construction needs, modern construction technologies, which emphasize on standardization and industrialization, are applied and significantly developed. Moreover, the political and ideological liberation in China contributes to a bloom of various architectural creation concepts. Architects have freed themselves from the bondage of nationalism and traditional cultures. New owners also prefer modern buildings. In a word, modernism has been widely disseminated and become the mainstream of domestic architectural creation.

3.2 Fall of regional architectural cultures

The ratio of works characterized with traditional regional architectural cultures has seen a decline from 64.71% down to 19.59%, and traditional regional architectural cultures, once the mainstream, accordingly become the minority. The ratios of buildings completely based on traditional concepts and those with combined traditional and modern characteristics

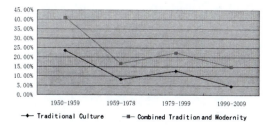

| | 1950-1959 | 1959-1978 | 1979-1999 | 1999-2009 |

-◆- Traditional Culture   -■- Combined Tradition and Modernity

Figure 2. The ratios of the two categories of prize-winning works with traditional characteristics to total works of the four periods (from the analysis of [2]).

have both been declining at similar rates in the various periods (see Figure 2).

The fall of regional architectural cultures results from multiple reasons. One of the most significant reasons is that the building space in traditional Chinese buildings is no longer able to meet the functional needs of a modern society. The relatively complicated decorations in traditional Chinese buildings require too much time and effort, making rapid mass production impossible and thus holding back the development of a modern society; it is the space and technical problems that lead to the fall of traditional architectural cultures. Another cause lies in our diffidence in traditional regional cultures. We tend to regard traditional architectural forms as being conservative and old-fashioned, while western buildings are seen as fashionable and luxurious. We worship western classic architectural culture blindly, and disdain traditional architectural cultures; that is how the "continental style" sweeps over China. In recent years, our blind worship for western cultures has been changed with the development of China's economy. Traditional Chinese symbols have been applied in some commercial projects and "Chinese-style" buildings have begun have begun to appear in the construction of commercial buildings in commercial buildings. However, the traditional regional cultures used in these projects are seen from the angle of consumption concept, and show characteristics of segment and collage, whether it is right is subject to further verification by time. Yet it is still hard for "Chinese-style" buildings to lead the field because, except in some important public buildings, most residences and hotels are still of the western style or the modern style.

3.3   Alienation of traditional western cultures

Among all the prize-winning works, there are not many buildings of a sheer western classic architectural style. Most buildings show mixing and compromising of various architectural cultures. The most typical case is in the "Top Ten Buildings" built in the early years of the new nation. In these buildings, we can see the architectural forms combining modernism and traditions, as well as colonnades and decorations of western parliament buildings. They are mixtures of architectural cultures, which is a common tendency in most

buildings constructed in the early years of the new nation, and related to the special context at that time like the political atmosphere and architectural creation context. With the rational development of nationalism and the liberation of architectural creation context since the reforms and opening up of China, such buildings have barely been seen. The national style featuring mixed eclecticism has become an historic label of an era.

Nonetheless, the limited number of prize-winning works based on classical western architectural culture does not mean it has disappeared. On the contrary, it has been disseminated over a wider range in another form. Since the 1980s, the real estate market has experienced a gradual rise in China, and commercial buildings featuring European and American cultures have been popular. Newly built residences and hotels usually show a Roman or continental architectural style. The "continental style" has spread from coastal cities to other parts of China, becoming a significant architectural style in domestic popular buildings. Nonetheless, none of the works in this style have been awarded this time, due to lack of innovation in design concepts and mechanical transplantation of cultural concepts. Obviously, these commercial buildings set in a traditional western cultural context have won popularity with ordinary people. but they are not recognized by the authoritative architectural theoretical circle, who has even criticized such buildings on other occasions. This shows the disagreement between the masses and the authoritative academic circle in their attitudes towards architectural cultures.

4   LINEAGE OF CREATION CONCEPTS OF REGIONAL TRADITIONAL BUILDINGS

Modern architectural thoughts have been constantly spread within China and architects have never stopped thinking about the relationship between traditions and modernity, which, with the constant development of the economy, society, and culture, shows different characteristics during different periods. Based on the 300 prize-winning architectural works, we can make a summary of the inheritance of regional architectural cultures and determine their characteristics in the various architectural periods, as well as the close relationship between their subtle changes and social, humanistic, and political concepts. In brief, the inheritance context of traditional architectural cultures falls into two aspects as follows.

4.1   Inheritance of classical architectural techniques

Traditional architectural concepts have always been applied to one kind of architectural design of buildings with regional cultural characteristics. According to the statistical data, a certain number of buildings of this sort can be seen in every period. Architectural innovation, space layout, and the façade of such buildings all follow traditional architectural rules to present

21

Figure 3.   Memorial Hall of Ven. Jian Zhen in Yangzhou.

Figure 4.   Beijing train station.

Figure 5.   National agricultural exhibition centre.

classical architectural forms, express the lasting appeal of traditional regional buildings, and realize its final goal of eternal historic architectural beauty. The scale and measure of such buildings also follows traditional architectural rules. The creation thought focuses on how to reconstruct ancient Chinese buildings and how to restore, protect, and repair traditional regional buildings. These works cover reconstruction and restoration of massive heritage buildings, the repair and conservation of historic buildings, and the renewal and revitalization of ancient villages and historic blocks. Their creation concepts even extend to the protection of the relevant intangible cultural heritage. Research achievements in anthropology, archaeology, and architectural history are applied to these buildings, which have been constructed in strict accordance with classical construction technical standards to carry on traditional regional architectural cultures. The Memorial Hall of the Ven. Jian Zhen in Yangzhou is a typical example (see Figure 3). To sum up, classical construction techniques crystallize the wisdom of our ancestors and serve as the most important lineage of regional architectural cultures.

4.2   *Inheritance of traditional regional cultures in modern buildings*

Another significant issue is how to express traditional regional cultures in modern buildings. Since the foundation of China, we have been in different phases and experienced some changes in our concepts. Accordingly, there have been different architectural creation techniques that show various styles and characteristics. Multiple techniques rather than only one are applied in each phase. However, there is always one that dominates. For instance, in the early years since the foundation of China, architects attached importance to the application of traditional regional cultural symbols and often expressed traditional architectural cultures in way of concretization. However, architects in later years laid emphasis on formal abstraction, which means the natural combination of modernity and traditions in forms. They did not apply architectural symbols or traditional architectural design methods indiscriminately, but rather, stressed the similarity in spirit rather than in appearance. Nowadays, traditional materials like black bricks and black tiles, and construction methods like wooden structures and masonry are emphasized in new buildings. Such materials and methods are applied in combination with modern technologies to express the cultural characteristics of traditions and create a traditional Chinese

architectural atmosphere. According to relevant analysis, the inheritance of traditional regional cultures in modern buildings roughly falls into three phases.

4.3   *Emphasis on formal meaning*

In the early years of the new nation, full-scale reconstruction was undertaken Influenced by the political atmosphere at that time, architects absorbed various construction techniques and presented an architectural style with mixed characteristics, which was determined by the social environment at the time. Firstly, some architects, with western education backgrounds, were significantly influenced by traditional western buildings. Therefore, they absorbed many traditional western elements like colonnades and decorations. Buildings at that time were also influenced by the culture of the Soviet Union. Soviet-style facades with vertical lines were often adopted in buildings, showing a touch of solemnity. Moreover, right after the end of the Maoism. Chinese people were full of nationalism sentiment and attached great significance to traditional architectural style, which was characterized by the frequent application of traditional large roofs.[3] Lastly, these large public buildings in relatively concise forms were constructed with modern technologies to achieve new functions. Influenced by the previously mentioned factors, buildings at that time presented various characteristics, traditional, western, Soviet, or modern. They are mixtures of various styles, characterized with eclecticism. The expression of formal meaning is quite important for such buildings, which means that the forms of the buildings are related to cultural concepts like political culture, national culture, etc. Beijing Train Station and the National Agricultural Exhibition Centre serve as good examples (see Figures 4 and 5).

4.4   *Emphasis on formal abstraction*

Back at the start of China's reforms and opening up, some large public buildings, including hotels, museums, etc., shouldered the responsibility of carrying

Figure 6. Fragrant hill hotel.

Figure 7. Jin Mao Tower.

Figure 8. Suzhou Museum.

Figure 9. Jia Pingwa Gallery of Literature and Art.

Figure 10. Luyeyuan Stone Carving Museum.

on regional architectural cultures by combining traditional and modern technologies.[4] Compared to the inheritance of traditional architectural culture before the reforms and opening up that was restricted by those conservative techniques, this period was obviously much more unconstrained. Architects no longer limited themselves to those heavy architectural symbols like large roofs and brackets. They paid more attention to the expression of formal abstraction with modern technologies. Besides, the architectural forms were no longer restricted to the three-segment design or the classical design, but began to present diversity. Concretization was not often applied to express architectural forms. On the contrary, traditional cultures were expressed in a more abstract and obscure manner so that regional architectural characteristics were more distinct and diversified. Obviously, architectural forms still attracted most attention in this period, although regional cultures were inherited in a more abstract manner and became closer to modern space. Regional cultures in this period evolved into influential concepts and generated a batch of world-renowned architectural works that led the trend of the period, like the Fragrant Hill Hotel and the Jin Mao Tower (see Figures 6 and 7).

4.5 Multidimensional development trend

In the past twenty years, China's economy has flourished. Accordingly, buildings of a great number have been constructed on a large scale. Although the ratio of buildings based on regional architectural cultures to all modern buildings is falling, their absolute number is on a constant rise, which brings great creation opportunities.[5] There are two different attitudes towards the expression of regional cultures in this context.

One still sticks to the solemn creation that advocates the expression of traditional cultures at deeper levels. The emphasis on form in design, which was a practice in the previous period, was somewhat changed by a batch of young architects in this new period, who tried to apply new ways to express regional cultures. They tried to reflect the characteristics of traditional buildings from aspects of artistic conception, space, and materials. They applied modern methods to express traditional cultures, including formal abstraction, abstract artistic conception, and construction expression. These architects made their own way, and were recognized worldwide by their peers. One representative is Wang Shu who has applied many local construction materials to his architectural works based on the traditional construction method, and expresses traditional regional cultures in modern buildings in a deep and implicit manner. He won the Pritzker Architecture Prize in 2012, which was seen as affirmation of his achievements in the exploration of traditional regional architectural cultures of this period. Typical buildings include the Suzhou Museum, the Jia Pingwa Gallery of Literature and Art, and the Luyeyuan Stone Carving Museum (see Figures 8 to 10).

The other attitude refers to the design with traditional regional cultures from the angle of consumerism. Buildings of this type mainly include hotels and residences, which have become "Chinese-style" buildings in the real estate field. This design concept serves the market and is usually determined by market demands. Therefore, masonry and retro commercial designs have been usually applied to the buildings of this period, catering to ordinary people, advertising traditional architectural concepts, and enhancing the masses' understanding of local cultures. However, from the angle of architectural creation, buildings of this style do not really impress people. Traditional architectural concepts may even become commercialized and vulgarized when applied too frequently to these buildings, which is a phenomenon we should be aware of.

5 DILEMMAS AND WAYS OUT

Though many persuasive examples could be listed to prove that we are still carrying on our local cultures, we could not overlook the destruction of traditional architectural cultures by architects during ongoing city construction. From Hutongs in Beijing to arcades in Guangzhou, from Zhouzhuang in River South to Miao Villages in the southeast of Guizhou Province, as well as batches of traditional architectural settlements removed during urbanization, we have to admit that we have done so much more to destroy traditional regional architectural cultures than to protect them. We have put the inheritance of traditional regional architectural cultures in multiple dilemmas.

Culture is the first dilemma we are faced with. The fall of traditional architectural cultures results from the whole community's concentration on economic benefits rather than on intangible cultures. We adore western cultures while belittling our own traditional cultures. This is how our cultural values become distorted and create an adverse social and cultural environment for the dissemination of regional architectural cultures. On what basis can we protect and carry forward our traditional architectural cultures when we lack the confidence in our cultures, the concept of regional cultural inheritance, or the consensus to conserve traditional buildings? Fortunately, there are still things we can do for improvement. We could make efforts to understand more of our traditional architectural cultures, to summarize and study them systematically, to strive for complete academic achievements, and to advertise and carry forward traditional architectural cultures. We could popularize the literal education of architecture and reshape the values of traditional architectural cultures through various approaches to win the masses' confidence in local cultures.

Form is the second dilemma. Our nation boasts a wide expansion of land and variety of traditional regional architectural cultures, but these treasures also restrict many architects to applying traditional forms and symbols in their work, like the fire walls and large roofs, because they find it hard to abandon them. The problem is that such forms and symbols sometimes are not compatible with modern life and technologies, and therefore, are unable to achieve the expected effects. To solve this problem, we should divert our emphasis on forms to materials, construction, and even artistic concepts to break free from the formal restriction and to understand traditional buildings on a deeper level. Only in this way can we make new breakthroughs.

Content is the third dilemma Traditional architectural cultures are closely related to traditional

lifestyles. However, the content of city construction is no longer the same with that of ancient times. Therefore, the overemphasis on classical architectural concepts in the design of modern buildings only makes them incompatible with modern life. To take the conservation and renewal of historic cultural buildings as an example, much as we want to preserve traditional structures and their styles, these traditional buildings are not as compatible with the modern life we are living. This is a miniature of the conflict between traditional buildings and modern life. How can we reconcile the conflict? Obviously, we should start by accommodating our inheritance of traditional regional architectural cultures with the changes of modern life and combine the two closely to make a natural fusion. To be specific, to get out of this dilemma, we should no longer replicate traditional cultures indiscriminately but prepare for sublation, adaptation, and transition.

6 CONCLUSIONS

Although regional architectural cultures have been constantly impacted by modernism, there are still ninety pieces of prize-winning works out of the 300, accounting for 30 percent of the total, that carry forward these cultures. Therefore, it is self-evident how important regional architectural cultures are for domestic modern buildings. Seen from the whole development history, the inheritance of regional architectural cultures, once focusing on application of symbols, abstraction of forms, and construction of materials, will divert to the creation of artistic conception in order to express traditional architectural cultures. It is true that we are faced with multiple dilemmas. However, if breaking free from the restriction of forms and symbols and exploring more of the characteristics of traditional cultures, we will find a way to express the artistic conception of traditional architectural cultures and raise them to a new level.

ACKNOWLEDGEMENT

This paper is from the Study on the Creative Thought of Contemporary Regional Architecture in Lingnan, as the Youth Program in 2011 of Guangdong Provincial Philosophy & Social Science Research Program, during the twelfth Five-Year Plan, Program No.: GD11YYS01).

ABOUT THE PICTURES

Table 1/Figure 1/Figure 2 are from the analysis of reference 2.
Picture 3 is from http://www.xinli110.com/luyoi/mjxz/jsly/yz/201209/322103.html
Picture 4 is from: http://httpzh.wikipedia.orgwiki Beijing
Picture 5 is from: http://zjbwg.cdstm.cn/index.php?m=sevenbooks&a=showall&placeid=144&typeid=75

Picture 6 is from: http://www.panoramio.com/photo_explorer#view=photo&postion=529&with_photo_id=23484631&order=date_desc&user=3322710

Picture 7 is from Lu Huihua: http://citylife.house.sina.com.cn/detail.php?gid=66528###

Picture 8 is from zrd512: http://citylife.house.sina.com.cn/detail.php?gid=64529

Picture 9–10 are from Chen Changyong.

REFERENCES

[1] Chen Changyong, Xiao Dawei; Study of New Trends of Domestic Regional Architectural Practice Based on Lingnan Buildings; Architectural Journal, 2010 (02)

[2] Award-winning Works in Architectural Design Creation Competition of Architectural Society of China; Architectural Journal, 2009 (09)

[3] Zhou Denong; Architectural Progress in Storms and Waves – the First Thirty Years of Architecture in New China (From 1949 to 1978); Architectural Journal, 2009 (09)

[4] Ma Guoxin; From 1979 to 1999: Old Days of Architecture in Two Decades; Architectural Journal, 2009 (09)

[5] Cui Kai; Review of Chinese Architectural Creation during 1999 and 2009; Architectural Journal, 2009 (09)

Bridge engineering

Advances in Civil Engineering and Building Materials IV – Chang et al (eds)
© 2015 Taylor & Francis Group, London, ISBN: 978-1-138-00088-9

The effects of moisture content fluctuation on a dynamic resilient modulus of compacted clay

Dongxue Li
Jilin Provincial Communication Scientific Research Institute, Changchun City, Jilin Province, P.R. China

Zongyuan Sun
Northeast Electric Power Design Institute of China Power Engineering Consulting Group, Changchun City, Jilin Province, P.R. China

ABSTRACT: Subgrade soils are important materials to support highways. For the past few decades, the resilient modulus (M_R) has been used for characterizing the stress-strain behaviour of subgrades subjected to repeated vehicle loadings. It is well known that the moisture content has an important impact on M_R of clay. Furthermore, the moisture content of subgrade does not keep stable but fluctuates seasonally. In this paper, the impact of moisture content fluctuation on M_R of compacted clay was investigated by laboratory experiment. It is shown that M_R of compacted clay decreases significantly with the moisture content fluctuation cycles, and the bigger the fluctuation amplitude is, the lower M_R is. Therefore, the reduction factor of moisture fluctuation is proposed to reflect this impact on the long-term performance of subgrade.

Keywords: cohesive subgrade soil, resilience modulus, moisture content, moisture cyclicity

1 INTRODUCTION

Subgrade is one of the most important parts of highway structures, and has a great impact on pavement performance and traffic safety. The 1993 AASHTO (Association of State Highway and Transportation Officials) Design Guide introduced resilient modulus (M_R) as an important subgrade material property (AASHTO, 1993). In the new AASHTO mechanistic-empirical design guide (NCHRP, 2004), M_R also plays a major role in representing the properties of the materials in various pavement layers. It is well known that the moisture content has an important impact on the M_R of subgrade soil. Relationships between stress state, water content or matric suction, and resilient modulus have been established by previous studies, and different adjustment methods have been proposed. Li and Selig developed a formula to estimate the M_R using water content and dry density as independent variables (Li & Selig 1994). Drumm et al. proposed that the value of M_R decreased significantly with the increase of water content for specific soils (Drumm, Reeves, & Madgett 1997). Recently, Khoury found that the M_R-moisture relationship is hysteretic and is vital in predicting the response of pavement structure because of changes in climate (Khoury 2009). In fact, the moisture of subgrade does not keep stable but fluctuates seasonally. More importantly, the moisture fluctuation beneath the pavement is not completely dry to fully saturated, but in a certain water content range near the Optimum Moisture Content (OMC). Some scholars used the strength reduction factor of the freezing-melting cycle to correct the influence on subgrade resilient modulus. Similarly, it is worth investigating whether the limited range of moisture fluctuation causes M_R reduction, and how the impact of moisture fluctuation on M_R. is best described. Therefore, in this paper, the moisture fluctuation simulation method was introduced, and then the impact of moisture content fluctuation on the M_R of compacted clay was investigated by laboratory experiment.

2 MATERIAL AND TESTING METHOD

2.1 Soils and samples preparation

The soils used in this study were obtained from in-service subgrades, and described as CL (Low liquid limit clay) according to the Unified Soil Classification System of the American Society of Testing and Materials (ASTM. 2006). The liquid limit was 33.7, the plastic limit was 21.6, the percentage of fine ($P_{0.075}$) was 75%, the OMC was 13.6%, and the maximum dry density was 1.82 g/cm^3.

The specimen was statically compacted to target relative compaction (ρ_d/ρ_{dmax}) with OMC. In this study, the relative compaction was chosen as 94%. The prepared specimen for moisture fluctuation simulation

Working condition	EMC %	AMC %	Number of times
A	13.6	—	—
B-0	18	3	0
B-2	18	3	2
B-4	18	3	4
B-6	18	3	6
B-8	18	3	8
C-0	18	4	0
C-2	18	4	2
C-4	18	4	4
C-6	18	4	6
C-8	18	4	8
D-0	22	2	0
D-2	22	2	2
D-4	22	2	4
D-6	22	2	6
D-8	22	2	8

Table 1. Experimental scheme.

Table 2. Stress loading sequence of fine-grained subgrade soil specimen.

Loading sequence	Confining pressure σ_3 (kPa)	Cyclic deviator stress σ_d (kPa)	Loading times
Preload	30	55	500
LS-1	30	30	100
LS-2	30	30	100
LS-3	30	30	100
LS-4	30	30	100
LS-5	30	55	100
LS-6	25	55	100
LS-7	25	55	100
LS-8	25	55	100
LS-9	25	75	100
LS-10	25	75	100
LS-11	20	75	100
LS-12	20	75	100
LS-13	20	105	100
LS-14	20	105	100
LS-15	20	105	100

and repeated loading triaxial test was 100 mm in diameter and 200 mm in height.

2.2 Testing method

The maximum moisture content in the sampling location is about 10% higher than its OMC, the EMC (Equilibrium moisture content) in this experiment was set to 18% and 22%, and the amplitude of moisture content (AMC) was set to 2%, 3%, and 4%. Resilient modules of specimens were measured after moisture fluctuation cycles. The experimental scheme is shown in Table 1.

The Dynamic servo-hydraulic material testing system, UTM-100, was used to conduct repeated loading triaxial tests. Air pressure was used as the confining pressure and half-sinusoid was used as the loading waveform, whose loading frequency was 1 Hz, the loading duration was 0.1 s and the interval was 0.9 s. The deviator stress was chosen as 25 kPa, 30 kPa, 35 kPa, 40 kPa, and 105 kPa, and the confining pressure was set to 20 kPa, 25 kPa, and 300 kPa (Luo, 2007). The number of loading cyclic of each loading level was 100.

2.3 Simulation of moisture content fluctuate

The great mass of subgrade soil at the scene is unsaturated. The moisture content of subgrade soil decreases during the dry season and increases when the season becomes rainy. Usually, it keeps the moisture content within the range of OMC to OMC + 7% such as in the humid region of China (Li, 2007). Furthermore, this phenomenon occurs repeatedly during the lifetime of a highway subgrade. Therefore, the moisture content of specimens' changing process means, 'EMC→EMC + AMC→EMC-AMC→EMC', and is defined as a moisture content fluctuation path in this paper, as shown in Figure 1. The figure also illustrates

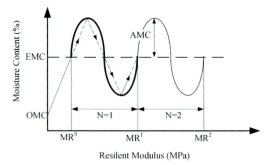

Figure 1. Moisture content fluctuation.

the testing time of the resilience modulus when moisture content changes. To investigate the influence of the variation of moisture on M_R in a laboratory, it is critical to control precisely the moisture content fluctuation (Zemenu & Martine 2009, Alhomoud & Basma 1995, Dif 1991, Day 1994, Basma & Alhomoud 1996, Rao & Reddy 2001). The drying process was carried out in an oven, and the wetting process was simulated by an absorption control device, which was composed of a rubber membrane, filter paper, porous stone, and an elastic band, as shown in Figure 2. The rubber membrane was made with a half-open hollow cylinder to seal the soil sample. Because the specimen dimensions in the triaxial experiment were 100 mm × 200 mm, the rubber membrane cylinder was 100mm in diameter and 250 mm in height; the filter paper and porous stone were both 100 mm in diameter. To provide enough confining ability to encase the soil sample, the thickness of the rubber sleeve was 2 mm.

The several steps of the wetting process are as follows. Step 1: The prepared specimen was put on the operation platform and then packed by a half-open

30

Figure 2. Experiment device.

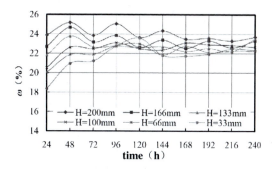

Figure 3. Relationship between moisture content and height and time.

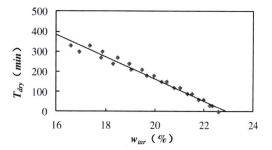

Figure 4. Relationship between moisture content and dring time.

rubber sleeve from top to bottom; Step 2: The filter paper was soaked and put on the exposed bottom circle of the specimen, then a soaked porous stone was put on the soaked filter paper; Step 3: After making sure that there was no bubble and fold in each contact surface, the entire specimen, filter paper, and porous stone were packed into the half-open rubber membrane cylinder; Step 4: The specimens were placed into a container, and a predetermined amount of water was added from the top of the specimen to enable it to reach the target moisture content; Step 5: Following Step 4, water was added to each assembled specimen and numbered, then the container was sealed to make sure the moisture content reached its equilibrium.

2.4 Wetting equilibrium time

At the beginning of wetting, the distribution of the moisture content along the height of the specimen was not equal. A rational equilibrium time was determined in order to ensure that the moisture content of each layer inside the specimen reached EMC and was equal before the dynamic resilient modulus test. For this purpose, ten specimens were carried out wetting process firstly. Specimens were prepared at OMC and at 96% of their compaction; the target moisture content was 23%. The moisture content along the height (H) was tested between one and ten days, respectively. The results are shown in Figure 3. In the initial stage of adding water, the difference between the maximum and minimum moisture content of the specimen was 5.5%. After seven days, the difference between the maximum and minimum moisture content of the specimens was about 1.5%. Between days eight and ten, the difference of the moisture content did not vary significantly. If the time continued to increase, the water would keep moving downwards, which would cause the moisture content on the top of the specimen to be low, whilst at the bottom it would be high. Therefore, the absorption time was determined as seven days in this working condition study. The equilibrium time of the other working conditions in this study can obtained using the same method.

2.5 Drying time

An oven-drying method was used to dehydrate specimens. The oven was set at 50°C with no air blow. To determine the drying time, specimens were taken to conduct a drying experiment and the results of those specimens were used to make comparisons and analyses. Two specimens were prepared at OMC and 96% of their compaction. Water was added until the moisture content reached 23%. After seven days, the dehydration experiment started and was tested continually for 1440 minutes. In the first 450 minutes, specimens were weighed every 30 minutes to calculate their moisture content.

The results show that when the dehydration time was 450 minutes, the total dehydration mass of the specimen was about 210 g, and the moisture content of the specimen decreased to about 15%. When the dehydration time was 1440 minutes, the total dehydration mass of the specimen was about 490 g, and the moisture content of the specimen decreased to about 5%. The dehydration velocity of the specimen was slow in the earlier stage and fast in the later stage. The reason for this phenomenon may be as follows. At the initial stage of drying, the difference between the temperature inside and outside was relatively large, so the water loss came mainly from the surface of the specimen. At the later stage, the inside and outside temperature of the specimen were gradually closed and reached oven temperature Therefore, water moved fast in the high temperature condition, and the increase of the gas

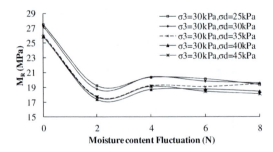

Figure 5. Relationships between MR and Number of fluctuation times (working condition C).

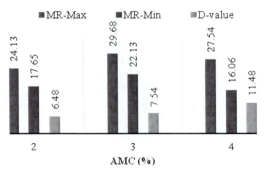

Figure 6. Relationship of MR and AMC.

phase inside the specimen accelerated the movement
of the water. Parts of experiment data was picked out
used for fitting in order to control the drying time, as
shown in Figure 4. It can be seen that the linear fitting,
as Equation 1, is relatively satisfactory. The drying
time of the other working conditions in this study can
obtained using the same method:

$$T_{dry} = 1150 - 50 \times \omega_{tar} \quad (R^2=0.98) \tag{1}$$

where, ω_{tar} = target drying moisture content (%);
T_{dry} = drying time (min).

3 RESULTS AND DISCUSSION

3.1 Influence of fluctuation cycles on M_R

When the EMC was 18% and the AMC was ±4%,
the relationship between the resilient modulus and
fluctuation cycles is shown in Figure 5. It can be
seen that the resilient modulus decreased by 25–32%.
There are some significant conclusions from these
curves. The first two fluctuations contribute greatly
to reducing the modulus of resilient. The test data then
reflects a rebound of resilient modulus and then continue to reduce. Therefore, fluctuation is one of the
causes of resilient modulus decrease; the greater the
number of fluctuations, the smaller the resilient modulus becomes. However, this decrease is not linear but
curvilinear.

3.2 Influence of fluctuation amplitude on M_R

M_R decreased with the moisture content fluctuation, as
shown in Figure 6. For condition B, M_R decreased from
24.13 to 17.65 MPa after eight times of moisture content fluctuation. For condition C, M_R decreased from
29.68 to 22.18 MPa. For condition D, M_R decreased
from 27.54 to 16.06 MPa. Comparing conditions B, C,
and D, where AMC are 3%, 4% and 2%, it can be found
that, M_R was sensitive to the amplitude of the moisture
content the greater the AMC was, and the higher the
D-value of M_R was. In this condition, D-value means
the difference between M_R-Max value and M_R-Min.
On the other hand, when comparing condition D and
condition B, the D-value is slightly lower, though the

AMC is increased. It reveals that, although M_R shows
a linear correlation to the AMC, the results would be
more complex with the EMC also being considered.
In spite of this, the AMC is clearly an important factor
of M_R under moisture content fluctuation.

4 CONCLUSION

In this paper, a moisture fluctuation simulation method
was proposed, a soaking equilibrium time and drying
curve were determined for precise control of the moisture variation, and then the impact of moisture content
fluctuation on M_R of compacted clay was investigated.
The experimental results reveal that the tested compacted clay is extremely sensitive to moisture content
fluctuation and M_R decreases continuously with the
repeated cycles. The greater the fluctuation amplitude,
the lower M_R is. Significant fluctuation amplitude
with low equilibrium moisture content has a similar
unfavourable impact on the long-term performance of
clay subgrade as a high equilibrium moisture content.
Finally, the reduction factor of moisture fluctuation is
proposed to reflect this impact.

REFERENCES

AASHTO. 1993. AASHTO guide for design of pavement
structures, Washington, D.C. AASHTO.
Alhomoud, A. S. & A. A. Basma, et al. 1995. Cyclic
swelling behavior of clays. Geotechnical Engineering-
ASCE, (121)7: 562–565.
ASTM. 2006. Standard practice for classification of soils
for engineering purposes (Unified Soil Classification
System). ASTM standard D2487. American Society for
Testing and Materials, West Conshohocken, PA.
Basma, A. A. & A. S. Alhomoud, et al. 1996. Swelling-
shrinkage behavior of natural expansive clays. Applied
Clay Science, 11(2–4): 211–227.
Day, R. W. 1994. Swelling-shrink behavior of compacted clay.
Geotechnical Engineering-ASCE, 120(3): 618–623.
Dif, A. E. 1991. Expansive soil under cyclic drying and
wetting. Geotechnical Testing. 14(1): 96–102.
Drumm, E. C., Reeves, J. S., Madgett, M. R. & Trolinger,
W. D. 1997. Subgrade resilient modulus correction for
saturation effects. Geotechnical and Geoenvironmental
Engineering 123(7): 663–670.

Khoury, C. K. & Khoury, N. K. 2009. The effect of moisture hysteresis on resilient modulus of subgrade soils. 8th Int. Conf. Bearing Capacity Roads, Railways, and Airfields, Univ. of Illinois–Urbana-Champaign, Champaign, IL. 71–78.

Li Cong. 2007. The Study on resilient modulus taking into account theoretic of unsaturated soils. Ph.D. Thesis, Chongqing Jiaotong University, Chongqing,

Li, D. & Selig, E. T. 1994. Resilient modulus for fine-grained subgrade soils. *Geotechnical and Geoenvironmental Engineering.* 120(6): 939–957.

Luo Zhigang. 2007. Research on dynamic modulus parameters of subgrade and aggregates layer. Ph.D. Thesis, Tongji University, Shanghai.

NCHRP. 2004. Guide for mechanistic-empirical design of new and rehabilitated pavement structures. Final Report. No. NCHRP 1-37A. National Research Council, Transportation Research Board. Washington, D.C.

Rao, S. M. & B. V. V. Reddy, et al. 2001. The impact of cyclic wetting and drying on the swelling behavior of stabilized expansive soils. *Engineering Geology*, 60(1–4): 223–233.

Zemenu, G. & A. Martine, et al. 2009. Analysis of the behavior of a natural expansive soil under cyclic drying and wetting. *Bulletin of Engineering Geology and the Environment.* 68(3): 421–436.

Advances in Civil Engineering and Building Materials IV – Chang et al (eds)
© 2015 Taylor & Francis Group, London, ISBN: 978-1-138-00088-9

Research on seismic response and fragility of lock-up device

Zhanzhan Tang, Jie Liu & Xu Xie
Zhejiang University, Hangzhou, Zhejiang, China

Yixin Jin
Chengdu Alga Engineering New Technology Development Co. Ltd., Chengdu, Sichuan, China

Yan Wang
Zhejiang Sci-Tech University, Hangzhou, Zhejiang, China

ABSTRACT: In order to analyze the seismic response and fragility of lock-up device (LUD), this paper selected 100 historical strong ground motions to study the influence of ground motion types on the seismic response of it, and generated another 100 ground motions randomly as near-field ground motions to study the seismic fragility of it further. The results show that, LUD has a good seismic performance under the ground motions; the velocity pulse in near-field ground motions has a great impact on the response of LUD; there is a satisfactory logarithmic-linear relationship between the response of LUD and the ground motion intensity index *SI*, which can reflect the seismic safety of LUD well when introduced in the seismic fragility curve.

Keywords: lock-up device, seismic response, fragility, velocity pulse, near-field ground motion

1 INTRODUCTION

LUD is a seismic protection device filled with viscous fluid medium. The damping force produced by it has an exponential relationship with relative velocity, therefore it is able to release displacement under normal service condition and restrict displacement under

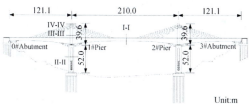

(a) Elevation view

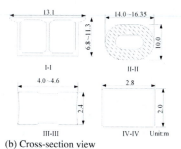

(b) Cross-section view

Figure 1. Geometry of the Super-span Railway Bridge.

seismic action. The application of LUD in engineering structures has been studied for a long time, and there are lots of essays and reports on it (Park et al. 2012, Zhang et al. 2012, Kong et al. 2012). However, the failure characteristics of LUD itself have rarely been studied. The results of fragility analysis can provide designers basis for a comprehensive judgment about the seismic performance of structures (Liu et al. 2010, Wang et al. 2012). Substantial research have revealed the relationship between seismic damage probability and ground motion intensity, helping make assessment about the safety of structures from the perspective of failure probability.

This paper studied the influence of ground motion types on the seismic response on LUD and its seismic fragility. The results by this paper provide a calculation method for the seismic reduction and isolation design of railway bridges and the reasonable application of LUD.

2 BRIDGE MODELING AND DYNAMIC CHARACTERISTICS

2.1 *Bridge modeling*

Figure 1 is the design scheme of a long-span railway bridge. It is a PC extradosed bridge with a main span of 210 m. The abutments and piers are equipped with spherical bearings, of which 1# is fixed. There are 8 LUDs on 2# pier and 4 LUDs for each on

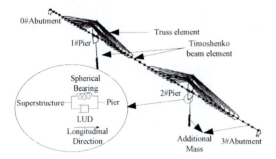

Figure 2. Calculation model of the bridge.

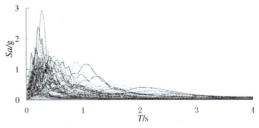

Figure 3. Response spectrum of ground motions.

0# abutment and 3# abutment. The design resistance peak ground acceleration is 0.16 g, site type is II, and the characteristic period of response spectrum is 0.45 s.

The friction coefficient μ of the spherical bearing is 0.02. The relationship between load F_d (the damping force) and relative velocity of a single LUD is

$$F_d = cv^2 \tag{1}$$

where, $c = 4000\,kN/(m/s^2)$ is the damping coefficient. The maximal design carrying capacity of a damper is 4000 kN.

A beam model shown in Figure 2 was employed for the simulation of the response of the bridge. The calculation was conducted by FEM package ABAQUS 6.10. Geometric nonlinearity and nonlinear reaction of the spherical bearings and the LUDs were taken into account.

2.2 The free-vibration characteristics

Since the equivalent stiffness of bearings and LUDs is relative to the structure's seismic response, the fundamental frequency calculated from the original stiffness under the completion state is not accurate to reflect the dynamic characteristics of the bridge. Therefore, when calculating the structure with free-vibration, it was assumed that δ was 70% of the average relative displacement of the bearings at the time when the maximal relative velocity of LUD was achieved. δ was used as the effective displacement to calculate the secant stiffness of the bearings for analyzing the free-vibration characteristics of the bridge. The first three natural frequencies were 0.557 Hz, 0.704 Hz and 1.300 Hz respectively.

3 THE DYNAMIC RESPONSE OF LUD UNDER SEISMIC ACTION

3.1 Input of ground motions

According to the site type and the design seismic resistance level of the bridge, 100 historical strong ground motion records from PEER were selected to conduct the nonlinear time-history response analysis. Figure 3 shows the response spectrum Sa with a damping ratio of 5%. The results indicate that the predominant periods of the ground motions range between 0.3 s and 1.5 s, which are significantly less than the free-vibration period of the bridge.

3.2 Seismic response analysis of LUD

Traditional theories always measure the ground motion with peak ground acceleration (PGA), spectral acceleration value (SA) or spectral velocity value (SV) corresponding to the fundamental frequency of the bridge, and SI, which is calculated according to the integral of the spectrum of velocity. SI, proposed by Housner, can be expressed as:

$$SI = \int_{0.1}^{2.5} S_v(\xi, T)dT \tag{2}$$

Figure 4 shows the logarithmic relationship between the maximal velocity response v_m of LUD and the ground motion intensity, which is represented respectively by PGA, SA, SV and SI as mentioned. The results indicate an effective logarithmic linear fitting of v_m and the ground motion intensities, and that the determination coefficients R^2 are all above 0.75. The fitting line has the highest relevancy with the samples when SI is taken as the independent variable, and R^2 reaches 0.95.

The results of the historical ground motions indicate that under only two ground motions, the velocity response of LUD can meet the condition of ln $(v_m/[v]) > 0$, where $[v]$ is the permissible velocity of LUD. The SI values of these two ground motions are above 3.0, while those of other ground motions are all under 2.5, which means that these two ground motions have higher spectral velocity values and possess some features of near-field ground motions. According to the research by Loh et al. (2002), the higher the ratio of PGV to PGA of an earthquake, the more obvious the near-field effects. Figure 5 shows the relationship between the ratio of PGV to PGA and the ratio of v_m to $[v]$. It can be inferred that the more obvious the near-field effects, the greater the response of LUD.

Figure 6 indicates more clearly that near-field ground motion has a greater influence on LUD, and the possibility of damage induced by near-field ground motions is greater accordingly.

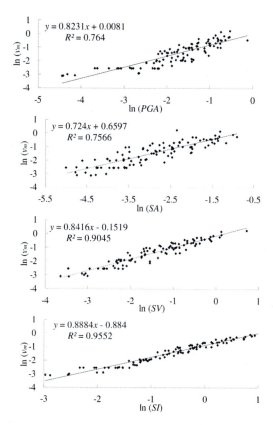

Figure 4. The relationship between seismic response of LUD and the ground motion intensity.

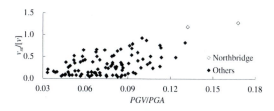

Figure 5. Relationship between PGV/PGA and the response of LUD.

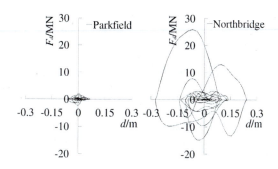

Figure 6. Hysteretic curve of LUD.

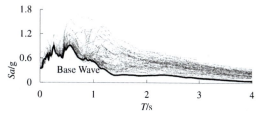

Figure 7. The acceleration response spectrum of near-field ground motions.

4 THE RESPONSE OF LUD UNDER ARTIFICIAL NEAR-FIELD GROUND MOTIONS

4.1 The generation of artificial near-field ground motions

This paper adopted the method proposed by Makris & Chang (2000), simulating velocity pulses by trigonometric function. For the sake of randomness of artificial ground motions, the Monte Carlo method was adopted to generate 100 groups of random numbers, and three basic parameters of velocity pulse were distributed randomly in a certain interval. There are three kinds of velocity pulse: progressive type, round-trip type and cyclic type, but only progressive type was adopted in this paper on account of the similar characteristics these three kinks of pulses share. The progressive-type velocity pulse can be expressed as:

$$v_e(t) = v_g(t) + \frac{V_p}{2} - \frac{V_p}{2}\cos(\frac{2\pi}{T_p}t) \tag{3}$$

$$t_0 \leq t \leq t_0 + T_p \tag{4}$$

where, $v_e(t)$ is the velocity of artificial ground motion, $v_g(t)$ is the velocity of base ground motion, V_p is the equivalent velocity peak, ranging between 0.8 m/s and 1.6 m/s, t_0 is the time when velocity pulse is inserted, ranging between 0.5 s and 8.5 s and T_p is the duration of velocity pulse, ranging between 1 s and 3 s.

Base ground motion wave (abbreviated as base wave in the following) is a record of a far-field ground motion named El Centro occurred in 1940, the velocity wave can be acquired by the integral of acceleration wave and the correction of datum line. Figure 7 shows the acceleration response spectrum of 100 ground motions generated randomly. As can be seen, the artificial ground motions cause stronger vibration during the moderate long period, reflecting the long-period vibration characteristics of near-field ground motion. Figure 8 indicates that the ratios of PGV to PGA of artificial ground motions are all above 0.2, so the artificial ground motions can reflect the near-field characteristics of real ground motions well.

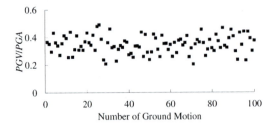

Figure 8. *PGV/PGA of artificial ground motions.*

4.2 *The response of LUD under artificial near-field ground motions*

In this section, 100 artificial near-field ground motions were input, and analyze about the response of LUD was conducted, together with the calculation results of 100 historical ground motions as mentioned. Figure 9 shows the relationship between the maximal velocity response of LUD and the intensity of 200 ground motions. A logarithmic linear fitting was conducted to the response of LUD. The results indicate that as the amount of ground motion records increasing to 200, the logarithmic-linear rule is more obvious, and R^2 is higher. The distribution of the statistics proves that LUD has a greater response under near-field ground motions, and the corresponding damage is more obvious.

5 FRAGILITY ANALYSIS OF LUD

5.1 *Demand probability theory of LUD*

The fragility curve reflects the probability that the response of the structure exceeds the capacity of a given damage state defined as a function of a parameter representing the ground motion. The probability of exceeding the capacity can be expressed as following formula:

$$p_f = P_f\left(\frac{D_c}{D_d} \leq 1\right) \qquad (5)$$

where, D_c and D_d represent the capacity and the demand respectively, and are assumed to obey normal distribution. Therefore, P_f can be expressed as following as well:

$$P_f = \phi\left[\frac{\ln\left(\frac{D_d'}{D_c'}\right)}{\xi}\right] \qquad (6)$$

where, Φ is the standard normal distribution function; D_c' and D_d' are the mean value of seismic capacity and demand, corresponding to the threshold of the damage state and the logarithmic regression function of the demand respectively; ξ is the standard deviation of the damage probability.

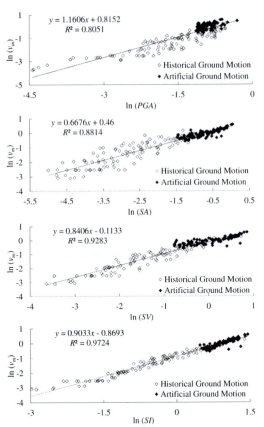

Figure 9. The relationship between the response of LUD and the intensity of 200 ground motions.

Table 1. Damage states based on failure criterion of velocity.

Damage State	Threshold
No Damage	$v \leq [v]$
Slight Damage	$[v] \leq v \leq 1.2[v]$
Medium Damage	$1.2[v] \leq v \leq 1.5[v]$
Complete Damage	$v \geq 1.5[v]$

5.2 *Results of fragility analysis*

Due to the lack of relative specifications, damage indexes are not unified, so this paper divided the damage of LUD into four states: no damage, slight damage, medium damage and complete damage state according to the failure criterion of velocity. Table 1 shows the division of all the damage states:

Among all LUDs in this bridge, the one with the greatest response was taken as the object of fragility analysis. The fragility curve was generated according to formula (5)–(6). Figure 10 shows the fragility curves of LUD vs different parameters of ground motion. As can be seen from the figures, as ground motion intensity increasing, the probability of damage

38

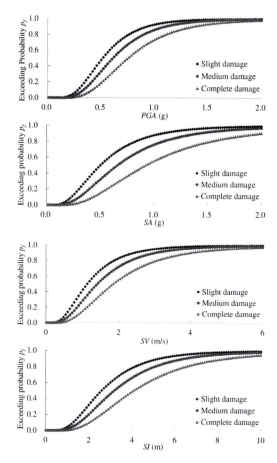

Figure 10. Seismic fragility curves of LUD under near-field ground motions.

increases. When *PGA* and *SA* reach rare ground motion intensity 0.32 g, the probability of complete damage is still under 4%, which means LUD can remain in the in-service state basically. By far, specifications have not provided any rules that measure ground motion intensity by *SV* or *SI*. However, according to the study of this paper, analysis based on ground motion intensity represented by multiple indexes is conducive to a more comprehensive understanding of the seismic response and safety performance of LUD.

6 CONCLUSIONS

Through the calculation and analysis, several conclusions can be drawn:

(1) The results of historical ground motions indicate that LUD has greater response under near-field ground motions. The displacement and velocity of LUD both reach the maximum when the velocity component of the ground motion reaches the maximum.

(2) The results of artificial ground motions prove further the destructive power of near-field ground motions on LUD. The maximal response of LUD has a logarithmic-linear relationship with ground motion intensity, and the determination coefficients of linear regressions are all above 0.8. The fitting effect based on ground motion intensity represented by *SI* outperforms than others, but ground motion intensity can be reflected effectively by *PGA* and *SA* as well, which is also consistent with the indexes adopted to measure ground motion intensity in current specifications.

(3) According to the results of fragility analysis, along with the increase of ground motion intensity, the probability of damage increases. Designers can assess the safety performance of LUD more comprehensively through the fragility curves produced by analogous method.

This paper provides a method to analyze the fragility of LUD, but if relative technical statistics of LUD can be accessible and taken into consideration, more rigorous seismic fragility curves can be produced by this method, and a direct reference for practical engineering can be provided.

ACKNOWLEDGEMENT

The author would like to appreciate the financial support of National Natural Science Foundation of China (51378460).

REFERENCES

Alavi B, Krawinkler H. 2004. Behavior of moment-resisting frame structures subjected to near-field ground motions. *Earthquake Engineering & Structural Dynamics*. 33(6): 687–706.

Charles Menum, Qiang Fu. 2002. An analytical model for near-fault ground motion and the response of SDOF systems. *7th US National Conference on Earthquake Engineering. Boston. September 2002*. Boston: Mira Digital Publishing.

Jingbo Liu, Yangbing Liu, Heng Liu. 2010. Seismic fragility analysis of composite frame structure based on performance. *Earthquake Science*, 23(1): 45–52.

Kidong Park, Daeyoung Kim, Donghyun Yang et al. 2012. A comparison study of conventional construction methods and outrigger damper system for the compensation of differential column shortening in high-rise buildings. *International Journal of Steel Structures*, 10(4): 317–324.

Lingjun Kong, Yanbei Chen, Peng Li et al. 2012. The anti-seismic analysis of the lock-up device on the arch bridge. *Advanced Materials Research*. 368: 1047–1050.

Loh C H, Shiuan Wan, Wen-I Liao. 2002. Effect of hysteretic model on seismic demands: Consideration of near-field ground motions. *The Structural Design of Tall Buildings*. 11(3): 155–169.

Makris Nicos, Shih-Po Chang. 2000. Effect of viscous, viscoplastic and friction damping on the response of seismic isolated structures. *Earthquake Engineering & Structural Dynamics*. 29(1): 85–107.

Qiang Wang. Ziyan Wu, Shukui Liu. 2012. Seismic fragility analysis of highway bridges considering multi-dimensional performance limit state. *Earthquake Engineering and Engineering Vibration*. 11(2): 185–193.

Shinozuka Masanobu, M. Q. Feng, Jongheon Lee et al. 2000. Statistical analysis of fragility curves. *Journal of Engineering Mechanics*. 126(2): 1224–1231.

Shuang Li, Lili Xie. 2007. Progress and Trend on Near-field Problems in Civil Engineering. *Acta Seismologica Sinica*. 20: 105–114.

Yamazaki Fumio, Motomura Hitoshi, Hamada Tatsuya. 2000. Damage assessment of expressway networks in Japan based on seismic monitoring. *12th World Conference on Earthquake Engineering; Proc.intern. symp., Auckland, 30 January to 4 February 2000.*

Yan Wang. 2013. Research on Seismic Performance and Fragility of Isolated Railway Bridges. Hang Zhou: Zhejiang University.

Yongliang Zhang, Xingchong Chen, Wenjie Hou et al. 2012. Comparative study on schemes of seismic response reduction for the railway extradosed cable-stayed bridge. *Advanced Materials Research*. 383: 6103–6109.

Advances in Civil Engineering and Building Materials IV – Chang et al (eds)
© 2015 Taylor & Francis Group, London, ISBN: 978-1-138-00088-9

A study on the deterioration of hangers with corroded wires

Xiaoyu Pan, Xu Xie & Xiaozhang Li
Department of Civil Engineering, Zhejiang University, Hangzhou, Zhejiang, China

Hanhua Zhu
Highway Administration Bureau of Zhejiang Province, Hangzhou, Zhejiang, China

ABSTRACT: The remaining capacity of hangers on arch bridges is found to be sensitive to corrosion. In order to analyse the corrosion effect on the mechanical properties of a hanger, wires used for more than a decade on an arch bridge are employed. An accelerated corrosion test and artificial notch simulation are conducted on the wire. The ultimate strength and ductility of wires have been investigated through a tension test and finite element analysis. The study indicates that the elastic modulus and ultimate strength of the wire are not sensitive to pitting. However, the ductility of the corroded wire decreases sharply. Stress concentration is the cause of the deterioration, and the ultimate strain of the wire is significantly influenced by the depth and opening angle of the pits. A classification of corroded wires is proposed and a probability model of the hanger is established. The result of statistical research on the remaining strength shows that the effect of corrosion on hangers is mainly the deterioration of ductility. However, the ultimate strength is not seriously affected, which is the same as for the wires.

Keywords: hanger, corroded wires, finite element analyses, classification method, remaining capacity

1 INTRODUCTION

Half-through or through arch bridges have been widely used in China over the last two decades. However, the durability problem of a hanger which leads to the reduction of service life is exposed during the service process. Bridge accidents due to hanger failure have been frequently reported in recent years (Gao, 2011).

Corrosion is the fatal cause which leads to the degradation of the mechanical properties and service life of a hanger. Studies on the mechanism of corroded high-strength steel wires have been carried out. It is well recognized that the tensile strength of wire decreases linearly with the loss of the effective section, due to corrosion. The actual tensile strength does not decrease, but ductility and fatigue strength significantly decrease with the aggravation of corrosion. The ductility is controlled by the roughness of the surface geometry of the wire.

On the other hand, the reliability analysis of wire bundles is traditionally based on series-parallel system for the capacity of a cable. Faber et al. (2003) use statistical models of parallel wire cables to give a reliability-based assessment of parallel wire cables, both for design and for assessment of cables in existing structures. The length effect and Daniels effect on the ultimate tensile strength of the model are also considered.

For the purpose of evaluating the corrosion effect on strength and ductility, tension tests on accelerated corroded wires and artificial notched wires are

performed. With the help of finite element analysis, a classification of corroded wires is proposed. A probability model of hanger ultimate capacity is also established according to the series-parallel system. Individual wires with different levels of corrosion are simulated by the Monte Carlo method. With the theory of fiber bundle, the remaining capacity of a hanger with corroded wires is evaluated.

2 MECHANICAL PROPERTIES AND CLASSIFICATION OF CORRODED WIRES

2.1 *Tensile properties of accelerated corroded wires*

Wire samples are obtained from the same bridge. To access wires with different corrosion levels, accelerated corrosion methods are employed. The wires are covered with wet gauze and placed in an enclosed box to keep them wet, which simulates the humid environment of the hanger (Suzumura & Nakamura, 2004). The tension test is carried out according to the code. The stress-strain relationships of different wires are plotted in Figure 1. The wires in group A were placed in the natural environment for one and a half years. Wires in group B were under an accelerated corrosion environment for one year, and wires in group C were under an accelerated corrosion environment for one and a half years. Original wires are taken as the control group. It is noted that the ductility of corroded

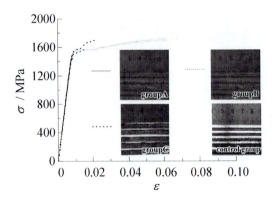

Figure 1. Results of tension test of corroded wires.

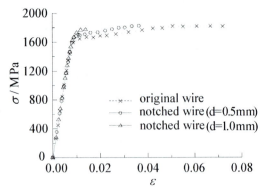

Figure 3. Stress-strain relationship of notched wires.

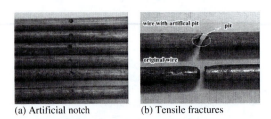

(a) Artificial notch (b) Tensile fractures

Figure 2. Morphology of tension test samples.

wires shows a remarkable decrease, while the elastic modulus yields strength and the ultimate strength of the wires do not change.

2.2 Tension test of wires with an artificial notch

It is known that the stress concentration on a pitting area can lead to the change of mechanical properties (Betti et al., 2005 & Mayrbaurl et al., 2001). Pitting corrosion is one of the corrosion phenomena which affect the strength and ductility of high-strength wires. Since the surface roughness of the wire is mainly caused by pitting, this research is focused on the corrosion pits. To evaluate the influence of pits, an artificial method (Okamoto et al., 2010) is used. Different shapes of pits are simulated by electrical cutting on the surface of the wire. The appearance of hemispherical pits are shown in Figure 2(a), the depths of the pits are 0.5 mm and 1.0 mm, respectively. Figure 2(b) makes a comparison of the tensile fractures between an artificial notched wire and an original wire. It confirms that the tensile fracture of the notched wire is similar to the corroded wire, which means that the artificial pits can be used to analyse the impact of pitting corrosion. It also indicates that stress concentration on a pitting area causes a reduction in ductility.

The stress-strain relationship of a notched wire and an original wire is shown in Figure 3, with the tensile stress σ as the ordinate, and the strain ε as the abscissa; d means the depth of the pit. The curves indicate that the ductility of wires decrease with the growth of the depth, while the elastic modulus and ultimate strength do not decrease. It is therefore believed

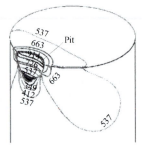

(a) Von Mises stress on pitting area

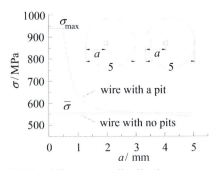

(b) Von Mises stress distribution on cross-section

Figure 4. Results of elastic finite element analysis.

that the elastic modulus and ultimate strength are not sensitive to artificial pits.

2.3 Classification of corroded wires

As mentioned previously that stress concentration is the fatal cause of the change of tensile properties, a elasto-plastic finite element model of a 30 mm long wire with a pit on the surface is established using the finite element software ABAQUS. The material properties are used from the tension test of the original wires, and the Von Mises criterion is used as the yielding criterion.

Figure 4(a) shows the stress distribution on the pitting area, and Figure 4(b) is a comparison of stress

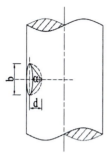

Figure 5.   Diagram of pit geometry parameters.

distribution on a cross-section between a wire with a pit and a wire with no pits. It is apparent that the pitting area has an obvious stress concentration, and stress on the centre of the pit reaches the maximum, while the stress far from the pitting area towards to a constant.

To estimate the corrosion degree of wire, the stress concentration factor K_t (Pilkey, 1997) and ultimate strain factor η are defined as follows:

$$K_t = \sigma_{max} / \bar{\sigma} \qquad (1)$$

$$\eta = \varepsilon_u / \varepsilon_{u0} \qquad (2)$$

where σ_{max} and $(\bar{\sigma})$ are given in Figure 4, and ε_u is the ultimate average strain of corroded wire, ε_{u0} is the ultimate average strain of uncorroded wire.

According to the results of tension tests on notched wires, the relationship between the stress concentration factor and the ultimate strain factor can be expressed by the following equation:

$$\eta = -0.3992K_t^2 + 0.7462K_t + 0.653 \qquad (3)$$

The ultimate strain of wire can be obtained by equations (1)–(3). Through a large amount of finite element models, the ultimate strain is expressed by:

$$\varepsilon_u = 0.0425 + 0.0065\ln(\alpha/d - 1.7193) \qquad (4)$$

where α and d are parameters representing the angle and depth of the pit as shown in Figure 5.

The coefficient of determination (R^2) is 0.9584, which means this method has a good correlation.

By comparing the stress distribution of wires with different amounts of pits, it is seen that the ductility of wire is mainly affected by the worst corroded pit. A classification for corroded wires is proposed according to this research, as shown in Table 1.

3   REMAINING CAPACITY ASSESSMENT OF HANGERS

3.1   Corrosion distribution in hanger

The level of corrosion in a hanger should be investigated before the evaluation of the mechanical properties of the hanger. A dismantled hanger from an arch

Table 1.   Classification for corroded wires.

Grade	Description	Ductility reduction %
I	Wire with metallic lustre or a small amount of white rust.	0
II	Wire with a small amount of ferrous rust; no pits are observed on the surface.	5
III	Wire is covered with ferrous rust, and d_{max}^* is below 0.2 mm.	20
IV	d_{max} is between 0.2 mm and 0.5 mm, and b/d^* exceeds 4.	40
V	d_{max} is between 0.2 mm and 0.5 mm, and b/d is below 4.	50
VI	d_{max} exceeds 0.5 mm, and b/d exceeds 4.	70
VII	d_{max} exceeds 0.5 mm, and b/d is below 4.	100

$^*d_{max}$ is the maximum depth of the pits; b/d is the ratio of the width and depth of the pits.

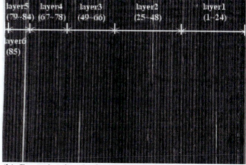

(a) Number of wires

(b) Corrosion level of each layer

Figure 6.   Corrosion distribution in a hanger.

bridge is observed, and the corrosion level of each wire in different layers is shown in Figure 6. Visual inspection demonstrates that the wires in the hanger are almost the same grade if the outer sheath is intact, and the distance between the wire and the cable centroid is irrelevant to the corrosion distribution as assumed by Cremona (2003) and Elachachi et al. (2006). Therefore, the corrosion level of a hanger can be divided by the grade of wires, and considering the worst situation, all the wires in a hanger are of the same grade.

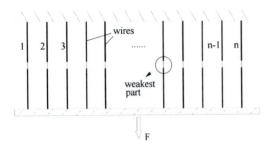

Figure 7. Simplified model of a hanger.

Table 2. Ultimate strain of each wire.

No.	ε'_u	No.	ε'_u	No.	ε'_u
1	0.0652	8	0.0737	15	0.0695
2	0.0850	9	0.0827	16	0.0705
3	0.0750	10	0.0787	17	0.0700
4	0.0809	11	0.0739	18	0.0748
5	0.0686	12	0.0762	19	0.0707
6	0.0815	13	0.0685		
7	0.0651	14	0.0679		

3.2 Hanger model for remaining strength assessment

The strength theory of a fiber bundle is widely used to calculate the remaining strength of a cable. A series-parallel link model is employed, which assumes that the cable is composed of independent parallel wires of the same length, while the wires are made up of a system of weakest link structural systems (Xu & Chen, 2013). Hangers are represented in this research by a simplified model, as shown in Figure 7; the strength of a wire is determined by the weakest part of the wire, and the remaining capacity of a hanger can be obtained by elasto-plastic calculation of the individual parallel wires.

The weakest part in each wire has a random corrosion level, and this randomness can be simulated by the Monte Carlo simulation. Taking the bilinear model as the material constitutive relation, all the wires have the same elastic modulus and ultimate strength. However, the ductility reduces according to the corrosion level. While a hanger is under an axial load, the weakest wire breaks earlier than the others do. After its break, the hanger redistributes its stress. With the increase in axial deformation, more wires break and eventually the hanger fails.

In order to investigate the effect of the corrosion distribution, a variety of situations are analysed. In addition, the reliability assessment of hangers is also obtained by a statistical method.

3.3 Numerical analysis

3.3.1 Model inspection

Considering a hanger with 19 wires, the corrosion level of each wire is a discrete random variable, and the ductility is reduced to grade IV in Table 1. The ultimate strain of each wire is listed in Table 2, where ε'_u is the ultimate strain of the wires after random reduction in the range of 60% to 80%.

By increasing the end displacement of the hanger, wires will break one by one. After the failure of a wire, the redistributed stress of the unbroken wires is in proportion to the distance from the broken one. These wires are still in service until they break. The number of broken wires and the strength of the hanger on the overall process are shown in Table 3, where N is the

Table 3. Failure process of a hanger.

N	ε	F	Diagram	N	ε	F	Diagram
0	0.064	672.6		13	0.075	216.9	
2	0.066	602.3		14	0.076	180.8	
5	0.069	506.2		15	0.079	144.6	
6	0.070	470.1		17	0.082	72.3	
9	0.071	361.6		18	0.083	36.2	
11	0.074	289.3		19	0.086	0	

number of broken wires in the process, and ε and F are the average strain and remaining strength of the hanger, respectively.

As shown in Table 3, the remaining strength reaches its maximum value before the first wire breaks, and it decreases with the increase in the number of broken wires. In addition, the failure time of each wire is relevant to its ultimate strain; the wire with the lowest ultimate strain breaks first.

3.3.2 Statistical analysis of the remaining capacity of hangers

For hangers of different grades, the bearing capacity will be affected by corrosion. It can help to understand the effects by using a statistical analysis of hangers with wires at different corrosion levels.

It is noted in Table 1 that for wires of grade VI and VII, the maximum depth of the pit exceeds 0.5 mm; the loss of the section area due to pitting corrosion cannot be neglected. Therefore, the former five grades are taken into account in this research. 500 group stochastic simulations for each grade are conducted

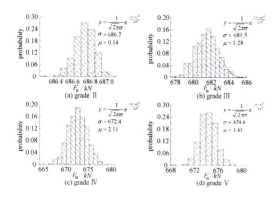

Figure 8.   Probability distribution of ultimate capacity.

Table 4.   Simulation results of different hangers.

Grade	F_{u-mean} kN.m	ε_{u-mean}
I	687.0	0.1000
II	686.7	0.0999
III	681.5	0.0941
IV	672.4	0.0775
V	674.4	0.0568

to calculate the remaining capacity of the hanger. Figure 8 gives this probability distribution for 500 group simulations of each grade, where F_u is the ultimate capacity of the hanger. The mean and the standard deviation based on 500 simulations are also shown.

As plotted in Figure 8, the ultimate capacity of the hanger can be modelled as a normal distribution with a small discreteness.

The probability distribution of the maximum plastic strain is similar to the results shown in Figure 8. By using the Monte Carlo method, the ultimate capacity and the maximum plastic strain of every grade of hanger are calculated, as listed in Table 4, where F_{u-mean} and ε_{u-mean} represent the mean value of the ultimate capacity and maximum plastic strain respectively.

It is shown in Table 4 that the ultimate capacity of the hanger shows a downward trend from grade I to grade V, but it is not significantly affected by corrosion. However, the ultimate plastic strain dramatically decreases as the grade rises. This fact indicates that corrosion has a significant influence on the ductility of the hanger. Moreover, when the grade of hanger is lower than III, the ductility is rarely affected by corrosion as well as the ultimate capacity, but if the grade reaches V, the ductility reduces by a half, and, therefore, the safety performance of the hanger is under threat. According to the hanger assessment, grade V can be the limit to determine whether the hanger should be replaced or not.

4   CONCLUSION

The deterioration of corroded wires and hangers is studied in this paper. Some conclusions are drawn:

(1) The elastic modulus and strength of wires are not sensitively influenced by corrosion; however, the ductility is significantly affected by corrosion. Pits on the surface of the wires cause the reduction of ductility.

(2) Pitting corrosion deteriorates the tensile properties because of the stress concentration on the surface. Therefore, it is reasonable to simulate pitting corrosion by an artificial notch on the wire.

(3) A classification for corroded wires is proposed. The simplified model of a hanger is established and random reduction of the ductility of wires is considered, therefore enabling a numerical analysis of hangers to be carried out.

(4) Corrosion has a significant influence on the ductility of a hanger. For assessing the remaining capacity of a hanger, the decrease of ductility due to corrosion should be taken into consideration.

This research mainly focuses upon the effects of corrosion. In reality, hangers are also affected by manufacturing error and transverse deformation, etc. Further research will be conducted in the future on these problems.

ACKNOWLEDGEMENT

The author acknowledges and appreciates the financial support of the National Natural Science Foundation of China (51378460).

REFERENCES

Barton, S.C., Vermass, G.W., Duby, P.F., West, A.C. & Betti, R. 2000. Accelerated corrosion and embrittlement of high-strength bridge wire. *Journal of materials in civil engineering* 12(1): 33–38.

Betti, R., West, A.C., Vermass, G.W. & Cao, Y. 2005. Corrosion and embrittlement in high-strength wires of suspension bridge cables. *Journal of Bridge Engineering* 10(2): 151–162.

Cremona, C. 2003. Probabilistic approach for cable residual strength assessment. *Engineering Structure* 25(3): 377–384.

Daniels, H.E. 1945. The statistical theory of strength of bundles of threads, part I. *Proceedings of the Royal Society of London*, Series A 183: 405–435.

Elachachi, S.M., Breysse, D., Yotte, S. & Cremona, C. 2006. A probabilistic multi-scale time dependent model for corroded structural suspension cables. *Probabilistic Engineering Mechanics* 21(3): 235–245.

Faber, M.H., Engelund, S. & Rackwitz, R. 2003. Aspects of parallel wire cable reliability. *Structural Safety* 25: 201–225.

Gao, X. 2011. Analysis methods for suspender damage and system reliability of existing concrete filled steel tubular arch bridge. Harbin: Harbin Institute of Technology.

General Administration of Quality Supervision, Inspection and Quarantine of the People's Republic of China, 2002. Steel wires for the prestressing of concrete. Beijing: China Standard Press.

Mayrbaurl, R.M. & Camo, S. 2001. Cracking and fracture of suspension bridge wire. *Journal of Engineering* 6: 645–650.

Nakamura, S., Suzumura, K. & Tarui, T. 2004. Mechanical properties and remaining strength of corroded bridge wires. *Structural Engineering Internationa* 14(1): 50–54.

Okamoto, Y., Nakamura, S. & Suzumura, K. 2010. Measurement of corrosion roughness of galvanized bridge wires and fatigue strength of wires with artificial pits. *Journal of Structural Mechanics and Earthquake Engineering* 66(4): 691–699.

Pilkey, W.D. 1997. *Peterson's Stress Concentration Factors*. New York: John Wiley & Sons.

Suzumura, K. & Nakamura, S. 2004. Environmental factors affecting corrosion of galvanized steel wires. *Journal of Materials in Civil Engineering* 16(1): 1–7.

Xu, J. & Chen, W.Z. 2013. Behavior of wires in parallel wire stayed cable under general corrosion effects. *Journal of Constructional Steel Research* 85: 40–47.

Advances in Civil Engineering and Building Materials IV – Chang et al (eds)
© 2015 Taylor & Francis Group, London, ISBN: 978-1-138-00088-9

A method of hanger tension identification of an arch bridge considering the effect of boundaries

Tingting Zhang, Xu Xie & Xiaozhang Li
Zhejiang University, Hangzhou, Zhejiang, China

ABSTRACT: Hanger tension plays a significant role in the management of the structural state of arch bridges and health monitoring. The vibration frequency method is most widely used for measuring the tension force of hangers during their service term. On the purpose of improving the accuracy of measured tension force, a multi-frequency method for hanger tension measurement, realized by a genetic algorithm, is proposed. It combines the generation algorithm and the FEM (Finite Element Model) of hangers, and utilizes two or more natural frequencies that can be picked, consecutive or not, to solve the inverse eigenvalue problems, thus identifying the hanger tension. By comparing the numerical simulation results and field measurement, the efficiency and accuracy of the proposed method is testified and verified.

Keywords: tension identification, equivalent dynamical model, genetic algorithm, frequency-based method

1 INTRODUCTION

The vibration frequency method is a principle approach in measuring the tension force of in-service hangers, but once applied, the method with single frequency neglects the effect of flexural rigidity and boundary rotational constrains, and the accuracy of the estimated tension force is not ideal. In order to improve the precision of the vibration frequency method, a variety of approaches has been proposed. For example, H Zui (Hiroshi Zui et al., 1996), W-X Ren (Wei-XinRen et al., 2005) gave empirical formulas taking the effects of flexural stiffness and sag of a cable into account. Z. Fang (Fang Zhi & Wang Jianqun, 2012) proposed a practical formula to estimate the cable tension based on the transverse vibration equation of a cable with consolidation boundaries under axial force, and by using a curve-fitting technique in a simple explicit form, in which the bending stiffness is concluded and the sag effect is ignored. Besides, some other researchers estimate tension force by using natural frequencies of low-order vibration modes, and various methods were posed in order to solve reversely the eigenvalue equations. Among them, Sagues (A. A. Sagüés et al., 2006) tried the trial algorithm. S. Lagomarsino (S. Lagomarsino & C. Calderini, 2005) used the first three natural frequencies of the tie-rod to identify tensile force and the rotational restrict at both ends, based on a minimization procedure of a suitable error function to obtain the characteristic equation of the structural system. Kim

(B. H. Kim & T. Park, 2007) figured up a finite element model that can consider both sag-extensibility and flexural rigidity, and applied a frequency-based, sensitivity-updating algorithm to identify the model. Ceballos (M. A. Ceballos & C. A. Prato, 2008) presented an explicit analytical expression for the natural frequencies, taking both the bending stiffness of the cable and the rotational restriction at the ends into account. Xu Xie (Xu Xie & Xiaozhang Li, 2014) utilized multiple natural frequencies, through genetic algorithms to solve inverse eigenvalue equations, and thus searched and identified the hanger tension force.

However, the mentioned methods all need determined boundary conditions, such as boundary or damper locations, which are unknown in some hangers of existing arch bridges after long-term service. In order to exclude the effect of uncertain parameters related to boundary conditions, this paper developed a method that uses multiple combinations of measured natural frequencies to estimate hanger tension with unknown boundary locations. By numerical simulation and field measurement on an arch bridge, the accuracy and efficiency of the proposed method is verified.

2 BASIC PRINCIPLES AND METHOD

2.1 *Principles of hanger tension identification*

Figure 1(a) shows an actual in-service hanger. The vibration frequency method is widely used to measure

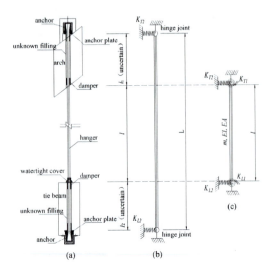

Figure 1. Calculating model based on the vibration characteristic of hangers.

tension based on the relationship between natural frequencies and the tension force of a hanger. In order to build up the correspondence between frequencies and tension force, the general solution is to take all other parameters, such as the bending stiffness, boundary conditions, and the length of hangers, except tension force, as known quantities, based on the taut string vibration theory as shown in Figure 1(b), to determine the hanger tension. Indeed, some hangers are equipped with dampers at both the upper and lower end, whose location and stiffness significantly affect the accuracy of the measured tension force through the calculating process, according to measured natural frequencies. Thus, the location of dampers is of great importance in hanger tension measurement.

To avoid the effect on the measurement of the tension force caused by unknown parameters at anchorage, such as location of dampers and stiffness of restricts, this paper proposes a new method for calculating the tension force of hangers according to the equivalent dynamic model as shown in Figure 1(c), in which the length of the hanger is defined with its exposed part. To equate the stiffness of those buried in the arch rib and tie beam, elastic lateral and rotational constraints on both ends of the exposed part of the hangers are set, thus simulating the vibration characteristics of the actual hanger as shown in Figure 1(c). In Figure 1(c), E represents the elastic modulus of a hanger in kN/m², I is the section moment of inertia in m⁴, A is the sectional area in m², and m is the mass per unit length in t/m. Meanwhile, T is the hanger tension in kN, K_{T1} and K_{L1} are the rotational stiffness of the upper and lower ends, respectively, and K_{T2} and K_{L2} are the lateral stiffness of each end. The parameters are as mentioned, except that l and m are all uncertain.

The natural frequencies of a hanger shown in Figure 1(c) can be determined by using the following

eigenvalue equation, which describes the relationship between the hanger tension T and the natural frequency f:

$$\left| \mathbf{K} + T\mathbf{K}_G - (2\pi f)^2 \mathbf{M} \right| = 0 \tag{1}$$

where $\mathbf{M}$ = mass matrix; $\mathbf{K}$ = elastic stiffness matrix; $\mathbf{K}_G$ = geometric stiffness matrix generated by unit tension; f = natural frequency. The matrix $\mathbf{M}$, $\mathbf{K}$, and $\mathbf{K}_G$ can be obtained according to the theory of FEM.

The stiffness matrix $\mathbf{K}$ is assembled by structural parameters including the unknown bending stiffness EI, and the rotational restrict stiffness K_T and K_L. Thus, the above equation contains six unknown parameters: bending stiffness (EI), boundary restrict stiffness both on the upper $(K_{T1}$ and $K_{T2})$ and lower end $(K_{L1}$ and $K_{L2})$, as well as the tension force of a hanger (T).

By focusing on measuring the tension force of hangers with field-measured natural frequencies, the followed eigenvalue equations are applied.

$$\begin{cases} \left| \mathbf{K} + T\mathbf{K}_G - (2\pi f_i)^2 \mathbf{M} \right| = 0 \\ \left| \mathbf{K} + T\mathbf{K}_G - (2\pi f_j)^2 \mathbf{M} \right| = 0 \\ \left| \mathbf{K} + T\mathbf{K}_G - (2\pi f_r)^2 \mathbf{M} \right| = 0 \\ \dots \end{cases} \tag{2}$$

where f_i, f_j, f_r represent any order of natural frequencies of vibration modes by field measurement.

However, it is difficult to determine the variables in Equation (2), as the solution of inverse eigenvalue equations is also a problem of solving transcendental equations. On this basis, this paper utilizes Genetic Algorithm (GA) to solve the inverse eigenvalue equations, thus identifying the tension force of the hangers.

Figure 2 shows the flow chart of GA in solving transcendental equations, thereby searching and identifying the most optimal tension force of a hanger. Firstly, input m, l, and field-measured natural frequencies as the known parameters, and randomly generate initial populations with combinations of the six variables: T, EI, K_{L1}, K_{T1}, K_{L2}, and K_{T2}, in the related regions. Secondly, according to each sample's input parameters, the GA calculates their theoretical frequencies through solving eigenvalue equations, and by comparing these with field-measured natural frequencies, each sample's fitness value, with which all the samples are sorted in descending order, can be determined Thirdly, male parents are selected by an adaptive selection operator to generate a new individual through an arithmetic crossover operator. Once the newly generated individual has a higher fitness value compared with the worst sample, that has the least fitness value, in the elder generation, survive the fittest. Meanwhile, the elite individuals in the elder generation are retained into the next generation in case the good genes are lost. Additionally, a mutation is implemented to promote the diversity of populations. When the frequency precision of the best individual in one generation reaches

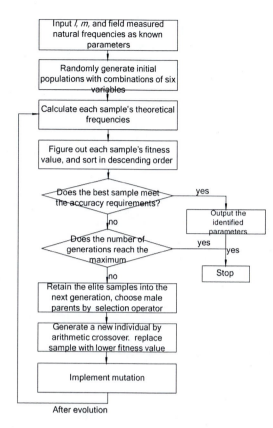

Figure 2. Flow chart of GA for hanger tension identification.

The flow chart contains the following boxes:

Input *l*, *m*, and field measured natural frequencies as known parameters

Randomly generate initial populations with combinations of six variables

Calculate each sample's theoretical frequencies

Figure out each sample's fitness value, and sort in descending order

Does the best sample meet the accuracy requirements? — yes → Output the identified parameters

no ↓

Does the number of generations reach the maximum — yes → yes → Stop

no ↓

Retain the elite samples into the next generation, choose male parents by selection operator

Generate a new individual by arithmetic crossover. replace sample with lower fitness value

Implement mutation

After evolution

Table 1. Simulating parameters of hangers and relative theoretical natural frequencies.

Pa^*	Unit	$l = .0$ m	3.0 m	4.0 m	5.0 m
A	cm^2	20	20	20	20
EI	kN.m^2	50.0	50.0	50.0	50.0
EA	MN	410	410	410	410
T	kN	500	500	500	500
m	kg/m	15.7	15.7	15.7	15.7
K_{T1}	kN/m	500	500	500	500
K_{T2}	MN.m/rad	5.0	5.0	5.0	5.0
K_{L1}	kN/m	500	500	500	500
K_{L2}	MN.m/rad	5.0	5.0	5.0	5.0
f_1	Hz	70.387	39.487	27.272	20.825
f_2	Hz	168.495	88.601	58.848	43.900
f_3	Hz	305.955	152.918	97.741	70.990
f_4	Hz	485.769	234.807	145.636	103.262
f_5	Hz	708.975	335.289	203.418	141.428
f_6	Hz	976.010	454.848	271.563	185.918

*Pa represents parameter.

Table 2. Identification results of hanger tension with different combinations of theoretical frequencies.

Case	f (Hz)	T (kN) $l = 2.0$ m	3.0 m	4.0 m	5.0 m
1	f_1	1269.92	899.27	762.59	694.79
2	f_1–f_2	499.99	500.00	500.00	500.00
3	f_1–f_3	500.00	500.00	499.99	499.99
4	f_1–f_4	499.48	500.00	500.00	500.00
5	f_1–f_5	499.74	500.00	499.99	500.00
6	f_1–f_6	499.85	500.00	500.00	499.99
7	f_1, f_3, f_5	499.99	499.99	499.99	500.00
8	f_1, f_2, f_4	499.33	500.00	499.99	500.00
9	f_1, f_3, f_4	499.54	500.00	500.00	500.00
10	f_1, f_2, f_6	499.99	500.00	500.00	499.99
11	f_1, f_3	499.99	499.99	500.00	500.00
12	f_1, f_4	499.40	500.00	499.99	500.00
13	f_1, f_5	500.00	500.00	499.99	500.00
14	f_1, f_6	500.00	499.99	500.00	500.00

the required value, or the number of generation reaches the maximum value, end computing.

2.2 Efficiency verification of the proposed method by theoretical simulation

To testify the feasibility and accuracy of the measured tension force of the proposed method, a numerical simulation of tension identification is built. On this focus, the method uses theoretical frequencies calculated by simulated data of hangers as shown in Table 1, by solving inverse eigenvalue problems to identify the optimal tension force, as shown in Table 2. Then, through a comparison with the simulated tension data, the feasibility of the method is verified. Table 1 displays the input parameters of four simulated hangers, 2 m, 3 m, 4 m, and 5 m in length respectively, and each hanger's theoretical frequencies. In Table 1, *EA* represents the tensile stiffness, and f_1 to f_6 are natural frequencies from the 1st to the 6th vibration mode.

Table 2 shows the results of the identified tension force with GA according to different combinations of the 1st to the 6th frequencies of vibration modes. In particular, $f_j - f_k$ means utilizing frequencies from the *j*th to the *k*th vibration mode while f_j, f_k means only the two frequencies are used in searching for the optimal

tension force. In Table 2, up to fourteen cases of various combinations of frequencies are used in tension identification. It can be indicated that the precision of the estimated tension force, based on string theory with only fundamental vibration frequency (f_1), is negative. Actually, its calculated tension is significantly larger than the simulated tension force (500 kN) for figuring frequencies, as mentioned earlier. In addition, with the increase of the length of the hangers, its calculated hanger tension, based on string theory to 500 kN. In contrast, the proposed multi-frequency method with GA gives ideal results for identified tension, whose accuracy is quite acceptable as shown in Table 2, whether using consecutive or inconsecutive frequencies.

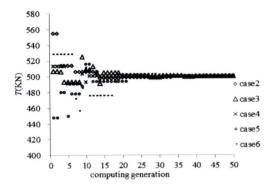

Figure 3. Hanger tension searching process in computing generation of GA with different combinations of natural frequencies.

The tension force searching process in computing generation of GA with different cases of combinations of natural frequencies is illustrated in Figure 3. When the number of computing generation reaches about 30, the identified tension force keeps stable at about 500 kN. Besides the high precision, the multi-frequency method also improves the efficiency in the tension searching process.

3 FIELD TEST VALIDATION FOR APPLICATION

To verify further the validation of the proposed method, field tests are conducted on some in-service hangers of an arch bridge in Zhejiang, China, which is a three-span, half-through concrete arch bridge, with 30 m, 40 m, and 50 m spans, respectively. As is shown in Figure 4, the bridge is of a floating system structure. The hanger consists of 85 parallel steel wires of 5 mm diameter. The mass per unit length is 13.10 kg/m. According to the deck load that each hanger bears, the tension force of a hanger can be deduced to be about 500 kN.

Five short hangers are selected as research objects to testify the feasibility of the proposed method. The relative measured parameters are listed in Table 3, where the length is determined by the exposed part from the lower end of the arch rib to the upper end of the deck. The natural frequencies of the hangers are picked by DH5901 (DongHua testing technology Co. Ltd.), a hand-held dynamic signal analyser, which can display two valid data after the decimal point of the frequency. Table 3 lists the 1st to the 6th natural frequencies of the vibration modes of each hanger, where the slash signal (/) means the frequency in that order cannot be picked by DH5901.

Table 4 shows the identification results of the hanger tension with different combinations of field-measured natural frequencies. In particular, they are divided into four groups, with the first two, three, or four vibration modes of field-measured natural frequencies to identify the hanger tension respectively, while the last

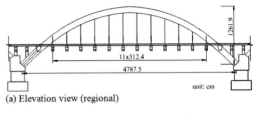

(a) Elevation view (regional)

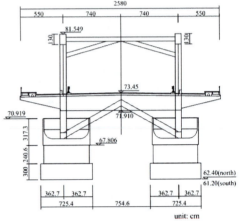

(b) Deck section

Figure 4. General conditions of the arch bridge.

Table 3. Field measured natural frequencies of in-service hangers.

	Measured frequency (Hz)					
l (m)	1st	2nd	3rd	4th	5th	6th
2.32	51.56	117.34	183.13	285.78	390.63	/
2.69	48.59	110.16	191.41	294.38	235.00	313.13
3.78	30.63	66.72	112.03	167.97	/	/
4.29	26.69	57.06	93.38	137.63	199.94	/
5.89	18.44	38.81	62.31	89.81	122.50	159.88

column lists the estimated hanger tension based on string theory with fundamental frequency. It can be illustrated that the identified hanger tension by the proposed method is consistent with the value dispersed on each hanger that is calculated from the total deck load, due to the floating system of the structure, which strongly verified the feasibility and accuracy of the proposed method. However, the result of hanger tension based on string theory is significantly deviates from the theoretical tension force. Furthermore, with the increase of the hanger length, the calculated tension force by string theory becomes more precise, which is in accordance with the numerical simulation mentioned earlier.

Figure 5 shows the identification process in computing generation of the shortest hanger ($l = 2.32$ m) with three combinations of natural frequencies. It can

Table 4. Identification tension results of in-service hangers with different combinations of frequencies (unit: kN).

Theoretical T	with $4f$	with $3f$	with $2f$	with $1f$
500	489.1434	488.7726	487.4463	765.0803
500	491.7536	487.4878	484.3241	913.4895
500	452.9454	451.2974	452.3248	716.7749
500	482.1987	481.8556	476.9812	700.9983
500	464.8315	461.4525	456.2175	630.7505

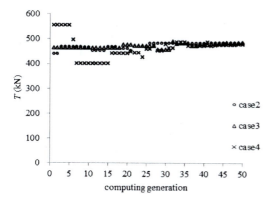

Figure 5. Identification process in computing generation of the shortest hanger ($l = 2.32$ m) with three combinations of natural frequencies.

be illustrated that by adopting two or more natural frequencies, the identified hanger tension keeps stable within about 40 computing generations, which promises the efficiency of the method.

4 CONCLUSIONS

In this study, a multi-frequency method applied with GA is proposed for tension measurement of short hangers. With the focus on solving the problem of the boundary uncertainty of in-service hangers, the paper suggests the exposed length of the hangers, instead of the exact one, as an input parameter, and uses an equivalent FEM model of the elastic supporting hanger to identify the tension force. The accuracy and efficiency of the proposed method are verified through both theoretical simulation and field tests. In addition, it also has a better applicability due to easy operation in field measurement.

ACKNOWLEDGEMENT

The authors would like to thank the Natural Science Foundation of China (51378460), for the financial support for this project.

REFERENCES

A. A. Sagüés, S. C. Kranc, M.ASCE and T. G. Eason, 2006. Vibrational tension measurement of external tendons in segmental posttensioned bridges. Journal of Bridge Engineering 11(5), 582–589.

B. H. Kim, T. Park, 2007. Estimation of cable tension force using the frequency-based system identification method. Journal of Sound and Vibration 304, 660–676.

Fang Zhi, Wang Jian-qun, 2012. Practical formula for cable tension estimation by vibration method. Journal of Bridge Engineering 17(1), 161–164.

Hiroshi Zui, Tohru Shinke and Yoshio Namita, 2006. Practical formulas for estimation of cable tension by vibration method. Journal of Structural Engineering 122(6), 651–656.

Marcelo A. Ceballos, Carlos A. Prato, 2008. Determination of the axial force on stay cables accounting for their bending stiffness and rotational end restraints. Journal of Sound and Vibration 317, 127–141.

Sergio Lagomarsino, Chiara Calderini, 2005. The dynamical identification of the tensile force in ancient tie-rods. Engineering Structures 27, 846–856.

Wei-XinRen, Gang Chen and Wei-Hua Hu, 2005. Empirical formulas to estimate cable tension by cable fundamental frequency. Structural Engineering and Mechanics 20(3), 363–380.

Xu Xie, Xiaozhang Li, 2014. Genetic algorithm-based tension identification of hanger by solving inverse eigenvalue problem. Inverse Problems in Science and Engineering 22(6), 966–987.

Advances in Civil Engineering and Building Materials IV – Chang et al (eds)
© 2015 Taylor & Francis Group, London, ISBN: 978-1-138-00088-9

Dynamic responses of an aluminium alloy footbridge under stochastic crowd loading

Jingru Zhong & Xu Xie
College of Civil Engineering and Architecture, Zhejiang University, China

ABSTRACT: Characterized by high strength-to-weight ratio, strong corrosion resistance, as well as good formability, aluminium alloy is a novel material which is used more and more widely for footbridges in China. However, since pedestrian loading takes a much higher proportion of the structural quality, and the elastic module of aluminum alloy is one third of that of steel, the dynamic features of aluminum alloy structures must be quite distinct from that of steel or reinforced concrete structures. Given the vibration serviceability problem which occurs in many footbridges, and the study shortage on this kind of footbridge, this paper analyses the dynamic responses of an aluminum alloy footbridge under stochastic crowd loading to give a reference for the approaching design of this bridge type.

Keywords: aluminum alloy footbridge, dynamic, stochastic crowd, foot force

1 INTRODUCTION

Since the 1930s, aluminium alloy has emerged as a substitution for steel, especially in Europe, America, and Japan, with the advantages of having low maintenance and management costs, and being lightweight with a high strength-to-weight ratio. Recently, aluminium alloy footbridges are developing rapidly in China, where ten aluminium alloy footbridges have been constructed in the last eight years. The rise of this kind of footbridge has attracted some attention in China, but reference documents for this kind of bridge are lacking. Therefore, a sensible specification for aluminum alloy footbridge design is urgently required.

The structural design of a footbridge must ensure its safety and serviceability. Most notably, the dynamic serviceability problems of modern footbridges have become a primary research topic in recent decades (Živanović S, 2005), which mainly focus on the simulation of the foot force model and the vibration perception threshold (Parsons, 1988). The present Chinese footbridge specification (CJJ 69–95, 1995) sets the serviceability standard based on the structure's natural frequency limit, regardless of different materials or structure features. It has been widely applied to steel or reinforced concrete bridges, but has not been tested on aluminum alloy bridges.

Given the dynamic serviceability issues and the emerging development of the aluminium alloy footbridge in China, this paper provides a comprehensive analysis on the dynamic responses of the aluminum alloy footbridge under stochastic crowd loading, as an example for the reference of the approaching design of this bridge type.

2 MODEL OF BRIDGE AND STOCHASTIC CROWD LOADING

2.1 *Bridge structure and its finite element model*

The example used is the Fuxin footbridge, a footbridge in Hangzhou, for the dynamic responses analysis of the aluminum alloy footbridge under stochastic crowd loading. The bridge has the girder of a prefabricated aluminium alloy truss structure, with continuous spans of $15 + 19$, and a simple supported span of $26\,\text{m}$ (shown in Figure 1). Here, we only calculate the $26\,\text{m}$ span.

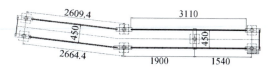

Figure 1a. General plane layout.

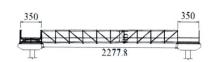

Figure 1b. Elevation.

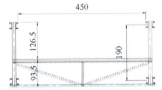

Figure 1c. Cross section. Unit: (cm).

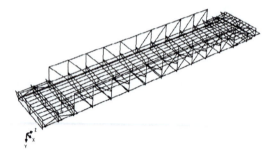

Figure 2. Finite element model of the Fuxin footbridge.

Table 1. Value and mode shape type of the first five natural frequencies.

Mode No.	Frequency Hz	Mode Shape Type
1	6.015	Lateral 1st order symmetry bending
2	8.056	Vertical 1st order symmetry bending
3	9.152	Torsion
4	11.407	Lateral 2nd order symmetry bending
5	12.782	Handrail locally symmetry bending

A finite element model with three-dimensional beam elements is used for this calculation, and pedestrian are randomly located on the deck.

The Rayleigh damping set in this model is employed and described as:

$$C = \alpha_0 M + \beta_0 K \tag{1}$$

where, C, M, and K indicate the damping, mass, and stiffness matrix respectively. Coefficients α_0 and β_0 are both equal to 0.02, of which the corresponding damping ratio is around 0.5%.

The first five natural frequencies and the corresponding mode shape types are shown in Table 1. The first model shape is 1st order symmetry bending in a lateral direction. The frequency of the 1st order symmetry bending in a vertical direction reaches to 8.056 Hz, far beyond the normal pacing frequency, which means vibration renounce is unlikely to happen.

2.2 Single foot force model

Since the 1970s, many scholars have researched pedestrian foot force (Racic, 2009). Pedestrian locomotion can be defined as walking, running, and jumping,

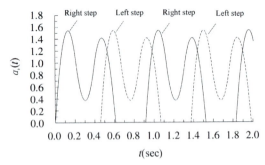

Figure 3. Shape of the vertical DLF of 2.0 Hz pacing frequency.

Table 2. Stochastic pedestrian parameter.

Parameters	Average	Ranges of random number
Weight	G_0 (=0.65 kN)	$G_i \in [0.95, 1.05]G_0$
Walking Width	W_0 (=0.2 m)	$W_i \in [0.95, 1.05]W_0$
Walking Length	L_0	$L_i \in [0.95, 1.05]L_0$
Frequency	f_0 (=1.5 ~ 2.0 Hz)	$f_i \in [0.95, 1.05]f_0$
Location	Y_0	$Y_i \in [-1.0, 1.0]Y_0$
Delete Time	T_total	$T_i \in [0,1]T\_total$

and the force model is developed in time domain and frequency domain. Reports by Andriacchi, Wheeler, etc. (Wheeler, 1982; Andriacchi, 1977) are considered as the earliest comprehensive studies on the force-time history. Following these, the Fourier trigonometric series was proposed to simulate the foot force in frequency-domain (Giakas, 1997). Nowadays, a biomechanical force model (Qin, 2013) is gradually being developed for a more accurate simulation.

Based on the characteristic shape of footfall force and many different force-time models, a typical shape of vertical foot-force Dynamic Loading Factor (DLF) varying from different walking frequencies is referring to fib (fib, 2005). This DLF shape is used to simulate the single footfall, which is exerted alternatively in accordance with the given velocity. For instance, Figure 1 shows the shape of the vertical dynamic load factor (DLF) $\alpha_v(t)$ of 2.0 Hz pacing frequencies.

2.3 Stochastic crowd loading

Pedestrians are distributed randomly on the bridge in actual cases, including the pedestrian's weight, pacing frequency, location, step length, etc. To obtain a more reasonable dynamic response result, this paper calculates the vibration responses in a probabilistic way, by setting stochastic pedestrian parameters (in Table 2). According to Bachmann and Venuti (Bachmann, 1987; Venuti, 2012), the average pedestrian weight G_0 and bipedal width W_0 are set as 0.65 kN and 0.2 m respectively; the mean pacing frequency f_0 is set as 1.5–2.0 Hz, with which the average walking length

l_0 is related. These four parameters are randomly generated by the Monte Carlo method varying from 95% to 105% of the mean value. In addition, Y_0 means half of the deck width, and the pedestrian could be located in the range of $-Y_0$ to Y_0; T_total means the total time that a single pedestrian passes through the deck. The time into the deck (T_i) is randomly set as 0 to T_total.

3 DYNAMIC RESPONSES

3.1 Dynamic response under a single pedestrian

In order to check the reliability of this calculation program, a single pedestrian is firstly simulated and located on the middle girder passing through the bridge at 1.8 Hz (a normal pacing frequency). The time-history responses (with a peak of 0.047 m/s²) and the FFT spectral analysis results in midpoint are shown in Figure 4. The spectral picture shows that harmonic vibration mainly concludes the integral multiply of the extra excitation, which is not consisted of the structural natural frequencies.

3.2 Dynamic response under stochastic crowding loading

3.2.1 Time-history response and spectral analysis of one random case

Figure 5 shows the time-history response of a crowd stream with stochastic walking parameters. Pedestrians can be recycled continuously, and the stream will be stable after twice as long as the full passing period. In this case, the pedestrian density is 0.5 person/m², and the mean pacing frequency is assumed as 2.0 Hz. Similarly, the spectral analysis indicates that the extra excitation is the main component of the structure vibration. However, owing to the frequency randomness, the bandwidth of each peak frequency becomes greater.

3.2.2 Probability character of structure dynamic responses

Large numbers of stochastic samples, like the one shown in section 3.2.1, are generated by the Monte Carlo method to obtain the dynamic probability character. Research by Bachmann, Živanović, and fib, etc. propose that the mean pacing frequency of normal adults is in the range of 1.5 Hz to 2.5 Hz, and the pedestrian density is generally between 0.1 person/m² to 1.5 person/m². In order to simplify things, this paper set the pacing frequencies of 1.5, 1.75, and 2.0 Hz, and the pedestrian densities of 0.5 to 1.0 person/m² as the calculation cases.

The average maximum acceleration of fifty random cases of every different walking condition is finally extracted (shown in Figure 6). They indicate that a rise in either the pedestrian density or the pacing frequency will cause a stronger dynamic response and, comparably, the pacing frequency is of more consequence.

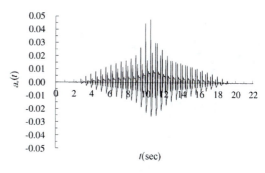

Figure 4a. Vibration response under single pedestrian (m/s²).

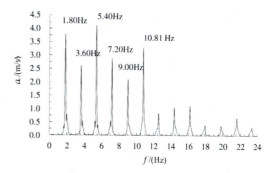

Figure 4b. Spectral analysis for the response of a single pedestrian passing through (m/s²).

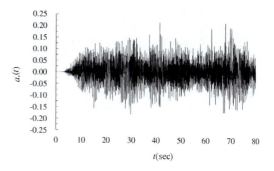

Figure 5a. Vibration response under stochastic crowd (m/s²).

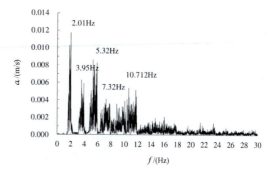

Figure 5b. Spectral analysis for the response under stochastic crowd stream (m/s²).

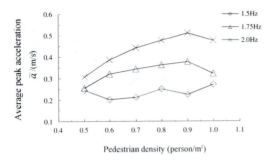

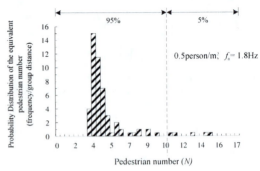

Figure 6. Mean value of the maximum acceleration corresponding to different walking conditions.

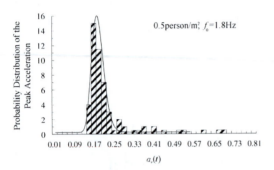

Figure 7. Frequency histogram of the maximum accelerations under the condition of 1.75 Hz, 0.8 person/m².

One more interested thing based on a specific walking condition, is how the dynamic responses are distributed. As an example, 100 random cases under the condition of 1.8 Hz, 0.5 person/m² are generated. The frequency histogram of the maximum accelerations is shown in Figure 7. It indicates that the histogram is similar with the shape of lognormal distribution. Almost all the maximum responses are less than 0.8 m/s², which is in the range of medium comfort level according to the EN03 specification (EN03, 2007).

3.3 Maximum equivalent pedestrian number

3.3.1 Reference on equivalent pedestrian number
As for the convenient engineering application, some researchers intended to describe the structure maximum response by the single pedestrian-induced maximum response. According to the stochastic vibration theory, Matsumoto earlier reported a calculation equation of the equivalent pedestrian number, assuming a Poisson distribution of pedestrian arrival probability (Matsumoto, 1978):

$$N_p = \sqrt{n} \qquad (2)$$

where n is the whole pedestrian number at any given time.

Based on Matsumoto, Bachmann (1987) proposed that for structural frequencies between 1.8 Hz to

Figure 8. Frequency histogram of the maximum equivalent pedestrian number under the condition of 1.75 Hz, 0.8 person/m².

1.6 Hz, and between 2.2 Hz to 2.4 Hz, this equivalent number should be reduced linearly.

In 2006, the French Ministry of Transport provided a concept of the maximum equivalent number (Setra, 2006), which was aimed at a specific vibration mode and only applied to the case when people density is less than 1.0 person/m². In the sense of probability, the pacing frequencies and phases are set to have a normal and uniform distribution respectively. Certain numbers of random samples are generated by the Monte Carlo method and this equivalent value should be larger than 95% of the samples:

$$N_p = 10.8\sqrt{m \cdot \xi_i} \qquad (3)$$

where m is the whole pedestrian number at any given time, and ξ_i is the damping ratio of the ith mode.

On the foundation of the calculation methods above, a similar concept of maximum equivalent pedestrian number is put forward. Random pedestrian models mentioned in section 2.2 are generated, and different equivalent numbers corresponding to the maximum dynamic responses will have a frequency histogram by consequence. The maximum equivalent pedestrian number that ranks in the 95% percentile of all samples could be described as:

$$\left. \begin{aligned} N_{p,i} &= A_{max.i} / a_{max} \\ P\{N_{p,i} \le N_{p,max}\} &= 95\% \end{aligned} \right\} \qquad (4)$$

where $N_{p,max}$ is the maximum equivalent number; $N_{p,I}$ is the equivalent number of stochastic case i; $A_{max,i}$ is the peak acceleration of stochastic case i; a_{max} is the peak acceleration under single pedestrian excitation.

3.3.2 Maximum equivalent pedestrian number
Based on the probability histogram depicted in section 3.3.2 and the peak acceleration under single pedestrian (0.047 m/s²) shown in section 3.2, we divide the 0.5 person/m²-induced peak acceleration by that of the single excitation. Consequently, the probability histogram of the equivalent pedestrian number is developed. The value ranking in the 95 percentile of

the whole number is defined as the maximum equivalent pedestrian number. In this case, the whole number of pedestrians with density of 0.5 person/m^2 is 68 people, and the maximum equivalent number is calculated as 10.4 people, which is larger than the result obtained by the Matsumoto method ($68^{1/2} = 8.2$ people).

4 CONCLUSION

In order to obtain a better understanding of the dynamic features of aluminium alloy footbridges, this paper goes in depth into the study on the structure dynamic responses of an actual aluminum alloy footbridge with stochastic crowd flow excitation. Pedestrian parameters like pacing frequency, weight, time into deck, etc. are randomly generated by the Monte Carlo method.

Fifty cases with every different condition from 0.5 person/m^2 to 1.0 person/m^2 pedestrian density, with a mean pacing frequency of 1.5 Hz, 1.75 Hz, and 2.0 Hz are calculated. Results show that either a rise in the pedestrian density or pacing frequency will cause a stronger dynamic response; when comparing these, pacing frequency is of more consequence.

The responses of 100 cases of a specific walking condition are obtained. The corresponding probability histogram shown in the paper has a typical shape of lognormal distribution. The calculation results indicate that the maximum acceleration of an aluminum alloy footbridge under stochastic is generally less than 0.8 m/s^2, a figure that belongs to the medium comfort level and could meet the serviceability requirement.

In addition, based on the response probability distribution, the concept of a maximum equivalent pedestrian number is put forward. Under the same traffic conditions, the equivalent pedestrian number obtained in this paper is larger than the number provided by the Matsumoto method.

This paper does not consider the effect of pedestrian loading weight on the structure's natural frequency, which has been considered as an essential influence on the dynamic character of an aluminium alloy structure, and which still requires further investigation.

ACKNOWLEDGEMENT

The author would like to show appreciation for the financial support of the National Natural Science Foundation of China (51378460).

REFERENCES

Andriacchi T P, Ogle J A, Galante J O. Walking speed as a basis for normal and abnormal gait measurements[J]. Journal of biomechanics, 1977, 10(4): 261–268.

Bachmann H, Ammann W. Vibrations in structures: induced by man and machines [M]. Iabse, 1987.

EN03, 2007 (Human induced vibrations of steel structures): Design of footbridges, Bachground document, September 2008.

Fédérationinternationale du béton (fib). Task Group 1.2, International Federation for Structural Concrete. Guidelines for the Design of Footbridges: Guide to Good Practice [M]. FIB-Féd. Int. du Béton, 2005.

Giakas G, Baltzopoulos V. Time and frequency domain analysis of ground reaction forces during walking: an investigation of variability and symmetry [J]. Gait & Posture, 1997, 5(3): 189–197.

Matsumoto Y., Nishioka, T., Shiojiri, H., Matsuzaki, K. Dynamic design of footbridges [M]. IABSE Proceedings, No. P-17/78, 1978, pp. 1–15.

Ministry of Housing and Urban-Rural Development of the People's republic of China. ≪Technical Specifications of Urban Pedestrian Overcrossing and Underpass≫ (CJJ69-95), 1995

Okura I. Application of aluminium alloys to bridges and joining technologies [J]. Welding international, 2003, 17(10): 781–785.

Parsons K C, Griffin M J. Whole-body vibration perception thresholds [J]. Journal of Sound and Vibration, 1988, 121(2): 237–258.

Qin J W, Law S S, Yang Q S, et al. Pedestrian–bridge dynamic interaction, including human participation [J]. Journal of Sound and Vibration, 2013, 332(4): 1107–1124.

Racic V, Pavic A, Brownjohn J M W. Experimental identification and analytical modelling of human walking forces: literature review [J]. Journal of Sound and Vibration, 2009, 326(1): 1–49.

Schlaich M. Guidelines for the design of footbridges [J]. GUIDE TO GOOD PRACTICE-BULLETIN 32, 2005.

Setra. Assessment of Vibrational Behavior of Footbridges under Pedestrian Loading, 2006.

Siwowski T. Aluminium Bridges Past, Present and Future [J]. Structural engineering international, 2006, 16(4): 286–293.

Venuti F. Lively footbridges: overview and perspectives for the development of crowd and structure monitoring systems [J]. International Journal of Lifecycle Performance Engineering, 2012, 1(1): 92–113.

Wheeler J E. Prediction and control of pedestrian-induced vibration in footbridges [J]. Journal of the Structural Division, 1982, 108(9): 2045–2065.

Advances in Civil Engineering and Building Materials IV – Chang et al (eds)
© 2015 Taylor & Francis Group, London, ISBN: 978-1-138-00088-9

Effects of lateral loading patterns on seismic responses of reinforced concrete bridge piers

Shanwen Zhang & Xiuxin Wang
Key Laboratory of C & PC Structure of the Ministry Education, Southeast University, Nanjing, China

ABSTRACT: Bridge pier models were chosen as research objects, in which different section forms were considered, in order to investigate the effects of lateral load patterns in pushover analysis on the structural responses. A series of nonlinear time history analyses under six earthquake waves and pushover analyses with three common lateral load patterns were performed. The statistical results of the time history analyses were taken as the benchmark for comparison research. The results indicate that the analysis errors of the static loading method are the biggest. Compared with the time history method, the modal method and the constant acceleration method underestimate the ultimate displacement and overestimate the base shear. The analysis results are almost the same for both the square and the circular double-column piers, but the circular forms give a slightly superior performance in energy dissipation than the square forms.

Keywords: lateral load pattern, pushover analysis, structural shape, pile-soil-structure interaction (PSSI), numerical analysis

1 INTRODUCTION

The pushover method was initially applied to the seismic design and research of building structure, which is introduced into the seismic analysis of bridge structure by the earliest Imbsen and Penzien. In recent years, the United States, Japan, and other countries have applied this method to the bridge seismic design specifications. It is mentioned briefly in Chinese updated bridge specifications.

Pushover analysis is a diagram calculation method that combines static nonlinear analysis with seismic design response spectra; structural performance can be clearly reflected in the diagrams. The purpose of pushover analysis is to get the values of calculation which displacements and ductility factors of bridge structure are, and in the comparison with the values of demand; find the weak parts of the structure, and carry out the safety assessment. Pushover analysis was performed with different section forms and different lateral loading patterns, in order to obtain a more reasonable loading method concerning double-column piers.

2 MATERIAL PARAMETERS AND ANALYSIS METHODS

2.1 Material parameters and pier models

Analysis models have two structural forms when considering the pile-soil-structure interaction (PSSI), and the square and the circular double-column piers.

Based on the equal cross-sectional areas and the same reinforcement ratio, the square piers are 1.4 m long and 1.4 m wide, and the circular piers have a diameter of 1.58 m, a height of 10 m, and a center spacing of 4.0 m. The foundation is made up of circular piles with a diameter of 1.1 m; the pile caps are 6.9 m long, 4.9 m wide, and 2.0 m high. The pier concrete grade is C40, the pile and pile cap are C30, and the steel grade is HR335. The longitudinal reinforcement ratio is 1.256%. The axial compression ratio of the analysis model was 0.10, 0.12, 0.16, 0.20, 0.24, 0.28, and 0.30. The intensity of the earthquake is 8 degrees, the site type is II class, the design ground motion peak acceleration is 0.2 g, and the characteristic period is 0.45 s. The bridge pier loading pattern and reinforcement drawing is shown in Figure 1.

Midas/Civil software is used to establish the FEA model; the steel material constitutive model is the Menegotto-Pinto type; the constrained concrete and unconstrained concrete is the Mander type; the Winkler beam model is used for the pile foundation; the spring stiffness coefficient of the soil mass used the "m" method, and the "m" value of the dynamic calculation is 2.5 times more than the static calculation.

2.2 Pushover analysis method

2.2.1 Establish capacity spectrum curves
The load-displacement curves of were transformed into an acceleration-displacement response demand spectrum curve by Formulas (1) and (2);

$$S_a = v_b / M^k \tag{1}$$

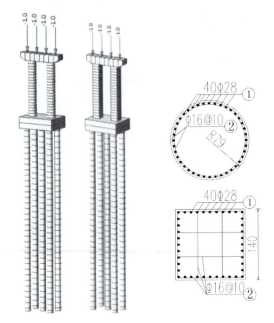

Figure 1. Schematic of a double-column model.

$$S_d = D_N / (\Gamma^k \cdot \phi_{Nk}) \qquad (2)$$

where S_a = proposed acceleration spectra; S_d = proposed displacement spectra; V_b = pier bottom shear; D_N = pier top displacement; and N = the number of nodes.

The k-order vibration modal mass is calculated as follows:

$$M^k = (\sum_{j=1}^{N} m_j \phi_{jk})^2 / \sum_{j=1}^{N} m_j \phi_{jk}^2 \qquad (3)$$

The k-order vibration modal mass is calculated as follows:

$$\Gamma^k = \sum_{j=1}^{N} m_j \phi_{jk} / \sum_{j=1}^{N} m_j \phi_{jk}^2 \qquad (4)$$

where m_j = the lumped mass of node j; φ_{jk}^k = the amplitude of k order vibration mode in the node j.

2.2.2 Establish demand spectrum curves

The single degree demand spectrum curves (damping ratio $\zeta = 0.05$) were turned into $S_a - S_d$ relation curves by Formula (5).

$$S_d = S_a T_n^2 / 4\pi^2 \qquad (5)$$

The multi-degree demand spectrum curves need to be reduced based-on the equivalent damping ratio. The principle that energy dissipation of a vibration period is equal, whatever the elastic-plastic system or the equivalent linear system is, is shown in Formula (6);

$$\xi_{eq} = E_D / 4\pi E_s \qquad (6)$$

where E_n = hysteresis damping energy dissipation, and E_s = the maximum strain energy.

2.2.3 Determining the performance points

Capacity spectrum curves and demand spectrum curves were put in the same coordinate system, and the intersection of two curves was called the performance point. It shows the values of ultimate bearing capacity and deformation capacity if there is a point of intersection. If there is no point of intersection of the two curves, it indicates that the seismic capacity is insufficient, and the structure needs to be redesigned.

3 INTRODUCTION OF MODELS

3.1 Lateral loading patterns

3.1.1 Static loading method

This is regarded as a lateral loading pattern to the static load condition of the coefficient adjustment. In the case of a bridge, the superstructure mass loading acts on the top of the piers. Therefore, the lateral force is located on the center of the pier cap in a pushover analysis.

3.1.2 Constant acceleration method

The constant acceleration lateral force is only associated with the mass distribution of the bridge; it is equal to the product of the constant acceleration and corresponding distribution mass.

3.1.3 Modal method

The modal lateral force is equal to the product of the defined mode and the square of the circular frequency of the mode and corresponding distribution mass. The eigenvalue analysis uses the Ritz vector method; the defined mode of vibration is the second order for pushover analysis.

When the axial compression ratio of the square double-column piers is 0.20, the load-displacement capacity curves, spectral curves, and performance points are shown in Figures 2 and 3.

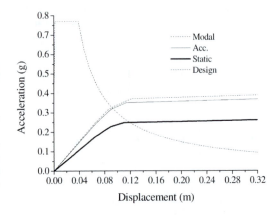

Figure 2. Load-displacement capacity curves.

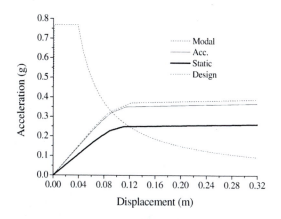

Figure 3. Spectral curves & performance points.

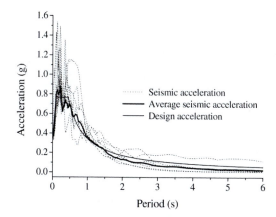

Figure 4. The normalized acceleration response spectrum.

Table 1. Ultimate displacement and pier bottom shear of the square double-column piers.

Axial compression ratio	Displacement (m)	Shear (kN)
0.12	0.0949	4094
0.16	0.1019	4378
0.20	0.1154	4679
0.24	0.1229	4983
0.28	0.1329	5290

Table 2. Ultimate displacement and pier bottom shear of the circular double-column piers.

Axial compression ratio	Displacement (m)	Shear (kN)
0.12	0.0965	4111
0.16	0.0864	3673
0.20	0.1053	4130
0.24	0.1174	5013
0.28	0.1527	6008

3.2 Time history method

The nonlinear time history method is also known as the direct integral method, which directly solves the integral equation of the input ground acceleration record and obtains the structural seismic response of the total time history. There is a spectacular disparity of seismic responses of structure due to the different seismic waves input. In order to reduce the error of seismic wave choice, five actual seismic records and an artificial seismic wave are selected as the seismic inputs for the nonlinear time history analysis by the Midas Building software. They are the Taf earthquake wave in 1952, the San Fernando wave in 1971, the Imperial Valley wave in 1979, the Tal-280 wave in 1985, and the Northridge wave in 1994. The normalized acceleration response spectra are shown in Figure 4.

The structural basic frequency changes largely with the axial compression ratio; hence, the suitable ratios are 0.12, 0.16, 0.20, 0.24, and 0.28 through comparative analysis. The pier bottom shear force values in time history analysis are the corresponding numerical values when the control points reach ultimate displacement. The statistical results of six time history analyses

were taken as the benchmark for comparison research. The calculated values are listed in Tables 1 and 2.

4 NUMERICAL ANALYSIS

Based on the equal cross-sectional areas and the same reinforcement ratio for the square and circular double-column piers, the pushover analysis and time history analysis indicate that the calculation displacements and pier bottom shear values have certain differences, due to the different lateral loading patterns. The statistical results of the time history analyses were taken as the benchmark for comparison research. The calculated values of top displacement and bottom shear are shown in Figures 5 and 6. The results of the time history analysis are nearly identical under the two structural forms; the horizontal force of the square piers passed from the bridge pier to the pile foundation is slightly larger than that of the circular piers.

The results of the pushover analysis indicate that the errors of static loading method are the greatest. The maximum displacement values were overestimated compared with the values of the time history method, which is not safe for engineering design. The errors in the modal method in the specified mode are the smallest. The distribution trends of the calculated values were similar in the two cases with the time history method. However, the error, which is associated using the modal participation mass and the characteristic period, cannot be ignored. The difference will be improved by increasing the modal participation mass. The errors of the constant acceleration method fall in between the two patterns above, and the calculated values are closest to the time history method when the

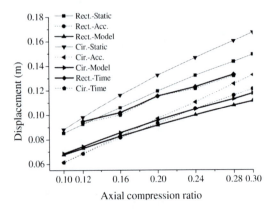

Figure 5.   Displacements of the pier top.

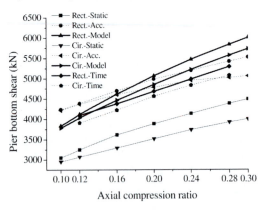

Figure 6.   Shears of the pier bottom.

axial compression ratio is 0.2. The errors of the top displacements are smaller and the errors of the bottom shear values are larger when the axial compression ratio is greater than 0.2.

Through four methods of analysis, excepting the static method, the ultimate displacements show almost the same errors; the results show that the structural seismic response has little difference between the square piers and the circular piers. Compared with the time history method, in the modal method analysis, the top displacements were underestimated and bottom shear values were overestimated. This is favorable for seismic engineering design to the safety of structure. Therefore, the calculation results of the modal method are better than the other lateral force loading patterns.

5   CONCLUSION

A series of nonlinear time history analyses under six earthquake waves and pushover analyses with three common lateral load patterns were performed using square and circular double-column piers. In addition, the statistical results of the time history analyses were taken as the benchmark for comparative research. The numerical examples verify the following conclusions and suggestions.

The results indicate that it is of vital importance to select lateral loading patterns correctly; there are differences under the three common lateral loading patterns on the structural responses.

The largest errors were caused when the static loading method was adopted.

Compared with the time history method, in the modal method analysis, the top displacements were underestimated and the bottom shear values were overestimated. This is favorable for seismic engineering design to the safety of structure.

The capacity for energy dissipation in the circular double-column piers is better than that of the square piers, based on the equal cross-sectional areas and the same reinforcement ratio.

The errors of the modal method are smallest than the other methods; the results under different axial compression ratios are in good agreement with the time history method. No matter what the displacement or shear, the modal method achieves more accurate results than the other lateral loading patterns for the different structure forms.

By comprehensive comparison, the modal method is recommended in the actual engineering analysis. Correspond with the values of the nonlinear time history analysis method; this method obtains results that are more reliable.

REFERENCES

Building seismic safety council. 2000. NEHRP Guidelines for the seismic rehabilitation of buildings. FEMA-356. Washington DC: Federal Emergency Management Agency.
Chopra, A.K., Geol, R.K. 2002. A modal pushover analysis procedure for estimating seismic demands for buildings. Earthquake Engineering and Structural Dynamics, 31: 561–582.
Code for Seismic Design of Urban Bridges. 2011. Beijing: China Architecture & Building Press, 2011.
Hou, S. & Ou, J.P. 2004. Study of load pattern selection of pushover analysis and influence of higher modes. Earthquake Engineering and Engineering Vibration 24(3): 89–97.
JTG/TB02-01-2008 Guidelines for seismic design of highway bridges. 2008. Beijing: China Communications Press.
Krawinkler, H. & Seneviratna, G.D.P.K. 1998. Pros and Cons of a pushover analysis of seismic. Engineering Structures 20(4): 452–464.
Ma, Q.L., Ye, L.P. & Lu, X.Z. 2008. Comparative Evaluation of Correctness Between MAP and Pushover Analyses. Journal of South China University of Technology (Natural Science Edition) 36(11): 121–127.
Priestley, M.J.N. 2000. Performance based on seismic design. Bulletin of the New Zealand National Society for Earthquake Engineering 33(3): 325–346.
Tatjana, I., Pompeyo, N.L.M. & Matej, F. 2008. Applicability of pushover methods for the seismic analysis of single-column bent viaducts. Earthquake Engineering and Structural Dynamics 37(8): 1185–1202.
Ye, A.J. & Lu, C.A. 2010. A seismic performance analysis approach based on pushover analysis for group pile foundations. Civil Engineering Journal 43(2): 88–94.

Advances in Civil Engineering and Building Materials IV – Chang et al (eds)
© 2015 Taylor & Francis Group, London, ISBN: 978-1-138-00088-9

Dynamic analysis of a PC composite box girder bridge with corrugated steel webs

Baotong Shi & Xiangxing Kong
The First Highway Survey and Design Institute of China Communications Construction Company Ltd., Xi'an, China

ABSTRACT: Based on a long-span PC composite box girder bridge with corrugated steel webs, the dynamic characteristics and seismic behaviour of finite element model built by Midas Civil is analysed through response spectrum method. The results show that the deformation of the bridge under the function of E1 and E2 earthquake is mainly for vertical deformation, and the bending moment distribution has similar regularity.

Keywords: dynamic analysis, Girder Bridge, corrugated steel webs, displacement, and inner force.

1 INTRODUCTION

The PC composite box girder with corrugated steel webs was a new type of composite structure which emerged in the 1980s (Nie, 2011). The PC composite box girder with corrugated steel webs has been applied widely, and in great progress in China in recent years (Xu et al., 2009). It was a new type of steel-concrete composite structure in which the traditional concrete webs were replaced by corrugated steel webs in the prestressed concrete box girder. As the research on the prestressed concrete composite box girder with corrugated steel webs is advancing and its application gradually becomes mature, this structure will get a wide range of applications in the bridge engineering in our country, therefore it is necessary to study its dynamic characteristics and seismic behaviour.

2 ENGINEERING BACKGROUND

H-bridge is a prestressed concrete composite box girder bridge with corrugated steel webs, and with a continuous span arrangement of 15 m + 30m + 15 m, whose bridge roadway width is 19.9 m. The bridge is a variable cross-section prestressed concrete continual structure, whose main girder is a single box with four cells sections. Beam depth changes from midspan to root as the form of quadratic parabola, whilst the depth of root and mid-span are respectively 2.2 m and 1.5 m. Vase piers and ribbed plate type abutments are also used in the bridge. Bored piles are used as the foundation. Tub bearings are used to transfer force from the superstructure to substructure.

3 FINITE ELEMENT SIMULATION

The simplified beam model is simulated by Midas Civil 2012, where bearings are simulated by elastic connection (Wang, 2009). Shifting bearing can be simulated by ideal elastic-plastic model, and the restoring force model, which is shown in Figure 1.

It is unnecessary to simulate the abutments, because their stiffness is much higher than the main beam and piers respectively. A Winkler foundation beam model is used to simulate the structure-earth effect (Kong et al., 2005). A three-dimensional finite element dynamic model was built by Midas Civil 2012, including 215 nodes and 190 elements as shown in Figure 2.

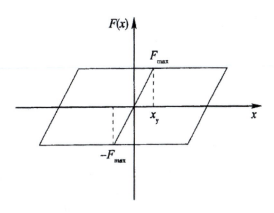

Figure 1.   Restoring force model of shifting bearing.

Figure 2.   Finite element model.

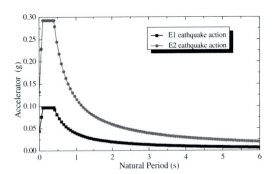

Figure 3.   Horizontal response spectrum curve of E1 and E2 earthquake action (g = 9.81 m/s²).

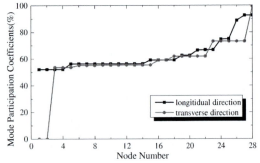

Figure 4.   Mode participation coefficients of the first 28 modes.

4   RESPONSE SPECTRUM ANALYSIS

The principles of two-level fortification and two-level seismic design are used according to Guidelines for Seismic Design of Highway Bridges. The characteristic period in the area is 0.40s and the site condition is II kind of subsoil, therefore, the characteristic period of the seismic response spectrum is 0.40s. The seismic importance coefficient for earthquake action E1 and E2 is 0.43 and 1.3 respectively. Site coefficient is 1.0 in the condition that the seismic fortification intensity is VII in the II kind of subsoil. Damping ratio of the structure is 0.05, whilst damping adjustment coefficient is 1.0. Horizontal seismic peak ground acceleration in this condition is 0.10 g (g is acceleration of gravity). The designed horizontal acceleration of the response spectrum is respectively 0.09675 g and 0.2925 g at E1 and E2 earthquake action. The horizontal response spectrum curve of E1 and E2 earthquake action is

shown in Figure 3. (note: Y direction means vertical direction and Z direction means transverse direction).

The mode participation coefficients of the first 28 modes in three directions are more than 90% as shown in Figure 4, which meets the demands of related standard. The method of CQC combination is applied for the modes composition while the method of SRSS combination is applied for spatial combination.

5   RESULTS AND DISCUSSION

5.1   Displacement analysis

Based on the response spectrum analysis of the long-span PC composite box girder with corrugated steel webs, the displacement of H bridge acted on by E1 and E2 earthquake is shown in Figure 5-a and Figure 5-b.

It can be inferred from Figure 5-a, and Figure 5-b, that the deformation of the bridge under the function

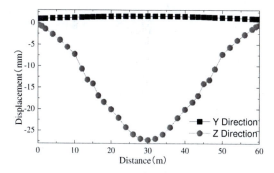

Figure 5a. Displacement under E1 earthquake action.

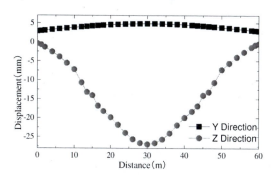

Figure 5b. Displacement under E2 earthquake action.

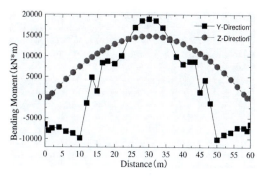

Figure 6a. Bending moment under E1 earthquake action.

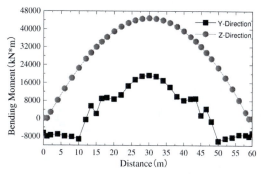

Figure 6b. Bending moment under E2 earthquake action .

of E1 and E2 earthquake is mainly for vertical deformation whilst the transverse deformation is smaller respectively, and the maximum is about 5 mm, for the vertical stiffness of the structure is lower than the transverse stiffness. The vertical displacement under E1 and E2 earthquake actions is basically unchanged, which will be illustrated in the next part.

5.2 Inter-force analysis

Based on the response spectrum analysis of the long-span PC composite box girder with corrugated steel webs, the inter-force distribution of H bridge acted on by E1 and E2 earthquake is shown in Figure 6-a, and Figure 6-b.

From Figure 6-a, and Figure 6-b, the bending moment distribution has similar regularity that the bending moment at mid span ranks the most. The vertical bending moment is basically unchanged in both earthquake actions, and this explains why the vertical is basically unchanged. The transverse bending moment greatly changes from E1 earthquake action to E2 earthquake action.

6 CONCLUSIONS

The dynamic characteristics of a long-span PC composite box girder bridge with corrugated steel webs are analysed as follows: the deformation of the bridge under the function of E1 and E2 earthquake is mainly for vertical deformation whilst the transverse deformation is smaller respectively. The bending moment distribution has similar regularity that the bending moment at mid span ranks the most, which should be considered seismic design according to the relevant regulations.

REFERENCES

Kong Deshen, Luan Maotian, Yang Qing. Review of dynamic Winkler model applied in pile soil interaction analyses. World Earthquake Engineering, 2005, 21(3), 12–17

Nie Jianguo. Steel-Concrete Composite Bridges. China Communications Press, Beijing, 2011: 313–315

Wang Kehai. Bridge Seismic Research. China Railway Publishing House, Beijing, 2007: 237–256

Xu Qiang, Wan Shui. Design and Application of PC Composite Box girder Bridge with Corrugated Steel Webs. China Communications Press, Beijing 2009: 3–12

Building materials performance analysis

Advances in Civil Engineering and Building Materials IV – Chang et al (eds)
© 2015 Taylor & Francis Group, London, ISBN: 978-1-138-00088-9

Mechanism analysis of different ageing road asphalt by RTFOT and IR methods

Yihua Nie, Jie Ding & Dingling Li
School of Civil Engineering, Hunan University of Science and Technology, Xiangtan, P.R. China

ABSTRACT: In order to study the ageing performance of road asphalt, different ageing degree asphalts were made by RTFOT (Rolling Thin Film Oven Testing) method. The three indexes of penetration, softening point, and ductility were tested at five different ageing times, and IR (Infrared Radiation) parameters were tested by the Fourier transform infrared spectroscopy method. The research showed that the relationship between each of the three indexes of asphalt and different ageing times is exponential function, and the relationship between each of IR parameters and different ageing times is power function.

Keywords: road asphalt, ageing, microanalysis, mechanism.

1   INTRODUCTION

Three indexes of asphalt are the key performance factors. So studying asphalt performance usually starts from studying its three performance indexes. Asphalt is inevitably affected by water, high temperature, low temperature, force during the process of mixing, construction and usage, which accelerate asphalt's ageing, and speeding up the reduction in its performance and shorten its lifetime. However, we often questioned, only a short time ago, the ageing test (such as heating test of 163° with RTFOT) and it was not accurate enough to simulate asphalt's ageing degree over different ageing times.

For microstructure analysis in petroleum asphalt, infrared absorption spectrum is more prominent, especially in the identification of the structure of asphalt. In addition, in determining the structure type, aromatic side chain length and number can be involved. This paper analyses different ageing degree mechanisms of asphalt by the Fourier transform infrared spectroscopy method and provides important theoretical reference.

2   EXPERIMENTAL SECTION

2.1   Asphalt sample preparation

RTFOT is used to heat No. 70 road asphalt in this paper, and six groups of ageing time tests were done to simulate asphalt ageing with different heating times: 0 min, 40 min, 85 min, 180 min, 240 min and 300 min. With penetration, softening point, and ductility, three indexes were respectively measured. Mathematical statistical methods were used to analyse

and compare performance indexes under different ageing times, getting their changing regularity to study the mechanism of asphalt ageing.

Reverse mould temperature of specimens in experiments in this paper is 135–145°C. A kind of asphalt reverse mould was used only once and was not heated twice. Experiments of asphalt penetration and ductility were done strictly as procedureT0604-2011 and T0605-2011, RTFOT was usually heated only four to six hours before it was used with heated asphalt. The other operations were all done as per the procedure T0610-2011.

2.2   Infrared spectrum test

2.2.1   Infrared spectrum instrument

Thermo Fisher Scientific Fu Liye infrared spectrometer was adopted, its resolution being 4 cm^{-1} scanning for thirty-two times, and its test range being 4000–400 cm^{-1}. An instrument infrared spectrometer and operation should be installed with air conditioning, and the instrument room temperature should be controlled at 17–27°C. Water vapor has a heavy effect on measurement of sample absorption peak, which has a lot of absorption peaks in the far infrared spectral region and a big interference with the sample absorption peak. So the sample needs to keep dry by lighting and heating before being tested.

2.2.2   Infrared spectrum test

Take a small part of asphalt on the agate bowl, then add potassium bromide (potassium bromide and asphalt ratio is about 3:1), with a pestle fully grinded, pressured into about 1 mm of circular section, the sample scanning and background scanning, and the

69

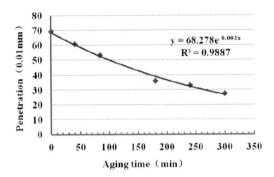

Figure 1. The changing curve of asphalt penetration under different ageing time.

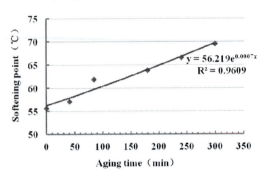

Figure 2. The changing curve of asphalt softening point under different ageing time.

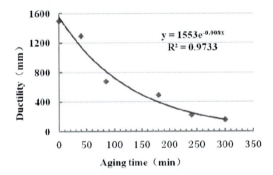

Figure 3. The changing curve of asphalt ductility under different ageing time.

sample scanning from environmental background, the IR spectra obtained curve required.

3 MECHANISM ANALYSIS

3.1 Three performance indexes analysis

No. 70 asphalt three performance indexes changing curves and fitting equations under different ageing time are shown in Figure 1–Figure 3.

From Figure 1, it is known that the penetration of No. 70 road asphalt reduces roughly as an index, and the data fitting is very precise with its correlation

coefficient being 0.989. It is also shown in the Figure 2 that the softening point of No. 70 road asphalt increases as an approximant as the index of its ageing time is increasing and its correlation coefficient is 0.961. It is indicated in the Figure 3 that the ductility of No. 70 road asphalt reduces as its ageing time increases, and ductility reduces as an index of the ageing time increasing and its correlation coefficient is 0.973. From comparison between Figures 1–3, we can know that the amplitude of variation in ductility is much larger than that of the penetration and the softening point.

3.2 IR indexes analysis

3.2.1 Ageing mechanism analysis

Ageing asphalt is often expressed as two types. The first type is the mass loss caused by saturate and aromatic light component volatilization, the second type is mainly oxygen combined with carbonyl and sulfoxide chemical functional groups. Ketones and carboxylic acid are the weakest parts of ageing asphalt (or the most active part). At the same time, aromatic benzene methyl and aliphatic side chain, easily combined with oxygen, remaining in the ketone, sulfoxide or other oxygen functional groups, strengthening the polarity of asphalt which promoted interaction asphalt molecules and promoted the asphalt stiffness greatly. Asphalt stiffness increment amplitude is related to size, number of polar functional groups and dispersive power of non polar functional groups.

3.2.2 IR wavelength range

The oxygen absorption of asphalt can be obtained through infrared spectrum experiment analysis. Infrared spectra wavelength range is between 0.8–1000 um. The characteristic absorption peaks can be used in spectral analysis of polymers. Existing research shows that C-H wavenumber is between $2800-3000\,cm^{-1}$, and non-aromatic C=C wavenumber is between $900-1000\,cm^{-1}$, and carbonyl C=O wavenumber is between $1650-1800\,cm^{-1}$, and acid anhydride wavenumber is between $1650-1700\,cm^{-1}$, and ester wavenumber is between $1740-1750\,cm^{-1}$, and aldehyde wavenumber is between $1720-1750\,cm^{-1}$, and ketone wavenumber is between $1720-1750\,cm^{-1}$, and aromatic wavenumber is concentrated in about $1600\,cm^{-1}$.

3.2.3 IR test results analysis

Asphalt IR absorption curves under different ageing time are given in Figure 4. As aromatic components can reflect the strength of ageing asphalt, wave peak ratios of $1700\,cm^{-1}$ and $1459\,cm^{-1}$ can reflect oxidation of the aromatic C=C, and wave peak ratios of $1033\,cm^{-1}$ and $1459\,cm^{-1}$ can reflect the oxidation of S elements.

Asphalt carboxyl and sulfoxide characteristic absorption peak heights are shown in Table 1 under different ageing time. The horizontal vertical line comes from the absorption peak top, and the distance from the intersection between the horizontal vertical line

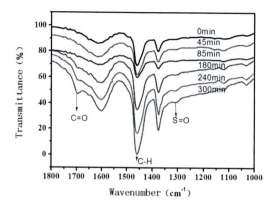

Figure 4. Asphalt IR absorption curves under different ageing time.

Table 1. Asphalt IR indexes under different ageing time.

Ageing time (min)	C=O Peak height	C=H Peak height	H=O Peak height	Carboxyl index CL	Sulfoxide index SI
0	0.941	21.280	0.602	0.044	0.028
40	1.508	21.025	0.744	0.072	0.035
85	0.823	14.396	0.582	0.057	0.040
180	2.128	31.755	1.391	0.067	0.044
240	2.465	39.120	2.311	0.063	0.059
300	3.095	46.383	5.232	0.067	0.113

and baseline to peak top is the peak height of C=O, C-H, S=O peak value.

Limited by sample test operations, the thickness of the coating film is different, the quantitative calculation of absorption peak intensity exists with calculation errors. Therefore, an internal standard method is adopted to eliminate the effect of film thickness on absorbance, and the saturated C-H bending vibration absorption peak is used to calculate the change of carbonyl and sulfoxide groups' relative content after ageing, of which the absorption peak height ratio is respectively between the carbonyl (wavenumber $1700 \, cm^{-1}$), sulfoxide (wavenumber $1033 \, cm^{-1}$) absorption peak height, and saturated C-H (wavenumber $1459 \, cm^{-1}$).

Carbonyl index (CI) and sulfoxide index (SI) calculation formula:

$$CI = \frac{H_{C=O}}{H_{C-H}} \qquad (1)$$

$$SI = \frac{H_{S=O}}{H_{C-H}} \qquad (2)$$

where: $H_{C=O}$ = carbonyl absorption peak height, and $H_{S=O}$ = sulfoxide absorption peak height, and H_{C-H} = absorption peak height.

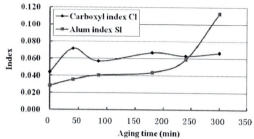

Figure 5. Asphalt IR index curves under different ageing time.

Table 2. Asphalt fitting formula for carboxylic index and sulfoxide index under different ageing time.

The fitting formula	Formula symbols	Correlation coefficient
$y_{羧酸} = 0.05x^{0.1793}$	$y_{羧酸}$-Relative index of carboxylic	$R = 0.6816$
$y_{亚砜} = 0.26x^{0.6296}$	$y_{亚砜}$- Relative index of sulfoxide	$R = 0.8539$

Asphalt IR index curves under different ageing time are shown in Figure 5. From the Figure 5 it is shown that the carbonyl index and the sulfoxide index increase as power function with ageing time is increasing. The added speed for sulfoxide index is faster than the carbonyl index. The result shows that the ageing process of asphalt is generated in aldehyde, ketone, carboxylic acid, ester or oxygen containing components. In addition, sulfoxide base asphalt after ageing near wavenumber $1033 \, cm^{-1}$ is increasing gradually, which means that sulphur has an important effect on the asphalt ageing process. During the asphalt ageing process, the asphalt molecule effect becomes harder, and asphalt stiffness became bigger, due to the functional groups being combined with oxygen to form carbonyl, sulfoxide polarity or amphoteric functional groups.

4 CONCLUSIONS

(1) To different age levels of road No.70 asphalt, the penetration, softening point, and ductility three performance indexes of road No.70 asphalt have exponential function with different ageing time, and the amplitude of variation of ductility is much larger than that of penetration and softening point.
(2) To different age levels of road No.70 asphalt, asphalt IR carbonyl index and the sulfoxide index increase as power function with ageing time increasing. The added speed for the sulfoxide index is faster than the carbonyl index.

ACKNOWLEDGEMENTS

This research was supported by Hunan National Science Found (Project No. 11JJ4037).

REFERENCES

Chen, H.X., Chen, X.F., Wang, B.K. Ageing Behavior and Mechanism of Base Asphalt [J]. Journal of Shandong University (Engineering Science), 2009, 38(1), 13–19.

Cui, Y.N., Xing, Y.M., Wang, L., et al. Micro-Structure and Rh eological Behavior of Composite Crumb Rubber Modified Asphalt [J]. *Polymer Materials Science and Engineering*, 2012, 28(2).

Geng, J.H. The mechanism of asphalt ageing and regeneration technology research [D]. Xi'an Chang'an University, 2009.

Jun W., Zhu Bensong. IR Spectrum analysis of changes in molecular structure of pitch pre-oxidized fiber during carbonization [J]. Synthetic Fiber Industry, 1995, 18(5), 30–33.

JTG F40-2004, Standard Test Methods of Bitumen and Bituminous Mixtures for Highway Engineering [S].

Liu, P., Nie, Y.H., et al. Performance changing regularity of No.70 heavy traffic road asphalt with different ageing time [J]. *Journal of Hunan University of Science and Technology*, 2013, 28(2):56–59.

Ou Yang, C.F., Li, Y.S., Jia, R.P. Research Progress of Base Asphalt and Polymer-modified Asphalt Ageing [J]. *Synthetic Materials Ageing and Application* [J]. 2007, 36(2): 22–28.

The ministry of transport highway research institute. Standard test methods of bitumen and bituminous mixtures for highway engineering [S]. Beijing: *China Communication Press*, 2011.

Weng, S.P., et al. Fourier transform infrared spectrometer. Beijing: *Chemical Industry Press*, 2005.

Xie, J.G., Qian, C.X., Xiao, Q.Y., et al. Performances and mechanism analysis of SEBS modified asphalt [J]. *Journal of Southeast University (Natural Science)*, 2004, 34(1), 96–99.

Xu, J., Hong, J.X., Liu, J.P. Asphalt ageing mechanism were reviewed [J]. *Petroleum Asphalt*, 2011, 25(4), 1–7.

Zhang, J.S., Zhang, Y.Y., Xia, X.Y., et al. Bituminous Materials. Beijing: *Chemical Industry Press*, 2009.

Advances in Civil Engineering and Building Materials IV – Chang et al (eds)
© 2015 Taylor & Francis Group, London, ISBN: 978-1-138-00088-9

Prediction of chloride diffusivity of concrete with different coarse aggregates using artificial neural network

L.J. Kong & X.Y. Chen

School of Materials Science and Engineering, Shijiazhuang Tiedao University, Shijiazhuang, China

ABSTRACT: Coarse aggregates occupy half of the volume of concrete, and can influence the concrete durability to some extent. Therefore, it is necessary to choose the aggregate reasonably and further optimize the design of the concrete structure. In this paper, an artificial neural network (ANN) model is applied to determine the influence of the particle size, water absorption, density and strength of coarse aggregate, water-cement ratio and curing age on the chloride diffusivity of concrete. The model depends on 108 data sets from experiments, and an appropriate choice of architecture and learning processes. It is developed, trained and tested by using a multi layer back propagation method. Then the model predicted values are compared with actual test results, and the average relative error of prediction of test results is found to be 4.44% and the coefficient of determination (R^2) is 0.99. The results indicate that the developed model is successful in learning the relationship between the different input and output parameters, and can predict the chloride diffusivity with adequate accuracy required for practical design purpose. ANN procedure provides guidelines to select appropriate coarse aggregate for required chloride diffusivity of concrete and will reduce the number of trial and error, save cost and time.

Keywords: concrete; chloride diffusivity; coarse aggregate; prediction; artificial neural network.

1 INTRODUCTION

Coarse aggregates for concrete production, occupy nearly half of its volume. Desired characteristics of concrete, such as density, strength and abrasion, are determined mainly by the properties of aggregates (Kilic et al. 2008). However, the influence of aggregate on concrete durability has been given little attention, since it is often assumed that most rock aggregates used in concrete are dense and chemical inert. In fact, the aggregate can influence the formation and development of the microstructure of the interfacial transition zone (Elsharief et al. 2003, Tasong et al. 1999), which is very important to concrete durability, especially concrete impermeability (Torgal & Gomes 2006, Pereira et al. 2009). Therefore, it is necessary to choose the aggregate reasonably and further optimize the design of the concrete structure. Moreover, it is better to find a method, which can predict the durability of concrete with different coarse aggregate more convenient and quickly.

In recent years, with the development of computer and life science, artificial neural network (ANN) theory and the model has been developed rapidly and successfully used in many fields of engineering (Ji et al. 2006, Bal & Buyle 2013, Khan 2012). The neural network modeling approach is simpler and more direct than traditional statistical methods, particularly when modeling nonlinear multivariate interrelationships, thus it has attracted more and more attention. The aim of this study is to construct an ANN model

to predict the influence of coarse aggregate on the chloride diffusivity of concrete, which can reduce the experimental workload greatly. For this purpose, the coarse aggregates with different particle sizes, water absorptions, densities and strengths were used to prepare concrete with different water-cement ratios, and the chloride diffusion coefficients of concrete at different ages were tested. The data collected from the experiments are used to develop the neural network model, which is expected to provide a predictive tool in the optimization of coarse aggregate for the preparation of high durability concrete.

2 PREPARATION OF ANN SAMPLES

The first step in developing the network is to obtain good and reliable training and testing examples. To obtain the data for developing the ANN model, three groups of concrete were prepared with water-cement ratios of 0.32, 0.40 and 0.49, the mix proportions of concrete are shown in Table 1. Grade 42.5 Ordinary

Table 1. Mixture proportions of concrete/kg/m³.

w/c	Water	Cement	Sand	Crushed stone	Fly ash	UNF
0.32	160	425	650	1061	98	7.32
0.40	160	340	669	1128	78	3.34
0.49	160	278	687	1154	64	1.54

Portland Cement and Class I fly ash were used in all mixtures. Three different mineralogical coarse aggregate types were selected for this study, include: limestone from Shijiazhuang (SHY-1) and Fuping (SHY-2), granite from Nanjing (HGY) and basalt from Lianyungang (XWY), and for each type, three different particles sizes of coarse aggregate (5–10 mm, 5–20 mm and 5–30 mm) were used, the physical properties of the aggregates are summarized in Table 2. Fine aggregate used was river sand with a fineness modulus of 2.8 and an apparent density of 2610 kg/m³. A naphthalene-based superplasticizer was used. After demolding, the specimens with dimension of $100 \times 100 \times 50$ mm were placed in a standard curing room until the day of testing (7d, 28d and 56d). Before the test, specimens were immersed in the sodium chloride solution with a concentration of 4 mol/L, and saturated by vacuum pumping. The chloride diffusion coefficients of specimens were shown in Table 1, which obtained using a rapid chloride penetrant method named NEL (Li et al. 1998).

3 APPLICATION OF ANN

In this study, the ANN was developed and performed under MATLAB programming. To simplify the learning process of the back propagation (BP) neural network and to reduce the time required for training, the learning algorithm adopted to train the network model in this study is the Levenberg-Marquardt algorithm [7]. The ANNs model is developed, trained and tested by using a total of 108 data sets. To test the reliability and accuracy of the models, 20% of the 108 data sets were randomly selected as testing sets, while the remaining 86 samples were used to train the network. In order to eliminate the magnitude difference between the variables, a normalization method is used to deal with the input and output data in this study as follows, which scale all the data in the 0 to 1 range.

$$\overline{x_i} = \frac{x_i - x_{min}}{x_{max} - x_{min}} \tag{1}$$

Table 2. Sample database.

No.	Aggregate	Normal size/mm	R/mm	W_a/%	ρ/(kg/m³)	f/MPa	w/c	$D_{Cl}^-/(\times 10^{-8} cm^2/s)$ 7d	28d	56d
1	SHY-1	5-10	6.27	0.9	2640	65	0.32	4.15	3.43	1.89
2	SHY-1	5-20	8.04	0.9	2640	65	0.32	3.09	2.57	1.56
3	SHY-1	5-30	8.59	0.9	2640	65	0.32	3.02	2.51	1.34
4	SHY-2	5-10	6.78	0.5	2780	80	0.32	4.06	3.11	2.18
5	SHY-2	5-20	8.16	0.5	2780	80	0.32	3.11	2.23	1.49
6	SHY-2	5-30	8.95	0.5	2780	80	0.32	3.05	2.08	1.21
7	HGY	5-10	7.12	0.3	2860	148	0.32	2.97	2.18	1.22
8	HGY	5-20	8.41	0.3	2860	148	0.32	2.85	2.12	1.18
9	HGY	5-30	9.33	0.3	2860	148	0.32	2.56	2.05	1.13
10	XWY	5-10	6.86	0.4	2730	112	0.32	2.39	1.77	1.17
11	XWY	5-20	7.94	0.4	2730	112	0.32	2.31	1.60	1.15
12	XWY	5-30	8.92	0.4	2730	112	0.32	2.25	1.58	1.09
13	SHY-1	5-10	6.27	0.9	2640	65	0.40	5.58	5.03	3.48
14	SHY-1	5-20	8.04	0.9	2640	65	0.40	4.63	4.69	2.65
15	SHY-1	5-30	8.59	0.9	2640	65	0.40	5.37	4.76	3.28
16	SHY-2	5-10	6.78	0.5	2780	80	0.40	5.49	4.75	3.82
17	SHY-2	5-20	8.16	0.5	2780	80	0.40	4.81	3.95	2.87
18	SHY-2	5-30	8.95	0.5	2780	80	0.40	5.12	4.26	2.94
19	HGY	5-10	7.12	0.3	2860	148	0.40	5.37	4.88	3.79
20	HGY	5-20	8.41	0.3	2860	148	0.40	4.88	4.24	3.06
21	HGY	5-30	9.33	0.3	2860	148	0.40	5.21	4.67	3.64
22	XWY	5-10	6.86	0.4	2730	112	0.40	3.95	3.16	2.45
23	XWY	5-20	7.94	0.4	2730	112	0.40	3.56	2.69	1.94
24	XWY	5-30	8.92	0.4	2730	112	0.40	4.03	3.01	2.58
25	SHY-1	5-10	6.27	0.9	2640	65	0.49	7.32	6.63	5.13
26	SHY-1	5-20	8.04	0.9	2640	65	0.49	6.29	5.71	4.29
27	SHY-1	5-30	8.59	0.9	2640	65	0.49	7.87	7.25	5.52
28	SHY-2	5-10	6.78	0.5	2780	80	0.49	6.75	6.28	5.08
29	SHY-2	5-20	8.16	0.5	2780	80	0.49	6.54	5.92	4.81
30	SHY-2	5-30	8.95	0.5	2780	80	0.49	7.32	6.76	5.43
31	HGY	5-10	7.12	0.3	2860	148	0.49	6.85	6.22	4.98
32	HGY	5-20	8.41	0.3	2860	148	0.49	6.97	6.36	4.93
33	HGY	5-30	9.33	0.3	2860	148	0.49	7.48	6.83	5.91
34	XWY	5-10	6.86	0.4	2730	112	0.49	5.75	5.04	3.96
35	XWY	5-20	7.94	0.4	2730	112	0.49	5.53	4.83	3.77
36	XWY	5-30	8.92	0.4	2730	112	0.49	6.07	5.42	4.73

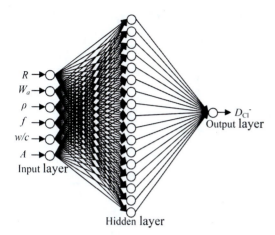

Figure 1. Chart of ANN for predicting concrete permeability.

where x_i = input or output data, x_{min} = the minimum data, x_{max} = the maximum data, $\overline{x_i}$ = data after normalization.

To predict the chloride diffusivity of concrete with different coarse aggregates, take the six factors that influence the chloride diffusivity of concrete as the input layer neurons: particle sizes (R), water absorptions (W_a), densities (ρ) and strengths (f) of coarse aggregate, water-cement ratio (w/c) and curing age (A), take the chloride diffusion coefficient of concrete (D_{Cl}^-) as the output layer neuron. Currently, there is no rule for determining the optimal number of neurons in the hidden layer or the number of hidden layers, except through experimentation. A single hidden layer has been found to be satisfactory for many problems. The architecture of prediction model for the chloride diffusivity of concrete consists three layers as shown in Figure 1.

The selection of the number of neurons in the hidden layer is the most challenging part in the total network development process. For determine the optimal number of hidden layer nodes, neural networks with different number of hidden layer neurons were trained.

The data from the training set was used to determine the number of neurons in the hidden layer, which resulted in the least error between the neural network output and the experimental data. The hidden layer neuron number was varied and the resulting mean-square-error (MSE) between the network outputs and the corresponding experimental outputs were determined and plotted in Figure 2. It can be seen that the error is minimum when the number of neurons is 17.

4 RESULTS AND DISCUSSION

The training of a neural network is stopped when the error falls below a user-specified level, or when the user-defined number of training iterations has been reached. In this case, 1000 iterations were planned for the final training process, as this was found adequate

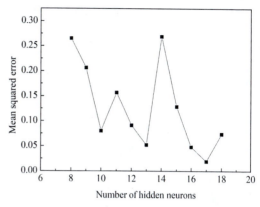

Figure 2. Error generated by different number of neurons.

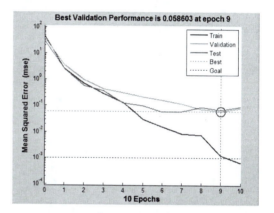

Figure 3. Reduction in error over training time.

in a series of test runs. From Figure 3 it can be seen that after 10 times of the network training, the error had been reduced to the specified level of 0.001, and the training of the neural network was stopped.

The predicted values obtained using ANN for the testing samples were listed in Table 3, and the relative error (E) between predicted value and measured value of concrete chloride diffusion coefficient were calculated as follows:

$$E = \left| \frac{Y - S}{S} \right| \times 100\% \qquad (2)$$

where Y = predicted value, S = measured value.

From the relative error results as shown in Table 3, it can be seen that the maximum error of the prediction is 9.40%, and the minimum error is 0%, and the predicted value is very close to the measured value. Moreover, the average relative error is 4.44%, which meets the requirement of engineering.

The performance of the ANNs model for predicting the total chloride diffusion coefficients of training and testing sets is illustrated in Figure 4. It shows the experimental results on the horizontal axis and the predicted results on the vertical axis, and can be seen that there is a good correlation between experimental

Table 3. Comparison of measured and predicted value of concrete chloride diffusion coefficient.

No.	w/c	R/mm	W_a/%	ρ/(kg/m³)	f/MPa	A/d	Measured value/ ($\times 10^{-8}$cm²/s)	Predicted value/ ($\times 10^{-8}$cm²/s)	E/%
1	0.32	8.04	0.9	2640	65	7	3.23	3.14	2.79
2	0.32	9.33	0.3	2860	148	7	2.56	2.37	7.42
3	0.32	8.59	0.9	2640	65	28	2.51	2.73	8.76
4	0.32	7.94	0.4	2730	112	28	1.60	1.51	5.63
5	0.32	8.16	0.5	2780	80	56	1.49	1.63	9.40
6	0.32	9.33	0.3	2860	148	56	1.13	1.22	7.96
7	0.40	8.59	0.9	2640	65	7	5.37	5.27	1.86
8	0.40	8.16	0.5	2780	80	7	4.81	4.85	0.83
9	0.40	7.12	0.3	2860	148	7	5.37	5.37	0.00
10	0.40	7.94	0.4	2730	112	7	3.56	3.79	6.46
11	0.40	7.12	0.3	2860	148	28	4.88	4.73	3.07
12	0.40	8.59	0.9	2640	65	56	3.28	3.08	6.09
13	0.40	8.95	0.5	2780	80	56	2.94	2.93	0.34
14	0.40	8.92	0.4	2730	112	56	2.58	2.72	5.43
15	0.49	8.95	0.5	2780	80	7	7.32	7.60	3.82
16	0.49	7.94	0.4	2730	112	7	5.53	5.30	4.16
17	0.49	8.16	0.5	2780	80	28	5.92	5.91	0.17
18	0.49	6.86	0.4	2730	112	28	5.04	4.65	7.74
19	0.49	8.59	0.9	2640	65	56	5.52	5.61	1.63
20	0.49	8.16	0.5	2780	80	56	4.81	5.18	7.69
21	0.49	9.33	0.3	2860	148	56	5.91	5.84	1.18
22	0.49	8.92	0.4	2730	112	56	4.73	4.99	5.49

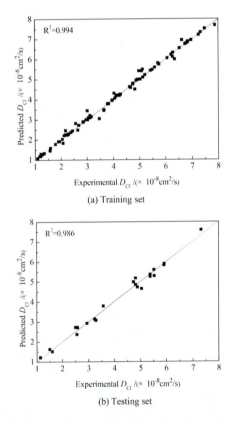

(a) Training set

(b) Testing set

Figure 4. Performance of the ANNs model for predicting the chloride diffusion coefficient of concrete.

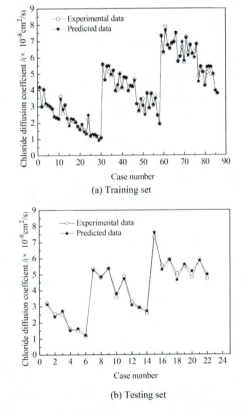

(a) Training set

(b) Testing set

Figure 5. Comparison of experimental and predicted chloride diffusion coefficient of concrete.

and predicted values for complete data set. The results illustrate that the ANN model is successful in learning the relationship between the different input and output parameters, and show the ability of the network to predict the influence of coarse aggregate on the chloride diffusivity of concrete with high precision.

Figure 5 plots the experimental and the predicted chloride diffusion coefficient for the total chloride diffusion coefficients of training and testing sets. It can be observed that the predictions made by neural network closely matches with the experimental data.

5 CONCLUSIONS

(1) Coarse aggregates, which occupy half of the volume of concrete, will certainly have some influence the concrete durability. ANN procedure provides guidelines to select appropriate coarse aggregate for required chloride diffusivity of concrete and will reduce the number of trial and error, save cost and time.

(2) A multi-layer back propagation method has been adopted for the development of the model. Based on the 108 data sets from the experiments of the influence of the particle size, water absorption, density and strength of coarse aggregate, water-cement ratio and curing age on the chloride diffusivity of concrete, the model is developed, trained and tested. Moreover, the model predicted values are compared with actual test results.

(3) The average relative error of prediction of test results is found to be 4.44% and the coefficient of determination (R^2) is 0.99. Such error levels are considered positive and acceptable in the field of engineering. The results indicate that the developed model is successful in learning the relationship between the different input and output parameters, and can predict the chloride diffusivity with adequate accuracy required for practical design purpose.

ACKNOWLEDGEMENTS

This work was financially supported by the National Nature Science Foundation of China (51108282), and Excellent Youth Scholars of University Science and Technology Research of Hebei Province (Y2011111).

REFERENCES

Bal, L. & Buyle-Bodin, F. 2013. Artificial neural network for predicting drying shrinkage of concrete. *Construction and Building Materials* 38(1): 248–254

Elsharief, A., Cohen, M.D. & Olek J. 2003. Influence of aggregate size, water cement ratio and age on the microstructure of the interfacial transition zone. *Cement and Concrete Research* 33(11): 1837–1849

Ji, T., Lin, T.W. & Lin, X.J. 2006. A concrete mix proportion design algorithm based on artificial neural networks. *Cement and Concrete Research* 36(7): 1399–1408

Khan, M.L. 2012. Mix proportions for HPC incorporating multi-cementitious composites using artificial neural networks. *Construction and Building Materials* 28(1): 14–20

Kilic, A., Atis C.D., Teymen, A. & O. Karahan, O. 2008. The influence of aggregate type on the strength and abrasion resistance of high strength concrete. *Cement and Concrete Research* 30(4): 290–296

Li, C.L., Lu, X.Y. & Zhang, H.X. 1998. Rapid test method for determining chloride diffusivities in cementitious materials. *Industrial Construction* 28(6): 41–43.

Parka, K.B., Noguchi, T. & Plawsky, J. 2005. Modelling of hydration reactions using neural networks to predict the average properties of cement paste. *Cement and Concrete Research* 35(9): 1676–1684

Pereira C.G., Gomes, J.C. & Oliveira. L.P. 2009. Influence of natural coarse aggregate size, mineralogy and water content on the permeability of structural concrete. *Construction and Building Materials* 23(2): 602–608

Tasong, W.A., Lynsdale, C.J. & Cripps, J.C. 1999. Aggregate-cement paste interface: Part I. Influence of aggregate geochemistry. *Cement and Concrete Research* 29(10): 1019–1025

Torgal, F.P. & Gomes, J.C. 2006. Influence of physical and geometrical properties of granite and limestone aggregate on the durability of a C20/25 strength class concrete. *Construction and Building Materials* 20(10): 1079–1088

Advances in Civil Engineering and Building Materials IV – Chang et al (eds)
© 2015 Taylor & Francis Group, London, ISBN: 978-1-138-00088-9

A mix design and strength analysis of basic magnesium sulphate cement concrete

Xiangchao Zeng & Hongfa Yu
Department of Civil Engineering, Nanjing University of Aeronautics and Astronautics, Nanjing, China

Chengyou Wu
Qinghai Institute of Salt Lake, Chinese Academy of Sciences, Xining, China

ABSTRACT: Basic Magnesium Sulphate Cement (BMSC) has many advantages. In order to explore the difference in performance between basic magnesium sulphate cement concrete and ordinary cement concrete, the experiments of basic magnesium sulphate cement concrete including mix design, measuring slump, apparent density, and compressive strength tests were conducted. Cube and prism compressive testing and some analysis were also conducted. We studied the development of the strength of basic magnesium sulphate cement concrete, which was compared with that of ordinary cement concrete. The results were in accordance with the current national product standards and test methods standards for Portland cement, from which we can know the advantages of basic magnesium sulphate cement concrete.

Keywords: basic magnesium sulphate cement, concrete, mix design, compressive strength, axial compressive strength, modulus of elasticity, Poisson's ratio.

1 INTRODUCTION

Magnesium oxychloride cement was invented by the French chemist S. Sorel, having the advantages of quick setting, early strength, high strength, low alkali, abrasion resistance, high bonding strength, and brine corrosion resistance. On the other hand, its weaknesses include poor water resistance, high moisture absorption, high deformability, and a tendency to corrode reinforcing bars.

Magnesium oxysulphate cement is a cementious material similar to magnesium oxychloride cement. Due to its lower strength, however, it does not have many applications (Urwong L. et al., 1980).

Magnesium phosphate cement is an air-setting cementious material made from dead-burned magnesium oxide powder, soluble phosphate, and a retarder. Its advantages are its quick-setting, early strength, high strength, frost resistance, high temperature resistance, good volume stability, high bond strength, high wear resistance, and good rust protection for reinforcing steel bars. However, it is very expensive and requires a great deal of phosphorus. Its properties make it an ideal material for quick repairs.

Magnesium oxysulphate cement is the most promising of the new magnesium-based cementious materials and has the highest chance to replace magnesium

oxychloride cement and occupy a greater share of the construction industry (Robert J. et al., 2008). In order to exploit the strengths of magnesium oxysulphate cement and overcome its weaknesses of low strength and insufficient hydration, academics have been working to use additives to improve its performance. It was found that by adding certain additives magnesium oxy sulphate cement exhibited a new, unknown basic magnesium sulphate cement crystallization phase, such that its physical strength increased dramatically. Under comparable conditions its resistance strength and flexural strength are higher than magnesium oxychloride cement. A German-Chinese cooperative venture successfully analysed its chemical composition and crystal structure (Runčevski T. et al., 2013). This new type of magnesium oxy sulphate cement with primary hydration product the 517 phase was named basic magnesium sulphate cement. The advantages of Basic Magnesium Sulphate Cement (BMSC) are its quick setting, early strength, high strength, water resistance, and corrosion resistance. Its physical-chemical and mechanical properties are similar to ordinary Portland cement (Chengyou Wu, 2014). Therefore, reinforced concrete members can be made with BMSC. Furthermore, thermal insulation mortar and thermal insulation blocks developed with basic magnesium sulphate cement has wide application

Table 1. The mixture proportions of basic magnesium sulphate cement concretes kg/M^3.

Number	Fine aggregate	Coarse aggregate	water	MSC	Composition of BMSW MgO:MgSO$_4 \cdot$7H$_2$O:Fly ash: Admixture
Nm1	704	1080	182	362	50.9%:22.9%:25.4%:0.8%
Nm2	740	1110	166	372	40.6%:18.3%:40.6%:0.5%
Nm3	745	1118	195	374	33.7%:15.2%:50.6%:0.5%
Nm4	697	1046	161	349	28.9%:13.0%:57.7%:0.4%
Nm5	748	1121	163	373	50.9%:22.9%:25.4%:0.8%
Nm6	866	1312	108	261	42.2%:14.8%:42.2%:0.8%

prospects in the construction industry (Chengyou Wu et al., 2012).

This paper introduces some new experimental results concerning BMSC.

2 EXPERIMENTAL

2.1 Materials

In this experiment, the primary material is BMSC with a specific surface area of 2500 cm^2/g, which is made of MgO, MgSO$_4$.7H$_2$O, fly ash admixture. A kind of locally available crushed gravel with a maximum size of 25 mm and specific density of 2690 kg/m^3 was used as the coarse aggregate. A locally available natural river sand with a fineness modulus of 2.44 and specific density of 2600 kg/m^3 was used as the fine aggregate. The mixture's proportions of basic magnesium sulphate cement concretes are given in Table 1.

2.2 Experiments

Based on the Chinese standard for test methods of the mechanical properties of ordinary concrete (GB/T 50081, 2002), the compressive strength of the concrete was measured. The specimen sizes were 100 × 100 × 100 mm for a compressive strength test, 100 × 100 × 300 mm for modulus of elasticity, and Poisson's ratio and axial compressive strength tests. All concrete specimens were cast in a standard laboratory at 20 ± 5°C and with 50 ± 5% relative humidity. After 24 h, all the specimens were demolded. They were then cured in water at room temperature for up to 3 days, 7 days, 14 days and 28 days, respectively.

According to the standards for test methods of the mechanical properties of ordinary concrete, the compressive strength, the modulus of elasticity and the Poisson's ratio of basic magnesium sulphate cement concrete were tested.

3 RESULTS AND DISCUSSION

3.1 Relationship between cube compressive strength and axial compressive strength

From Table 2, we can see that the strength grades of specimens were about C10~C20. The number one series, number two series and number five series

Table 2. The cube compressive strength of MSW concrete.

Curing age/d Number	3	7	14	28
1	10.9	15.9	19.2	21.4
2	10.3	12.5	17.7	20.2
3	7.1	10.2	10.9	12.9
4	6.1	10.0	11.2	11.6
5		18.9	21.5	23.2
6	2.1	4.2		5.9

of MSW concrete have higher axial compressive strengths, which is can reach 10 MPa. The reason is that these series of MSW concrete have less total water and lower mole ratio of MgSO$_4 \cdot$ 7H$_2$O. However, the number three series, number four series and the number six series of MSW concrete have lower axial compressive strength, which is no more than 6MPa. The reason is these series of MSW concrete have a lower mole ratio of MgSO$_4 \cdot$ 7H$_2$O. Therefore, the total water and mole ratio of MgSO$_4 \cdot$ 7H$_2$O have an effect on the axial compressive strength of MSW concrete. If the total amount of water and the mole ratio of MgSO$_4 \cdot$7H$_2$O are in a certain range, the axial compressive strength of basic magnesium sulfate cement concrete increase with the decrease of the total amount of water and the mole ratio of MgSO$_4 \cdot$7H$_2$O.

Figure 1 shows the relationship between the cube compressive strength and axial compressive strength. The date of the cube compressive strength and the axial compressive strength is from the code for design of concrete structures (GB 50010-2010).

From Figure 1 we can see that the relationship between the cube compressive strength and the axial compressive strength is in line with that of Portland cement concrete.

3.2 Static modulus of elasticity

According to the standard for test methods of the mechanical properties of ordinary concrete, the test of the modulus of elasticity is done. The results are shown in Figure 2.

From Figure 2, we can conclude:

(1) The relationship between the modulus of elasticity and compressive strength is in line with the law of concrete;

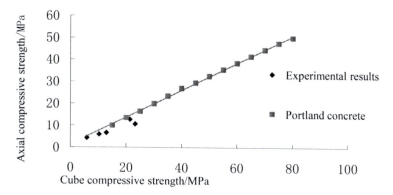

Figure 1. Relationship between cube compressive strength and axial compressive strength.

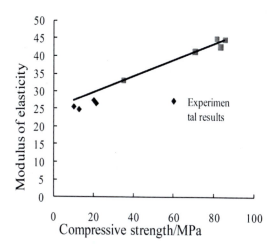

Figure 2. Curve of relationship between Modulus of elasticity and compressive strength.

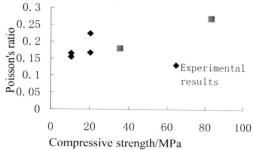

Figure 3. Poisson's ratio of basic magnesium sulphate cement concrete.

(2) Its modulus of elasticity is slightly lower than that of ordinary cement concrete of the same grade, which shows that its strain is larger than that of ordinary cement concrete. So it can resist larger deformation.

3.3 Poisson's ratio of BMS concrete

From Figure 3 we can see that the Poisson's ratio fluctuates up and down about 0.2, which is in line with that of Portland cement concrete.

4 CONCLUSIONS

Strength development of BMS concrete has been studied. Because of the unqualified coarse aggregate and fine aggregate, the results of compressive strength are lower than expected value. So, another strength experiment will be done in a few days. From the experiments above, we can conclude that:

(1) Its cube compressive strength and axial compressive strength gradually increases with age. The lower extent of fly ash in MSW, the quicker is the development of the concrete's strength.
(2) MSW concrete has a higher early strength than that of Portland cement concrete.
(3) Its modulus of elasticity is slightly lower than that of ordinary cement concrete of the same grade, so it can resist larger deformation.
(4) Poisson's ratio of basic magnesium sulphate cement concrete is in line with that of Portland cement concrete.
(5) Its relationship between cube compressive strength and axial compressive strength conforms to the design standard of concrete structures.

REFERENCES

Chengyou, Wu, 2014. Fundamental theory and civil engineering application of basic magnesium sulfate cement [D]. Nanjing: Nanjing University of Aeronautics and Astronautics.
Robert, J.M. Martian Cements. 2008. 10th Annual International Mars Society Convention, August 30,

2007 [DB/OL]. CEMENTSPORTLAND-CHEMISTRY-PROCESSING-LETS-TECHNOLOGY-MARS-Entertainment-ppt-power-point, 2008-01-27

Runčevski, T., Wu Chengyou, Yu Hongfa, Yang Bo, and Dinnebier R. E. 2013. Structural Characterization of a new Magnesium Oxysulphate Hydrate Cement Phase and its Surface Reactions with Atmospheric Carbon Dioxide [J]. Journal of the American Ceramic Society, 96(8): Article first published online: 30 AUG 2013, DOI: 10.1111/jace.12556

Urwong, L., Sorrell, C.A., 1980. Phase relations in magnesium oxysulfate cements [J]. Journal of the American Ceramic Society. 63(9–10): 523–526

Advances in Civil Engineering and Building Materials IV – Chang et al (eds)
© 2015 Taylor & Francis Group, London, ISBN: 978-1-138-00088-9

A strength developing law for an epoxy asphalt mixture for airport runways

Jianming Ling, Erchun Kong & Hongduo Zhao
Key Laboratory of Road and Traffic Engineering of the Ministry of Education, Tongji University, Shanghai, China

ABSTRACT: This paper analysed the chemical reaction mechanism of epoxy asphalt and the strength growth characterization of an epoxy asphalt mixture. Splitting tensile strength was chosen to characterize the strength of the epoxy asphalt mixture. Based on the increasing rules of viscosity of an epoxy asphalt system in a solidification process, a time-temperature model of strength was built. According to an equivalent temperature cure experiment, strength equation parameters were solved and the strength of an epoxy asphalt mixture after a natural cure was predicted by using the strength equation. The results show that the theoretical prediction curve agrees well with the actual strength growth process: increase of viscosity of a combined material system results in a change of mixture strength; the resolved strength equation can be used to predict the strength of epoxy asphalt runways in airports and determine the practical cure period and the opening time of traffic.

Keywords: epoxy asphalt mixture, strength growth mechanism, strength equation, discrete prediction, airport runway.

The Epoxy Asphalt Mixture (EAM) is composed of graded aggregates, fillers, and an epoxy asphalt binder. It is the kind of material with the dual advantages of cement concrete and asphalt mixture, which provides good corrosion resistance, high temperature stability, moisture susceptibility and an anti-fatigue property (Huang, W. 2006). Due to its excellent service performance, the EAM has been successively used in an airport at east China in 2013.

Current domestic research on EAM is mainly focused on the properties of this material as an adhesive layer for steel bridge decks (Cong, P. et al. 2009, Cong, P. et al. 2010), disease forms and disease mechanism (Chen, T. 2006, Zhang, H. et al. 2013), while the strength of epoxy asphalt's growth law or curing model is less involved.

In the mixing and curing process, epoxy's ring-opening reaction occurred under the effect of the functional groups of the curing agent, resulting in a space network system. The ring-opening reaction will occur immediately just after mixing the two components together. However, this chemical compound was not completed instantaneously and was greatly impacted by time and temperature in reaction, during which a higher temperature makes a faster reaction and a longer time makes a higher degree reaction (Lu, W. & Guo, Z. 1996). Thus, the strength growth of EAM shows a gradual process, in which the intensity was low at the initial stage and would gradually grow with time. Moreover, the growing strength of the EAM was greatly influenced by the outside temperature, mainly as the higher the ambient temperature was, the faster the strength grew; the lower the ambient temperature was, the slower the strength grew.

In order to effectively guide the EAM production and the large-scale construction at airports and to determine the opening traffic time of airport runways, this article will examine the strength growth mechanism of EAM, reveal the process of developing its strength, and draw the relationship between the EAM's strength and its durability and temperature.

1 MATERIALS

1.1 Raw materials

In research, the adopted epoxy asphalt binder which was produced by a chemical factory in Jiangsu province in China consists of two cement components: component A (epoxy resin) and component B (base asphalt and curing agent). This kind epoxy asphalt in which the weight of component B was 2.9 times that of component A was HLJ-2910. Basalt stone chips collected from Baoshan in Shanghai were used as an aggregate. The properties of the raw materials used in this paper all met the required specifications.

1.2 Aggregate gradation

The aggregate gradation used in an airport runway in east China was adopted for the research. The

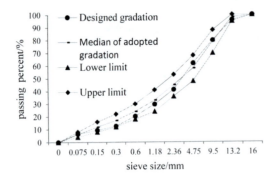

Figure 1. Adopted and designed aggregate gradation in test.

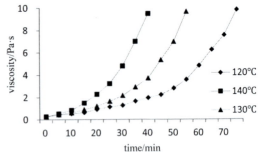

Figure 2. Rising curves of viscosity under different temperatures.

nominal maximum aggregate size was 13.2 mm and the gradation curve is shown in Figure 1.

1.3 Optimum binder content

The Marshall Mixture Design procedure was used to produce the optimum binder content. Considering the skid resistance of airport runway, the target air void of the mixture ranged from 3% to 5%, while the target air void of EAM used for steel deck runways was controlled with less than 3% which was based on the consideration of anti-flooding performance. And the Marshall stability of the cured specimen is more than 40 kN. According to the test results, a binder content of 5.0% is found to be the optimum for the EAM in airport runways.

2 STRENGTH GROWTH MECHANISM

2.1 Viscosity growth law of epoxy asphalt

The ring opening reaction of epoxy asphalt was a gradual process. In the first stage, the microgel emerged unevenly in the system in the core of the macromolecular reactants (epoxy resin polymer) which had a faster rate of chemical reaction; then the microgel grew gradually until it reached the point of a large gel while the curing reaction was going on. After which the whole system developed towards the direction of three-dimensional network until the reaction end. This change can be expressed by viscosity in the apparent. Rising curves of viscosity of epoxy asphalt under different temperatures are shown in Figure 2. From Figure 2, it is obviously possible to plot the characteristics of the viscosity growth in the epoxy asphalt under three different temperatures: in the beginning stage, the viscosity grew slowly and the amount of growth was small; then the growth rate grew gradually faster and then the viscosity increased rapidly in the final stage. In addition, a sharp increase in viscosity appeared in a short time at a higher temperature, while it required a little longer time at a lower temperature (Min Zhaohui et al. 2007).

2.2 Strength growth law of epoxy asphalt mixture

2.2.1 Intensity parameter selection

During curing, the aggregate in the EAM remained unchanged; the only change which was embodied in the change of viscosity was in the binder. It was well to know that the essence of the strength growth of the EAM is a chemical reaction between component A and component B. The binder that really worked in the EAM was the cured epoxy asphalt, and it bonded the discrete aggregate together which indicated the strength source of the EAM, especially its tensile properties, and that the strength of the EAM would stop increasing when the reaction ended. So, among the numerous material strength parameters, tensile strength would be the best choice for characterizing the nature of intensity of EAM. Considering the convenience of the test, the splitting tensile strength which was based on Marshall Specimens was finally selected as the intensity parameter.

2.2.2 Strength growth process

At 120°C, the needed Marshall Specimens were built under the optimum binder content (5.0%) with a seventy-five times double-sided hammer. Specimens were divided into three groups, two of which were respectively placed in 120°C and 60°C oven, and the rest group was placed outdoors under natural conditions.

Before the splitting test, the specimens need thermal insulation so that their internal temperature and external temperatures were consistent with each other. Taking into account the time-temperature curing properties of epoxy asphalt, split test should be carried out at low temperatures, which could reduce the effects of experimental holding time on the strength of epoxy asphalt mixture. According to the literature research and experimental test, 15°C would be ideal.

The splitting tensile strength of two group specimens cured in isothermal conditions were measured at 15°C. The results are shown in Figure 3 and Figure 4.

As was seen from the above two figures, the properties of strength growth curve match well with those of viscosity growth curve. With the curing time increasing, the intensity increased. In addition, the strength

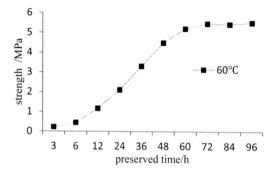

Figure 3. Strength-preserved time curve at 60°C.

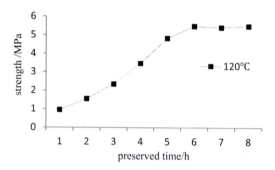

Figure 4. Strength-preserved time curve at 120°C.

growth rate at a higher temperature was much faster than that at a lower temperature. The ultimate strength at different curing temperature was maintained at the same level.

3 STRENGTH GROWTH EQUATION

Being different from general modified asphalt, the rheological properties of epoxy asphalt during the curing process was related to not only temperature but also time (Guo, Z. et al. 2004). The viscosity of epoxy asphalt system at a certain temperature and time could be written as following (Theriault, R.P. et al. 1999):

$$\eta(t) = \eta_0 \exp[k(T)t] \tag{1}$$

$$k(T) = A \exp[-\frac{E_a}{RT}] \tag{2}$$

where η_0 = viscosity at zero time; $k(T)$ = the reaction rate constant; t = time; E_a = the activation energy; R = the gas constant; T = the absolute temperature; A = coefficient.

In addition, the relationship between viscosity at time zero time and temperature history can also be described by the Arrhenius relationship:

$$\eta_0 = \eta_\infty \exp[\frac{E_\eta}{RT}] \tag{3}$$

Table 1. Solution of the parameters in strength equation.

S_∞	0.53509 MPa
E_η	0.83999 kJ/mol
A	108572.8 h^{-1}
E_a	40.77599 kJ/mol

where η_∞ = the Arrhenius pre-factor; E_η = flow activation energy.

Then the viscosity of epoxy asphalt system with time and temperature could be expressed by the following new equation:

$$\eta(t,T) = \eta_\infty \exp[\frac{E_\eta}{RT}] \exp[k(T)t] \tag{4}$$

A new form could be obtained by taking the logarithm of both sides of the equation:

$$\ln \eta(t,T) = \ln \eta_\infty + \frac{E_\eta}{RT} + At \exp(-\frac{E_a}{RT}) \tag{5}$$

Considering the relationship between the viscosity of epoxy asphalt and the strength of the epoxy asphalt mixture expounded in the second section of the article, the viscosity in equation could be replaced as the strength:

$$\ln S(t,T) = \ln S_\infty + \frac{E_\eta}{RT} + At \exp(-\frac{E_a}{RT}) \tag{6}$$

where $S(t, T)$ = the splitting tensile strength; S_∞ = the Arrhenius pre-factor. Equation 6 was the dual-Arrhenius model of strength growth of the EAM, and could be used for predicting the strength growth of EAM after the parameters in this model were parsed, being based on the measured data.

4 STRENGTH PREDICTION UNDER NATURAL CONDITION

4.1 Strength parameter parsing

According to the measured splitting tensile strength of two groups of Marshall Specimens, cured under equivalent temperature, two sets of data from each of the two groups would be used in the strength equation and the parameters could be obtained as shown in Table 1.

4.2 Discrete prediction

The daily temperature was not necessarily the same under natural conditions. In research, the average daily temperature during natural curing was shown in Figure 5.

So, the formula 6 could not be used for continuous prediction but for discrete prediction instead. According to the time-temperature equivalence principle, the

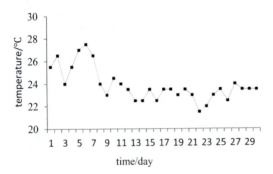

Figure 5. Average daily temperature in natural curing period.

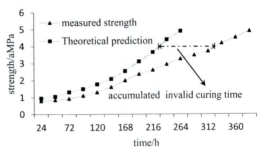

Figure 6. Strength of theoretical prediction and field test.

strength of EAM at different temperatures obeys the following rule:

$$S = S(t_i, T_i) = S(t_j, T_j) \quad (i \neq j) \tag{7}$$

where S, T_i and T_j were already known, then:

$$t_i = t(T_i, S(t_j, T_j)) \tag{8}$$

If $S(n)$ represent the first n days strength, Δt represent the time interval (actually equal to one day or 24 hours under natural conditions), then a recursive relationship can be expressed as follows:

$$S(1)=S(\Delta t, T_1) \quad (t_1 = \Delta t) \tag{9}$$

$$S(2) = S(t(T_2, S(1)) + \Delta t, T_2) \tag{10}$$

...

$$S(j) = S(t(T_j, S(j-1)) + \Delta t, T_j) \tag{11}$$

...

Since the daily average temperate was known already, the predicted value of strength for each day during which the strength of the EAM was growing can be found according to this process.

4.3 Analysis of prediction result

The predicted strength and the measured strength of Marshall Specimens under natural conditions were shown in Figure 6. The value of theoretical prediction was always larger than that measured, and the difference between them grew over time. It seemed that the prediction was a complete failure, while the result was in fact normal and could be explained.

In the process of prediction, the interval time was one day (an entire 24 hours); but during the interval, the actual curing time of the specimen did not reach the entire length of the interval which meant that there was some invalid time during which the strength of the specimen did not grow. All of this is because of the average temperature. To make it clearer, there was always a period of time in which the temperature

stayed at a lower level, especially the average daily temperature at the latter part of the study was lower. For epoxy asphalt, the curing reaction was temporarily stopped when the temperature was below a certain critical temperature.

Even so, the strength equation can still be used to predict the behaviour of strength growth of EAM under certain temperature conditions, and the invalid curing time ought to be removed from the interval if needed.

5 CONCLUSION

The strength of EAM is affected by the time and the temperature of the curing process, mainly that the strength of EAM grows with the growth of time, and the growth rate accelerates when the temperate rises. The essence of the strength growth of EAM is the growth of epoxy asphalt.

The dual-Arrhenius model of strength growth of EAM is built on the base of the viscosity growth rule of the epoxy asphalt system. The parameters in the strength equation are parsed, based on the measured splitting tensile strength of the two groups of Marshall Specimen cured at two different temperatures. The strength equation can be used to predict the behaviour of strength growth of EAM under certain temperature conditions. The predicted result of the strength under natural condition only agrees with the measured strength at the initial stage of the process because of the existence of invalid curing time which is inevitable at a lower temperature. The test result and prediction method can be used as a reference to the traffic opening time and further research at the next stage.

REFERENCES

Chen, T. 2006. Research on crack behavior of long-span steel bridges deck epoxy asphalt runway. Jiangsu: Southeast University.
Cong, P. et al. 2009. Investigation on influence factors of epoxy asphalt and its mixture properties. Journal of Wuhan University of Technology, 31(19):7–10.
Cong, P. et al. 2010. Effects of epoxy resin contents on the rheological properties of epoxy-asphalt blends. Journal of Applied Polymer Science, 118(6): 3678–3684.

Guo, Z. et al. 2004. Cure kinetics and chemorheological behavior of epoxy resin used in advanced composites. Acta Materiae Compositae Sinica, 21(4): 146–151.

Huang, W. 2006. Theory and method of deck paving design for long-span bridges. Beijing: China Construction Industry Press: 47–62.

Lu, W. & Guo, Z. 1996. Compounding of the high strength asphalt concrete and its properties. China Journal of Highway and Transport, 9(1): 8–13.

Min, Z. et al. 2007. Dependency of time-temperature of epoxy asphalt mixture strength. Chinese Journal of Highway, 03: 1–4.

Theriault, R.P. et al. 1999. A numerical model of the viscosity of an epoxy resin system. Polymer Composites, 20(5): 628–633.

Zhang, H. et al. 2013. Research on performance of crack seal materials for long-span steel deck epoxy asphalt runway [J]. Highway, 12: 36–39.

Advances in Civil Engineering and Building Materials IV – Chang et al (eds)
© 2015 Taylor & Francis Group, London, ISBN: 978-1-138-00088-9

Research on the low-temperature cracking resistance of epoxy asphalt mixtures for airport runways

Jianming Ling, Zhifeng Gu, Guoxi Liang & Erchun Kong
Key Laboratory of Road and Traffic Engineering of the Ministry of Education, Tongji University, Shanghai, China

ABSTRACT: The relationship between the temperature stress of epoxy asphalt concrete and its tensile strength were measured under different initial temperatures and rates of cooling at low temperatures. On this base, the low-temperature cracking resistance of epoxy asphalt mixtures was evaluated, which showed that the limit temperature of low-temperature cracking was roughly −23°C and the relevant limit strain was approximately 2000 micro strains. The results showed that epoxy asphalt mixture had good performance under the conditions of low temperature and satisfied the performance of epoxy asphalt runways in cold areas when there was a sudden drop in temperature.

Keywords: airport runways, epoxy asphalt mixture, temperature stress, the low temperature performance.

1 INTRODUCTION

Epoxy asphalt is composed of epoxy resin and asphalt according to a certain proportion. The epoxy asphalt mixture has been used in airports in recent years because of its stability at high temperatures. At present, research into epoxy asphalt has progressed a lot, but the study of epoxy asphalt concrete in runways of airports is less involved. Due to the limitation of the layer and its own, thermal stress without relaxation will appear in epoxy asphalt runways because of the sudden change in temperature, and low-temperature cracking will occur when the thermal stress exceeds the allowable stress of epoxy asphalt mixture (Zhu Yiming, 2006). Different temperature drop range and cooling rate will directly affect the low-temperature cracking of epoxy asphalt concrete.

Methods currently used to study and evaluate low temperature cracking resistance of asphalt mixture can be divided into three categories: the estimated cracking temperature of the mixture; evaluation of asphalt mixture temperature deformation or stress relaxation ability; The evaluation of the fracture energy of asphalt mixture (Zhou Xiaohua et al. 2006). In order to effectively study the airport low-temperature performance of epoxy asphalt concrete runways for airports, the limit cracking temperature will be estimated by calculating the thermal stress under different conditions in base of low-temperature bending test and contraction coefficient tests (Lu Ruitie, 1998).

2 MATERIALS

2.1 Grading

The aggregate gradation used in an airport runways in East China was adopted for the research. The nominal maximum aggregate size was 13.2 mm and the gradation curve was shown in Fig. 1.

2.2 Raw materials

In the research, the adopted epoxy asphalt binder which was produced by a chemical factory in Jiangsu province in China consists of two cement components: component A (epoxy resin) and component B (base

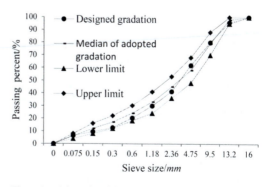

Figure 1. Adopted and designed aggregate gradation in test.

Table 1. Epoxy asphalt major technical indicators.

Experiment items	Technical indicators	Technical Requirements
Marshall test	Marshall stability/kN	≥40
	Voidage/%	3~5
	Residual stability/%	≥80
Trabecular bending test	Flexural tensile strain (15°C)/10^{-3}	≥2
Rutting test	Dynamic stability (70°C, 1.4 MPa)/number	≥5000
Freeze-thaw splitting test	TSR/%	≥80

Table 2. Different temperature ranges of linear shrinkage factor.

Temperature range/°C	Linear shrinkage coefficient/°C^{-1}
5~0	1.84×10^{-5}
0~−5	1.78×10^{-5}
−5~−10	1.71×10^{-5}
−10~−15	1.63×10^{-5}
−15~−20	1.52×10^{-5}

asphalt and curing agent). This kind epoxy asphalt in which the weight of component B was 2.9 times that of component A was HLJ-2910. Basalt stone chips collected from Baoshan in Shanghai were used as aggregates. The properties of the raw materials used in this paper all met the required qualifications.

2.3 Best binder content

The Marshall Mixture design procedure was used to design the optimum binder content. Considering the skid resistance of airport runways, the target air void of the mixture ranged from 3% to 5%, while the target air void of EAM used for Steel deck runways was controlled at less than 3% which was based on the consideration of anti-flooding performance. And the Marshall stability of the cured specimen is more than 40 kN. According to the test results, a binder content of 5.0% is found to be optimum for the EAM of airport runways. The epoxy asphalt major technical indicators are shown in Table 1.

3 THE EPOXY ASPHALT TEST

3.1 The test of linear shrinkage coefficient

The test uses prismatic specimens with the size of 20 mm × 20 mm × 200 mm, and the group of specimens were at least three. Contacts were kept at both ends of the specimen and the dial gauge when measuring, ensuring the end portion of the dial gauge could be freely contracted. The specimen was fixed with a temperature regulator that set the initial temperature of 5°C, control temperature according to the cooling rate of 5°C per hour, then read change in length ΔL after the specimen temperature stability indicator, calculating the epoxy asphalt mixture linear shrinkage coefficient at different temperatures. The results were shown as follows.

As the temperature decreases, the linear shrinkage coefficient of the epoxy asphalt mixture becomes lower, It is not a constant.

3.2 Bending test of small beam at low temperature

The stiffness modulus and the maximum tensile strain of the epoxy asphalt mixture were got according to

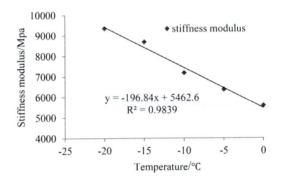

Figure 2. The change rule of the stiffness modulus as the temperature changed.

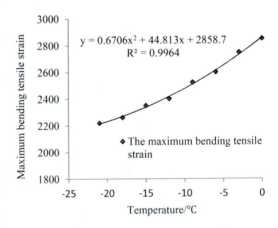

Figure 3. The change rule of the stiffness modulus as the temperature changed.

the bending test of a small beam with the size of 30 mm × 35 mm × 250 mm under different temperature conditions (Zhan Zhanjun et al. 2012). The three-point loading method on the MTS testing machine was used, adopting the loading rate of 50 mm/min. The experimental temperature was adjusted by the temperature control box. The focus force was applied to the middle of the small beam until there was fracture failure. The recorded data, and the results were as follows.

As the temperature decreases, the stiffness modulus of the epoxy asphalt mixture increases linearly, the maximum tensile strain decreases gradually less, and

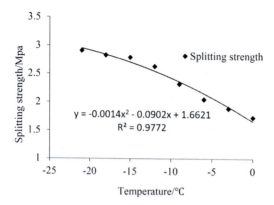

Figure 4. The change rule of the splitting strength as the temperature changed.

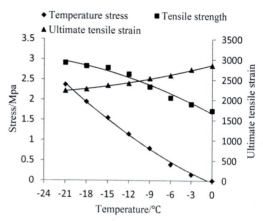

Figure 5. The comparing temperature stress and tensile strength at different temperatures and the estimation of the ultimate tensile strain.

tends to be stable. The stiffness modulus is inversely proportional to low-temperature deformation of the epoxy asphalt mixture. When there was a sudden drop in the temperature, the increase in the stiffness modulus leads to temperature stress increase. The comparisons between the temperature stress and tensile strength will show a more intuitive reaction of the epoxy asphalt mixture resistance to low temperature performance, and the maximum tensile strain can be used as low-temperature epoxy asphalt mixture performance control indicators to evaluate its crack resistance performance.

3.3 The splitting test of low temperature

The tensile strength of epoxy asphalt is generally measured by the splitting test. The low-temperature splitting test can be used to evaluate the epoxy asphalt cracking resistance performance. (Tan Yiqiu et al. 2010) Use the standard Marshall compaction method molding the specimen with the diameter of 101.6 mm, the height of 63.5 mm, the specimens were tested on the MTS testing machine after they were completely cured with the loading rate of 1 mm/min, the test temperature was controlled through the temperature control box, ranging the test temperature of $0° \sim -20°$ The test results were shown as follows.

The splitting strength of the epoxy asphalt mixture gradually increased with the temperature decreases, but, the rate of increase became smaller, and it can be inferred that, when the temperature dropped to a certain extent, the splitting strength tended to be stable.

The specimen's splitting tend to brittle failure as the temperature decreased, and the splitting strength of epoxy asphalt mixture exist limit.

3.4 Temperature stress

The epoxy asphalt temperature stress generated within the constraints could not catch up with the relaxation as the temperature suddenly dropped, when exceeding the allowable stress values of the epoxy asphalt mixture, and cracks would occur (Shang Yapeng, 2008).

As to the top layer of epoxy asphalt concrete runways of the airport, the effects of temperature gradients was not taken into consideration, According to the constraints specimen temperature stress test (TSRST), the temperature stress calculation of asphalt concrete runway's temperature stress is now a commonly used formula as that made by Hills and Brien (Tian Xiaoge et al. 2002), and on this basis, slightly amended to:

$$\sigma_x(T) = \sum_{T_0}^{T_n} \alpha \ (\Delta T) \ S(\Delta T) \cdot \Delta T \qquad (1)$$

where $\sigma_x(T) =$ Cumulative temperature stress, MPa; $\alpha(\Delta T) =$ linear shrinkage coefficient among ΔT, 1/°C; $S(\Delta T) =$ stiffness modulus among ΔT, MPa; $\Delta T =$ Temperature interval, °C; $T_0 =$ Initial temperature, °C; $T_n =$ final temperature, °C.

According to China's meteorological climate information, the night cooling rate could be up to 4°C/h~7°C/h, and in some areas could even reach 10°C/h, in northern of China at the time of the arrival of a cold wind, and the maximum temperature reduce can reach −15°C, and extreme low temperatures of −20°C can cover the northern regions of China (Zhu Ling et al. 2013).

The creeping test showed that, when the temperature was higher than 0°C, the temperature stress of the epoxy asphalt mixture could release its stress through stress relaxation, on the basis of equation (1), where the initial temperature is setting at 0°C, and then the temperature is adjusted, to compare the tensile strength and temperature stress of the epoxy asphalt mixture at different temperature. Finally, the test estimated the cracking temperature of the epoxy asphalt mixture when the temperature suddenly dropped and the result was shown as follows.

As shown in the graph, the temperature stress increases faster than the tensile strength of the epoxy asphalt with the decrease of cooling rate. According to a temperature stress and tensile strength parabola formula, the cracking temperature of the epoxy asphalt is reached when the temperature sudden dropped to about $-23°C$, and in this cooling rate, the limit of the tensile strain on the epoxy asphalt is 2000×10^{-6}. When the When the strain exceeds the ultimate tensile strain, shrinkage cracks occurred.

4 CONCLUSIONS

The limit of the cracking temperature of epoxy asphalt concrete is estimated and the low temperature performance when temperature dips is evaluated by comparing the temperature stress and tensile strength under different cooling rates on the basis of tests and analysis.

Taking into account that the epoxy asphalt mixture linear shrinkage coefficient varies at different temperature ranges, the method of temperature stress superposition at different temperature ranges is used to accurately calculate the thermal stress.

According to extreme climate changes nationwide, the limit of the cracking temperature is estimated under extreme conditions of temperature stress, and the result shows that the epoxy asphalt mixture can be used in large areas of northern China.

The study of the limit cracking temperature and the relevant ultimate tensile strain are conservative because the stress relaxation effect is not be considered.

REFERENCES

Lu, R. 1998. Study on evaluation index of low temperature properties of asphalt mixture. Petroleum Asphalt, 01:23–35t + 22.

Shang, Y.P. 2008. Study on temperature shrinkage of semi-rigid base asphalt runways and the effect on temperature stress. Chang'an University.

Tan, Y.Q. 2010. An evaluation of several kinds of asphalt mixtures' low-temperature performance based on TSRST. Highway, 01:171–175.

Tian, X.G. 2002. Tests on thermal stress in asphalt cement sample and its numerical simulation.stress. Civil Engineering Journal, 03:25–30.

Zhan, Z.J. 2012. Research on bending performances of epoxy resin asphalt mixes with different crosslink densities at low temperatures. China Journal of Highway and Transport, 01:35–39.

Zhou, X.H. 2006. Research on low temperature performances of epoxy asphalt mixture. Highway, 01:179–182.

Zhu, Y.M. 2006. Research on the performance of domestic epoxy asphalt mixture. Southeast University.

Zhu, L. 2013. Study on the temporal and spatial distribution of the extreme low temperature. Wind Energy, 09:74–78.

Advances in Civil Engineering and Building Materials IV – Chang et al (eds)
© 2015 Taylor & Francis Group, London, ISBN: 978-1-138-00088-9

Comparative performance of two commercially available superplasticizers on the flow behaviour of cement paste incorporating mineral admixtures

J. Chakkamalayath, M. Abdulsalam, S. Al-Bahar & F. Al-Fahad
Construction and Building Materials Program, Energy and Building Research Centre,
Kuwait Institute for Scientific Research, Kuwait

ABSTRACT: High performance concrete incorporates several mineral and chemical admixtures to achieve the desired properties. The incompatibility between admixtures and cements can be avoided by the appropriate selection of admixtures at their optimum dosage. This paper discusses the combined influence of cement, mineral admixture, and superplasticizer on the flow behaviour of cement paste. Cement paste was prepared with a w/c of 0.35, with ordinary Portland cement, two types of superplasticizers and two types of mineral admixtures. The saturation dosages of the superplasticizer for different combinations were determined, using mini slump test immediately after mixing and after 60 minutes. High performance concrete was prepared with the dosage of the superplasticizer obtained from paste studies and its mechanical properties were studied.

Keywords: Superplasticizer, mineral admixture, high performance concrete, mini slump, compressive strength.

1 INTRODUCTION

Superplasticizers are commonly used with other mineral admixtures for the production of High Performance Concrete (HPC). In addition to providing good workability, the Superplasticizer (SP) also influences the mechanical and microstructural properties of concrete. Also, it reduces the permeability of concrete and improves its durability (Sakai et al., 2006).

However, the performance of a superplasticizer in concrete depends on the physical and chemical characteristics of cement, type and dosage of the superplasticizer, type and dosage of the mineral admixture under the prevailing environmental conditions (Aïtcin, 1998; Aïtcin et al., 2001, Rixom and Mailvaganam, 1999). An incompatible combination can result in rapid slump loss, retardation in setting and reduction in strength (Ramachandran, 2002).

Sulphonated Naphthalene Formaldehyde (SNF) and Polycarboxylates (PCE) are commonly used superplasticizers in Kuwait. The local construction industry demands the use of HPC for its mega projects and is facing the challenge of selecting a suitable type of admixture as there are different families of SPs and brands of SP, and a wrong selection sometimes results in incompatible combination of cement and SP. In addition, the incorporation of mineral admixtures, especially natural Pozzolana, may result in additional incompatibility problems. Therefore, the selection of an appropriate type of superplasticizer at optimum dosage, incorporating natural Pozzolana and a conventional mineral admixture under the prevailing

environmental conditions in Kuwait, is the objectives of the study.

The interaction between the cement and other admixtures governs the flow behaviour of concrete and can be studied by understanding the flow behaviour of cement paste as the paste phase gives fluidity to concrete (Jayasree and Gettu, 2008). The flow behaviour of superplasticized cement paste is studied with mini slump and the dosage obtained from paste study is used to produce concrete with adequate workability.

2 MATERIALS

Ordinary Portland Cement (OPC) Type I, two types of superplasticizers (Sulphonated Naphthalene Formaldehyde (SNF) and Polycarboxylate (PCE) and two mineral admixtures, Volcanic Ash (VA) and Ground Granulated Blast Furnace Slag (GGBS) were used in the study. In the cases of mixes with VA and GGBS, OPC was replaced with 20% VA (by mass) and 40% GGBS respectively. The chemical composition and physical properties of VA were determined as per ASTM C618 to check its suitability to be used as a Pozzolanic material in concrete. The specific gravity of SNF based SP is 1.245 and that of PCE based SP is 1.110. The solid content of both superplasticizers were determined as per ASTM C 494, and it was observed that SNF had a solid content of 41% and PCE had a solid content of 45%. The SP was added as a percentage of cement/cementitious material weight.

Figure 1. Mini slump test.

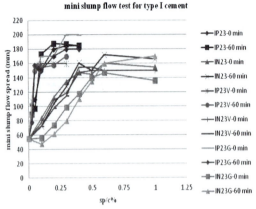

Figure 2. Mini slump test result for Type I cement.

Table 1. Saturation dosage of superplasticizer for different mixes.

Designation	Saturation dosage for constant flow	Saturation dosage from intersection of 0 and 60 min flow curve
IP23	0.2	0.3
IN23	0.4	0.5
IP23V	0.1	0.2
IN23V	0.4	0.5
IP23G	0.1	0.2
IN23G	0.5	0.5

3 EXPERIMENTAL METHOD

The mini-slump test developed by Kantro (1980) was used to study the flow behaviour of superplasticized cement paste. In this test, a mould in the shape of a truncated cone, with dimensions proportional to the Abram's cone, was filled with cement paste, and the spread diameter was measured after lifting the mould (Fig. 1). Also, visual examination helped to evaluate the bleeding and segregation of the paste.

A consistent mixing procedure was adopted throughout the study. The spread was observed with mini slump, immediately after mixing and after 60 minutes, in order to understand the influence of SP on loss of fluidity. The saturation dosage was taken as the dosage corresponding to a constant flow before bleeding. Mixes were designated by a code, the first letter representing type I cement (I); second letter representing the type of SP; P for PCE type and N for SNF type superplasticizer; third and fourth letter representing the temperature at which the study was conducted (23°C). The last letter represents the type of mineral admixture, volcanic ash designated as V and GGBS is designated as G. The superplasticizer dosages given in this work are all in terms of a solid content; the water contained in the superplasticizer has been accounted for in the water content of the mixes. The flow behaviour of the cement paste was observed at 0 and 60 minutes and the intersection point of 0 and 60 minutes flow curve was also noted.

In order to correlate the results obtained from paste tests to the behaviour of concrete, the tests were conducted on concrete with two superplasticizers. The saturation dosage of the superplasticizer obtained from the paste study was used as a guideline to get a concrete of medium workability. A designed concrete mix was used in this study and the fresh and hardened properties of the concrete were studied.

4 RESULTS AND DISCUSSION

4.1 *Flow behaviour of superplasticized cement paste*

The mini-slump test results (Figure 2) show that the spread increases with an increase in dosage of the superplasticizer and it remains constant or shows only a slight difference in spread with the further addition of the superplasticizer. This is taken as the saturation dosage of the SP for that particular mix. However, bleeding was observed in the paste whenever the saturation superplasticizer dosage was exceeded. It was also visually observed that the paste mixes of PCE and SNF at an over dosage with VAshowed setting at 60 minutes later,more than that with GGBS.

Table 1 shows the saturation dosage of the superplasticizer obtained for various mixes; saturation dosage, taken as one with constant flow before bleeding and the other as point of intersection of flow curve at 0 and 60 minutes.

The test results show that the type of superplasticizer has the predominant effect on the saturation dosage than the type of mineral admixture. The performance of both mineral admixtures are comparable and shows well-defined saturation dosage. It can be clearly seen that the PCE based SP has a stronger

Table 2. Compressive and splitting tensile strength of concrete mixes.

Mix	Compressive strength (MPa)		Splitting tensile strength (MPa)		Percentage of Voids (%)
	7 day	28 day	7 day	28 day	
IP23	51.33	61.17	3.92	4.21	8.42
IN23	51.67	59.50	3.93	4.83	9.71
IP23V	46.73	59.50	3.49	3.66	10.94
IN23V	41.73	51.83	3.18	3.87	11.22
IP23G	50.47	52.67	3.48	4.19	9.80
IN23G	47.50	58.83	3.34	3.98	10.07

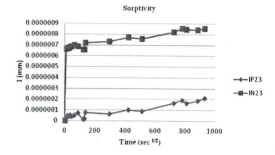

Figure 3. The sorptivity performance of different mixes.

effect on the mixtures, as only low concentrations of SP were required, ranging from 0.1 to 0.3% by weight of cement, as expected, whereas similar mixtures with SNF, required higher dosages, usually 0.4–0.6%. This is due to the fact that the PCE action is based on both electrostatic repulsion and steric hindrance for dispersion, while the SNF action is based on electrostatic repulsion only, as reported earlier (Ramachandran, 2002). The mixes with dosages higher than the saturation dosage had shown a setting after one hour of mixing. Even though the saturation dosage for mixes with both mineral admixtures were comparable, GGBS had shown better fluidity at lower dosages. The dosage obtained from paste studies were used to prepare concrete with adequate workability.

4.2 Effect of superplasticizer on the compressive and the splitting tensile strength of concrete

A concrete mix was prepared with the following proportions; Cement – 480 kg, coarse aggregate – 1170 kg, fine aggregate – 570 kg and w/c −0.35. OPC was replaced with 20% VA and 40% GGBS for the respective mixes. The saturation dosage of the superplasticizer, obtained from paste study based on the intersection of zero and sixty minute flow curves, was used to produce concrete of different combinations to get a medium slump of 80 to 100 mm.

In addition to slump test, the compressive strength and splitting tensile strength after 7 and 28 days were evaluated and the studied mixes had achieved the designed strength after 28 days. The compressive strength values of various mixes are given in Table 2. The percentage of voids was determined after 28 days as per ASTM C642. It was observed that the compressive strength varies with the combination of superplasticizer and mineral admixture.

In both cases of PCE and SNF, the compressive strength and splitting tensile strength of concrete decreases with the addition of mineral admixture. The compressive strength result shows that cement-PCE-VA combination and cement-SNF-GGBS combinations give comparable strength results to the control mix of IP23 and IN23. The combination of cement-SNF-VA and cement-PCE-GGBS combinations give

reduced strength compared to control mixes, which may be because of the higher percentage of voids in the respective concrete mixes.

4.3 Effect of superplasticizer on the rate of absorption of water

The rate of absorption (sorptivity) of water by the concrete was determined by measuring the increase in the mass of the specimen due to the absorption of the water as a function of time when only one surface of the specimen was exposed to water. Assuming a constant water supply at the inflow surface, Sorptivity (S) is related to the cumulative absorbed volume of water per unit area of the inflow surface (i) and elapsed time (t) by the relationship,

$$I = A + S\sqrt{t} \qquad (1)$$

The constant term (intercept at $t = 0$) represents the effect of initial water filling on the concrete surface.

In this study, the test was conducted to understand the influence of PCE and SNF on the rate of absorption of water. Cylinders of size 100 mm diameter and 50 mm height was cut from the previously prepared cylinder samples. Two specimens were taken from each mix. Vacuum saturation of samples was done as per ASTM C1202 without epoxy coating. The test was conducted as per ASTM C 1585. The surface of concrete cylinder, which was parallel to the direction of casting was placed in a tray containing water and rested on 5 mm glass rods to permit the free access of water to the inflow surface. The water level was maintained at 1 to 3 mm above the base of the cylinder. The absorption of water was measured at intervals as mentioned in the ASTM standard.

The comparative performance of different mixes on sorptivity is given in Figure 3. The test results show that the rate of absorption of water varies with the type of superplasticizer. It was observed that the mixes with PCE based SP show a lower rate of absorption of water compared to that with SNF. The sorptivity is significantly influenced by the type of SP. The reduction in sorptivity, while using PCE, indicates the presence of relatively smaller capillary pores in concrete. The reduced sorptivity indicates finer pore structure, which in turn inhibits the ingress of aggressive elements in the pore system.

5 CONCLUSION

A well-defined saturation dosage was obtained for various combinations of superplasticized cement paste indicating compatibility among the studied cement, SP and mineral admixture. The study concludes that the dosage of the superplasticizer obtained from paste study can be used as a guideline for selecting the dosage required for producing concrete with an adequate workability. The hardened properties of concrete show that the performance of SP varies with the type of SP and the type of mineral admixture. As far as the compressive strength is concerned, the cement-PCE-VA combination and the cement-SNF-GGBS combinations are giving better performances. The sorptivity test result shows that there is a higher rate of absorption of water initially with SNF based SP and the results are supporting the higher compressive strength obtained with PCE based SP.

ACKNOWLEDGEMENT

This paper is based on the project EU085K funded by the Kuwait Institute for Scientific Research (KISR) The authors are grateful for the support of KISR to accommodate all the rising challenges during the execution of the project's tasks. The authors also wish to acknowledge the support extended by the technical staff of Concrete and Material Testing Laboratory while performing experiments.

REFERENCES

Aïtcin, P.C. 1998. High performance concrete. London, E & FN Spon.

Aïtcin, P.C., Jiang, S., Kim, B.G., Nkinamubanzi, P.C., & Petrov, N. 2001. Cement/superplasticizer interaction: The case of polysulphonates. *Bulletin Des Laboratoires Ponts et Chausses*, July–August: 89–99.

Jayasree, C. & Gettu, R. 2008. Experimental study of the flow behaviour of superplasticized cement paste. *Materials and Structures*, RILEM. Vol. 41: 1581–1593

Kantro, D.L. 1980. Influence of water-reducing admixtures on properties of cement paste-A miniature slump test. *Cement, Concrete and Aggregates*. Vol. 2: 95–102

Rixom R & Mailvaganam. N. 1999. *Chemical admixtures for concrete*. London: E& FN Spon.

Ramachandran, V.S. 2002. *Concrete admixtures handbook*. New Delhi, India: Standard Publishers,

Sakai, E., Kasuga, T., Sugiyama, T., Asaga, K. and Daimon, M. 2006. Influence of superplasticizers on the hydration of cement and the pore structure of hardened cement. *Cement and Concrete Research* Vol. 36: 2049–2053.

Advances in Civil Engineering and Building Materials IV – Chang et al (eds)
© 2015 Taylor & Francis Group, London, ISBN: 978-1-138-00088-9

An assessment of sand quarrying waste in Kuwait

S.B. Al-Fadala, M. Abdulsalam, O. Hamadah & S. Ahmed
Construction and Building Materials Program, Energy and Building Research Centre,
Kuwait Institute for Scientific Research, Kuwait

ABSTRACT: Sand is the only locally sourced material in Kuwait and sand quarrying waste is causing an environmental problem and economic loss to the quarrying industry. The assessment of the sand wastes generated from sand quarrying in Kuwait is needed in order to investigate the potential applications for its reuse and recycling. The results of research into the particle size of sand waste, using hydrometer analysis, showed that clay was found to represent more than 50% of the sand waste samples. The study also revealed the high degree of plasticity of clay present in the sand waste. The mineralogical characterization of sand waste using XRD (X-Ray Diffractometer) indicated the presence of clay but the exact type and nature of the clay mineral can only be identified through further studies of more samples. This study will divert the reuse of sand quarrying waste from concrete and pavement applications towards its incorporation into brick manufacture or the production of manufactured aggregates.

Keywords: engineering evaluation, clay minerals, mineralogical analysis, plasticity of microfines, sand quarrying wastes.

1 INTRODUCTION

Sustainable development, global warming, and energy conservation are growing issues that governments are addressing in order to prove their sustainable development responsibility towards their people. During the last few decades, the construction industry has spurred much attention towards an efficient use of resources, reducing environmental impact, and increasing its profitability by waste reuse. Concrete is the most commonly used structural material in the state of Kuwait and its importance cannot be underestimated in any modern society. The State of Kuwait is a small developing country with limited resources of raw materials. Sand is the only locally sourced material in Kuwait, though not all available desert sand is suitable for concrete production due to its deficient gradation and salt contaminations. Quarrying operations including sieving and washing are mainly performed to upgrade the quality of sand for the use in concrete by lowering the amount of microfines and reducing salt contamination. Sand quarrying waste is the fractions passing 150 mm with the majority passing 75 mm. The composition of the waste will vary depending on the locality, but will be composed of Ca, Mg, Al silicates. The stockpiling and disposal of sand quarrying waste are causing an environmental problem and economic loss to the quarrying industry. Globally, efforts are being pursued to find opportunities for the promotion of reusing and recycling waste materials through technically and economically feasible technologies. Among the most

common uses of quarrying waste are in concrete, road construction, or the manufacture of calcium silicate products. The comprehensive characterization of the waste is an essential step towards the development of specifications and procedures suitability evaluation of reusing and recycling. This paper presents the results of an engineering evaluation of the physical, chemical, and mineralogical properties of the sand wastes generated from sand quarrying in Kuwait.

2 SAND QUARRYING IN KUWAIT

At present, several private sand quarries are located in the northwest area of Kuwait and they are the only sand suppliers of sieved and washed sand to the construction market. Sand quarrying operations including sieving and washing are mainly performed to lower the amount of microfines and reduce salt contamination. In sieving fixed shakers with 10 mm to 18 mm sieves are used in sieving the sand. Locally, three washing techniques are used to wash sand with fresh and recycled water. The most recently introduced technique is the hydro-cyclone washing technique. Details of washing techniques are as follows:

- Hydro-cyclone washing technique. This is the latest technique introduced to the sand mining industry. The sand is washed with pressured water and sieved within a tank. Contaminated water is drained to a collecting pond where flocculent chemicals are used to settle any water contamination and the recycled water is used again for washing.

- Double screw washing technique. This is the most common technique. Sand is transferred by conveyor belts to a water tank operated by double screws and sieved. Drained water is collected in a collecting pond where flocculent chemicals are used to settle water contamination and the recycled water is used again for washing.
- Bucket machine-washing technique. Sand is sieved and transferred by conveyor belts to rotating buckets with a sieved opening of 5 mm. The water is drained to a waste pond.

3 MATERIALS AND EVALUATION PROGRAM

In collaboration with one of the well-known sand quarries, microfines samples were collected between May to October 2013. The sampling procedures and the number of samples collected followed ASTM D 75 and ASTM C 702.

The evaluation mainly focused on qualifying the microfines by determining the particle size gradation in accordance with ASTM D 422-2007, using an hydrometer and investigating the plasticity by Atterberg limit tests in accordance with ASTM C4318-2010, and the methylene blue according to the European Standard EN 933-9:2013 entitled 'Tests for geometrical properties of aggregates. Part 9: Assessment of fines: Methylene blue test'. In addition, a mineralogical analysis of sand microfines was performed by X-Ray Diffractometer (XRD) to determine the type of mineral present in the sample. The chemical composition of sand microfines was also analysed by X-Ray Fluorescence spectrometry (XRF).

4 RESULTS

4.1 Microfines Grading

The grading of the microfines is determined by the use of hydrometer and the clay content was identified by particles, which were smaller than 0.005 mm, and the silt by the sample portion, which had particle sizes between 0.074 to 0.005 mm according to the ASTM particle size specifications. In all microfine samples, the clay was found to represent more than 50% of the microfines samples and ranged between 53 to 88%. Details are presented in Table 1. The grading graph is presented in Figure 1.

4.2 The Plasticity of microfines

The plasticity index was calculated by Atterberg limit tests and it ranged between 35 to 42. The average PI calculated is 38, which indicated a high degree of plasticity.

The methylene blue test is widely accepted in Europe as an effective method to quantify the clay content both in natural sand and manufactured sand.

Table 1. Grading of microfines.

	particle size	Minimum (%)	Maximum (%)	Average (%)
Silt	0.074 to 0.005 mm	10.5	35.7	25.9
Clay	smaller than 0.005 mm	52.6	88.3	67.5

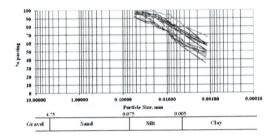

Figure 1. Grading of sand waste.

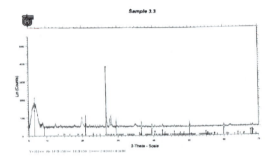

Figure 2. A typical XRD diffractrogram of sand waste.

High MB value indicates the presence of a high amount of clay (swelling) and vice versa. This test is newly introduced to Kuwait and there are no data on the MB_F values of Kuwait sand. Results of methylene blue test performed on the microfines (passing 75 Micron) showed that the methylene blue values vary between 60 and 228 g/kg. The average MB_F calculated is 109 and indicates a high swelling clay content

4.3 Mineralogical analysis

The XRD of microfines showed predominant peaks of Quartz and a peak of aluminosilicates at 5–10° indicating the traces of clay. This is in agreement with the results of the methylene blue test and the hydrometer analysis tests showing the presence of clay in the microfines. In addition to quartz and aluminosilicates, the samples of microfines had also shown the presence of calcite. The identification of the exact type and nature of clay mineral can be identified only through further studies on more samples of quarry waste. A typical XRD diffractrogram of microfines is presented in Figure 2.

Table 2. The chemical composition of Microfines and sand.

Oxide	Microfines	Sand
SiO_2	51.76875	88.048
Al_2O_3	12.06875	3.4457
Fe_2O_3	14.1565	2.0027
K_2O	2.71875	3.2343
CaO	13.31175	1.9274
MgO	2.0825	0.1929

4.4 Chemical analysis.

The chemical composition of microfines is presented in Table 2. When compared with the chemical composition of sand quarried from the same location, the results of microfines samples showed higher Ca, Al_2O_3 and Fe_2O_3 oxides as compared with the sand samples

5 CONCLUSIONS

The study of the particle size of sand waste, using a hydrometer analysis, showed that clay was found to represent more than 50% of the sand waste samples. The study also revealed the high degree of plasticity of the clay present in the sand waste. The mirological characterization of sand waste using XRD showed the presence clay, which is in agreement with the results of the methylene blue test and the hydrometer analysis tests. The exact type and nature of clay minerals can be identified only through further studies on more samples. This study will divert the reuse of sand quarrying waste from concrete and pavement applications towards an incorporation into brick manufacture or the production of manufactured aggregates.

REFERENCES

ASTM C117.2013. Standard Test Method for Materials Finer than 75-μm (No. 200) Sieve in Mineral Aggregates by Washing. *American Society for Testing and Materials, Philadelphia, Pennsylvania.*

ASTM D75.2009. Standard Practice for Sampling Aggregates. *American Society for Testing and Materials, Philadelphia, Pennsylvania*

ASTM C702.2011. Standard Practice for Reducing Samples of Aggregate to Testing Size. *American Society for Testing and Materials, Philadelphia, Pennsylvania.*

ASTM D422.2007 Standard Test Method for Particle-Size Analysis of Soils. *American Society for Testing and Materials, Philadelphia, Pennsylvania.*

ASTM D4318.2010 Standard Test Methods for Liquid Limit, Plastic Limit, and Plasticity Index of Soils. *American Society for Testing and Materials, Philadelphia, Pennsylvania.*

EN 933. 2013. Test for geometrical properties of aggregates-Part 9-Assessment of fines-Methylene Blue test. *The British Standards Institution. United Kingdom.*

KSS GSO 1809.2008. Natural aggregates used for concrete. Standardization Organization for G.C.C (GSO). *Public Authority for Industry, State of Kuwait.*

Hudson, B. Discovering the lost aggregate opportunity. *Parts 1 to 9, Pit and Quarry, 2002–2003.*

Quiroga, P.N. The effect of the aggregates characteristics on the performance of Portland cement concrete. *Ph.D. Dissertation, University of Texas at Austin, 2003.*

Ahn, N. (2000) An Experimental Study on the Guidelines for Using Higher Contents of Aggregate Microfines in Portland Cement Concrete. *PhD Dissertation, University of Texas at Austin.*

Nikolaides, A. Manthos, E. and Sarafidou, M. Sand equivalent and Methylene Blue value of aggregates for Highway Engineering. *Foundations of Civil and Environmental Engineering, Publishing House of Poznan University of Technology, Poznań, No. 10, 2007.*

Advances in Civil Engineering and Building Materials IV – Chang et al (eds)
© *2015 Taylor & Francis Group, London, ISBN: 978-1-138-00088-9*

Evaluation of concrete treated with a migrating corrosion inhibitor

S. Abdulsalam, A. Husain, R. Al-Foraih & A. Abdul Jaleel
Energy and Building Research Centre, Kuwait Institute for Scientific Research, Kuwait

ABSTRACT: The aim of this study is to evaluate the performance of concrete treated with Migrating Corrosion Inhibitor (MCI). A suitable testing program was suggested and applied in this study. This program included different types of tests on reinforced concrete of two different strengths with regular steel rebars serving under various conditions. The MCI was introduced into these concrete mixes as a corrosion protection agent for the steel rebars. The results of these mixes were compared with other concrete control mixes that had no corrosion inhibitor and with concrete mixed with another commercial inhibitor. The overall Electrochemical Impedance Spectroscopy (EIS) results indicated that MCI is an efficient corrosion inhibitor. When it is mixed with concrete, it enhances the concrete's performance and shows an equivalent strength to a conventional concrete corrosion inhibitor and is even better in some cases.

Keywords: MCI, corrosion inhibitor, compressive strength, electrochemical impedance.

1 INTRODUCTION

There are various methods for reducing the corrosion of steel reinforcement in concrete. One major method is the use of corrosion inhibitors that protect steel reinforcements in concrete from corrosion induced by chloride and atmospheric attack. A testing program was designed and performed to evaluate the performance of The Migrating Corrosion Inhibitor (MCI) and Calcium Nitrite (CN) when they are mixed with concrete. The suitability of The Migrating Corrosion Inhibitor (MCI) in reducing corrosion of steel reinforcements in concrete was investigated, using the compressive strengths, setting time, and AC Impedance tests.

The testing program included different types of concrete of two different strengths with regular steel rebars under various conditions. The MCI corrosion inhibitor was introduced into these concrete mixes as a corrosion protection agent for the steel rebars. The results of these mixes were compared with other concrete control mixes without any corrosion inhibitor and the concrete was mixed with another commercial inhibitor (CN).

The overall Electrochemical Impedance Spectroscopy (EIS) results indicate that MCI is an efficient corrosion inhibitor; when it is mixed with concrete; it enhances the concrete performance and shows an equivalent strength to calcium nitrite. The results from a Corrosion Rate Meter and a Great Dane Instrument complied with the EIS results.

2 EXPERIMENTAL METHODS

The MCI corrosion inhibitor was applied to two different concrete mixes: control mix number 1 (1C) of 400 kg/cm² target strength and control mix number 2 (2C) of 250 kg/cm² target strength. The composition of the above mentioned mixes are given in Table 1. The MCI was added to mix 1C with a dosage of 0.6 l/m³ and identified as mix 1 MCI; then it was added to mix 2C with the same dosage and identified as mix 2 MCI. Another two mixes were prepared with calcium nitrite added to them. They were identified as mix 1 CN and 2 CN consecutively.

The concrete mixture proportions of the samples prepared for this experiment, Table 1, were designed to yield an ordinary range of transport properties and are based upon identical concrete mixture proportions. The cement was of Type I. The concrete mixes were used to cast specimens for the following testing:

2.1 Setting time

The standard procedure of ASTM C-403 was applied on all the above concrete mixes using a penetrometer that indicates the setting time when its pressure reading reaches 3.5 MPa.

2.2 Compressive strength

Concrete cube specimens of $100 \times 100 \times 100$ mm were prepared from the above-mentioned mixes and

Table 1. Composition of concrete control mixes.

| | Target Slump (mm) | W/C | Weights Per Cubic Meter of Concrete (kg) | | | |
			Cement	Sand	Age. 3/4″	Age. 3/8″
Mix 1C	100	0.45	450	600	770	380
Mix 2C	100	0.60	300	700	770	380

Table 2. Results of concrete setting time test.

Mix No.	Setting Time (hr:min)
1 C	4:45
2 C	5:30
1 MCI	5:45
2 MCI	6:30
1 CN	3:45
2 CN	4:30

then they were aged in fresh water and seawater. The compressive strength of these cubes was determined at different ages.

2.3 AC impedance

The impedance behaviours of the designed specimens were studied using a special corrosion cell designed to accommodate the test specimens of concrete reinforcement with two steel bars. This was done in order to characterize qualitatively the EIS spectrum for each system in terms of Nyquist and Bode plots, as well as in terms of the general equivalent circuit elements.

The concrete mixtures were prepared according to the procedures in ASTM standards. The samples were cast in 380 mm × 200 mm × 76 mm moulds made of Perspex materials, covered, and at 24 hours of age, the specimens were removed from the moulds and stored in a humidity chamber for curing until they were submerged in 3% NaCl solution. These specimens were tested periodically at intervals of every day in the first week after casting, then every week for the rest of testing period. Although no temperature controls were used, the laboratory temperature could be characterized by the interval $20 \pm 2°C$.

AC impedance measurement and data acquisition was controlled using Potentiostat/Galvanostat interfaced with Frequency Response Analyser (RA) to provide sweep frequency measurement. Both instruments were interfaced with computer for data logging, storage, and analysis. The applied potential amplitude was in the range of (10–100 mV) in the nominal frequency range of 100 kHz to 10 MHz. The AC impedance test data were obtained at the open circuit potential.

The impedance tests were performed with a three-electrode configuration:

(a) Saturated silver/silver chloride electrode as a reference electrode.
(b) Test specimen reinforced steel bar in concrete (350 mm in length) as a working electrode.
(c) Test specimen reinforced steel bar in concrete (350 mm in length) as a counter electrode.

Three identical concrete test specimens were designed and constructed for this study and the steel specimen selected for test was mounted horizontally. The exposed area of the test specimen (working electrode) and counter electrode were identical and equivalent to 153.94 cm^2.

Table 3. Results of compressive strength after ageing in fresh water.

| Mix No. | Compressive Strength (kg/cm^2) | | | | |
	7 Days	28 Days	60 Days	120 Days	180 Days
1 C	353	506	606	522	622
2 C	205	294	409	409	413
1 MCI	484	585	716	7.14	705
2 MCI	282	373	465	460	485
1 CN	375	482	588	522	602
2 CN	231	314	397	386	445

Table 4. Results of compressive strength after aging in sea water.

| Mix No. | Compressive Strength (kg/cm^2) | | |
	60 Days	120 Days	180 Days
1 C	517	527	580
2 C	322	343	360
1 MCI	616	649	683
2 MCI	399	425	428
1 CN	515	527	577
2 CN	338	336	354

3 RESULTS AND DISCUSSION

3.1 Setting time

Table 2 indicates that the setting time for concrete mixed with MCI is longer than that for plain concrete and, unlike the concrete mixed with CN, it has a shorter setting tine than plain concrete.

3.2 Compressive strength

From the results of the compressive strength testing, shown in Tables 2 and 3, it is concluded that MCI concrete has the best compressive strength in different environments and after different lengths of time.

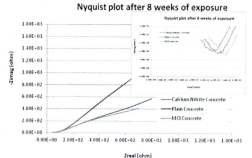

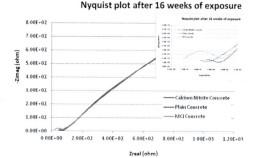

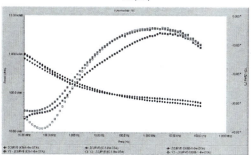

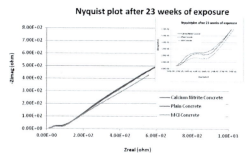

Figure 1. Nyquist and Bode plots for EIS after 8 weeks exposure showing impedance values for MCI concrete with respect to CN concrete and ordinary concrete.

Figure 2. Nyquist and Bode plots for EIS after 16 weeks exposure showing impedance values for MCI concrete with respect to CN concrete and ordinary concrete.

3.3 AC impedance

Figures 1, 2 and 3 represent EIS measurement results in the form of Nyquist and Bode plots for the mortar concrete cement specimens taken after 8, 16, and 23 weeks respectivelyof salt water exposure process in awater bath.

The impedance results have shown that the laboratory prepared ordinary concrete specimensduring casting exhibited the lowest impedance and capacitance values ofthe rest of the specimens. This is shown clearly in Figure 1 after 8 weeks exposure. In other words, the presence of a Warburg diffusion response indicated that the slow diffusion processes of the water species through porosity in the material and formation of passive film product was the rate-controlling factor for the mechanism of concrete corrosion protection. The natural process of alkaline passive formation and build up for this particular sample affects this process. Similarly, water saturated concrete tends to show resistivity of the order of a few Kohmcm, while oven-dried concrete can approach resistivity values typical of an electrical insulator as is reflected on the EIS spectrum taken before exposure toother concrete samples.

As is expected from the protective nature of the CN concrete specimen, the inhibition process of CN concrete slowed down the diffusivity of the chloride species to reach the steel substrate. This is very much related to the mechanism of passivation coalescence and interaction of the inhibitor cement element on the surface of the rebar steel embedded in the structure of the concrete whilst in the dehydration process.

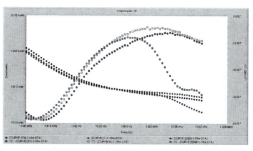

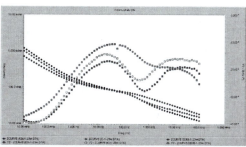

Figure 3. Nyquist and Bode plots for EIS after 23 weeks exposure showing impedance values for MCI concrete with respect to CN concrete and ordinary concrete.

In contrast to the ordinary concrete and the calcium nitrite, the MCI inhibited concrete showed the highest impedance and capacitive behaviour. This is due to the low rate of permeability and adsorption or destruction of the chloride corrosive species to the steel substrate surface. The EIS test results after 16

103

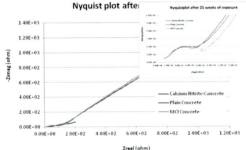

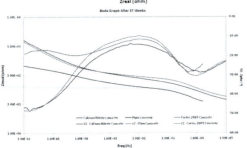

Figure 4. Nyquistand Bode plots for EIS after 37 weeks expo-sure showing impedance values for MCI concrete with respect to CN concrete and ordinary concrete.

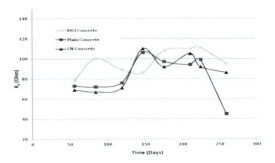

Figure 5. EIS impedance Polarization resistance (Rp) profile for MCI concrete with respect to CN concrete and ordinary concrete.

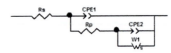

Element	Freedom	Value	Error	Error %
Rs	Free(+)	332	N/A	N/A
CPE1-T	Free(+)	1.8679E-5	N/A	N/A
CPE1-P	Free(+)	0.23714	N/A	N/A
Rp	Free(+)	2.388E8	N/A	N/A
CPE2-T	Free(+)	1.8679E-5	N/A	N/A
CPE2-P	Free(+)	0.23714	N/A	N/A
W1-R	Free(+)	6.4352E5	N/A	N/A
W1-T	Free(+)	3.5758E5	N/A	N/A
W1-P	Free(+)	0.23721	N/A	N/A

Data File:
Circuit Model File:
Mode: Run Simulation / Freq. Range (0.001 - 1000000)
Maximum Iterations: 100
Optimization Iterations: 0
Type of Fitting: Complex
Type of Weighting: Calc-Modulus

Figure 6. Example of an equivalent circuit model with curve fitting analysis for capacitance and impedance elements for ordinary concrete and concrete with MCI concrete and metal/solution interface under study.

and 23 weeks has determined that the improper application of an inhibitor concentration and or a variation in specimen solution submersion or curing may cause the corrosion process to become even worse than in ordinary concrete (C). The hypothesis behind thisun-expected behaviour of the inhibited concrete sample was more or less related to the improper film cover-age and adsorption of the nitrite species to the passive oxide or hydroxide layer of the steel surface with the subsequent formation of unprotected micro-corrosion cells.

If chloride ions arrive first as a contaminant to the surface of the steel, the surface tends to become active when the molar ratio of Cl^- to Nitrite and hydroxyl ions in the pore solution reaches a critical value that exceeded 0.9. Microscopically, the concrete acts as a non-homogeneous electrolyte, which varies greatly with the overall moisture content, in addition to the degree of diffusivity of corrosive and inhibited ions such as oxygen, chloride and nitrite.

In the case of the MCI concrete specimens an almost purely capacitive and diffusion response was observed and has been reflected as straight spectra line in the Nyquist and Bode plot, shown in the same Figures 2 and 3, and even after 37 weeks of exposure, as shown in Figure 4. There is little apparent change during this time, suggesting that the system is relatively stable and that there is no more water uptake and the material is acting as some sort a die-electric component. In many cases, this system has exceeded the CN concrete and has shown better performance and impedance values as has been indicated on the Nyquist plot that the spec-trum for MCI concrete always shown trends to be on

the right position of the plots. This means that MCI has given an additional support to the concrete, which will prolongs its life by protecting the reinforcement steel bar.

In the case of the ordinary concrete, it is expected that the Nyquist spectrum will reach its semi-circular shape and will become smaller with time. Similarly, the same EIS response will be expected from the CN concrete specimen.

Both specimens have shown a greater tendency toward reaching its charge transfer resistance and polarization resistance at a faster rate (corrosion reac-tion) once exposed to the environment. It can be seen from Figure 5 that concrete specimens inhibited with the MCI inhibitor has attained a constant stable profile of impedance behaviour showing higher impedance values with respect to the other concrete specimens. Similar behaviour can be deduced from capacitance plot of Figure 6 where the same inhibited concrete shows low water uptake with respect to the other concrete specimens.

The equivalent parallel R-C circuit in Figure 6 explains and represents quite well the frequency impedance response of the concrete mix design systems used in this study.

4 CONCLUSION

- The MCI enhances the compressive strength of concrete.
- Electrochemical Impedance Spectroscopy (EIS) technique is a useful tool that highly complements the ASTM tests method in order to follow and anticipate the real concrete behaviour with respect to environmental conditions.
- In general the decrease in steel natural alkaline surface film of corrosion resistance with exposure suggests that solution ingress permanently damages the passive film locally at discrete pathways, increases with time. The corrosive effect is enhanced by the presence of surface contaminants such as sodium chloride, sulphide and other environmentally corrosive species that compete with each other to breakdown the passive film.
- In general the highest concrete mix design performance was observed in this experimental study for MCI concrete with respect to the CN concrete and plain concrete.
- Upon application of concrete with NaCl addition, the reaction sequence would follow one of diffusion controlled (passivation reactions) and changed to charge transfer control (polarization reactions).

ACKNOWLEDGEMENT

Authors would like to extend their gratitude to The Gulf Supplies General Trading and Contracting Company for their financial support of this research study. Authors also are grateful for the support of KISR to accommodate all the rising challenges during the execution of the project's tasks.

REFERENCES

Andrade, C., C. Alonso, and J.A. Gonzalez. 1989. *Proceedings of the 3rd International Symposium on Electrochemical Methods in Corrosion Research. Materials Science Forum. Edited by B. Elsnor. Zurich, Switzerland, 1988, Vols. 44 & 45, pp. 330–335.*

Arya, C., N.R. Buenfeld, and J. B. Newman. 1990. Factors influencing chloride-binding in concrete. *Cement and Concrete Research 20: 291–300.*

ASTM 403/C403M-08. Standard test method for time of setting of concrete mixtures by penetration resistance. *American Society for Testing and Materials, Philadelphia, Pennsylvania.*

Ha-Won Song and Velu Saraswathy. 2007. Corrosion Monitoring of Reinforced Concrete Structures – A Review. *International Journal of Electrochemical Science. 2. pp. 1–28.*

Hakkarainen, T.J. 1989. *Materials Science Forum. Vol. 44 & 45, pp. 347–356.*

Hepburn, B. J., Gowers, K. R. and Scantlebury, J. D. *British. Corrosion. Journal. Vol. 21, p. 105, (1986).*

Shamsad Ahmad. 2003. Reinforcement corrosion in concrete structures, its monitoring and service life prediction – a review. *Cement and Concrete Composites. Vol. 25. Issues 4–5. pp. 459–471.*

V. Feliiu, J. A. Gonzalalez, C. Andrade and S. Feliu. 1998. Equivalent circuit for modeling the steel-concrete interface. I. experimental evidence and theoretical predictions. *Corrosion Science. Vol. 40. Issue 6. pp. 975–993.*

Advances in Civil Engineering and Building Materials IV – Chang et al (eds)
© 2015 Taylor & Francis Group, London, ISBN: 978-1-138-00088-9

Numerical models of end-plate assembling bolt connections of CHS profiles

Anežka Jurčíková
Department of Structural Mechanics, Faculty of Civil Engineering, VSB-Technical University of Ostrava,
Czech Republic

Přemysl Pařenica & Miroslav Rosmanit
Department of Structures, Faculty of Civil Engineering, VSB-Technical University of Ostrava, Czech Republic

ABSTRACT: The aim of this work was to create numerical models of the common truss-type assembling joints of CHS (Circular Hollow Section) profiles. Two different models were created – basic (simplified) model of joint and more accurate model which corresponds to the experimental specimens in preparation. Models with different end-plate thicknesses and consequently with different failure modes were solved. The results obtained from numerical models were compared with the analytical solution of such joints using the Eurocode procedure recommended in EN 1993-1-8. These results are planned to be verified and further developed based on planned experiments.

Keywords: ANSYS; FEM; assembling joint; prying; equivalent T-stub flange in tension.

1 INTRODUCTION

In practice, for the construction of halls and for spanning large distances, lattice girders or truss frames made of either hollow or open sections are often used conveniently. Main advantages of such structures are their good static effect, sufficient stiffness while preserving lightweight structure and also their aesthetic appearance. However, for large span structures, it is necessary to ensure proper assembling connection of individual components, for which the end-plate bolt joints are used often.

Such connections, if under tensile stress, should be designed taking into account also the possible influence of prying. Design methods given by Eurcode procedure are complicated and for some types of joints are not even exactly described. Behavior of connections using equivalent T-stub flange in tension theory which allows to predict the influence of prying is still actual and many of the world research teams are dealing with that [1, 2].

This work is aimed to create numerical models of end-plate assembling bolt connections of CHS profiles which will reflect the expected behavior of the joint. These models will be subsequently verified and further precise by the experiments which are currently in preparation. The goal of this research is simplification and improvement of available analytical solutions.

2 GEOMETRY OF THE RESEARCHED JOINT

The researched joint (Fig. 1) connects two CHS profiles (TR 89x8) and consists of two circular end-plates,

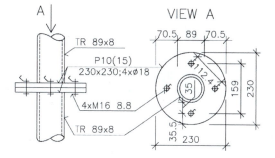

Figure 1. Basic geometry of analyzed assembling joint.

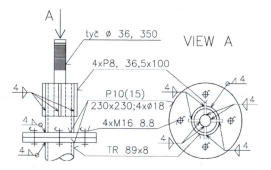

Figure 2. Geometry of the specimen in preparation.

with thicknesses $t_{f,1} = 10$ mm or $t_{f,2} = 15$ mm respectively, which are connected by four bolts M 16, structural grade 8.8. Fig. 2 shows scheme of the prepared specimen.

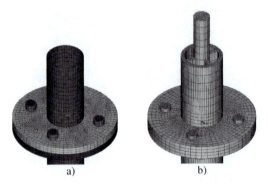

a)   b)

Figure 3. Numerical model of joint with the FE mesh a) simplified basic model; b) updated model.

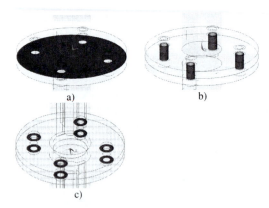

a)   b)

c)

Figure 4. Contact pairs between the surfaces of a) end-plates; b) bolts and the holes in the plates; c) end-plates and nuts.

3 ASSESSMENT OF THE EQUIVALENT T-STUB FLANGE ACCORDING TO EC3

For CHS profiles and end-plates the steel S235, with the yield stress $f_y = 235$ MPa was considered. M 16 structural grade 8.8 bolts with tensile strength $f_{ub} = 800$ MPa and the tensile stress area $A_s = 157$ mm^2 were used.

First design of the joint was done using the equivalent T-stub flange in tension theory recommended by Eurocode [3]. The resulting load-bearing capacities were primarily used for initial estimation of the joints behavior. The lowest load-bearing capacity for three possible failure modes of T-Stub should be considered:

1. Complete yielding of the flange
2. Bolt failure with yielding of the flange (large influence of prying forces)
3. Bolt failure

When using standardized formulas, load-bearing capacity of the joint with end-plate *thickness* $t_{f,1} = 10$ mm resulted in value $F_{Rk} = 91.5$ kN (failure mode corresponds to complete yielding of the

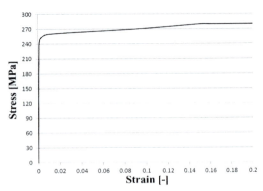

Figure 5. Multilinear stress-strain material properties of the end-plates and CHS profiles.

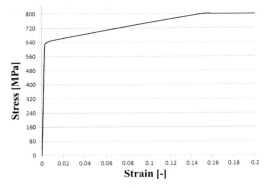

Figure 6. Multilinear stress-strain material properties of the bolts.

flange) and for the joint with end-plate *thickness* $t_{f,2} = 15$ mm the resulting load-bearing capacity was $F_{Rk} = 205.8$ kN (bolt failure with yielding of the flange is decisive for joint capacity). Note that original recommended EC3 procedure had to be modified because of the type of connection.

4 NUMERICAL MODELS

Two different models were developed – simplified basic model of the joint and more accurate model of the prepared experimental specimen. Both numerical models (Fig. 3 a), b)) were created in the FEM software ANSYS 12.0 using the finite elements enabling nonlinear calculations (both plastic behavior of materials and influence of large deformations). In the simplified model for modeling the CHS profiles the shell finite element SHELL 43 was used and for modeling the other components of the joint (steel plates, bolts and nuts) 3D SOLID 45 finite element was used. For an updated model the entire joint was modeled using the SOLID 186 element. For creation the contact pair between the end-plates' surfaces and between both the surfaces of bolts and the inner surface of the holes in

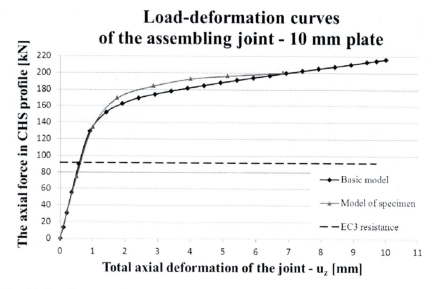

Figure 7. Load-deformation curves of end-plate bolt joint, thickness of the plate is 10 mm, with the characteristic load-bearing resistance according to EC3 [3].

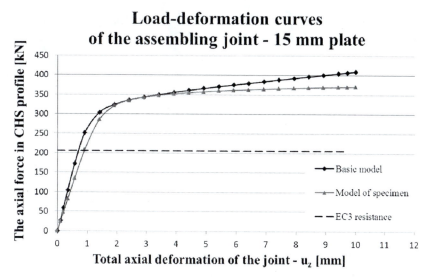

Figure 8. Load-deformation curves of end-plate bolt joint, thickness of the plate is 15 mm, with the characteristic load-bearing resistance according to EC3 [3].

the end-plates, the contact elements TARGE 170 and CONTA 174 were used – Fig. 4 a), b), c).

The following material properties were assigned to the finite elements (similar to [4], [5]): Young's modulus of elasticity $E = 210$ GPa and Poisson's ratio $\nu = 0.3$. Both physical and geometrical non-linear aspects were considered within the calculation (plastic behavior considering the large deformations). In the simplified model the elasto-plastic behavior of the material was expressed by a bilinear diagram (similar to e.g., [6]) with the value of yield stress $f_y = 235$ MPa and about 1% hardening (i.e., with value of the tangent modulus $E_2 = 2$ GPa) for the

end-plates and for CHS profiles and with the value of yield stress $f_y = 640$ MPa and with the tangent modulus $E_2 = 10$ GPa for bolts. In the other, more accurate model, the multilinear diagrams were used (Fig. 5, 6).

First model is simplified in many ways, because the connections between the bolts and nuts and then between the nuts and plates were modeled as a rigid, which does not exactly match the real situation, but is sufficient for capturing the behavior of the joint. In the second model the connections between end-plates and nuts were also performed by using the contact elements.

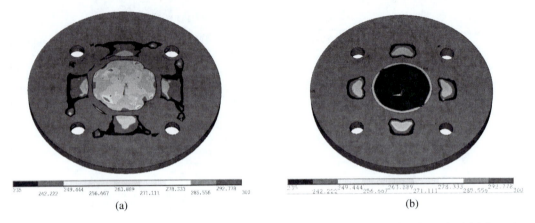

(a)   (b)

Figure 9.   Evolution of von Misses stress [MPa] beyond the yield stress value *on the plates* when the total axial deformation is 0.9 mm (estimated beginning of plastification) a) Model with 10 mm end-plate (axial force *129 kN*); b) Model with 15 mm end-plate (axial force *252 kN*).

(a)   (b)

Figure 10.   Evolution of von Misses stress [MPa] beyond the yield stress value *on bolts* when the total axial deformation is 0.9 mm (estimated beginning of plastification) a) Model with 10 mm end-plate (axial force *129 kN*); b) Model with 15 mm end-plate (axial force *252 kN*).

5   RESULTS OF NUMERICAL MODELS

Both numerical models of the CHS assembling joint were studied in two variants – with the end-plate thickness $t_{f,1} = 10$ mm or $t_{f,2} = 15$ mm respectively. All models were used to observe the dependency of the CHS total axial deformation (u_z) on the axial force in the profile and thus the overall behavior of the joint – Fig. 7 and 8. The course of these load-deformation curves was also compared with the load-bearing capacity value according to the Eurocode. At the same time the evolution of stress on the steel plates and bolts were observed (Fig. 9–10) to specify the failure mode.

model with the end-plate thickness of 10 mm and for the joint with the plate thickness of 15 mm the *bolt failure with yielding of the flange* (prying of bolts) should decide. Numerical models have confirmed these assumptions.

The resultant load-bearing capacities obtained by the analytical calculations did not fully match the findings from the numerical modeling, as expected. Thus it would be appropriate to refine the analytical formulas for the calculation of load-bearing capacity of this type of joints.

Further refinement of numerical models based on the planned experiments will be performed and subsequently eventual modifications to the existing analytical relations will be proposed.

6   CONCLUSIONS

The numerical models of end-plate assembling bolt connections of CHS profiles were created, which fits the expected behavior of individual joints with different thickness of the plates. These assumptions were based on the analytical evaluation according to Eurocode [3], under which the *complete yielding of the flange* should be decisive failure mode within the

ACKNOWLEDGEMENT

This research was financially supported by Student Grant VSB-TUO, Reg. No. SP2014/168.

REFERENCES

EN 1993-1-8, Eurocode 3: Design of steel structures – Part 1–8: Design of joints. 2005.

Hantouche, E., G., Rassati, G., A., Kukreti, A., R., Swanson, J., A.: Built-up T-stub connections for moment resisting frames: Experimental and finite element investigation for prequalification. In *Engineering Structures,* Volume 43, 2012, p. 139–148. ISSN 0141-0296

Hantouche, E., G., Kukreti, A., R., Rassati, G., A.: Investigation of secondary prying in thick built-up T-stub connections using nonlinear finite element modeling. In *Engineering Structures,* Volume 36, 2012, p. 113–122. ISSN 0141-0296

Jurčíková, A., Rosmanit, M.: FEM Model of Joint Consisting RHS and HEA Profiles. In: *Steel structures and bridges 2012: 23rd Czech and Slovak International Conference.* Podbanské, September 2012. Procedia Engineering, Volume 40, 2012, p. 6. ISSN 1877-7058

Katula, L., Márai, P.: Study the prying effect on bolted base-plate connections. In *Periodica Polytechnica – Civil Engineering,* Volume 57(2), 2013, p. 157–172. ISSN 1587-3773

Yang, J.-G., Park, J.-H., Kim, H.-K., Back, M.-C.: A Prying Action Force and Contact Force Estimation Model for a T-Stub Connection with High-Strength Bolts. In *Journal of Asian Architecture and Building Engineering,* Volume 12(2), September 2013. ISSN 1346-7581

Advances in Civil Engineering and Building Materials IV – Chang et al (eds)
© 2015 Taylor & Francis Group, London, ISBN: 978-1-138-00088-9

The application analysis of fly ash in magnesium phosphate cement

Rui Huang & Xiangxing Kong

The First Highway Survey and Design Institute of China Communications Construction Company Ltd., Xi'an, China

ABSTRACT: The addition of fly ash into magnesium phosphate-base material will reduce the original compressive strength, the flexural strength, and the abrasion resistance of magnesium phosphate cement, and the expansion character of it will increase as the amount of fly ash increases. With the increase of the dosage of fly ash, the compressive strength of the pure magnesium phosphate cement will reduce after the first increase and the flexural strength and abrasion resistance of magnesium phosphate cement will decrease with the increase of the dosage of fly ash. However, the ultimate strength could meet the need of practical engineering if the dosage is appropriate. The addition of fly ash can help reduce the cost and improve the properties of phosphate cement-based repair material. Adding appropriate dosages of fly ash into the cement could help improve its liquidity, adjust the setting time, and improve the bonding degree of the magnesium phosphate cement with Portland cement concrete. Therefore, it could be better used in many areas such as patching material, oil well cementing, and waste disposal.

Keywords: application analysis, fly ash, magnesium phosphate cement, mechanical property.

1 INTRODUCTION

Magnesium phosphate cement is made up of burnt magnesia and phosphoric acid or acid phosphate, and some cementitious materials in a certain proportion. The chemical reactions between magnesia and phosphoric acid is very quick, usually within a few minutes of rapid condensation, whilst half an hour creates a certain strength, which contributes to its high early strength. In order to facilitate the construction, some certain retarder will be needed, which is mostly the boric acid or sodium borate. Magnesium phosphate cement hydration reaction is essentially an acid – alkali reaction, which will release a lot of heat. Therefore, magnesium phosphate cement can rapidly congeal at a low temperature.

In addition, magnesium phosphate cement also has peculiarities of wonderful bonding performance whilst splicing old concrete, good abrasion resistance, and freeze resistance as well as the dry shrink (Jiang, 2004). Based on the above characteristics of magnesium phosphate cement, it has a good prospect in application to both civil buildings and military facilities. It can be used in the mending and repairing of airport runway, bridges, roads and other civil and military engineering (Wang, 2005). In addition, it also can be used as a precision casting of coated materials, dental and bone cement binder, and so on (Chen, 2006, and Wu, 2006). Adding a suitable amount of fly ash to the magnesium phosphate cement can improve the performance of it whilst making use of a lot of fly ash, which used to be a waste. Fly ash has the advantages of low price and rich source, and adding it into magnesium phosphate cement can reduce the cost of the cement and improve its performance. After mixing with fly ash, the cost of the magnesium phosphate cement base material could be greatly reduced, which provides economic feasibility for the popularization and application of magnesium phosphate cement base material (Zhang, 2009).

2 INFLUENCE OF FLY ASH ON MAGNESIUM PHOSPHATE CEMENT BASE COMPOSITE MATERIAL'S PERFORMANCE

2.1 *Mechanical property*

Under similar conditions, with the increasing dosage of fly ash, the strength of the repaired phosphate cement base material basically showed a trend of decrease, but when the dosage of fly ash reached 25%, 1 d and 28 d, compressive strength of the magnesium phosphate cement reached 41.0 MPa and 63.2 MPa, which is enough to meet the engineering requirements. Due to the role of shape effect, fine grinding fly ash in repair material can rise to adjust the grading and close-grained filling effect, thus increase the compactness of cement mortar; it can also reduce the water cement ratio, increase strength, and its micro aggregate effect of fly ash also enhanced effect of phosphate cement-based repair material. At the same time, the addition of fly ash will "dilute" the paste, influence the crystal structure of the network formed between hydration product particles; the fly ash contains fairly coarse, porous and loose particles (such as carbon particles), which will have adverse impact on the liquidity and strength.

2.2 *Compressive strength*

Under the same age, with the increase of dosage of fly ash, the strength of magnesium phosphate cement net pulp first increases and then decreases, and when the dosage of fly ash reaches 10%, the compressive strength is the largest. With the increase of age, the strength of the same dosage of fly ash net slurry of magnesium phosphate cement increases gradually. When mixed with 10% of fly ash, the cement reaches its highest compressive strength. That is because of the micro aggregate effect of fly ash that, the particle could fill the wool stoma of the cement paste and refine the capillary porosity of the cement paste, which is beneficial to the increase of the compressive strength of the material. When the dosage of fly ash was increased further, a number of porous particles in fly ash can absorb part of the moisture so that the water needed for the reaction of the material will be reduced, inhibiting the development of magnesium ammonium phosphate crystal. Simultaneously, the fly ash and magnesium phosphate cement base material are not obviously chemically combined, making the adhesion strength of the fly ash with the magnesium phosphate cement base material poor. This affected the continuity of the magnesium ammonium phosphate crystal network, thus reducing the compressive strength of the magnesium phosphate cement base material.

2.3 *Flexural strength*

With the increasing dosage of fly ash, the flexural strength of the magnesium phosphate cement base material gradually decreases; with the increase of age, the flexural strength of the same amount of fly ash using material increases. Under the same age, the flexural strength of the material will decrease with the increase of the dosage of fly ash, and the main reason for the increase is that the chemical bond between fly ash and the magnesium phosphate cement matrix has not been formed so that the connection is very weak. Moreover, the increasing dosage of fly ash will seriously damage the crystallization of the matrix network. Therefore, under the action of shear force, its strength is on the decline.

2.4 *Wear-resisting performance*

As magnesium phosphate cement base material is often used to repair roads, the wear resistance of the materials is particularly important. The research based on that, taking the wear resistance of pure magnesium phosphate cement mortar as a measurement of that of the material, has found that the magnesium phosphate cement base material without fly ash had the least wear loss, and the higher the dosage was, the greater the wear loss was, and the worse the wear resistance was. Also, the speed of the wear resistance is gradually faster. This is related to the fact that the connection between fly ash and the magnesium phosphate cement matrix is weak, and the wear resistance of fly ash is not high in itself.

2.5 *Expansion properties*

With the increase of fly ash content, inflation rate increased gradually. After 7th d, the inflation rate of magnesium phosphate cement base material which contained 30% of fly ash, surpassed that with 40% of fly ash (Zhang, 2009). The expansion rate of the magnesium phosphate cement base material increases along with the growth of the age. This is related to the exothermic reaction of the magnesium phosphate cement base material, and the heat of the reaction is released mainly in its early stages, during which the temperature rises rapidly. Therefore, the inflation rate of the magnesium phosphate cement base material increased quickly before 7–14 d stage, but increased slowly later. The reason why the base material containing 30% and 40% of fly ash has a larger inflation rate, is that with large dosages of fly ash, the number of porous particles in the fly ash will increase, and after being mixed with water, these particles can absorb a large amount of moisture, which decreases the amount of the water involved in the reaction, and increases the concentration of Mg2+ in the solution. From the perspective of reaction dynamics, the increasing of the concentration of reactants will speed up the chemical reaction so that a large amount of heat is released in a short time, thereby making the material undergo a large amount of expansion.

3 TEST RESULTS OF FLY ASH ON MAGNESIUM PHOSPHATE CEMENT

The latest test showed that fly ash can significantly improve the interfacial bonding strength of magnesium phosphate cement, and ordinary cement. In the repairing of the concrete, it is a key factor to successful repair (Si 2009). When the fly ash content was 20%, the highest bonding strength can be obtained, and its 3rd d bonding strength can achieve 50 MPa. When the fly ash content is increased to 30%, the bonding strength will decrease.

The main reason for this phenomenon may be the reaction between the hydration products of ordinary cement mortar and MPC; second, the active SiO_2 and Al_2O_3 in fly ash react with $Ca(OH)_2$ in ordinary cement mortar to generate the C-S-H gel, which improves the bonding strength with both physical bond, and chemical bond. Along with the continuous increasing use of fly ash, hydration rate slowed, the hydration products decreased, and the strength decreased. Excellent adhesive performance between MPC mortar and ordinary cement mortar can make bonding strength reach 6.3 MPa after l day. The bonding strength reaches 10 MPa when the amount of fly ash is 10%, and reaches 5.6 MPa even if the amount of fly ash is as high as 50% after one day. Researchers have analysed and believed that three effects play a great role in enhancing the bonding strength. The micro aggregate effect makes fly ash effectively fill pores on the interface, and get the old and new interface region to become more compact. The micro fly ash particles filled within the MPC matrix uniformly, and the porosity decreases significantly make the interface zone structure more compact, and effectively improve the pore structure of the transition zone. Fly ash has a ball effect in slurry, which can decrease the internal frictional resistance of inter particles, reduce the porosity, improve the compactness, and effectively increase the interfacial bonding strength.

4 CONCLUSION

According to the research results above, it can be seen that although magnesium phosphate cement with the addition of fly ash would reduce the strength of the cement to some extent, the final strength does not affect its normal usage in the general engineering. Moreover, it can also improve other properties. The appropriate addition of fly ash can prolong the setting time of magnesium phosphate cement itself, increase the liquidity, reduce the unit volume weight itself, and improve the adhesion with ordinary cement products. These improvements can make better application of magnesium phosphate cement to the real work. We can choose the appropriate amount of fly ash according to the actual needs, in order to to make it better adaptable to the need of practical engineering, so that the material can be fully utilized and its cost can be reduced at the same time.

As an early strength rapid hardening cement, magnesium phosphate cement can be widely applied to roads, bridges and other civil and military engineering repair. Magnesium phosphate cement can also be used as coating materials, dental cement, and bone adhesive. As repairing materials, magnesium phosphate cement needs to have good adhesiveness with raw materials in engineering, and can improve its bonding strength with the ordinary cement products

after adding fly ash into it to meet the project requirements. Because of the high cost, magnesium phosphate cement is not suitable for large-scale application to general construction engineering. A large amount of fly ash used to mingle with the magnesium phosphate cement cannot only improve its property, but also greatly reduce the material cost. This largely broadens out the way of application of magnesium phosphate cement-based materials. Fly ash modified magnesium phosphate cement not only can be used for engineering repairs, but also can be used for the treatment of oil well cement, and toxic and radioactive waste [6]. For the improvement of the material property, or from an economic and environmental protection point of view, the addition of fly ash in magnesium phosphate cement is worth developing.

REFERENCES

HongTao wang, JueShi qian, JuHui cao, ChunPing tang. The Effect of Fly Ash on The Properties of Phosphate Cement-based Repair Materials [J]. Journal of new building materials, 2005 (12): 41–43.

HongYi jiang, HuanZhou, HuiYang. Super Fast Fepair Hard Phosphate Cement Hydration Hardening Hechanism Research [J]. Journal of wuhan university of technology, 2004 (4): 18–20.

LiuMing su, YiXiong huang, JueShi qian. Fly ash modified magnesium phosphate cement and Portland cement matrix adhesion performance research [C]. The third international symposium on high performance concrete: Hong Kong, macau, Taiwan, 78–85.

SiYu zhang, HuiSheng shi, ShaoWen huang. The Influence of the Dosage of Fly Ash Has on the Mechanics Performance of Magnesium Phosphate Cement Base Composite Material [J]. Journal of nanchanguniversity. 2009, 31(1):80–82.

SiYu zhang, HuiSheng shi. The Application of Fly Ash Modified Magnesium Phosphate Cement Base Material [J]. Journal of comprehensive utilization of fly ash, 2009(1): 54–56.

YuHong chen, YuanZhi tang. Magnesium Phosphate Coating Lithium Cobalt Nickel Manganese Oxide Synthesis and Performance Research [C]. Chinese chemical society academic conference paper sets the 25th session(I), 2006.

ZiZheng wu, JianZhang. The Experimental Study of Magnesium Phosphate Bone Cement Adhesive Fracture [J]. Chinese journal of reconstruction surgery, 2006(12): 912–915.

Building materials preparation

Advances in Civil Engineering and Building Materials IV – Chang et al (eds)
© 2015 Taylor & Francis Group, London, ISBN: 978-1-138-00088-9

Potential use of natural Pozzolana as a mineral admixture in concrete

S. Al-Bahar, J. Chakkamalayath, M. Abdulsalam & A. Al-Aibani
Construction and Building Materials program, Energy and Building Research Centre,
Kuwait Institute for Scientific Research, Kuwait

ABSTRACT: The deterioration of concrete structures due to the combined effect of chlorides and sulphates in the marine environment demand the use of pozzolanic materials for sustainable constructions in Kuwait. This paper presents the effective use of Volcanic Ash (VA), a regional natural Pozzolana, and evaluates and characterizes the physical, chemical, mineralogical and mechanical properties of cement matrix incorporating volcanic ash obtained from the Arabian Gulf region. Locally available Ordinary Portland Cement (OPC) was used for this research study and it was replaced with different percentages of VA. Its influence on the mechanical properties was assessed, as well as the microstructure and hydration behaviour. The study was conducted with 10–50% replacement of OPC with volcanic ash. The results show that the compressive strength of concrete decreases with the increase in volcanic ash content. The Strength Activity Index (SAI) was also determined to understand the pozzolanicity of volcanic ash and the results are discussed. Generally, the test results showed good potential for using volcanic ash showing lower heat of hydration and refinement of microstructure.

Keywords: volcanic ash, natural Pozzolana, strength activity index, microstructure characteristics, scanning electron microscopy, consistency and setting time.

1   INTRODUCTION

The extreme hot and cold climates and the marine environment of Kuwait have an adverse effect on the life of concrete structures and the construction sector always demands the use of innovative materials and methods. The deterioration of concrete structures in the existing climates demand the use of Pozzolanic materials for construction in Kuwait. Several studies have been conducted on blended cements incorporating artificial Pozzolana like silica fume, fly ash and slag in concrete production. Modifications in the pore structure and microstructure have been studied through these studies and the efficiency of blended cements against chloride and sulphate exposure has been explored (Yeau and Kim, 2005; Rahman et al., 2011). Even though, Gulf region has vast deposits of natural pozzolana like volcanic ash (VA), it is not explored fully in construction sector. Also, the literature on the study of blended cements with natural Pozzolana (NP) is limited. The studies are mainly concentrated on the hydration behaviour of cement paste with the addition of volcanic ash (Escalante-Garcia and Sharp, 1998, 2001; Hossain and Lachemi, 2006). A systematic study on the use of natural pozzolana in concrete is required for ensuring sustainability and durability in construction in Kuwait due to its extreme climates and marine environment. Characterization of the cement matrix incorporating volcanic ash will be carried out in this study to assess the physical, mechanical, and microstructure properties.

2   MATERIALS AND EXPERIMENTAL METHODS

Saudi Arabia, the nearby country to Kuwait in the Gulf region has abundant reserves of good quality volcanic ash (VA). Locally available ordinary portland cement and volcanic ash samples of average sizes of 40 μm and 20 μm obtained from Saudi Arabia was used in this study. Characterization of volcanic ash was carried out as per ASTM C 648. The compressive strength tests, and hydration tests on cement paste using isothermal calorimeter was conducted.

3   RESULTS AND DISCUSSION

3.1   *Characterization of the collected samples*

Preliminary investigation was conducted to understand the basic characteristics of the samples of volcanic ash. The physical properties like density, air content, compressive strength, fineness and setting time of samples were determined as per ASTM standard. The samples of both 40 μm and 20 μm volcanic ash was analyzed. A comparison of test results along with ordinary Portland cement (OPC) is given in Table 1.

It was observed from the test results that the density of volcanic ash is less than that of OPC and it is more than conventional mineral admixtures like fly ash and silica fume. The fineness of 20 μm volcanic ash is more than that of OPC where as it is less for 40 μm.

Table 1. Comparison of physical properties of OPC and VA.

Property	OPC	VA-20 μm	VA-40 μm
Air content (%)	6.6	6.5	7.0
Density (g/cm³)	3.15	2.75	2.78
Fineness (m²/kg)	367	396	296
Consistency (%)	23.8	24.3	24.5
Initial setting time (Minutes)	105	132	128
Final setting time (minutes)	175	202	198
Compressive strength of cement mortar with replacement of OPC with 20% VA (MPa)			
3 days	26.04	19.45	17.41
7 days	29.11	24.56	21.31
28 days	40.63	35.65	30.21

Table 2. Chemical composition of OPC and Volcanic ash.

Property (%)	OPC	VA-20 micron	VA-40 micron
CaO	62.27	9.09	9.17
SiO₂	16.63	38.89	37.72
Al₂O₃	3.63	13.00	12.41
MgO	1.22	6.16	6.03
SO₃	3.91	0.14	0.25
TiO₂	0.24	2.44	2.41
K₂O	0.61	1.31	1.33
Fe₂O₃	3.28	12.41	12.30
Na₂O	0.34	3.12	3.03

Consistency and setting time of cement paste with volcanic ash was more than that of control mix showing the retardation in hydration due to the addition of mineral admixture. The strength activity index (SAI) was determined as per ASTM C618 to understand the pozzolanicity of volcanic ash. It was observed that the strength activity index is 87% for cement paste with 20 μm volcanic ash which is more than the minimum required value of 75% specified by ASTM C 618. This confirms that the available volcanic ash of 20 μm have good pozzolanicity and can be used as microcement additive whereas 40 μm sample was not satisfying the criteria. Hence further studies were conducted with 20 μm VA samples.

3.2 Chemical characteristics

The chemical composition of OPC and volcanic ash was determined using XRF. The major compounds present are CaO, SiO₂, Al₂O₃ and the results confirms the low quantity of CaO and higher quantity of SiO₂ as in the case of other mineral admixtures. SO₃ content is lower in VA compared to OPC whereas MgO, Fe₂O₃ and alkali content is higher. The chemical composition of 20 and 40 micrometers is comparable and the results are given in Table 2.

Table 3. Consistency and setting time of different mixes.

Mix No.	Consistency	IST	FST
Control	22.5	145	185
V10*	23.4	160	205
V20	23.4	165	230
V25	23.5	180	225
V30	23.8	170	245
V40	24.0	175	230
V50	24.8	145	210

*V10 indicates replacement of OPC with 10% VA.

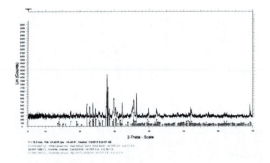

Figure 1. XRD Pattern of VA-20 micrometers.

3.3 Mineralogical analysis by XRD

The mineralogical analysis of VA by XRD was conducted. The volcanic ash diffractograph showed the presence of anorthite, albite, Augite, as major minerals along with traces of forsterite.

3.4 Consistency and setting time of cement paste

The tests carried out on the fresh cement paste included consistency and setting time using Vicat penetration test, to understand the influence of micro additives on hardening properties of cement paste (Refer Table 3).

The test results show that increase in the percentage of VA increases the water demand, even though the difference is marginal. Similarly the results show that replacement by 40% VA increases the initial setting time (IST) and final setting time (FST), though a trend in the reduction of setting time was observed when replacement of VA at 50% or above. These result has to be validated with further tests.

3.5 Hydration of cement paste

Preliminary study was conducted to understand the influence of volcanic ash on the hydration characteristics of cement paste. The test was conducted for OPC replacement with 10%, 20%, 30% and 40% of VA. The test result shows that the heat evolved decreases with increase in the amount of volcanic ash as shown in thermal power curve (see Fig. 2).

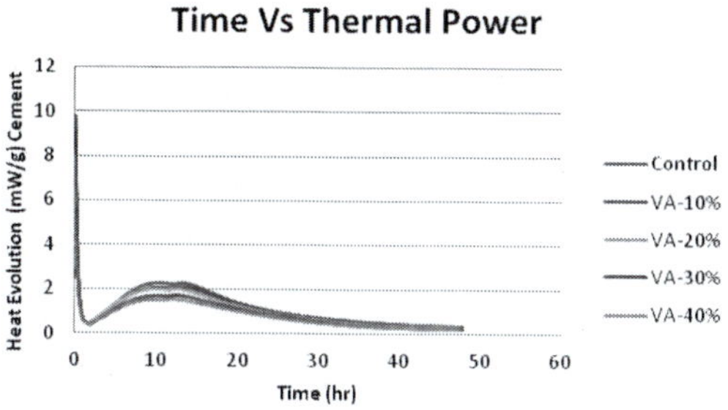

Time Vs Thermal Power

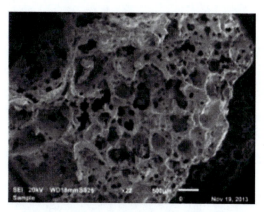

Figure 2. Thermal power curve for isothermal hydration of cement paste.

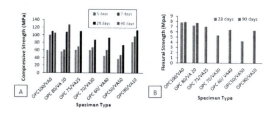

Figure 3. (A) Compressive and (B) Flexural strength of cement paste matrix.

3.6 Determination of compressive and flexural strength of cement paste

Compressive and flexural strength of cement paste with different percentages of volcanic ash was determined at a w/c ratio of 0.25.

Cement paste samples were prepared in 50 mm moulds for determining compressive strength and $40 \times 40 \times 160$ mm steel moulds for determining flexural strength. Addition of 10% VA shows an increase in compressive strength (Fig. 3. A) at 3 and 28 days compared to control mix whereas the compressive strength of cement paste decreases with increase in the percentage of VA from 20% to 50%. The tests will be continued up to 90 days to understand the influence of VA on late age strength. However, and as in Fig. 3. B, the flexural strength decreases with the addition of VA.

3.7 Microstructural analysis

A preliminary investigation on the influence of volcanic ash on the modification of microstructure was carried out and the SEM pictures are demonstrated below. Fig. 4 shows the SEM of VA-8 mm. SEM of highly vesicular glass shard of ash. Volcanic glass shards are fragments of molten part of magma that cools down and solidifies during eruption. Glass shards are remnants of tiny gas bubbles that develop and grow in size during the final ascent toward the surface with glassy phases in the amorphous material. Fig. 5 shows SEM of VA of 20 microns with clusters of droplet size particles of volcanic ash. Regular shaped particles are also present in the microstructure.

Fig. 6 shows the SEM of hydrated cement paste of 80% OPC and 20% VA at 0.25w/c after 28 days curing time, Ca(OH)$_2$ was observed and most of its densified,

Figure 4. SEM of VA-8 mm.

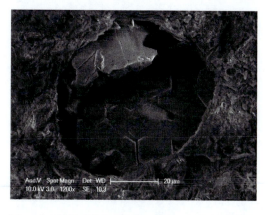

Figure 5. SEM of VA 20 microns.

Figure 6. SEM of Hydrated Cement Paste with OPC – 80% and VA – 20% for 28 days curing time.

No ettringite much because of the present of VA which reacts to form more C-S-H. On the other hand Fig. 7 shows SEM of hydrated cement paste of 80% OPC and 20% VA at 0.25w/c after 90 days curing time, classic C-S-H was found forming on VA, also the present of Ca(OH)$_2$ was observed. The ettringite is present as well mixed with the C-S-H.

Figure 7. SEM of Hydrated Cement Paste with OPC – 80% and VA – 20% for 90 days curing time.

4 CONCLUSION

This investigation was conducted on the collected samples of volcanic ash and OPC to understand its basic physical, chemical, hydration and microstructural characteristics. The results show that the heat of hydration decreases with increase in the dosage of VA. The physical and chemical characteristics of VA, and the basic test results show that VA can replace conventional pozzolanic materials in concrete construction. Further tests have to be conducted to make use of the material completely. These results can be utilized for understanding the properties of cement matrix in the micro level in the further studies.

ACKNOWLEDGEMENT

Authors would like to extend their gratitude to Kuwait Foundation for the Advancement of Sciences for their generous financial support of this collaborative research study between Kuwait Institute for Scientific Research (KISR) and MIT for the fulfillment of the project "Nano-engineered construction materials for durability in aggressive environments". Authors also are grateful for the support of KISR to accommodate all the rising challenges during the execution of the project's tasks.

REFERENCES

Escalante-Garcia, J. I., and J. H. Sharp. 1998. Effect of temperature on the hydration of the main clinker phases in Portland cements: Part II Blended cements. Cement and Concrete Research. 28(9): 1259–1274.
Escalante-Garcia, J. I., and J. H. Sharp. 2001. The microstructure and mechanical properties of blended cements hydrated at various temperatures. Cement and Concrete Research. 31: 695–702.
Rahman, A.A.; S. A. Abo-El-Enein; M. Aboul-Fetouh; and Kh. Shehata. 2011. Characteristics of Portland blast-furnace slag cement containing cement kiln dust and active silica. Arabian Journal of Chemistry. Article in press.
Yeau, K. Y., and E. K. Kim. 2005. An experimental study on corrosion resistance of concrete with ground granulate blast-furnace slag. Cement and Concrete Research. 35: 1391–1399.

Advances in Civil Engineering and Building Materials IV – Chang et al (eds)
© 2015 Taylor & Francis Group, London, ISBN: 978-1-138-00088-9

Design of asphalt Reclaimed Concrete Aggregate (RCA) mixtures based on performance evaluation of refusal density and moisture

A. Al-Aibani, S. Al-Fadala & A. Al-Arbeed
Construction and Building Materials Program, Energy and Building Research Centre,
Kuwait Institute for Scientific Research, Kuwait

ABSTRACT: This paper presents an assessment of Reclaimed Concrete Aggregate (RCA) in asphalt pavement as a material constituent of asphalt application offering profound and more sustainable engineering and economic advantages and a practical solution to the RCA waste managements. An optimized asphalt concrete mix design is developed from several trials of different percentages of RCA, as secondary aggregate of natural coarse aggregate, complying with road specifications requirements in Kuwait. Based on performance evaluation by measuring refusal density, the optimized selected asphalt mixture is an essential indicator and criteria for a more durable mixture. It has a prediction value of the ultimate densification and compaction that the asphalt concrete may withstand during service. Also in this study, the moisture effect on the asphalt mixture has been investigated in terms of Indirect Tensile (IDT) strength to assess the susceptibility of the pavement to moisture damage.

Keywords: recycled concrete aggregate, refusal density, Indirect Tensile (IDT) strength, asphalt pavement design.

1 INTRODUCTION

During the past decades, Construction and Demolition (C&D) waste issues have received increasing attention from both practitioners and researchers around the world (Weisheng Lu, 2011). Also with the increasing recognition of sustainable development as an important value, a diversity of industries, including the construction sector, have been taking action to promote sustainable development (Yuan, H. 2013). This raises the need for proper (C&D) waste management.Recycled concrete aggregate used in construction applications and especially in asphalt concrete mixtures is the focus of this study and can be considered asa type of engineering waste management. Several researches have been studying the implementation effects of adding the (RCA) as a material in pavement mixtures in order to produce more sustainable asphalt mixtures. In this study, a design of asphalt reclaimed concrete aggregate mixtures will be derived with an optimum percentage of (RCA) that meets local specifications and the asphalt institute manual MS-02 using the Marshall Mix design method.

Despite the fact that the approval of a Marshall designed asphalt concrete mixtures meets all the criteria and limits of the MS-02 manual, the roads still exhibit a premature permanent deformation under traffic loading. It has been observed that the roads in service are subjected to a secondary compaction where roads undergo further densification reducing the air void percentage to less than 3% leading to a very high risk failure from plastic deformation (rutting) (Smith HR, et al.). This form of failure results in large ruts caused by the flow of the surfacing material towards the edge of the wheel paths (TRL). This means that the Marshall method of compaction with 75 blows is not sufficient for the prediction of the resulting air voids of the road after service.The method of compaction needs to be modified to produce a refusal density where the mixture is compacted until refusal (BS 12697 part 32:2003) as an essential indicator and criteria for more durable mixture.This is a prediction value of the ultimate densification and compaction that the asphalt concrete may withstand during service.

Another important aspect should be the investigation for the final approval of the mix design through studying the asphalt performance, based on it moisture sensitivity. That the adhesion between asphalt and aggregate is reduced in the presence of water (Stripping) and that the cohesion within the asphalt binder itself deteriorates have been known topractitioners for a long time and it dates back to at least the 1920s (Solaiman, M. 2003).

The objectives of this research project are the assessment of reclaimed concrete aggregate in asphalt concrete mixtures in base course in order to develop an optimized pavement mixture with RCA, as one of the materials constituents, which can comply with the Kuwait road specifications, based on performance evaluations of refusal density and moisture effect in term of Indirect Tensile Strength (IDT).

2 EXPERIMENTAL INVESTIGATION

2.1 Asphalt recycled concrete aggregate mixture design

Coarse recycled concrete aggregate of sizes (1″, 3/4″, 3/8″) were used as a replacement of natural coarse aggregate with percentages of 20%, 30%, 40% for HMA Type I (base course). As a starting point and based on a designed control mixture (0% of RCA) with nominal maximum size aggregate of 1″, an asphalt mixtures with RCA were designed. The designed control mix met all the design requirementsof the Asphalt Institute Manual MS-02 criteria.

RCA was introduced as a replacement of the natural coarse aggregate of the designed control mix in different percentages of the total weight. All mixes were designed with a total coarse aggregate gradation proportions similar to the designed control mix. Table 1 shows the particles size distribution of the mixes, which comply with the specifications limits of Kuwait's roads for Type I asphalt mixtures.

Based on the MS-02 manual the MarshallMix design method was carried to finalize the selection of the optimized binder content associated with the preferred RCA percentage,conducted through different laboratory testing procedures and the volumetric properties of a compacted bituminous pavement mixtures, with different binder contents percentages, starting with 4% to 6.5% with 0.5 increments. Using bitumen 60/70 as a binder locally available inKuwait.

2.2 Performance evaluation

The optimized binder content with the selected RCA-percentage is considered as a preliminarily design and further performance investigations are required in terms of assessing the refusal density using the vibratory hammer as the compaction method to refusal, and the moisture effect using the Marshall-immersion test, and the moisture susceptibility test in terms of Indirect Tensile strength (IDT) were also performed.

2.2.1 Refusal density

The control and selected optimum percentage of RCA asphalt mixtures test samples of 150 mm diameter were prepared according to (BS: 12697 part 32, 2003). The samples were compacted using an electric vibrating hammer with 102 mm tamping foot for 120 seconds at each side of the samples. After completion of test, the samples were allowed to be cooled for 2 hrs. prior to de-moulding. Later the refusal density was determined according to (EN 12697 part 6:2003). The initial bulk density of a compacted asphalt specimen with the same asphalt mixture design was obtained using the Marshall method of compaction applying 75 blows, to investigate the effectiveness and the differences on the two types of approaches for the

Table 1. Particles size distribution of different RCA percentages.

Sieve-Size	Particles size distribution of different RCA%				Specification limits
	control mix	20%	30%	40%	
37.5	100.00	100.00	100.00	100.00	100
25	98.68	98.58	98.53	98.48	72–100
19	77.13	77.13	77.13	77.13	60–89
12.5	56.23	56.63	56.82	57.02	46–76
9.5	56.04	56.17	56.24	56.31	40–67
4.75	34.97	39.21	41.33	43.44	30–54
2.36	31.07	32.63	33.41	34.19	22–43
1.18	23.87	23.88	23.89	23.90	15–36
0.6	17.90	17.91	17.91	17.91	10–28
0.3	11.76	11.72	11.70	11.69	6–22
0.15	8.68	8.63	8.60	8.57	4–14
0.075	7.32	7.26	7.23	7.21	2–8

computation of the bulk specific gravity of an asphalt concrete specimen.

2.2.2 Moisture susceptibility

Six cylindrical samples of the optimized percentage of RCA asphalt mixtures, with 101.5 mm diameter and 63.5 mm height, were prepared to conduct the Marshall-immersion test. Samples were divided into two groups: each of three specimens so that the average bulk specific gravity of (group I) is similar to that of (group II). Conditioning procedures for the samples were performed according to ASTM D1075-11. Group I was stored dry at a 25°C, and group II was immersed in water for 24 hrs. at 60°C.

The Marshall stability was measured for each sample as a strength parameter. Then, the numerical index of resistance of bituminous mixtures to the detrimental effect of water was calculated, as a percentage of the original strength retained after the immersion period.

The ASTM D 6931-12 is also adopted to calculate the moisture damage by the ratio of the indirect tensile strength of moisture for the two conditioned and unconditioned subsets. IDT is considered a potential indicator of rutting or cracking for laboratory designed mixtures. The test is conducted by loading a cylindrical specimen with 101.5 mm diameter and 63.5 mm height across its vertical diametral plane at a rate of 50 mm/min at test temperature of 25°C. The two subsets were divided with each subset having similar average air voids.

3 RESULTS AND DISCUSSION

3.1 Asphalt recycled concrete aggregate mixture design

Following the Marshall Mix design method for the selection of the optimized RCA mix (4% air voids), it was noticed that blends with 30% RCA and 40%

Table 2. Mix Design Volumetrics of Control and 20% RCA Asphalt Mixes.

Volumetric Properties	Asphalt Mix	
	Control	20% RCA
Optimum asphalt content, %	4	4.7
Gse	2.704	2.628
Gmm	2.637	2.570
Gmb	2.521	2.459
Stability, N	9283	8485
Flow, 0.25 mm	13.25	14.3
AV, %	4.39	4.28
VMA, %	12.5	11.9
VFA, %	65.0	64.2

Table 3. Moisture Susceptibility Tests.

	Moisture Susceptibility Test	
	Marshall-immersion Test	Indirect tensile Strength Test
Asphalt Mix	Index of retained strength, %	Tensile Strength ratio, %
20% RCA	86	81

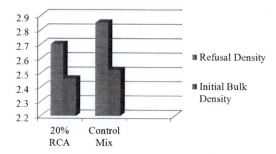

Figure 1. Initial bulk densities and refusal densities comparison.

RCA failed to comply with Marshall Mix design criteria. On the other hand, the 20% RCA blend mix when compared to the control mix, showed similar volumetric properties passing the MS-02 minimum limits, with higher optimum binder content than the control mix as shown in Table 2.

3.2 Refusal density

Figure 1 illustrate the comparison between the initial bulk densities and refusal densities test results for both control and 20% RCA asphalt mixtures at designed 4% air voids. The test results proving that bulk density computed from the Marshall Compaction method is not sufficient to simulate the maximum densification that the asphalt mixtures may withstand at service, which can be seen when compared to the higher refusal density amount.

The air voids calculated from the refusal density of 20% of the RCA mix was larger than the minimum limit of 3%, which predicted that the asphalt mix will not reach to air voids less than 3% after the application of the secondary compaction, and the mix is not likely going to exhibit a plastic deformation (rutting).

3.3 Moisture susceptibility

Table 3 shows the tests results percentages of the Marshall-immersion test and the IDT test for the selected optimized asphalt recycled aggregate mixture with 20% of RCA and a designed 4% air voids. Both test results met the minimum requirement 80%, but at the IDT test it can be notice that the result percentage indicates that the mixture has high moisture susceptibility when compared to control mixtures.

4 CONCLUSION

- Recycled concrete aggregate can be used as a material constituent of asphalt concrete mixtures for bass course, but in low percentages.
- High replacement percentages of recycled aggregate will decrease the stiffness and required high binder content percentages.
- The percentage of refusal densities has been proven to be an essential indicator and a predictor of the amount of densification the mixture will undergo after years of life, and is an important key for performance evaluation of the mixture prior to the final selection and approval of the mix design.
- The moisture susceptibility tends to increase whenever the RCA percentages increase in the pavement mix, thus reducing the life service of the pavement in its durability aspect.

REFERENCES

ASTM D 6926; 2010; Preparation of Bituminous Specimens Using Marshall Apparatus, *American Society for Testing and Materials, Philadelphia, Penn., U.S.A.*
ASTM D 1075; 2011; Standard Test Method for Effect of Water on Compressive Strength of Compacted Bituminous Mixtures, *American Society for Testing and Materials, Philadelphia, Penn., U.S.A.*
ASTM D 6931; 2012; *Standard Test Method for Indirect Tensile (IDT) Strength of Bituminous Mixtures, Philadelphia, Penn., U.S.A.*
ASTM D 6927; 2006; Marshall Stability and Flow of Bituminous Mixtures, *American Society for Testing and Materials, Philadelphia, Penn., U.S.A.*
BS EN 12697-6; 2012; Bituminous mixtures: Test methods for hot mix asphalt, Part 6: Determination of bulk density of bituminous specimens, *British Standard institute, London.*
BS EN 12697-9:2002; Bituminous mixtures: Test methods for hot mix asphalt, Part 9: Determination of the reference density, *British Standard institute, London.*

BS EN 12697-32:2003; Bituminous mixtures: Test methods for hot mix asphalt, Part 32: Laboratory compaction of bituminous mixtures by vibratory compactor, *British Standard institute, London.*

Lu, W. Yuan, H. 2011 A framework for understanding waste management studies in construction. *Waste Management.* 31: 1252–1260.

Ministry of Public Works (MPW) (2012), General specifications for Kuwait motorway/expressway system. *Roads Administration, Kuwait.*

Mix Design Methods for Asphalt concrete and other Hot-Mix Types, Asphalt Institute Manual series No. 2 (MS-2), Sixth Edition, 2012.

Smith H.R. & Jones C.R. (1997). Dense bituminous surfacings for heavily trafficked roads in tropical climates. TRIL *Annual Review 1996.*

Solaimanian, M. Harvey, J. Tahmoressi, M. & Tahmoressi, M. 2003. Test Methods to Predict Moisture Sensitivity of Hot-Mix Asphalt, Pavements: *A National Seminar; Proc, San Diego, 4-6 February 2003. TRB.*

Yuan, H. 2013 Key indicators for assessing the effectiveness of waste management in construction projects. *Ecological Indicators: 24: 476–484.*

Advances in Civil Engineering and Building Materials IV – Chang et al (eds)
© *2015 Taylor & Francis Group, London, ISBN: 978-1-138-00088-9*

Preparation of CIP/ZnO composites and design of λ/4 type EMWA coating in X-Band

Z.L. Zhang, M.M. Ming, Z.J. Xin, G.Y. Hou & L. Fan
China Building Materials Academy, Beijing, China

W.J. Hao
College of Materials and Chemical Engineering, Hainan University, Haikou, China

ABSTRACT: CIP/ZnO electromagnetic functional composites materials with core-shell structure were prepared by chemical precipitation method, and then characterized by X-ray diffraction, scanning electron microscopy and energy dispersive spectrometer. The electromagnetic parameters were measured in 8-12GHz by a vector network analyzer. In comparison with CIP, complex permittivity of CIP/ZnO composites had certain enhancement, but complex permeability decreased. Based with the Electromagnetic Wave Absorption (EMWA) Theory, the composites could prepare EMWA building coating by λ/4 type design, exhibiting excellent EMWA properties in 8–12 GHz. The reflection loss (RL) of −10 dB bandwidth was more than 75% and below −5 dB in the whole X band with the thickness of 3.5 mm, and the minimum RL of −20 dB was observed in 9.2 GHz. The magnetic loss of CIP and the dielectric loss of ZnO were the main mechanisms of EMWA for the CIP/ZnO composites, which could be used for architectural space electromagnetic radiation protection.

Keywords: core-shell structure, electromagnetic properties, λ/4 type coating, EMWA properties.

1 INTRODUCTION

With the rapid development of information and popularizing electronic equipment, architectural space electromagnetic radiation is more and more serious, electromagnetic radiation protection has been a hot problem of current international research[1−3]. Therefore, architectural space electromagnetic pollution control materials have aroused wide concern and deep study for researchers to reduce radiation[4−6]. Carbonyl iron (CIP) which exhibits excellent EMWA properties in GHz frequency range has superior conductivity and large saturation magnetization as a cheap and popular magnetic loss composite. However, the characteristics limit its application of which high density, narrow bandwidth, poor dispersion property and easy corrosion and oxidation[7−9]. The studies found that ZnO-based composites as EMWA materials have a fairly good EMWA performance, such as Fe/ZnO[10], Ni/ZnO[11] and ZnO/CoFe$_2$O$_4$[12], which are cheap, easy to prepare and have high temperature stability. At the same time, the structure, electromagnetic parameters and the impedance matching property have an influence on the absorption properties. In this paper, the CIP/ZnO composites with core-shell structure were prepared by chemical precipitation method innovatively, CIP as the core material, then grafted the ZnO shell. The micrography and structure of composites were preliminarily studied, and further discussed

CIP/ZnO special electromagnetic properties. We make the X-band absorption coating of λ/4 type design based on EMWA theory, in order to prepare new EMWA materials for architectural space radiation protection, and make the electromagnetic environment more safe and reliable.

2 EXPERIMENTAL

2.1 Materials and characterization

The micro-spherical carbonyl iron powders (CIP) were commercially purchased and its diameter was 2.0–7.0μm. Anhydrous ethanol and polyvinylpyrrolidone (PVP) from Beijing Yili Fine Chemical Co. Ltd.; Sodium hydroxide (NaOH) from Beijing chemical factory; Zinc nitrate hexahydrate (Zn(NO$_3$)$_2$) from Beijing Modern Oriental Fine Chemicals Co. Ltd., the above reagents were all A.R. level.

The micrography were characterized by Quanta 250 FEG Field emission environmental scanning electron microscope (SEM, FEI Company). The crystallite structure were confirmed by X-ray diffraction (XRD, Bruker). The relative complex permittivity ($\varepsilon_r = \varepsilon' - j\varepsilon''$) and permeability ($\mu_r = \mu' - j\mu''$) were measured by a vector network analyzer (AV3629D) in 8–12 GHz (X-band).

2.2 Preparation of CIP/ZnO Composites

The CIP/ZnO composites prepared by the chemical precipitation method were carried out as follows: because the micro-level CIP were easy to agglomerate, CIP should grind to spread out before reaction. Firstly, a certain amount of 1mol/L NaOH were added PVP, according to the rate mZnO:CIP = 1.2:1, and then mixed and dispersed evenly. Some CIP were added into the above mixture solution, and went on stirring 0.5h under the condition of 90°C. Then 0.5 mol/L Zn(NO₃)₂ was slowly added into the mixture and gone on stirring 5h. After the reaction, the final products were thoroughly washed with anhydrous ethanol for several times, and dried in 80°C for 6 h under the vacuum condition, then heat treatment for 1 h in 300°C.

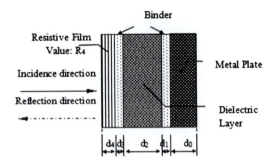

Figure 1. λ/4 type design.

2.3 Structural design of λ/4 absorber[13−14]

Figure 1. shows the λ/4 structural model, in which the thickness of dielectric layer is d_2. Dielectric layer and resistive film are bonded by binder, d_1, d_3 and d_4 are the thickness of binder and resistive film, respectively ($\leq$0.1 mm). The value of resistive film is R_4. Incidence direction of electromagnetic wave is from resistive film (hereinafter the same). It can design the EMWA building coating by the λ/4 model.

$\dot{Z}_{in1}$, $\dot{Z}_{in2}$, $\dot{Z}_{in3}$, $\dot{Z}_{in4}$ The impedance values of various layers ( ) can be calculated by the following formulas:

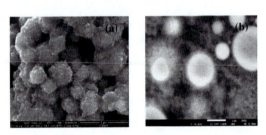

Figure 2. SEM spectrum of CIP/ZnO.

$$\dot{Z}_{ink} = Z_k \frac{\dot{Z}_{ink-1} + \dot{Z}_k \tanh(\dot{\gamma}_k d_k)}{\dot{Z}_k + \dot{Z}_{ink-1} \tanh(\dot{\gamma}_k d_k)} \quad (K \neq 4) \quad (1)$$

$$\dot{Z}_{in4} = \frac{\dot{Z}_4 \cdot \dot{Z}_{in3}}{\dot{Z}_4 + \dot{Z}_{in3}} \quad (K = 4) \quad (2)$$

Thereinto:

$$\dot{Z}_k = Z_0 \sqrt{\frac{\mu_{\gamma k}}{\varepsilon_{\gamma k}}} (K \neq 4) \quad (3)$$

$$RL = 20 \mathrm{Log}_{10} |\dot{\Gamma}| \quad (dB) \quad (4)$$

$$\dot{\Gamma} = \frac{\dot{Z}_{in4} - Z_0}{\dot{Z}_{in4} + Z_0} \quad (5)$$

Where RL(dB) refers to the reflection loss of λ/4 absorber, is the reflection coefficient of a vertical incidence wave, Z_0 is the impedance of the free space, ε_r and μ_r are the relative complex permittivity and permeability.

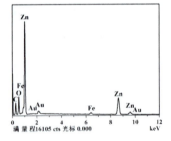

Figure 3. EDS energy spectrum.

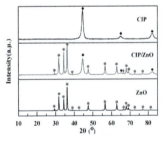

Figure 4. XRD spectrum.

3 RESULTS AND DISCUSSIONS

3.1 Morphological study and structural characterization

From Figure 2(a), it is observed that the surface of CIP/ZnO is very rough, and covered with a thin layer

of ZnO particles. The single micro-particle section can be observed the core-shell structure in Figure 2(b). In the reaction process, ZnO happen the heterogeneous nucleation on the CIP surface, in order to prepare CIP/ZnO composites with core-shell structure. From Figure 3, it can be seen that the composites exist Fe, Zn, O, C and Au element, Au is detected signal obviously for the spray-gold treatment before SEM. The strong detection signal of Zn and O indicates that the

CIP have coated with ZnO particles successfully. The XRD spectrum of CIP/ZnO composites exhibit three diffraction peak at 44.6°, 64.9° and 82.3°, which are associated with the (110), (200) and (221) planes cubic α-Fe[15] , respectively.When the CIP are coated with ZnO shell, the characteristic peaks of α-Fe become smaller, and the ZnO peaks are observed obviously, indicating that the composites exist ZnO phase and that CIP/ZnO were composed by α-Fe and ZnO.

3.2 Electromagnetic properties

Figure 5 shows the ε' and ε'' of the CIP, ZnO and CIP/ZnO composites in 8–12 GHz. For CIP, it is found that the real permittivity (ε') substantially keep constant and is about 3.3, and the imaginary permittivity (ε'') fluctuates from 0.10 to 0.04. The ε' of ZnO increases slowly from 2.8 to 3.1, the ε'' keep about 0.1. Coated with ZnO shell, both ε' and ε'' have been an increase, ε' keeps about 4.65 and ε'' shows the increasing trend. As a result of the ZnO shell, it enhances interfacial polarization between CIP and ZnO, so that the dielectric loss of CIP/ZnO composites has improved.

From Figure 6, it can be seen that the μ' all present the decreasing trend in 8–12 GHz with frequency. Among them, the μ' of CIP falls slowly from 1.35 to 1.26, but the μ'' increases from 0.24 to 0.29. And the μ' of ZnO and CIP/ZnO also decrease a little, the μ'' all exist a characteristic peak and the changing trend in the same way basically, and the μ'' of

CIP/ZnO is somewhere between CIP and ZnO. It is indicated that the ZnO cut internal CIP and make the eddy current loss be suppressed. Meanwhile, the ZnO shell reduce the volume fraction of CIP at the same filling amount of absorbing reagent. The μ'' decreasing is the comprehensive performance of the above reasons.

3.3 EMWA Properties

According to the electromagnetic properties, we can design the EM-absorber and simulate its absorption performance. The design of $\lambda/4$ type EMWA building coating can be carried out based on EMWA theory. First of all, supposing that the thickness is 3.5 mm, the value of resistive film is 380Ω/°C, and the average value of complex permeability is $\mu = 1.17 - j0.12$, the influence of ε'' on the EMWA can be designed, the results are showed in Figure 7. The RL values of $\lambda/4$ type absorber are all below -5 dB as a whole and -10 dB bandwidth is more than 75%, displaying the good absorption properties and indicating that the design is successful relatively. In general, the ε'' is smaller, the RL value is better.

By this, further explore the influence of complex permeability variation on EMWA. Since the trends of the μ' and μ'' are different, it can grasp integral absorption properties by studying the both sides variation and the average value. Based on the previous calculation, presuming the ε' constant, the ε'' average value

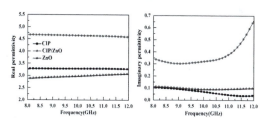

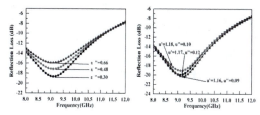

Figure 5. The permittivity of CIP/ZnO and CIP/ZnO.

Figure 7. The influence of electromagnetic parameters on RL.

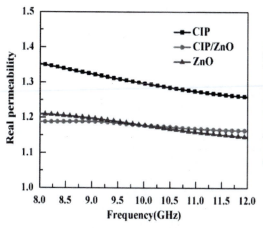

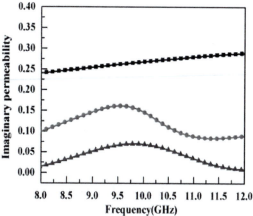

Figure 6. The permeability of CIP/ZnO and CIP/ZnO.

is 0.30, namely $\varepsilon = 4.65$–$j0.30$, simulative results are showed in Figure 7. It can be seen that the RL values are basically below -5 dB when the μ'' take the both sides' values, indicating that the new materials can prepare the high-performance absorbers. In particular, the RL value is below -5 dB when $\mu = 1.18$–$j0.10$ at 8 GHz, the minimum value is close to -20 dB. The RL value is about -7.5 dB at 12 GHz, but the terminal absorption is controlled by the complex permeability at 12 GHz, so the influence is little. Moreover, the RL value is below -5 dB when $\mu = 1.16$–$j0.09$ at 12 GHz, the minimum value is -20.3 dB, and the RL value around -14 dB at 8 GHz. But the front absorption is under the control of complex permeability at 8 GHz, as a result, the influence is little. Analyzed as a whole, the changing trends of the μ' and μ'' are different, so the ensemble absorption should be below -5 dB. It is proved by taking the average values of μ' and μ'' ($\mu = 1.17$–$j0.12$) to simulate, and find that the absorption are all below -5 dB.

As seen form the above calculation and simulation, the electromagnetic parameters of CIP/ZnO composites varies greatly, but it can be designed the high-performance $\lambda/4$ type EMWA building coating in X-band absolutely. The next job is carrying out the further experiments to validate.

4 CONCLUSION

The CIP/ZnO composites with the core-shell structure have been prepared successfully by the chemical precipitation method. After coating ZnO shell, the complex permeability of CIP/ZnO composites increases and the complex permittivity decreases. On the basis of EMWA theory, we can design the $\lambda/4$ type absorption building coating. By simulating calculation, it is

found that the minimum RL value can achieve -20 dB and the RL values are all below -5 dB in the whole X-band. Under the comprehensive effects of dielectric loss and magnetic loss, the CIP/ZnO composites show good electromagnetic wave absorption properties, and then the new materials can be used for the architectural apace electromagnetic radiation protection.

REFERENCES

Bala, H. et al. 2009. *Applied Surface Science* 255: 4050–4055.
Chen, W. 2011. *Synthesis of ZnO coated ferrite composite material and their electromagnetic properties.* Tianjin: Tianjin Universty.
Douglas, A.J. et al. 2002. *IEEE Transaction on Microwave Theory and Techniques* 50: 721.
Feng, Y.B. et al. 2006. *IEEE Trans. Magn.* 42:363–367.
Hao, W.J. 2004. *Science and technology of foreign building materials* 25: 74–76.
Hoshimoto, O. 1997. Introduction to Wave Absorber. *Morihoku Press*: 20–50.
Hoshimoto, O. 2004. Technologies and Applications of Wave Absorber. *CMC press*: 10–30.
Liu, S.H., Liu, J.M. & Dong, X.L. 2007. Electromagnetic wave shielding and absorbing materials. *Chemical Industry Press*: 20–40.
Liu, L.D. et al. 2010. *Magn. Magn. Mater* 332: 1736–1740.
Nayitou, Y. 1987. EMWave Absorber. *Omu press*: 30–50.
Qing, Y.C., Zhou, W.C., Luo, F. and Zhu, D.M. *J. Magn. Magn. Mater* Vol. 321(2009), p. 25–27.
Qing, Y.C. et al. 2011. *Physica B* 406: 777–780.
Simizu, T. 1989. Absorb and shield of electromagnetic wave. *Economic and technological's book:* 150–180.
Wang, S.H. et al. 2006. *Environmental Science and Technology* 29: 96–98.
Zhou, C. et al. 2012. *Journal of Magnetism and Magnetic Materials* 324:1720–1725.

Advances in Civil Engineering and Building Materials IV – Chang et al (eds)
© 2015 Taylor & Francis Group, London, ISBN: 978-1-138-00088-9

Design of high performance EMWA coating made of FeSiAl/PANI composites in S-band

Z.J. Xin, M.M. Ming, Z.L. Zhang, G.Y. Hou & L. Fan
China Building Materials Academy, Beijing, China

W.J. Hao
College of Materials and Chemical Engineering, Hainan University, Haikou, China

ABSTRACT: FeSiAl flaky powders were coated with polyaniline (PANI) to prepare FeSiAl/PANI electromagnetic wave absorption (EMWA) materials with core-shell structure by in situ polymerization. The electromagnetic properties of FeSiAl/PANI composites were studied in 2–4 GHz (S-band). Compared with the uncoated FeSiAl, the permittivity of FeSiAl/PANI composites increases, while the permeability decreases. Based with transmission line theory, it showed that PANI shell enhanced EMWA properties obviously. For monolayer coating absorber with a 4.8 mm thickness, the reflection loss (RL) of −5 dB bandwidth was more than 80% in S-band, and its minimum RL peak was −29.4 dB in 3.66 GHz, while the RL of FeSiAl almost cannot reach −20 dB in the same thickness. The magnetic loss of FeSiAl and the dielectric loss of PANI were the main mechanisms of EMWA for the FeSiAl/PANI composites, which could be used for architectural space electromagnetic radiation protection.

Keywords: core-shell structure, EM-properties, monolayer absorber, EMWA properties.

1   INTRODUCTION

In recent years, with the fast development of electrical communication technology and popularizing mobile communication and wireless network, architectural space EM- radiation is more and more serious, its protection has been a hot problem of current international research[1–3]. So, it is an important and emergent assignment to control architectural space EM-pollution and study the new EMWA materials that have aroused wide concern for researchers to reduce radiation[4–6]. Among all materials, FeSiAl flaky powders are especially focused on due to its excellent EMWA properties, but the characteristics limit its application of which poor chemical stability, easy corrosion and oxidation and bad impedance matching property[7–8]. And polyaniline (PANI) is one of the most popular conducting polymers as typical dielectric loss material[9–11]. Therefore, it is pressing to promote FeSiAl flaky powders application and prepare a new type lightweight materials with good EMWA properties that possess the comprehensive characteristics of two materials. In this paper, FeSiAl were coated with PANI shell by in situ polymerization to improve the EMWA performance of FeSiAl. The morphology and structure and electromagnetic properties of FeSiAl/PANI were also investigated in detail. As a absorbing reagent, we make the S-band monolayer coating absorber, in order to prapare a new type of EMWA material for

architectural radiation protection, and make architectural electromagnetic space more safe and reliable.

2   EXPERIMENTAL

2.1   *Materials and characterization*

The micro-level FeSiAl flaky powders were commercially purchased. Aniline monomer (ANI), which was distilled under reduced pressure before being used, and ammonium persulphate (APS) from Tianjin Guangfu technology development Co. Ltd.; Oleic acid from Tianijn Jinte chemical Co. Ltd.; Acetone and anhydrous ethanol from Beijing Yili Fine Chemical Co. Ltd., the above reagents were all A.R. Level. Dodecyl benzene sulfonic acid (DBSA) from Aladdin Industrial Corporation.

The micrography of samples were characterized by Quanta 250 FEG Field emission environmental scanning electron microscope (SEM, FEI Company). Crystallite structures were confirmed by X-ray diffraction (XRD, Bruker). The electromagnetic parameters were charactered by a microwave vector network analyzer (AV3629D) in 2–4 GHz (S-band), including the complex permittivity ($\varepsilon_r = \varepsilon' - j\varepsilon''$) and permeability ($\mu_r = \mu' - j\mu''$).

2.2   *Preparation of FeSiAl/PANI composites*

The FeSiAl/PANI composites were prepared by in situ polymerization in the presence of FeSiAl with APS

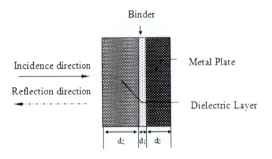

Figure 1. Monolayer type design.

as oxidant and DBSA as dopant, and carried out as follows: because FeSiAl were easy to corrosion in the process and used oleic acid to modify in order to avoid losses and obtain well compatibility. Firstly, DBSA (3.56 g) was dissolved in 200 ml of distilled water, and then aniline (1 ml) and FeSiAl (2 g) were added, according to the response rate mANI:FeSiAl = 1:2, Secondly, the mixture solution was mechanical stirred for 30 min under the ice-water bath, 20 ml of APS was slowly added into the solution and the reaction solution was vigorous mechanical stirred for 12 h under the ice-water bath. At last, the products were washed with acetone, anhydrous ethanol and distilled water for several times, and finally dried 24 hours in a vacuum drying oven at 60°.

2.3 Structural design of monolayer absorber

Figure 1 shows the monolayer absorber structural model, in which the thickness of dielectric layer is d_2. Dielectric layer and metal plate are bonded by binder, d_0 and d_1 are the thickness, respectively. Incidence direction of electromagnetic wave is from dielectric layer. It can design the EMWA building coating by the monolayer absorber model.

According to the transmission line theory, the reflection loss (RL) can be calculated from the ε_r and μ_r at a given thickness, described at the

$$Z_{in} = Z_0 \sqrt{\frac{\mu_r}{\varepsilon_r}} \tanh\left[j \frac{2\pi fd}{c} \sqrt{\mu_r \varepsilon_r} \right] \quad (1)$$

$$RL = 20Log_{10}\left|(Z_{in} - Z_0)/(Z_{in} + Z_0)\right| \quad (2)$$

$$Z_0 = \sqrt{\frac{\mu_0}{\varepsilon_0}} \quad (3)$$

where Z_{in} is the input impedance, Z_0 is the impedance of the free space, ε_r and μ_r are the complex permittivity and permeability. f is the frequency of microwave, d is the simulated thickness of the absorber and c is the velocity of the light in free space.

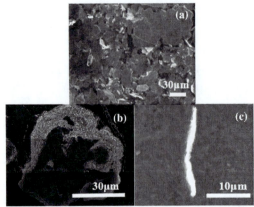

Figure 2. SEM spectrum of FeSiAl and FeSiAl/PANI.

3 RESULTS AND DISCUSSIONS

3.1 Morphological study and structural characterization

The SEM spectrum of FeSiAl/PANI and FeSiAl are showed in Figure 2. From Figure 2(a), FeSiAl are irregularity flake and its surface is smooth. After coated with PANI shell, it can be seen that the surface of FeSiAl/PANI composites were covered with a thin PANI layer and became rough from Figure 2(b), which could protect FeSiAl from oxidation and corrosion. The single micro-flake section can be observed the core-shell structure in Figure 2(c). The results indicate that FeSiAl/PANI composites with core-shell structure are successfully prepared.

From Figure 3, it can be seen that the composites exist Fe, Si, Al, C, O and S elements. The strong detection signal of C, O and S also indicates that FeSiAl have coated with DBSA doped PANI shell successfully.

In the XRD spectrum, pure PANI exist two broad peaks at 2θ of about 20° and 25°, which demonstrate that DBSA doped PANI is mainly of amorphous structure with a little degree of crystalline[13]. After coated with PANI, FeSiAl/PANI composites both remain the characteristic peaks of PANI and FeSiAl, while diffraction intensities of peaks decrease during the coating process. And it is indicated that the composites are composed by FeSiAl and DBSA doped PANI phases. Except for the characteristic peaks, there are no extra peaks for FeSiAl/PANI, indicating that there is no obvious chemical reaction between FeSiAl and polyaniline during in situ polymerization.

3.2 Electromagnetic properties

In Figure 5, it shows the ε' and ε'' of FeSiAl and FeSiAl/PANI composites in 2–4 GHz. For FeSiAl, the real permittivity (ε') substantially keep constant and is about 9.3, and the imaginary permittivity (ε'') increases from 0.07 to 0.26. At the same time, DBSA doped PANI is a dielectric loss EMWA materials. After the reaction, the ε' keep about 11.0, the ε'' increases

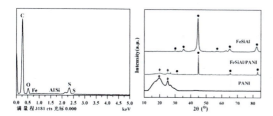

Figure 3. EDS energy spectrum.

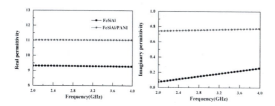

Figure 4. XRD spectrum.

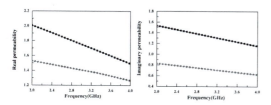

Figure 5. The permittivity of FeSiAl and FeSiAl/PANI.

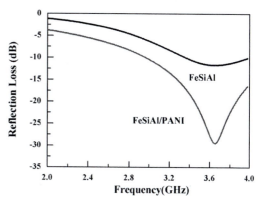

Figure 6. The RL values of monolayer absorber.

from 0.74 to 0.77 slowly. Compared with FeSiAl, the ε' and ε'' of FeSiAl/PANI composites all raise, not only the PANI shell enhances the interfacial polarization between core and shell materials, but also the dielectric loss component was added.

As a magnetic loss absorbing reagent, the complex permeability of FeSiAl are fairly large, the real permeability (μ') and the imaginary permeability (μ'') decrease from 2.00 to 1.49 and 1.53 to 1.15, respectively. After polymerization reaction, the μ' and μ'' of FeSiAl/PANI composites have a diminution in 2–4 GHz. And the μ' and μ'' all exhibit a decreasing trend. Among them, the μ' falls from 1.53 to 1.26, the μ'' decreases slowly from 0.83 to 0.62, and all show a good dispersion effect. The decreasing reason is mainly that the PANI shell reduce the volume fraction of FeSiAl at the same filling amount of absorbing reagent.

3.3 EMWA properties

According to the monolayer strucural model, we can design the EMWA monolayer absorber and simulate its EM-absorption performance. Supposing that the thickness d_2 is 4.8 mm, monolayer EMWA coating can be carried out based on transmission line theory. Figure 6 shows the calculated reflection loss for FeSiAl and FeSiAl/PANI composites in S-band. For FeSiAl flaky powders, the minimum value of -11.7 dB is obtained in 3.62 GHz and below -10 dB bandwidth is

about 50% in S-band. After coated with PANI shell, the minimum value decreases to -29.4 dB in 3.66 GHz, and the absorption peak position is not effected basically. Moreover, the RL value of -5 dB bandwidth is more than 80%, while the RL of FeSiAl cannot reach -20 dB in the same thickness. Under the comprehensive effects of dielectric loss and magnetic loss, the FeSiAl/PANI composites show excellent EMWA properties[14–15]. As seen from the above calculation and simulation, the FeSiAl/PANI composites can be designed the high-performance monolayer EMWA coating in S-band, and used for the architectural apace electromagnetic radiation protection. The next job is carrying out the further experiments to validate.

4 CONCLUSION

The FeSiAl/PANI composites with core-shell structure were fabricated by in situ polymerization. Compared with uncoated FeSiAl, the complex permittivity of FeSiAl with PANI shell increases, while the complex permeability decreases. In general, the FeSiAl/PANI composites had much better EMWA properties than FeSiAl. When the thickness of monolayer coating is 4.8 mm, its bandwidth of RL<-5 dB was more than 80%, and the minimum was nearly -30 dB. The enhanced EMWA properties can be attributed to the combined effect of both the magnetic loss of FeSiAl and the dielectric loss of PANI, and could be designed the monolayer coating absorber for the architectural space electromagnetic radiation protection.

ACKNOWLEDGMENT

This work was financially supported by the National Science and Technology support plan for Twelfth Five-years of China [No:2012BAJ02B04].

REFERENCES

Douglas, A.J. et al. 2002. *IEEE Transaction on Microwave Theory and Techniques* 50: 721.

Han, X. & Wang, Y.S. 2007. Effect of the contents of DBSA on the structures and electromagnetic properties of Fe_3O_4/conductive polyaniline nanoparticles. *Journal of Sichuan university (Engineering science edition)* 4(39): 90–93.

Hao, W.J. 2004. *Science and technology of foreign building materials* 25: 74–76.

Hoshimoto, O. 2004. Technologies and Applications of Wave Absorber. *CMC press*: 10–30.

Li, W.P. 2007. *The research on preparation and electromagnetic characters of polyaniline*. Dalian: Dalian university of technology.

Liu, S.H. & Liu, J.M. & Dong, X.L. 2007. Electromagnetic wave shielding and absorbing materials. *Chemical Industry Press*: 20–40.

Liu, X.D. et al. 2010. Study on preparation, electromagnetic property and microwave absorption property of doped polyaniline. *Plastic science* 4(38): 52–55.

Simizu, T. 1989. Absorb and shield of electromagnetic wave. *Economic and technological's book:* 150–180.

Toru, M. et al. 2004. *J. Magn. Magn. Mater*. 281: 195–205.

Wang, S.H. et al. 2006. *Environmental Science and Technology* 29: 96–98.

Wang, W. 2007. *The micro-structure and electromagnetic property study of anti-EMI FeSiAl soft magnetic flakes.* Hangzhou: Zhejiang University.

Wang, M.M. et al. 2014. Preparation of $CIP/SiO_2/PANI$ composites and design of electromagnetic wave absorption coating in X-Band. *Advanced Materials Research*. 846–847: 1905–1910.

Wang, M.M. et al. 2014. Preparation and Electromagnetic Properties of $CIP/SiO_2/PANI$ composites. *Advanced Materials Research*. 834–836: 187–190.

Zhang, Z.M. & Wan, M.X. 2003. *Synth. Met.* 132: 205–212.

Zhou, Y.D. 2004. *Studies on preparation, structures and characteristics of FeSiAl flaky powders*. Chengdu: University of Electronic Science and Technology of China.

Coastal engineering

Advances in Civil Engineering and Building Materials IV – Chang et al (eds)
© 2015 Taylor & Francis Group, London, ISBN: 978-1-138-00088-9

Risk evaluation of the navigational environment of Yingongzhou channel based on a single dimensional normal cloud model

Yanfei Tian & Liwen Huang

School of Navigation, Wuhan University of Technology (WUT), Wuhan, China
Hubei Inland Shipping Technology Key Laboratory, Wuhan, China

ABSTRACT: To make a fairer comprehensive risk evaluation, able to reflect the unity of fuzziness and random in the nature, or human knowledge, a weighted comprehensive risk evaluation model based on normal cloud concept was constructed, and its application procedures provided. The proposed model and procedures were successfully used to evaluation navigation environment risk of Yingongzhou Channel, which indicated themost possible quantitative risk gradewith a "HIGH" level on the whole. It showed the model was practicable and would substantially facilitate and benefitthe risk management.

Keywords: Normal cloud model, Yingongzhou Channel, navigation environment, comprehensive risk evaluation.

1 INTRODUCTION

Maritime risk is one of the most important focus areas for waterway transportation. Researchers and maritime authorities have been paying special attention to the risky areas where traffic density is high for a long time. The high density and high risk for transportation increase the complexity of navigation safety management (Jinfen, Zhang *et al.*, 2014).

Nowadays, lots of theories and models have been developed for maritime risk assessment. Cloud model, a conversion mode between qualitative and quantitative concepts, founded by Deyi Li, a Chinese scholar and anacademician, has been more and more frequently used because of its own advantages in dealing with fuzzy and random concepts arising in many fields (QiongYe *et al.*, 2011). This paper aimed to construct a weighted comprehensive risk evaluation model based on normal cloud concept and provide its application procedures. Comprehensive risk evaluation of navigation environment of YingongzhouChannel was made into an example research, which would offer support for a risk management decision-making process.

2 SINGLE DIMENSIONAL NORMAL CLOUD MODEL

Definitions of cloud and cloud droplet can be found in related literature (Qian Fu *et al.*, 2010). The numerical characteristicsof a cloud can be described by expected value, entropy and hyper entropy, and then a single dimensional cloud model can be expressed as

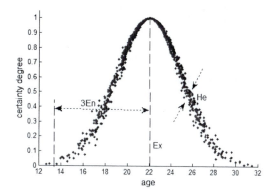

Figure 1. Normal cloud model of "YOUTH" with $C(22,3,0.2)$.

$C(E_x, E_n, H_e)$. Expected value (E_x) denotes the distribution expectation of a cloud droplet in the distribution domain, and also represents the most typical sample of this quantification concept. Entropy (E_n) denotes the uncertainty measure of qualification concept, which reflects the stochastic measure of qualification concept. Hyper entropy (H_e) is the uncertainty measure of entropy which is determined by both randomness and fuzziness of the entropy.

For example, the following YOUTH cloud indicates the uncertain expression "YOUTH" by basic single dimensional normal cloud with $C(22,3,0.2)$. For a qualitative concept, the quantitative values outside $[E_x - 3E_n, E_x + 3E_n]$ are ignored according to the "$3E_n$ rule" in normal cloud (Longquan Yong, 2011).

Table 1. Corresponding relations between grey classes and scores.

Grey classes	Scores (min, mid, max)
LOW	(0.0000, 0.1000, 0.2000)
LESS LOW	(0.2000, 0.3000, 0.4000)
MODERATE	(0.4000, 0.5000, 0.6000)
LESS HIGH	(0.6000, 0.7000, 0.8000)
HIGH	(0.8000, 0.9000, 1.0000)

3 RISK EVALUATION MODEL BASED ON NORMAL CLOUD

3.1 Model constructions

First, the 5 grey classes which qualitatively describe a risk grade (Yuehua Huang, 2014) are converted to scores varying from 0 to 1 for quantitative identification, whose corresponding relations are shown in Table 1.

There can be a lot of elements that have implications on safety. To indicate comprehensive risk expressed by cloud $CR(E_x, E_n, H_e)$ after evaluating the selected i indexes (factors) separately, whose individual weight cloud and risk grade (expressed by scores) cloud are respectively shown as $W_i(E_x, E_n, H_e)$ and $R_i(E_x, E_n, H_e)$, then the comprehensive risk (CR) can be calculated by:

$$CR(E_x, E_n, H_e)$$
$$= \Sigma W_i(E_x, E_n, H_e) * R_i(E_x, E_n, H_e) \tag{1}$$

where, supposing two clouds $CR(E_{x1}, E_{n1}, H_{e1})$ and $CR(E_{x2}, E_{n2}, H_{e2})$, the mathematical addition and multiplication operations are (Bingxiang Liu et al., 2009):

$$C_1 + C_2 = (E_x, E_n, H_e)$$

$$= \left(E_{x1} + E_{x2}, \sqrt{E_{n1}^2 + E_{n2}^2}, \sqrt{H_{e1}^2 + H_{e2}^2} \right) \tag{2}$$

$$C_1 * C_2 = \begin{bmatrix} E_x \\ E_n \\ H_e \end{bmatrix}$$

$$= \begin{bmatrix} E_{x1} \cdot E_{x2} \\ |E_{x1} \cdot E_{x2}| \times \sqrt{\left(\frac{E_{n1}}{E_{x1}}\right)^2 + \left(\frac{E_{n2}}{E_{x2}}\right)^2} \\ |E_{x1} \cdot E_{x2}| \times \sqrt{\left(\frac{H_{e1}}{E_{x1}}\right)^2 + \left(\frac{H_{e2}}{E_{x2}}\right)^2} \end{bmatrix} \tag{3}$$

3.2 Application procedures

There are mainly 6 steps used to carry out the proposed model.

Step 1: determine the evaluation indexes. This work can be done by literature consultation, expert investigation, etc.

Step 2: achieve (E_x, E_n, H_e) of each weight cloud. Expected weight (E_x) of each index could be obtained by means stated in (1), then specified values can be assigned to E_n and H_e.

In case of incomplete references, the Analytic Hierarchy Process (AHP) founded by T. L. Saaty (T. L. Saaty, 1980) is quoted here in order to obtain an index weight. Dealing with each of the N copies of questionnaires by AHP, N sets of weight data can be obtained as N droplets. With some specified value assigned to H_e, E_x and E_n of the weighted cloud will be approached through a backward cloud generator (Liyan Xing, et al., 2010, Guihua Liu, et al., 2007) by statistics of the droplets.

Step 3: approach individual risk assessment of each index indicated by $R_i(E_x, E_n, H_e)$.

Appropriate whiten weight function for each index is introduced to realize individual risk assessment (Yanfei Tian, 2013). Taking index i as an example, if the membership of 5 grey classes based on a certain index value is calculated as $(r_1, r_2, r_3, r_4, r_5)$ by the corresponding whiten weight function, then the score S representing its risk under the circumstance is:

$$S = 0.2r_1 + 0.4r_2 + 0.6r_3 + 0.8r_4 + 1.0r_5 \tag{4}$$

Then N sets of S can be obtained as N droplets by the above just mentioned means, because of N set of index values of index i existing in practice. With some specified value assigned to H_e, E_x and E_n of individual risk cloud will be approached through the backward cloud generator by statistics of the droplets.

Step 4: get a comprehensive evaluation. After the above steps, it is possible to calculate $CR(E_x, E_n, H_e)$ according to formulas (1), (2), and (3).

Step 5: generated droplets characterized by $CR(E_x, E_n, H_e)$ through forward cloud generator.

Step 6: plot the droplets, and make a result analysis.

Structures, roles, and algorithm implementations of forward and backward cloud generators among the steps are provided by previous works (Liyan Xing, et al., 2010, Guihua Liu, et al., 2007).

4 EXAMPLE RESEARCH

Comprehensive risk evaluation of navigation environment of Yingongzhou Channel on Yangtze River was made an example research in this segment.

4.1 Parameter settings

Parameters and their values were set as follows:

(I) parameter settings of evaluation cloud models
With E_n determined by '$3E_n$ rule', parameters of each cloud model of the five grey classes qualitatively describing risk grade are listed in Table 2 and the droplets are shown in Figure 1.

(II) Parameter settings of each index weight cloud model

138

Table 2. Parameter settings of each evaluation cloud.

Grey classes	E_x	E_n	H_e
LOW	0.0000	0.2000/3	0.0100
LESS LOW	0.3000	0.2000/3	0.0100
MODERATE	0.5000	0.2000/3	0.0100
LESS HIGH	0.7000	0.2000/3	0.0100
HIGH	1.0000	0.2000/3	0.0100

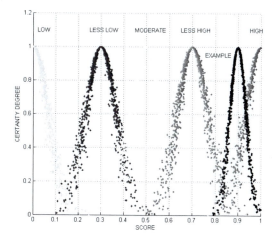

Figure 2. Risk cloud of navigation environment of Yingongzhou channel.

Original weight in the reference (Yanfei Tian, 2013) is assigned to E_x as the expected value. Subjectively make $E_n = 0.0100$ and $H_e = 0.0010$.

(III) Parameter settings of individual risk cloud model
The E_x of individual risk cloud is computed according to step 3. To agree with (I), set $E_n = 0.0333$ and $H_e = 0.0100$ for the cloud.

4.2 Simulation results and analysis

By the above means, a comprehensive risk of the navigation environment of Yingongzhou Channel was simulated as $CR(0.8986, 0.0357, 0.0036)$ shown in Figure 2.

The cloud CR distributes between "LESS HIGH" and "HIGH", revealing uncertainty of risk. The distribution expectation $E_x = 0.8986$ indicates the most possible quantitative risk grade, leading the cloud closer to 'HIGH', which is consistent with the result judged by the 'maximum membership degree' principle. The small E_n and H_e concentrating the droplets announce a higher certainty of the risk evaluation.

5 CONCLUSION

It can be seen that the proposed model based on cloud concept made a fair practice of comprehensive risk evaluation, which reflected unity of fuzziness and random in the nature or human knowledge.

The constructed model was successfully used to evaluate navigation environment risk of Yingongzhou Channel, indicating its most possible quantitative riskgrade with a "HIGH" level on the whole, which provided a visual assessment resultand substantially facilitated and benefitedthe risk management.

ACKNOWLEDGMENT

This work is supported by the Fundamental Research Funds for the Central Universities (Financially supported by self-determined and innovative research funds of WUT, Grant Nos. 2014-VII-026 and 2014-JL-010). The authors would like to thank the funds for the financial support for this project.

REFERENCES

Fu, Qian; Cai, Zhihua; Wu, Yiqi, 2010. A New Method for Reliability Evaluation Based on Cloud Model [C]. 2010 Second International Conference on Information Technology and Computer Science, 118–121.

Huang, Yuehua, 2014. Safety Assessment of Navigation Based Grey Fixed weight Cluster in Tianjin Port Areas [D]. Dalian Maritime University.

Liu, Guihua; Song, Chengxiang; Liu, Hong, 2007. Software Implementation of Cloud Generators [J]. Computer application research, (4):46–48.

Liu, Bingxiang; Li, Hailin; Yang, Libin, 2009. Cloud decision analysis method [J]. Control and Decision, 24(6): 957–960.

Saaty, T.L., 1980. The Analytic Hierarchy Process [M]. Mc.Graw:Hill International Book Company.

Tian, Yanfei, 2013. Establishment and Application of an Integrated Safety Evaluation Model Based on Connection Number [D]. Wuhan University of Technology.

Xing, Liyan; Shao, Chaohong, 2010. Risk assessment of project cost from the perspective of cloud [J]. Journal of Shandong Jianzhu University, 25(4):429–433.

Ye, Qiong; Li, Shaowen; Zhang, Youhua, et al., 2011. Cloud model and application overview [J]. Computer Engineering and Design, 32(12):4198–4201.

Yong, Longquan, 2011. A Cloudbased Model of Multi-attribute Decision Making [C]. Advances in Artificial Intelligence, 2011 International Conference on Management Science and Engineering, 6:438–444.

Zhang, Jinfen; Yan, Xinping; Zhang, Di, et al., 2014. Safety management performance assessment for Maritime Safety Administration (MSA) by using generalized belief rule base methodology [J]. Safety Science, 63:157–167.

Advances in Civil Engineering and Building Materials IV – Chang et al (eds)
© 2015 Taylor & Francis Group, London, ISBN: 978-1-138-00088-9

Development of a reflection coefficient diagram considering the effects of wave energy reduction due to riprap shapes and arrangement

Hayong Kim, Insang Yu, Keedong Kim & Sangman Jeong
Department of Civil and Environmental Engineering, Kongju National University, Chungcheongnam-do, Korea

ABSTRACT: The revetment structures constructed on the coastal areas of Korea are primarily made up of impermeable concrete material. The revetment structures are commonly arranged in a steep slope manner or in an upright position which are the so-called standard forms. However, the standard forms do not have the capacity to reduce the wave energy which results in the generation of reflective waves, and thus causes coastal erosion and bottom scouring to the revetment structures. The reflection coefficient, which is a design parameter required for the design of structures constructed in coastal areas, is difficult to obtain due to the complexity in the use of equations which are solely based on estimation or obtained by hydraulic experiments. There were a lot of studies that developed revetment structures for estimating design factor to deal with such problems however, shapes and arrangement of revetment were not considered in previous studies. In this study, artificial ripraps on the sloped coastal structures are arranged based on frictional area and void ratio. The effects on the wave energy changes of the artificial ripraps were analysed through a hydraulic model experiment, and the developed reflection coefficient estimated by graph was also used. Twelve (12) experimental revetment models having various frictional areas and void ratios were made for the hydraulic experiment. These consist of spherical, disk and cylindrical models. The hydraulic experiment was performed under regular wave conditions having wave heights of 2 cm, 4 cm, 6 cm, 8 cm, and 10 cm respectively as well as periods of 2 s, 4 s, 5 s, and 6 s respectively. A reflection coefficient diagram that can estimate reflection coefficients using height and period of wave was developed from the results of this study. Hydraulic experiment was carried out to verify the reliability of the diagram.

Keywords: wave energy, coastal erosion, reflection coefficient diagram, revetment.

1 INTRODUCTION

Beach erosion has long been a problem. In fact, countries like the US, Japan, the Netherlands, and other advanced countries considered it as a critical coastal engineering issue for decades. Scouring, one of the ways to damage coastal structures, was already gaining attention recently (Sumer et al., 2000). Besides, as the reflection coefficient is determined through the value of hydraulic experiments or estimation of wave force in designing ports, hydraulic experiments take a long time in addition to its uncertainty especially when estimation is used. There is a slight reduction of wave energy for the revetment structures built in the coast of Korea. In addition, the structures are collapsing due to the bottom scouring of revetment structures and the gravel layer is also forming due to beach erosion by reflective waves (Ministry of Land, Transport and Maritime Affairs, 2010).

As the population is growing and national infrastructures are being built along the coast, national issues to protect the coastal areas from various disasters as well as studies on ways to protect coastal facilities by efficiently controlling the effect of waves must be addressed. Thompson and Shuttle (1975) performed the study on stability of riprap structures for the first time, whilst Anderson et al. (1992), Van Djik (2001), Verhagen et al. (2003) and Van Gent and Pozueta (2004) reviewed the stability by checking breakaway of rear side cladding material depending on wave condition without considering the behaviour of overtopping. Kim, and Seonwoo (2000) drew hydraulic property of overtopping by run-up of wave dissipating block, whilst Shankar and Jayarathe (2003) showed that the combining effect of incident wave that affect overtopping and run-up and period explained wave steepness is an important index. Lee won (2008) studied frequency of overtopping by analysing overtopping property of sloped riprap revetment structures whilst Park Jingook (2003) analysed local applicability through overseas case studies on BMS technique to prevent beach erosion. Besides, Han Jaemyeong et al (2005, 2006) analysed reduction effect of coastal erosion by developing eco-friendly revetment blocks for coastal protection, and Kim Gwangjin (2007) developed eco-friendly CE blocks and evaluated their applicability. Kim Hayong (2012) developed artificial ripraps that have many frictional areas and pore space rate and

Table 1. Specifications of the wave flume.

Classification		Experiment facilities and equipment
Waterway Data		1.0 m (Wide) × 1.5 m (High) × 50.0 m (length)
Specification of wave machine	Length & Width	0.98 m × 1 wave boards
	Maximum wave height	0.4 m
	Period	0.5–5.0 sec
	Maximum Depth	1.0 m
	Operation Method	Electric servo piston wave maker

analysed the effect of frictional areas and pore space rate on wave energy under regular wave condition through hydraulic model test.

Studies over the effect of arrangement and shapes of ripraps on reduction of wave energy were rarely performed. Therefore, this study produced artificial ripraps with many frictional areas and pore space rate and analysed their effects on wave energy through hydraulic model test and developed reflection coefficients diagram using the result.

2 HYDRAULIC MODEL TEST

2.1 Experiment facility

Sectional wave test device used for hydraulic model test consists of rectangular waterway (1.0 m wide, 1.5 m high, and 50.0 m length) and electric servo piston wave maker. Irregular wave by spectrum function and regular wave that responds to given wave height and period can be produced from the system.

According to the result from the sectional water tank, data could be saved at 16 channels at the same time with wave height gauge, wave pressure gauge and current meter connected. Both sides of the waterway were built with reinforced glass to secure easy test observation. Furthermore, the control to absorb reflective waves is possible, based on the data measured from the wave height gauge, as the capacitance type wave height meter is fixed to the front side of the wave board, and as the multi-layers of porous wave absorber are installed at the other end of the wave board. Table 1 shows the specification of the wave flume used for the hydraulic model test of the study.

2.2 Model manufacturing

Greater friction (resistance) generally leads to greater loss in physics (force, energy, and velocity). Therefore, the artificial ripraps with the greatest frictional area (frictional area/volume) in a unit volume will be the most efficient way to reduce wave energy. When the diameter-height ratio of a circular cylinder is closer to 1, surface area-volume ratio becomes smaller. However, when the diameter of a plate (d/h > 1) is large and thin, and when a circular cylinder has d/h < 1, the frictional area-to-volume ratio can increase with smaller diameter and longer length. Figure 1 shows the

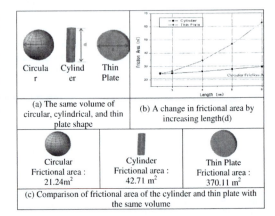

(a) The same volume of circular, cylindrical, and thin plate shape	(b) A change in frictional area by increasing length(d)

Circular Frictional area : 21.24m²	Cylinder Frictional area : 42.71 m²	Thin Plate Frictional area : 370.11 m²

(c) Comparison of frictional area of the cylinder and thin plate with the same volume

Figure 1. A frictional area variation of the configuration type.

frictional area increased when a sphere, circular cylinder, and thin plate that have the same volume became longer. Theoretically, as the circular cylinder is longer and the diameter of the plate is greater, the frictional area will increase infinitely. However, proper size is needed to apply in actual structures.

Therefore, the study considered manufacturability and strength of the thin plate and circular cylinder that are expected to affect the wave energy with increased frictional area at the same volume with spherical ripraps before manufacturing the model. The property of ripraps by shape is shown in Table 2 and the relationship between frictional areas and pore space rate by test model case is shown in Figure 2. The frictional area by riprap shape is planned to rise 50% and the pore space rate by 5%. By comparing the measured value, the effect on wave energy was analysed when the frictional area and pore space rate change.

2.3 Test condition

Ocean wave actually are very irregular having complex rhythms and force. However, such irregularity is the result of a combination of various regular waves. It is important to identify the effect of 2D regular waves that have steady periods and amplitude at a sea with constant depth on pore space and frictional area. Because of this, the study performed a hydraulic model test by using a regular wave that applied the same depth of 0.5 m for the entire test. To analyse the effect of various waves, five types of wave height (H) (2, 4, 6,

Table 2. Riprap characteristics.

Experimental cases	Case	Case 2	Case 3	Case 4	Case 5	Case 6	Case 7	Case 8	Case 9	Case 10
Slope	1 : 2.0									
Shape	Non Coastal	Circular			Thin Plate			Cylinder		
Schematic drawing										
Frictional area* (m²)	2.24	6.3	9.4	10.2	15.2	15.6	16.3	20.8	21.3	21.7
void ratio* (%)	-	39.8	36.8	34.3	34.3	39.7	44.9	34.3	39.3	44.8

*Frictional area and void ratio per unit volume (%)

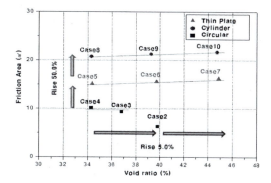

Figure 2. The relation between frictional area and void ratio.

8, and 10 cm) were tested and four periods (2, 4, 5, and 6 seconds) were used so that shallow sea areas and medium depth areas can be properly considered. When it was six seconds, the relative depth was 0.239 and when it was two, the relative depth was 0.775. Based on the results the test wave length specification of the hydraulic model test was presented in Table 3. As a result of wave length calculation, twenty test conditions were determined with five wave heights and four periods for Case 1, and mean value was used with three measured values for each case to reduce the error occurring from the measurement of the hydraulic model test. K represented wave number in the table, and it can be calculated with dispersion relationship shown below.

$$\sigma^2 = g\,k\,\tan h\,k\,h \qquad (1)$$

$$\sigma = \frac{2\pi}{T} \qquad (2)$$

where, σ = Angular frequency, h = Depth (m), T = Period (sec).

2.4 Test method

Three types of ripraps as shown in Table 2 were placed at 35m point of sea reach of 1.0 m × 1.5 m × 50.0 m, and four capacitance type wave height meters were installed to measure wave run-up and reflection coefficients. Wave height meter 1 was installed at the front

side of wave plate to identify water surface condition, and measure the reflected wave returning to the wave plate from model structure and used to measure input of reflective wave absorbing device. Wave height meter 2 and 3 were used for reflection coefficients, which was calculated with two point method proposed by Goda and Suzuki (1976), based on the data obtained from two wave height meters installed at the front side with random intervals. Goda and Suzuki (1976) calculate rates between component wave energies by using simultaneous wave, from the fact that each component wave of standing waves has constant phase difference within a given distance.

The interval of two wave height meters is adjusted depending on depth and periods to obtain wave shapes at each position, and then reflection coefficients are calculated from the wave data. In this case, reflection coefficients diverge if the interval is half the wave length, which should be avoided. Wave height meter 4 was installed in parallel with the slope of manufactured model, and the wave run-up of still water level was measured. Image analysis method was used which makes use of video for accurate and precise measurement to compare when there is measured data error. Besides, 2D current meter was installed at the joint of revetment structure and the bottom of waterway to analyse the measured current and the impact of riprap arrangement and revetment structure slope on the current. It is important for the revetment to measure and analyse overtopping among others; however, the study tested without allowing overtopping as wave energy is dissipated as much, and reflection coefficients is measured low when it is allowed. However, quantitative analysis on overtopping will be possible with wave run-up observation value by test models.

3 TEST RESULT AND ANALYSIS

There are higher wave results in greater maximum wave run-up using the measured data and will therefore end up facing its limit. Thus, the maximum run-up height measured through the hydraulic model test was analysed by making it dimensionless as the height of incident wave.

Table 3. Results from the calculation of wavelength (L), wavenumber (k), and relative depth based from the wave period (T).

Periods [T] (sec)	Depth [h] (m)	wavelength [L] (m)	wavenumber [k] (1/m)	Relative depth [kh] (Non-dimension)	Sea area
6	0.5	13.16	0.477	0.239	Shallow sea area
5	0.5	10.92	0.575	0.288	Shallow sea area
4	0.5	8.67	0.725	0.363	Middle depth
2	0.5	4.06	1.549	0.775	Middle depth

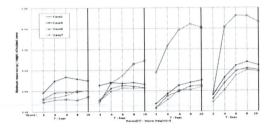

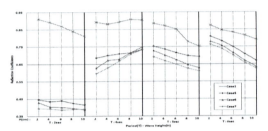

Figure 3. Non-dimension maximum wave run-up ratio about period (T)-wave height (H) in Case 5 · 6 · 7.

Figure 4. Reflection coefficient about period (T)-wave height (H) in Case 5 · 6 · 7.

3.1 Effect of increasing void ratio on wave energy

For Case 5 to Case 7, thin plate shape whose pore space rate rises by 5% with similar frictional area, and Case 8 to Case 10 that have a circular cylindrical shape, three tests were performed with four periods and five wave heights. Figure 3 to 6 compared dimensionless maximum run-up ratio and reflection coefficients by wave height depending on period of Case 5 to Case 10 with Case1 that has no riprap shape. Both shapes showed low reduction of maximum run-up in the short period. This is due to the fact that the water that can be stored among pores as pore space rate increases, and the wave comes back before the stored water becomes a run-up. A longer period and larger pore space rate produces lower maximum wave run-up and reflection coefficients, meaning that the larger pore space rate in a period have greater reduction effect on wave energy. Case 5 to Case 7, which is thin plate shaped marked a lower maximum wave run-up and reflection coefficients than Case 8 to Case 10, which are circular cylinder. This happens because water in the pores between thin plates is not stored but stirred up by hitting the plate, causing high run-up, lower energy reduction effect, and high reflection coefficients. Even when frictional area and pore space rate increase, the arrangement of ripraps should be efficient to reduce energy physically, in order to enhance wave energy reduction. When revetment structures are installed, it will be efficient to reduce wave energy if ripraps should be arranged in the vertical direction of incoming wave to reduce the area that the wave hits.

3.2 Effect of increasing frictional area on wave energy

Figure 7 to 8 show dimensionless maximum wave run-up ratio and reflection coefficients of Case 4 (riprap), Case 5 (thin plate), and Case 8 (circular cylinder)

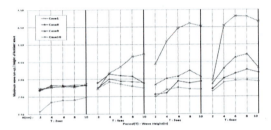

Figure 5. Non-dimension maximum wave run-up ratio about period (T)-wave height (H) in Case 8 · 9 · 10.

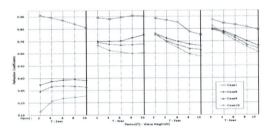

Figure 6. Reflection coefficient about period (T)-wave height (H) in Case 8 · 9 · 10.

whose frictional area increases 50% with the same pore space rate. Case 5 (thin plate) with the frictional area of 15.2 m² per unit volume recorded higher maximum wave run-up than Case 4 (ripraps) at 10.2 m². This happens, as mentioned above, because of the wave stirred up at pores between plates with vertical arrangement of ripraps that have thin plate shape. Case 8 with the greatest frictional area showed low maximum wave run-up and reflection coefficients, whilst Case 4 with the smallest frictional area showed the highest flexibility. This means that the rise of the frictional area is effective in wave energy reduction, and that the arrangement of ripraps is significant.

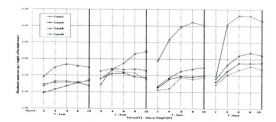

Figure 7. Non-dimensional maximum wave run-up ratio about period (T)-wave height (H) in Case 4 · 5 · 8.

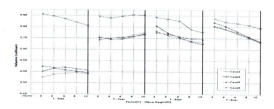

Figure 8. Reflection coefficient about period (T)-wave height (H) in Case 4 · 5 · 8.

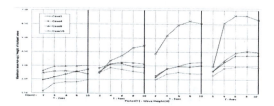

Figure 9. Non-dimension maximum wave run-up ratio about period (T)-wave height (H) in Case 4 · 6 · 10.

3.3 Effect of increased frictional area and void ratio on wave energy

Figure 9 to 10 presented dimensionless maximum wave run-up ratio and reflection coefficient graph of Case 4 (riprap), Case 6 (thin plate), and Case 10 (circular cylinder) whose frictional area increases 50% and void ratio 5%. Case 6 with greater frictional area and void ratio recorded higher maximum wave run-up than Case 4. Maximum wave run-up of Case 6 was the highest at a short period because of the riprap arrangement as described above. Case 10 has the greatest frictional area and pore space rate recorded, which presented a low maximum wave run-up and reflection coefficients. This means that the increase of frictional area and pore space rate is effective in reducing wave energy if the riprap arrangement is efficient in reducing energy.

4 DEVELOPMENT OF REFLECTION COEFFICIENTS DIAGRAM

As a result of the test, difference in the efficiency of reflection coefficient reduction of riprap arrangement where frictional area increases was merely 0.4% to 2.0%, whilst difference in the efficiency of reflection coefficient reduction of riprap arrangement where

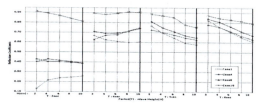

Figure 10. Reflection coefficient about period (T)-wave height (H) in Case 4 · 6 · 10.

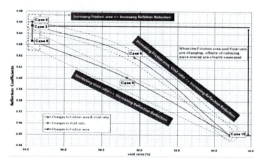

Figure 11. Changes in the reflection coefficient by variation of void ratio.

pore space rate increases was 3.7% to 9.8%, having a greater effect on wave energy than the former. Efficiency difference of reflection coefficient reduction of riprap arrangement where both frictional area and pore space rate increase was 8.1% to 11.4%. This said riprap arrangement has a great and remarkable effect on wave energy reduction. Figure 11 states that the measurement of reflection coefficients was clear with the change in the measured reflection coefficients by riprap arrangement. Therefore, the hydraulic model test result of Case 4 (riprap), Case 6 (thin plate), and Case 10 (circular cylinder) where frictional area and pore space rate have increased were used to develop revetment scale and reflection coefficients diagram.

In the graph, y axis means reflection coefficient and x is surf similarity parameter. The design condition of the study is 20 with five heights (H) and four periods (T). As one period has the measured value of five heights, simplification process is needed to show the conditions in one graph. Since Surf Similarity Parameter presents incident wave's height, period, and structural slope as one dimensionless coefficient, it is widely used to study wave run-up and reflection coefficients (Park Seunghyun, 2007). Surf Similarity Parameter (ξ) is calculated by Formula 3.

$$\xi = \frac{tan\alpha}{\sqrt{S}} = \frac{tan\alpha}{\sqrt{\frac{2\pi H}{gT}}} \tag{3}$$

where, α = angle of structure (θ), H = Wave Height (m), and T = Period (sec).

As a result of the test by riprap shape, two seconds recorded a big difference with measured reflection coefficients of other period, and different tendency from other reflection coefficients. Therefore, two types

145

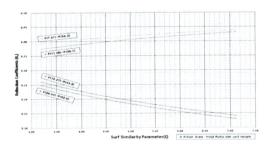

Figure 12. Reflection coefficient diagram (period = 2 sec).

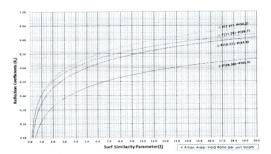

Figure 13. Reflection coefficient diagram (all periods, except for period 2 sec).

of revetment scales and reflection coefficient diagrams were developed, one using two seconds and the other, not using the two-second period. Figure 12 is the reflection coefficient diagram that can be applied to two seconds. In the graph, F of ① F (7.21)-P (34.3) means frictional area and F (7.21) represents the friction rate of 7.21 m². P is pore space rate and P (34.3) means the rate is 34.3%. ② F (11.28)-P (39.7), ③ F (15.27)-P (44.8) and ④ F (26.50)-P (52.6) also present frictional area and pore space rate. When the surf similarity parameter is the same, greater frictional area and pore space rate mean lower reflection coefficients from ① to ④. When surf similarity parameter of a certain coast is determined using the relationship curve of reflection coefficients and surf similarity parameter, whilst frictional area and pore space rate of riprap revetment are calculated, reflection coefficients can be estimated.

Inversely, the frictional area and pore space rate of riprap revetment can be determined and will allow having target reflection coefficients after wanted reflection coefficients is decided

Figure 13 is the curvilinear relationship of reflection coefficients and surf similarity parameter for all periods excluding two seconds. Similar to Figure 12, F means frictional area and F (7.21) represents that the friction rate is 7.21 m² whilst P is pore space rate, and P (34.3) means that pore space rate is 34.3%. In this graph as well, when surf similarity parameter of a certain coast is decided using the curvilinear relationship of reflection coefficients, and surf similarity parameter and frictional area and pore space rate of riprap revetment are calculated, reflection coefficients

(a) Before the installation of the anti-erosion revetment	(b) Three months after the installation of anti-erosion revetment

Figure 14. Before and after the installation of the anti-erosion revetment.

can be estimated. Inversely, frictional area and pore space rate of riprap revetment can be determined; target reflection coefficients will be obtained after the wanted reflection coefficient is decided.

Beach erosion and breakwater collapse resulting from wave and gravels generated by erosion have been a problem. Accordingly, the region, which calls for countermeasures to prevent coastal disasters, was selected as a test bed, and reflection coefficient diagram was used and riprap revetment developed by the study was designed and applied. Figure 14 shows pictures of before and after 3-month construction, showing no further development of erosion. Although it was difficult to reach a precise analysis and conclusion due to the short monitoring time, it is possible to conclude that quantitative evaluation is needed through regular monitoring of continuous coastal changes that occur due to erosion and sedimentation.

5 CONCLUSIONS

The study manufactured ten types of riprap models with various frictional areas and void ratio, analysed the effect of riprap arrangement and shapes on wave energy through the hydraulic model test and developed reflection coefficients diagram based on the analysis result.

(1) The analysis of the effect of increase in frictional area on wave energy when pore space rate is the same showed that Case 8 with the greatest area marked the lowest maximum wave run-up, and reflection coefficients for all periods and wave heights. Therefore, frictional area increase was effective in wave energy reduction.

(2) The analysis of the effect of increase in pore space rate on wave energy when frictional area is the same showed that Case 7 and 10 with the greatest pore space rate were evaluated to be the most effective in reducing the wave energy. However, when the period is short, water stored at a big pore space rate was analysed to strengthen the maximum wave run-up.

(3) The analysis of the effect of change in frictional area and pore space rate on wave energy showed that greater frictional area and pore space rate is effective in reducing wave energy. However, the riprap arrangement with small frictional area and pore space rate that has small vertical contact with wave direction is found to be more effective in reducing wave energy than the riprap arrangement that has big vertical contact with wave direction, even though it has big frictional area and pore space rate. This means that the riprap arrangement is also considered critical in wave energy reduction.

(4) The study used reflection coefficient diagram and performed the design. As a result, reflection coefficients and revetment scale that match the design condition could be determined without doing the hydraulic model test. It took less time for the design, as the design proceeded without the hydraulic model test, and uncertainty of design factors will also decrease as the diagram based on test observation value was used instead of estimation to calculate reflection coefficients.

ACKNOWLEDGMENT

This research was supported by a grant(13SCIPS04) from Smart Civil Infrastructure Research Program funded by Ministry of Land, Infrastructure and Transport (MOLIT) of Korea government and Korea Agency for Infrastructure Technology Advancement (KAIA).

REFERENCES

Anderson, O.H., Juhl, J. and Sloth, P (1992) Rear side Stability of Berm Breakwaters, Proc. 23rd Coast. Engrg. Conf., ASCE, pp. 1020–1029.

Choi, D.S. (2008) Wave Transformation and Overtopping Characteristics due to the Shape of Permeable Coastal Structures, Master thesis, Gyoengsang University, Jinju city, Gyoengsangnam-do, Korea.

Goda, Y. and Suzuki, Y. (1976) Estimation of incident and reflected waves in random Wave experiments. Proc. of 15th Int. Conf. on Coastal Eng., ASCE, New York, pp. 828–865.

Han, J.M., Kim, Y.M., Park, M.G. and Lee, J.S. (2006) Development of Wave Dissipation Block for Nature Friendly Structures, Korean Society of Civil Engineers regular workshop presentation file, pp. 245.

Han, J.M., Lee, J.S., Kim, H.S. and Sin, J.W. (2005) Development of Environmental Preservation Structures for Protection of Shoreline Erosion, Korean Society of Civil Engineers regular workshop presentation file, pp. 703–705.

Kim, K.J. (2010) Analysis of Efficiency and Improvement for Eco-Friendly Coastal Environment Block, Ph D. dissertation, Joongbu University, Geumsan county, Chungcheongnam-do, Korea.

Kim, S.W. (2000) Study on the Characteristics of Armor Block in Coastal Structure: In the Basis of Sloping Type Breakwaters, Master thesis, Yonsei University, Seoul, Korea.

Kim, I.C., Park, Y.W., Yoo, C.H. and Kweon, H.M. (2003) Experimental Study on Hydraulic Characteristics of Wave Dissipating New Armor Unit, International Journal of Ocean Engineering and Technology, 14, pp. 87–97.

Kim, H.Y., Han, K.J., Jeong, S.M. and Kim, K.D. (2012) Analysis of the Wave Energy Reduction Effect by Different Artificial Riprap Arrangement, Journal of The Korean Society of Hazard Mitigation, 12(5), pp. 241–249.

Lee, W. (2008) Wave Overtopping Characteristics of Rubble Mound Revetment, Master thesis, Hanyang University, Seoul, Korea.

Ministry of Land, Transport and Maritime Affairs (2010) Green construction technology for the Costal Zone.

Park, J.K. (2003) Overseas Case Study for Protection from Coastal Erosion with Beach Management System, Master thesis, Choongang University, Seoul, Korea.

Park, S.H., Lee, S.O., Jung, T.H., Cho, Y.S. (2007) Experimental Study on Reduction of Run-Up Height of Sloping Breakwater due to Submerged Structure, Journal of The Korean Society of Hazard Mitigation, 7(5), pp. 187–197

Ryu, Y.U., Lee, J.I. and Kim, Y.T. (2009) Behavior of Overtopping Flow of Caisson Breakwater with Dissipating Block: Regular Wave Conditions, International Journal of Ocean Engineering and Technology, 21(1), pp. 54–62.

Shankar, N.J. and Jayaratne, M.P.R. (2003) Wave run-up and overtopping on smooth and rough slopes coastal structure. Ocean Eng, pp. 221–238.

Sumer, B.M. and Fredose, J. (2000), "Experimental study of 2D scour and its protection at a rubble-mound breakwater" Coastal Engineering. Elsevier. Vol. 40. pp. 59–87.

Tompson, D.M. and Shuttler, R.M. (1975) Riprap design for wind wave attack, A laboratory study in random waves. Willingford, EX 707.

Van Dijk, B. (2001) The Rear slope stability of Rubble Mound Breakwaters. Ms thesis, Delft U of Tech.

Van Gent, R.A. and Pozueta, B. (2004) Rear-side stability of Rubble Mound Structures, Proc 29th Coast. Engrg. Conf., ASCE. pp. 3481–3493.

Verhagen, H.J., Van Dijk, B. and Nederpel, (2003) Riprap Stability on the Inner Slopes of Medium-Height Breakwaters. Proc. Coast. Struc. Portland. pp. 213–222.

Computer simulation and CAD/CAE

Advances in Civil Engineering and Building Materials IV – Chang et al (eds)
© 2015 Taylor & Francis Group, London, ISBN: 978-1-138-00088-9

Analysis of a construction scheme for neighbourhood metro tunnels with two different inner sections

Xiangxing Kong

The First Highway Survey and Design Institute of China Communications Construction Company Ltd., Xi'an, China

ABSTRACT: Based on the characteristics of different inner sections and small clear spacing, 2D-FEM was applied to simulate and analyse several construction schemes, of which the shield method was adopted in a left tunnel with a small inner section and the double-heading construction method or cross diaphragm method was adopted for the right tunnel with a large inner section. The results show that compared with the construction scheme with which the tunnel with a small inner section was firstly excavated, the method with which that the tunnel with a large inner section was first accomplished could evidently reduce the disturbance effect on the surrounding rock, and decrease the settlements of ground. In addition, compared with the center cross diagram method (CRD), the double-heading construction method has advantages in utilizing the self-supporting ability of surrounding rock and the supporting ability of preliminary lining, and also improving the safety margin of the second lining.

Keywords: construction scheme, neighbourhood metro tunnels, numerical simulation, mechanical analysis.

1 INTRODUCTION

Used in the city subway, underground tunnels are taken to two equal sections [1], only meeting special engineering needs it is designed with two unequal sections, for example crossover, contact lines and stop (deposit) car line and so on. The tunnel with parallel non-large section has carried out preliminary research. Through the establishment of a numerical model [2], the influence of the supporting mechanical behaviour and stability of the surrounding rock of an unequal span highway tunnel have been analysed; also, construction mechanics and surface settlement of a double line tunnel with antisymmetric sections have also been studied [3], however further study is still needed.

In order to meet the two lane and the right line stop line section expanding the engineering function, the left line of a small section tunnel with a diameter 6 m is constructed by the shield method, a large section tunnel right line of 9.31 m high and 11.24 m wide is constructed by the new Austrian tunnelling method. A hole for the shield and another hole for the new construction scheme of new Austrian tunnelling method (NATM) are selected. For solving the problem ofl linear programming, the small spacing tunnel are researched and applied. For example, Shenzhen rail transit technology park to Baidanzhou interval tunnel is about 0.5–1.2D (D as the tunnel span, similarly hereinafter) [4]. However, the engineering distance parallel to the tunnel is 4.247 m, only 0.38D, far less than the specified 1.0D current code for the design of the metro in China; this

will bring considerable engineering difficulties and risks.

2 GEOLOGICAL CONDITIONS

There are artificial fill, including quaternary late Pleistocene Aeolan loess soil, a new residual, late Pleistocene, Middle Pleistocene alluvial silty clay and sand soil. The interval tunnel passes through the upper layer. The new loess and paleosol are in the upper, clay soil and sand soil are in the lower, physical and mechanical properties of concrete are around the base soil.

3 NUMERICAL SIMULATION MODEL

The calculation program uses a two-dimensional plane strain model. It is assumed that the soil belong to Drucker-Prager strength criterion and isotropic hardening elastoplastic constitutive model of soil [5]. The shield segment is simulated by a beam element, and advanced grouting pipe is simulated by the quadrilateral element to improve the stability of the surrounding rock, and the second lining is simulated by beam element. In order to ensure accuracy, the dense units around the tunnel are used in the construction scheme of a finite element model as shown in Figure 1.

According to the Saint Venant's principle and the actual needs of the calculation model, the horizontal direction is 4 times diameter, vertical direction is 3

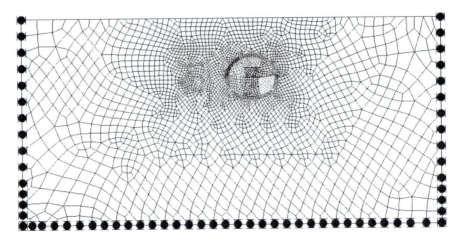

Figure 1. Mesh of finite element model.

Table 1. Parameters of support structures.

Soil/ Structure	Density/ (kN·m⁻³)	Elastic modulus/ MPa	Section area/ m²	Cohesion/ kPa	Inner grazing angle/(°)
Soil	18.6	10.3		36	24
Shield lining	25	34500, $\eta = 0.8$	0.3	–	–
Advanced grouting soil	23	22.5	–	54	37.5
Primary lining	25	23000	0.35	–	–
Geogrid shotcrete	25	34500	–	–	–
Secondary lining	25	30000	0.5	–	–

times diameter. Vertical direction is down to the ground surface. Model boundary conditions on the left and right sides are given X displacement constraint, and the bottom surface is given Y direction displacement constraint. According to the data of engineering geology, soil and calculation of related parameters are shown in Table 1.

4   CONSTRUCTION SCHEMES SIMULATION

The left line consisting of small section tunnel excavation diameter is 6.0 m, C50 concrete segment thickness is 0.3 m, the impermeability grade of S10, specific ring lining section shown in Figure 2. The large sections of the NATM tunnel excavation in the right line are 9.31 m high, 11.24 m wide, has a C25 lining thickness 0.35 m, and an HPB335 thick 0.25 m steel grille. In addition, in the loose loess shallow subway tunnel excavation, construction of a pre grouting pipe in the tunnel arch can improve the condition of the surrounding rock to ensure the stability of the tunnel face, and it is very important for increasing the stability of the tunnel. The specific scheme is diameter $\varphi 42 \times 3.5$

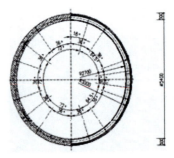

Figure 2. Segment cross-section of shield method tunnel.

(thickness 3.5 mm) grouting pipe, length 2.5 m, angle 15°, circumferential spacing 0.3 m, vertical spacing of 1.0 m; pressure injection of cement – water glass grout, grouting pressure control is 0.6–3.5 MPa.

The CRD method and the double side drift method are suitable for the large section and small spacing shallow buried loess tunnel. Following is the specific construction process of the dynamic simulation:

(1) the small section tunnel in the left line first then the right line of the large section tunnel by the double side heading method.

At first, the small section tunnel in left line is completed, simulation step of shield construction includes 2 steps. Then double side heading method construction process is simulated into 13 steps.

(2) large section of double side heading method right tunnel first then the shield small section of the left tunnel.

(3) the right line of large section tunnel by CRD method first then the small section tunnel in the left line.

5   RESULTS AND ANALYSIS

(1) mechanics analysis for large section tunnel in the right line.

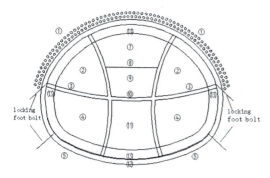

Figure 3. Construction sequence of double-heading method.

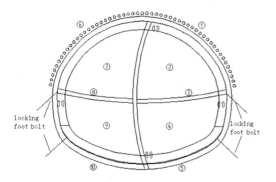

Figure 4. Construction sequence of CRD method.

Table 2. Results of preliminary lining axial stress (unit: kN · m).

Construction scheme	Left hance	vault	Right hance	Ratio/%		
				L. hance	vault	R. hance
Scheme 1	1263.0	761.3	1239.0	−6.4	−5.5	−5.4
Scheme 2	1350.0	835.1	1310.0	−	−	−
Scheme 3	1330.6	826.5	1295.0	−1.4	−0.6	−1.1

Ratio=(results of the selected scheme- results of Scheme 2)/ Scheme 2.

According to Table 2, Scheme 2 produced in construction of initial support axis force is bigger than Scheme 3, but the difference is small, more than Scheme 1. As the hole right arch waist, Schemes 1 and 3 respectively are 5.4%, 1.1%. The first through the NATM tunnel with large cross section of shield tunnel after excavation of small section of initial support axial force is bigger significantly than the first through hole after digging a hole, and the large section tunnel with double side heading method results in maximum. Therefore, Scheme 2 can maximize the tunnel lining support ability.

(2) mechanics analysis for small section tunnel in left line

According to Table 3, in scheme 2 segment lining negative moment is the smallest. The first segment lining moment through the NATM tunnel with large cross section is smaller, the results of large section

Table 3. Calculation results of segment moment (unit: kN · m).

Construction scheme	Left hance	vault	Right hance	Ratio/%		
				L. hance	vault	R. hance
Scheme 1	7.8	10.2	8.1	290.0	251.7	326.3
Scheme 2	2.0	2.9	1.9	−	−	−
Scheme 3	4.5	3.9	5.6	125.0	34.5	194.7

Table 4. Calculation results of ground settlement (unit: mm).

Construction scheme	Left hance	vault	Right hance	Ratio/%		
				L. hance	vault	R. hance
Scheme 1	9.2	12.1	15.6	16.5	14.2	13.0
Scheme 2	7.9	10.6	13.8	−	−	−
Scheme 3	8.3	11.1	14.2	5.1	4.7	2.9

tunnel with double side heading method produces are minimum. Therefore, Scheme 2 is most conducive to the safety of tunnel linings.

(3) analysis of surface displacement

The subway is usually built in the city population and built-up area, so the subway tunnel on the ground for the uplift and subsidence have strict requirements at the same time, more attention should be taken to and control of builders and managers. According to Table 4, the minimum ground surface settlement scheme, but the difference is small and less than 1. In the middle of the surface settlement, 1, 3, and 2 respectively are in the 14.2% scheme, 4.7%.

6 DISCUSSION

The tunnel is located in the loose loess strata, linking NATM tunnel Scheme 1 larger disturbance to the surrounding rock. It will have a larger surface deformation. Compared to that, after the excavation of the small section tunnel using the shield method in Schemes 2 and 3 through grouting, chemical grouting and reinforcement and enhancing the overall stability of the surrounding rock of the tunnel construction, measures such as freezing, compared to the former can be more effective, more timely control of surface deformation and ensure the stability of the intermediate soils.

7 CONCLUSION

The left small section tunnel constructed using the shield method and the right section tunnel constructed using the double side heading method or CRD method with small spacing parallel sequence is researched.

(1) constructed firstly by NATM tunnel with large section and small section tunnel boring shield

method construction scheme has less disturbance to the surrounding rock, and the first through hole after the excavation of shield method NATM hole scheme, will be more conducive to reducing the vault crown settlement, stress control surface deformation and middle soil body.

(2) CRD method is less safety and reliable than double side heading method first large section tunnel.

(3) The tunnel is located in the loose loess strata. Compared to NATM, after the excavation of small section tunnel by shield method, it can improves by chemical grouting to enhance the overall stability of surrounding rock in tunnel construction, compared to the former it will be more effective and have enough time to control surface deformation and ensure the stability of the intermediate soils.

REFERENCES

[1] Liu Bo, Tao Longguang, Ding Chenggang, et al. Prediction for Ground Subsidence Induced by Subway Tube Tunneling [J]. Journal of China University of Mining & Technology, 2006, 35(3): 356–361.

[2] Deng Jian, Zhu Hehua, Ding Wenqi. Finite Element Simulation of Whole Excavation Operation of a Unequal-span Double-arch Tunnel [J]. Rock and Soil Mechanics, 2004, 25(3): 477–480.

[3] Jiang Xiaorui. Finite Element Simulation of Construction Method of Crossover Tunnel with an Unequal-span Double-arch [J]. Railway Standard Design, 2009, 10: 103–105.

[4] Gong Jianwu. Study on Construction Mechanical Behavior of highway Tunnel with Large Section and Small Spacing [Ph. D. Thesis][D]. Shanghai: Tongji University, 2004.

[5] Kong Xiangxing, Xia Caichu, Qiu Yuliang, et al. Study on Construction Mechanical Behavior of Parallel Small Spacing Metro Tunnels Excavated by Shield and Cross Diaphragm (CRD) Method in Loess Region[J]. Rock and Soil Mechanics, 2011, 32(2): 516–524.

Advances in Civil Engineering and Building Materials IV – Chang et al (eds)
© 2015 Taylor & Francis Group, London, ISBN: 978-1-138-00088-9

Simulation of the vibratory soil compaction under the action of periodically changing elliptic external force

V.V. Mikheyev
Omsk State Technical University, Omsk, Russian Federation

S.V. Saveliev
Siberian State Automotive and Road Academy, Omsk, Russian Federation

ABSTRACT: This article is devoted to the model investigation of the interaction between the drum of vibratory roller and the soil to be compacted in the case of the periodic external force of an elliptic type.

It is shown that the usual harmonic vibratory force is not as effective for compaction as a periodic force with a more complex functional dependence on timing. The suggested approach is illustrated with a numerical analysis of a simple but non-trivial example of the plastic deformation of a soil specimen under an elliptic type of force. The simulation was performed using the approximation of lumped parameters. The growth of plastic deformation under an elliptic type of periodic force that leads to enhanced compaction is proven. The dependence of the plastic deformation of the soil on the parameters of an external elliptic force was also shown as a result of computer simulation.

1   INTRODUCTION

Soil compaction is a very important part of the preliminary stage of road construction. The significance of its quality is obvious from the number of researches (Kézdi, A., 1962, Forssblad, L., 1977, Townsend, F.C., Anderson, B., 2004). It was shown that durability and reliability of the road surface increases almost exponentially with an increase in achieved density of compacted soil. In relation to that two main problem arise.

1) How to compact the soil by an external dynamic action to the necessary level of density?
2) How to achieve that goal with the highest possible efficiency?

Vibratory soil compaction is nowadays one of the most frequently used types of soil modification in road building. Numerous researches were performed (Kloubert, H.J., 2001, Yoo, T.S., 1975, Massarsch, K.R.; Fellenius, B.H., 2002) to find out the ways to make vibratory compaction more effective, in other words, to achieve the necessary densities and mechanical properties of the soil while spending less time and resources.

That makes important the development of the model descriptions for the dynamic interaction between the working body of the roller and a mass of compacted soil. One must take into account that properties of the soil dramatically change during compaction so the feedback becomes an important issue, some type of adaptation of the external action to the properties of the soil becomes a necessary function.

The most common way of representing the periodic change in the force applied to the soil during compaction is a harmonic function.

$$F_{harm}(t) = F_0 \sin(\omega t + \varphi_0) + P \qquad (1)$$

Here we assume vertical direction of dynamic action. P is representing static force (weight of the machine usually). In most number of practical applications P is much less then F_0 because the mass of vibratory rollers are not as great as of static ones.

We will consider below the easiest possible model of a 'working body – compacted soil' interaction under the approximation of lumped parameters. In the framework of this suggested approach the interaction between the working body of the roller and mass of the soil is restricted to the so called 'effective volume' or 'effective mass' of the soil. The size and shape of the effective volume depend on the type of the soil, and basically on its viscosity, thickness of the layer and contact surface with the drum. All of these determine the amount of the soil that is forced into the motion right below the working body under the action of the force. The need for modelling is justified by the specific features of soil mechanics when an exact description of the processes in the depth of the soil is difficult. Lots of theoretical and experimental researches have been performed in order to investigate the properties of the soil during compaction and the distribution of stress in the depth of the soil (Seed, H.B.; Chan, C.K., 1959, Schmertmann, J.H. 1974, Schmertmann, J.H. 1985, Mooney et al. 2010).

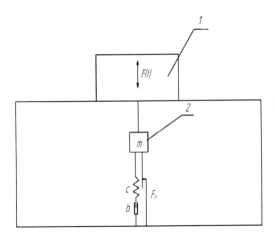

Figure 1. Graphic interpretation of a simple model of inter-action between working body 1 and a layer of the compacted soil 2 in the approach of lumped parameters (side view).

Therefore the total stiffness of the effective volume is c, total coefficient of the viscous friction is b and the mass which is considered to be a heavy point situated in the centre of gravity of the effective volume is m. Displacement of the mass under the action of periodic force is described by the equation

$$m\ddot{x} + b\dot{x} + cx = F(t) \qquad (2)$$

Here x presents displacement of the centre of mass 2 which depends on the displacement of the working body 1 and $F(t)$ is the periodic function that defines the type of dynamic action applied to the soil. Periodic harmonic force presented in (1) may be mentioned as an example. This is a model description because in reality the periodic force of the exciter is applied to the drum placed between the soil and the heavy frame of the roller. As a result the motion of the frame must be taken into consideration for a precise analysis. However, for qualitative consideration the motion of the roller's frame may be neglected since its weight exceeds the weight of the system 'drum – soil' by twenty to thirty times.

Compaction of the effective volume is obtained when displacements caused by the external force may be treated as a plastic deformation. The condition on contact stress which is required for that process is

$$\sigma_{pl} \le \sigma(t) \le \sigma_{destr}, \qquad (3)$$

where σ_{pl} is a yield stress (value of the contact stress that leads to the compaction with the start of the plastic deformation in the soil).

$$\sigma(t) = F(t)/s(t)$$

is a simultaneous value of contact stress and σ_{destr} is the contact stress that leads to the destruction of the soil specimen (ultimate compressive strength or rupture point). If that condition is not satisfied

$$\sigma_{pl} \ge \sigma(t)$$

then the mass of the soil under the roller's working body performs forced oscillations without compaction.

Even if the harmonic force can achieve the necessary values for the plastic deformation then the compaction takes place only during the part of the period. Also, if the value of F_0 is raised in order to satisfy condition (3) earlier in the period, then the higher limit in (3) may be overcome and the soil destruction may start. The effective part of the power of the vibratory working body that stresses soil and leads to the compaction, can be approximated by the expression for the work of the external force on the half-period of the vibration:

$$W_{plastic} = \int_{t_0}^{T/2 - t_0} (F(t) - F_{plastic})dx(t)$$

where $F(t)$ is a periodic force, for instance (1) and t_0 is found from the condition which determines the start of the plastic deformation $F(t) - F_{plastic} > 0$ and $x(t)$ is the value of plastic deformation. Here we assume that the maximum stress does not exceed σ_{destr}. For the periodic force of type (1) t_0 may be evaluated as

$$t_0^{harm} = \frac{1}{\omega} \arcsin(\frac{F_{plastic} - P}{F_0})$$

which gives the part of time in the semi-period when the soil is stressed according to condition (3) equal to

$$\frac{\pi}{\omega} - 2t_0^{harm} = \frac{1}{\omega}\left(\pi - 2\arcsin(\frac{F_{plastic} - P}{F_0})\right)$$

All of the above makes it possible to find the type of periodic force that can both achieve the necessary value of stress as fast as possible and maintain its value between limits required in (3) as long as possible during the period of the oscillation without exceeding ultimate compressive strength.

Harmonic function is not the best candidate for that because of the relatively slow rate of change in the beginning and the end of both of semi-periods and fast rate of change in the middle. In other words the necessary value of force is to be achieved slowly and sufficient stress is applied to the soil for a short period of time. Although traditionally harmonic function (1) is widely used because of easy technological realization (Mooney, M.A. et al. 2010, Yoo, T.S., 1975) in vibratory exciters.

Anyhow there are researches who prove a higher efficiency of the vibratory eccentric exciters of an elliptic type in comparison to purely harmonic case. The higher efficiency and lower energy consumption of such exciters were shown theoretically and proven by experiments (Kurita, Y. & Muragishi, Y. & Yasuda, H. 1998, Дудкин М.В. & Кузнецов П.С. 2005).

2 ELLIPTIC TYPE OF PERIODIC FORCE

Since harmonic functions as a realization of periodic force represent an easy, convenient, common but not the most effective way to obtain the most efficient compaction pattern of periodic action let us consider more general approach for obtaining this.

The obvious generalization of the harmonic function (on the unitary circle) by the second order curve on the plane may be given by a unitary ellipse (one of the semi-axis equals 1). The reasons for the choice are following:

1) The ellipse is a generalization of the circle (parameter $e = 0$)
2) There is a parameter e – eccentricity which may be varied in order to get the necessary view of the harmonic function and because of that to adapt the periodic force for the changing properties of the soil.

The equation of the ellipse in canonical Cartesian coordinates is

$$\frac{x^2}{a^2} + \frac{y^2}{b^2} = 1$$

and in polar coordinates (the centre of the ellipse coincides with the beginning of coordinate axis) looks like

$$\rho(\varphi) = \frac{b}{\sqrt{1 - e^2 \cos^2 \varphi}}$$

We will use the ellipse with parameter $b = 1$. That leads to the opportunity of choice for the time dependence of the periodic force applied in vertical direction to the soil as function

$$F(t,e) = \frac{F_0 \sin(\omega t)}{\sqrt{1 - e^2 \cos^2 \omega t}} + P \qquad (4)$$

The action of the force can be adjusted in order to achieve the necessary effect by the variation of the eccentricity parameter e ($e = 0$ transforms (4) to (1)). So one may obtain time dependence of the periodic force that is more appropriate for the efficient compaction. The graphs on Figure 1 present different patterns of function

$$f(\varphi,e) = \frac{\sin(\omega t)}{\sqrt{1 - e^2 \cos^2 \omega t}} \qquad (5)$$

depending on parameter e for the values of $e = \{0, 0.5, 0.75, 0.875, 0.9375, 0, 96875\}$ that demonstrates a higher rate of growth at the beginning and the end of the semi-period and a flatter top in the middle for greater values of the parameter e. That makes it possible, as it will be shown below, to prolong the time span for the soil to stay under the action of a compacting force of an almost constant value.

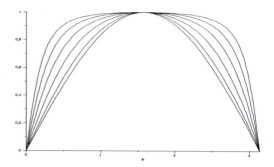

Figure 2. Graphic dependence of $f(\varphi, e)$ in (5) on $0 < \varphi < \varpi$ (inner graph represents $f(\varphi, 0) = \sin(\varphi)$) for $e = \{0, 0.5, 0.75, 0.875, 0.9375, 0, 96875\}$.

Time t_0 for elliptic type of force can be easily found after elementary calculations as

$$t_0^{ell} = \frac{1}{\omega} \arctan\left(\sqrt{\frac{(1 - e^2) F_0^2}{F_0^2 - (F_{plastic} - P)^2}}\right).$$

A comparison between a purely harmonic function (here we assume $\varphi_0 = 0$ without loss of generality) and an elliptic case of force gives the difference of the starting times

$$t_0^{harm} - t_0^{ell} = \frac{1}{\omega}\left(\arcsin(\frac{F_{plastic} - P}{F_0}) - \arctan\left(\sqrt{\frac{(1 - e^2) F_0^2}{F_0^2 - (F_{plastic} - P)^2}}\right)\right) =$$

$$= \frac{1}{\omega}\left(\arctan\left(\frac{\alpha e^2}{(1 + \sqrt{1 - e^2})(1 + \alpha^2\sqrt{1 - e^2})}\right)\right)$$

which shows that the time that is spent on compaction (plastic deformation) increases with transition from the purely harmonic function to the elliptic one.

Where parameter α is determined by the properties of the soil and the working body of the roller

$$\alpha = \sqrt{\frac{F_0^2}{F_0^2 - (F_{plastic} - P)^2}},$$

maximum amplitude of the periodic force F_0 may be found from the condition $(F_0 - P)/S \leq \sigma_{destr}$ or $F_{0max} = F_{destr} - P$. Here F_{0max}, F_{plasic} and P depend on the state of the soil and the properties of the roller. Since α must be evaluated explicitly for any particular case and after that an appropriate value of eccentricity e may be determined. For instance if the roller mass is $3000\,\text{kg}$ ($P = 30000\,\text{N}$) and the area of the contact surface is $0.1\,\text{m}^2$ then in case of loam with $\sigma_{destr} = 1000000\,\text{Pa}$ and $\sigma_{plasic} = 700000\,\text{Pa}$ we obtain $F_{destr} = 100000\,\text{N}$ and $F_{plasic} = 70000\,\text{N}$. That allows an estimation of $F_{0max} = 70000\,\text{N}$ and since that may be found $\alpha = (49/33)^{1/2} = 1.22$.

So one can find the difference between the times when the plastic deformation of the soil starts under purely harmonic and elliptic external forces.

The table shows that the time (in one period) when the soil is under the external elliptic force necessary for

157

Table 1. Increase of the time when plastic deformation starts in respect to purely harmonic case depending on an eccentricity parameter for $\alpha = (49/33)^{1/2} = 1.22$ (ω is a cyclic frequency of the periodic force).

e	$t_0^{harm} - t_0^{ell}$
0	0
0.5	$0.0868/\omega$
0.75	$0.2484/\omega$
0.875	$0.4192/\omega$
0.9375	$0.5861/\omega$
0.96875	$0.6844/\omega$

plastic deformation increases in respect to harmonic type from $0.1736/\omega$ to $1.3688/\omega$ while the eccentricity parameter grows from 0.5 to 0.96875.

3 NUMERICAL MODELLING OF THE ACTION OF THE ELLIPTIC PERIODIC FORCE ON THE SOIL

The time period when the plastic deformation might occur presents reasonable criteria for a qualitative consideration of the compaction process as it was discussed in Chapter 2. Further investigation of that problem requires a more detailed quantitative analysis. The exact solution of the differential equations that describe the compaction process is not simple because of the previously mentioned plastic deformations which change the initial conditions during compaction. So the numerical analysis can be used to reveal the specific properties of the compaction process even in the simplest cases which will be demonstrated below.

Let us consider the case of periodic force acting on the soil as suggested in (4) and on the effective volume of the soil which also can undergo plastic deformation, in a case when the necessary condition (3) is satisfied.

The model of soil behaviour that will be presented below has simple physical explanation.

Firstly the amount of soil which appears to be under the direct action of the internal force is restricted to the so called 'effective volume' in order to determine the stiffness and damping of the soil mass under the approximation of the lumped parameters.

Secondly the mass and mechanical properties of the effective volume depend on the shape of the volume and the parameters of the soil – elasticity modulus and viscosity.

Thirdly the plastic deformation takes place when the contact stress exceeds the necessary value and after that the elasticity force of the soil vanishes and a dry friction force replaces it. Therefore after the start of the plastic deformation the external force of any type meets the inertia force of the soil, with viscous damping and finally a dry friction instead of an elastic force.

After that, a computer simulation of the soil motion may be performed under following assumptions:

1) The contact surface between the roller drum and the soil is flat which is fairly good approximation even in case of cylindrical roller drum;
2) The simplest shape of the effective volume is a rectangular prism with dimensions $l \times d \times h$, where l, d are length and width of the contact spot, h is the depth of the soil down to the solid base which is considered absolutely rigid;
3) Lumped parameters of the effective volume are constant and do not depend on deformation. That is not absolutely correct because the stiffness of the soil grows and damping lessens with the process of compaction but, in case of small deformations, the rate of change for lumped parameters may be considered slow enough for that assumption to be right. Especially that may be considered correct for the one period of vibration.
4) The parameters are to be computed according to well-known formulas for stiffness of rectangular prism. Stiffness of the soil specimen is:

$$c = \frac{ES}{h} = E\frac{ld}{h},$$

damping (we assume the soil to be homogeneous media where the said prism moves without turbulence and the angle of internal friction is φ) is

$$b = \frac{\mu S}{z} = \mu\frac{l}{\tan\varphi},$$

and finally the mass of the prism is $m = \rho l d h$, where E – elasticity modulus of the soil, μ – coefficient of the dynamic viscosity and ρ is the density of the soil.

The resistance force of the soil to the external action is combined from the inertia force of the soil mass

$$F_{in} = m\ddot{x},$$

the damping force of the viscous friction caused by the movement of the mass in the depth of the soil

$$F_{damp} = b\dot{x},$$

and the elasticity force of the soil caused by deformation

$$F_{el} = cx$$

which is replaced by the force of dry friction $F_{fr} = F_{plasic}$ when the value of surface stress achieves the yield stress and plastic deformation starts as is shown on the diagram below. Here the stress is considered to be equal for any horizontal surface in the depth of the soil specimen. The response of the system against the external action is thought to be linear in respect of the coordinate x and its derivatives.

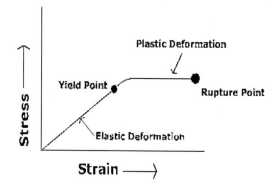

Figure 3. The strain-stress diagram used for the model representation.

Then the motion of the soil mass can be described with a well-known but slightly modified linear differential equation of a harmonic oscillator with damping and dry friction which represents the plastic deformation

$$m\ddot{x} + b\dot{x} + \theta(F(t) - F_{plastic})cx =$$
$$= F(t) - \theta(F(t) - F_{plastic})F_{plastic} + P \quad (7)$$

where $\theta(x)$ is Heaviside θ – function:

$$\theta(x) = \begin{cases} 1, x > 0, \\ 0, x \leq 0 \end{cases}.$$

Equation (6) describes the process taking into account the fact that after the force exceeds the limit necessary for the start of plastic deformation, and the stiffness of the soil vanishes and Hook's law does not work as it is shown in the well-known diagram below.

Here and below x is the displacement of the centre of gravity for the effective volume. Because of the choice of the shape oa the 'effective volume', the displacement of the soil surface x^{surf}, which is usually treated as the sign of compaction is connected by the simple relation

$$x^{surf} = 2x \quad (8)$$

since the centre of the mass is situated in the geometrical centre of the rectangular prism.

The external periodic force is determined by the formula (4) with parameters F_0, F_{plasic} and P which were discussed for example above. The eccentricity parameter remains optional in order to trace the influence of its value on the process of compaction.

The oscillating system will start some sort of forced oscillations with the frequency that coincides with the frequency of the external force. The difference from the 'normal' forced oscillations under the action of harmonic force lies in the moving equilibrium point so that even after external force is 'switched off' the soil will relax to position which differs from its initial position because of the retained plastic deformation.

Simple transformations allow us to rewrite the equation as

$$\ddot{x} + k\dot{x} + \theta(F(t) - F_{plastic})\alpha_0^2 x =$$
$$= f(t) - \theta(F(t) - F_{plastic})f_{plastic} + p \quad (9)$$

with new notations

$$k = \frac{b}{m}, \alpha_0^2 = \frac{c}{m}, f(t) = \frac{F(t)}{m}, f_{plastic} = \frac{F_{plastic}}{m}, p = \frac{P}{m}.$$

Here we assume $F(t)$ being the periodic force of elliptic type:

$$F(t,e) = \frac{F_0 \sin(\omega t)}{\sqrt{1 - e^2 \cos^2 \omega t}}$$

This equation cannot be solved exactly because of the special right side which makes it necessary to express the exact solution through elliptic integrals which may be a problem to compute.

Anyway the numerical solution of the equation may be easily performed for an arbitrary type of the external force at the right side. Standard transition to the finite differences gives

$$\frac{x_{i+2} - 2x_{i+1} + x_i}{(\Delta t)^2} + k\frac{x_{i+1} - x_i}{\Delta t} + \theta(F(t_i) - F_{plastic})\alpha_0^2 x_i =$$
$$= f(t_i) - \theta(F(t_i) - F_{plastic})f_{plastic} + p \quad (10)$$

with the initial condition $x_2 = 0$, $x_1 = 0$ which is equivalent to $x(0) = 0$, $\dot{x}(0) = 0$.

The number of steps is N, which determines Δt, may be found from the condition $\Delta t \ll 2\pi/\omega$.

In order to find the part of the power which is spent to compact the soil after the solution of equation (9) according to (10) one may perform numerical integration as follows

$$P(N\Delta t) = m \sum_{i=1}^{N-1} (x_{i+1} - x_i)\theta(F(t_i) - F_{plastic})(f(t_i) - f_{plastic})$$

Let us perform the calculations for the appropriate values of the parameters of the soil and the roller which are close to the real ones.

The graph represents the model process of plastic deformation for the layer of soil after ten periods of action. Different graphs correspond with the value of the eccentricity parameter e, showing increase in the accumulated deformation with a growth of eccentricity. In that model situation the speed of the roller appeared to be close 0.3 m/s. The value of the x-axis shows the number of steps (how many values of x_i were calculated).

It is obvious that the final plastic deformation is strongly influenced by the eccentricity of elliptic function in (5). Growth of the plastic deformation can be easily explained as a result of the prolonged action of the external force on the soil in each period. The

159

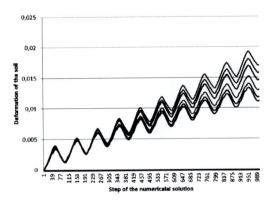

Figure 4. The evolution of the model soil volume's plastic deformation depending on the parameter of the external force (the lowest graph is obtained when $e = 0$, the graphs above it correspond to the value of e growing from 0.5 to 0.96875).

Table 2. Dependence of the power spent on the plastic deformation end efficiency coefficient on eccentricity parameter for ten periods of forced oscillations of the soil mass.

e	Δx_{plasic}, m	W_{total}, J	W_{plasic}, J	τ, %
0	0.0103	2193.36	226.63	10.33
0.5	0.0112	2284.35	252.68	11.06
0.75	0.013	2397.54	303.16	12.64
0.875	0.015	2663.71	367.89	13.81
0.9375	0.017	2705.64	428.19	15.83
0.96875	0.0185	2845.51	487.57	17.13

accumulated deformation shows 50% increase with a change in eccentricity from 0.5 to 0.96875.

Even more important is the issue of energetic efficiency. It is possible that the deformation growth was obtained with a non-proportional increase of the power being spent on it. We suggest that we compare total power with power that was spent on the plastic deformation in each of the cases considered above.

The table below presents the total energy W_{total}, energy that was spent on compaction W_{plasic} and efficiency coefficient τ for the different values of the eccentricity parameter.

The efficiency of the elliptic type of the dynamic action is obvious. It shows greater deformation and lower power that is spent on it. The problem still remains: how to make the working body of the roller act on the soil in the way described above?

That problem may be solved mechanically – by a change of the manner in which the vibration of the drum is exited. A common way of the excitation is the application of unbalanced rotational systems (eccentric weight) with inertia force of rotation on the axis being used as a force of dynamic action on the soil or a planetary type of the exciters. See for instance (Kurita Y. et al. 1998, Дудкин М.В. & Кузнецов II.С. 2005).

The other possible way is not that obvious and may imply the use of intelligent end electronically controlled vibratory excitation systems.

Table 3. The dependence of the degree of compaction on the eccentricity parameter for ten periods of forced oscillations of the soil mass.

e	Δx_{plasic}, m	κ
0	0.0103	0.9489
0.5	0.0112	0.9534
0.75	0.013	0.9626
0.875	0.015	0.9730
0.9375	0.017	0.9836
0.96875	0.0185	0.9917

The last question which has to be discussed here is the value of soil degree of compaction in the framework of the modelling presented above. The degree of compaction is defined as a ratio between the current density of the soil achieved during the compaction process and the recommended density

$$\kappa = \frac{\rho}{\rho_{rec}}.$$

A suggested representation of the effective volume of the soil gives an easy way to evaluate the current degree of compaction. Since the effective volume was chosen as a rectangular prism the degree of compaction depends only on the initial depth of the soil layer h and the current plastic deformation

$$\kappa = \frac{h_{rec}}{h - x^{surf}_{plastic}} = \frac{\rho_0}{\rho_{rec}} \frac{h}{h - x^{surf}_{plastic}},$$

where h_{rec} is the depth of the soil layer in the case of the recommended density ρ_{rec} being achieved starting from the initial density ρ_0 and where $x^{surf}_{plastic}$ is the retained displacement of the soil specimen surface under the roller which is connected to the plastic displacement of the centre of gravity of the effective volume by (8).

In example considered below the starting degree of compaction value is assumed to be equal 0.9 (density $\rho = 2000\,\text{kg/m}^3$), the initial depth of the soil layer is 0.4 m, the mass of the roller is 3000 kg ($P = 30000\,\text{N}$), and thearea of the contact surface is $0.1\,\text{m}^2$, $F_{plasic} = 70000\,\text{N}$ and the elasticity modulus is $E = 25000000\,\text{Pa}$. It gives the following results for the values of the degree of compaction after ten periods of vibration according to Table 3.

4 CONCLUSION

Vibratory compaction under additional surface weight is nowadays considered the most effective way of soil compaction in road and airfield construction. That method was realized in various types of machines, compaction systems and technologies.

The research that was done in this article has shown the advantages of the use of vibratory soil compaction types with periodic forces other than harmonic types. Despite some difficulties in the technical realization of elliptic types of the periodic force, it was proven to be more effective than the commonly used harmonic type of dynamic action for soil compaction.

Definitely, the example considered in this article is simple but illustrative. It was investigated to show the increased efficiency of soil compaction in cases where elliptic external force was applied to the soil even in the framework of the constant lumped parameters approach. The next step of more precise simulation must take into account the dynamic interaction between the roller drum, frame of the roller, and the effective volume of the soil with properties which change as compaction process continues. Also the modelling of the real situation must deal with the distribution of stress in the depth and on the surface on the soil (Schmertmann, J.H. 1974, Schmertmann, J.H. 1985) which can be found from the experimental data.

REFERENCES

Forssblad, L. 1977. "Vibratory Compaction in the Construction of Roads, Airfields, Dams, and Other Projects," Research Report No. 8222, Dynapac, S-171, No. 22, Solna, Sweden.

Kézdi, A. 1962. Erddrucktheorien. Springer Verlag, Berlin/Göttingen/Heidelberg.

Kloubert, H.J. 2001. "New intelligent compaction system for vibratory rollers" Report for IRF in Paris.

Kurita, Y.; Muragishi, Y.; Yasuda, H. Elliptical vibratory apparatus. US patent 6044710 A, 1998.

Massarsch, K.R.; Fellenius, B.H. 2002. Vibratory compaction of coarse-grained soils. Canadian Geotechnical Journal, 39(3) 695–709.

Mooney, M.A., Rinehart, R.V., Facas, N.W., Musimbi, O.M., White, D.J., Vennapusa P.K.R., 2010. Intelligent Soil Compaction Systems. NCHRP REPORT 67 6. Washington, D.C. 2010.

Schmertmann, J.H. 1974. Measurement of In-situ Shear Strength. Proceedings of American Society of Civil Engineers, ASCE, Geotechnical Division, Specialty Conference on In-Situ Measurement of Soil Properties, June 1–4, 1974, Raleigh, NC, Vol. 2. pp. 57–138.

Schmertmann, J.H. 1985. Measure and use of the in-situ lateral stress. In the Practice of Foundation Engineering, A Volume Honoring Jorj O. Osterberg. Edited by R.J. Krizek, C.H. Dowding, and F. Somogyi. Department of Civil Engineering, The Technological Institute, Northwestern University, Evanston, pp. 189–213.

Seed, H.B.; Chan, C.K. 1959. "Structure and Strength Characteristics of Compacted Clays", Journal of the Soil Mechanics and Foundations Division, American Society of Civil Engineers, Vol. 85, No. SM5, pp. 87–128.

Townsend, F.C.; Anderson, B. 2004 "A Compendium of Ground Modification Techniques," Research Report BC-354, pp. 16–60. Florida Department of Transportation (FDOT).

Yoo, T.S., 1975, "A Theory for Vibratory Compaction of Soil", The Dissertation for Degree of Doctor of Philosophy to University of New York at Buffalo.

Дудкин М.В.; Кузнецов П.С., 2005 Динамический анализ эллиптического планетарного вибровозбудителя для дорожных вибрационных катков. Журнал «Вестник ВКГТУ» № 1 / ВКГТУ, Усть-Каменогорск, 2005. – 7 с.

Construction technology

Advances in Civil Engineering and Building Materials IV – Chang et al (eds)
© 2015 Taylor & Francis Group, London, ISBN: 978-1-138-00088-9

A study of the propagation behavior of CFRP-concrete interfacial cracks under fatigue loading

K. Li & S. Cao
Department of Civil Engineering, Southeast University, Nanjing, China

X. Wang
Department of Civil Engineering, Zhengzhou University, Zhengzhou, China

ABSTRACT: The purpose of this paper was to investigate interfacial fatigue crack propagation behaviour in Carbon-Fibre-Reinforced Polymer (CFRP) reinforced concrete members, both experimentally and theoretically. The rate of energy release was chosen to describe the crack's growth. The theoretical method was proposed for calculating the energy release rate of the Fibre-Reinforced Polymer (FRP)-concrete interfacial crack. Modified beam tests were performed to capture the fatigue propagation law of CFRP-concrete interfacial cracks. The results showed that the evolution of interfacial fatigue cracks can be divided into three stages: namely, crack initiation and rapid propagation, steady propagation and instability propagation. The interfacial fatigue crack growth was described in a form of the Paris equation with the variable G/G_c. Combined with the experimental results, a semi-empirical formula was developed for the growth rate of interfacial cracks caused by flexural cracks in CFRP strengthened beams.

Keywords: CFRP, concrete, interface, fatigue crack propagation, energy release rate.

1 INTRODUCTION

It has become an important issue in recent years to extend the fatigue life of concrete structures, which are subjected to cyclic loading, by using FRP composites externally bonded to the structures (Kim & Heffernan 2008; Gheorghiu et al. 2006; Heffernan & Erki 2004). FRP-concrete interfacial debonding is one of the main failure modes of FRP reinforced concrete structures (Chen & Teng 2001). Therefore, it is of great importance to carry out studies on the interfacial crack propagation behaviour. However, the studies on the fatigue properties of FRP reinforced concrete structures have been limited. In particular, the studies on fatigue interfacial crack propagation behaviour of FRP reinforced concrete members were rarely reported (Diab et al. 2009; Ferrier et al. 2005; Nigro et al. 2011). In this paper, fatigue crack propagation behaviour of the CFRP-concrete interface was investigated experimentally and theoretically.

2 THE ENERGY RELEASE RATE OF FRP-CONCRETE INTERFACE CRACK

Let us first consider the end of a delamination in a FRP reinforced concrete flexural member as shown in Figure 1. The crack front is assumed to be originally at O on AB and propagated to O′ on CD, and AD is

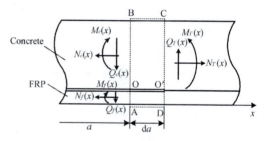

Figure 1. Loads at crack-tip.

the crack growth da. $M_T(x)$, $Q_T(x)$ and $N_T(x)$ denote respectively the bending moment, shear force and axial load applied on the section AB of the entire reinforced member before the crack reaches O. $M_c(x)$, $Q_c(x)$ and $N_c(x)$ denote respectively the bending moment, shear force and axial load applied on the section AB of only the concrete layer after the crack reaches O. $M_f(x)$, $Q_f(x)$ and $N_f(x)$ denote respectively the bending moment, shear force and axial load applied on the section AB of only the FRP layer after the crack reaches O.

The equilibrium equations can be obtained as follows considering that shown as ABCD in Figure 1:

$$Q_T(x) = Q_c(x) + Q_f(x) \tag{1a}$$

$$N_T(x) = N_c(x) + N_f(x) \tag{1b}$$

$$M_T(x) = M_c(x) + M_f(x) - N_c(x)h_1 + N_f(x)h_2 \tag{1c}$$

where h_1 is the distance from the centroid of the section of concrete layer to the centroid of the whole reinforced member section; and h_2 is the distance from the centroid of the section of concrete layer to the centroid of the whole reinforced member section.

Since the concrete layer is much thicker than the FRP layer, for simplicity, we assume that the layers on and beneath the interfacial crack have the same curvatures, then we have:

$$\frac{M_c(x)}{W_c} = \frac{M_f(x)}{W_f} \tag{2a}$$

$$\frac{Q_c(x)}{W_c} = \frac{Q_f(x)}{W_f} \tag{2b}$$

Considering the contour shown as ABCD in Figure 1, energy release rate G may be defined as:

$$G = \frac{1}{B}\left(\frac{dU_e}{da} - \frac{dU_s}{da}\right) \tag{3}$$

where B is the width of the FRP bonded to the concrete; U_e is the external work performed; and U_s is the strain energy for the contour.

The existing studies (Kim & Heffernan 2008) have showed that, in the case of a good bonding-property in the adhesive used, interfacial failure occurs mainly in the concrete side of the transition zone of concrete and adhesive resins. Thus, for simplicity, according to the relationship between U_e and U_s for isotropic homogeneous materials under constant external forces found by Anderson (1995), we can get:

$$G = \frac{1}{B}\frac{dU_s}{da}\bigg|_{\text{load constant}} \tag{4}$$

As Figure 1 shows, the change in the strain energy, resulting from the movement of an interfacial crack-tip from O to O', is caused by a bending moment, shear force, and axial load together. Referring to the literature by Williams (1988), the changes in strain energy, dU_s^M, dU_s^Q and dU_s^N, caused by bending moment, shear force, and axial load respectively, can be presented as follows:

$$dU_s^M = \frac{1}{2}\frac{M_c^2(x)}{W_c}da + \frac{1}{2}\frac{M_f^2(x)}{W_f}da - \frac{1}{2}\frac{M_t^2(x)}{W_t}da \tag{5}$$

$$dU_s^Q = \frac{3}{5}\frac{Q_c^2(x)}{S_c}da + \frac{3}{5}\frac{Q_f^2(x)}{S_f}da - \frac{3}{5}\frac{Q_t^2(x)}{S_t}da \tag{6}$$

$$dU_s^N = \frac{1}{2}\frac{N_c^2(x)}{D_c}da + \frac{1}{2}\frac{N_f^2(x)}{D_f}da - \frac{1}{2}\frac{N_t^2(x)}{D_t}da \tag{7}$$

Submitting equations 5, 6, and 7 into equation 3, the energy release rate G can be obtained as follow:

$$G = \frac{1}{B}\left(\begin{array}{l}\dfrac{1}{2}\dfrac{M_c^2(x)}{W_c} + \dfrac{1}{2}\dfrac{M_f^2(x)}{W_f} - \dfrac{1}{2}\dfrac{M_t^2(x)}{W_t} + \dfrac{3}{5}\dfrac{Q_c^2(x)}{S_c}da \\[2mm] + \dfrac{3}{5}\dfrac{Q_f^2(x)}{S_f} - \dfrac{3}{5}\dfrac{Q_t^2(x)}{S_t} + \dfrac{1}{2}\dfrac{N_c^2(x)}{D_c} + \dfrac{1}{2}\dfrac{N_f^2(x)}{D_f} - \dfrac{1}{2}\dfrac{N_t^2(x)}{D_t}\end{array}\right) \tag{8}$$

Table 1. Details of specimens and the test results.

Specimen	Test method	$\frac{P_{\max}}{P_u}$	$P_{\max}$ MPa	f_{cu} MPa	b_f mm	l_f mm	N
A-0	Static	1	29.17	62.2	50	160	1
A-1	Fatigue	0.8	23.34	62.2	50	160	2600
A-2	Fatigue	0.74	21.48	62.2	50	160	7800
A-3	Fatigue	0.65	18.96	62.2	50	160	168900
A-4	Fatigue	0.55	16.04	62.2	50	160	1550000
A-5	Fatigue	0.54	15.75	62.2	50	160	2380000

3 EXPERIMENTAL STUDY

3.1 Experimental program

Modified beam tests were performed to investigate the fatigue propagation behaviour of CFRP-concrete interfacial cracks caused by flexural cracks in CFRP reinforced concrete beams. Six beam specimens were used and tested. One of these specimens was tested under monotonic loading to determine the static capacity of all the specimens, and the others were tested under constant-amplitude fatigue loading. Details of the specimens are provided in Table 1.

Each specimen consisted of two same concrete blocks with inverted-T shapes joined by both a steel hinge at the top and two layers of CFRP sheets bonded to the bottom tension faces. The spacing of about 5 mm was set between the two concrete blocks at mid-span. A layer of CFRP sheet was installed on one concrete block transversely to make sure that the failure occurred at the other one (monitored side). The average 28-day concrete cube strength f_{cu} was tested and presented in table 1. The mean values of elastic modulus and tensile strength of CFRP sheets were 225 GPa and 4148 MPa respectively, resulting from tensile flat coupon tests. The nominal thickness of the CFRP sheets was 0.111 mm. As recommended by the manufacturer, the tensile strength and elastic modulus of the epoxy resin used for bonding the CFRP sheets was 47.2 MPa and 2857.7 MPa, respectively. In order to monitor debonding growth along the FRP-concrete interface, thirteen strain gauges were pasted to the surface of CFRP sheets along the midline (Fig. 2).

The beam was simply supported, and subjected to four point bending (Fig. 2). The monotonic load and the fatigue load were applied using a servo-hydraulic testing system. In the monotonic loading tests, loading was applied until the CFRP sheet peeled off from the concrete. Each fatigue test was performed under a constant-amplitude sinusoidal loading at a frequency of 2.5 Hz. The upper limit load $P_{\max}$, width of CFRP sheet b_f, bond length l_f and ratio of the upper limit load $P_{\max}$ to the static bond capacity P_u were listed in Table 1. The lower limit load ratio was 10 per cent of the static bond capacity for all fatigue specimens.

3.2 Experimental results

3.2.1 Failure mode

The number of cycles of failure N for each specimen is summarised in Table 1. Both the static and fatigue

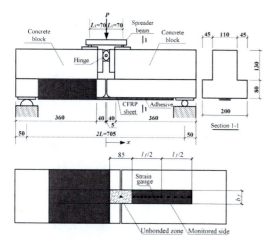

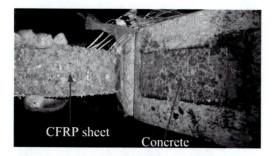

Figure 2. Specimen size, loading method and locations of strain gauges (unit: mm).

CFRP sheet

Concrete

Figure 3. The typical failure of fatigue specimens.

test specimens collapsed due to debonding of CFRP sheets. Figure 3 shows the typical failure mode. As shown in Figure 3, a thin layer of concrete with thickness of about 0.5–2 mm was peeled off by the CFRP sheet.

3.2.2 Crack growth of FRP-concrete interfaces

The crack growth of CFRP-concrete interfaces can be achieved based on development of strain distributions along the CFRP sheet. This method of obtaining the crack growth information was detailed by Diab et al. (2009). Figure 5 shows an example of the relationship between the crack length a and typical CFRP sheet strains along the bonded length. In Figure 4, a_{4080}, a_{10000} and a_{30000} represent the crack lengths at 4080, 10000 and 30000 load cycles respectively. Figure 5 shows the relationship between the number of load cycles and the crack length of specimen A-3 which was typical for all fatigue specimens. From the curve shown in Figure 5, three main stages can be detected, namely, the crack initiation and rapid propagation, steady propagation and instability propagation. In both the first and third stages, cracks grew quickly with an increasing number of cycles, corresponding to micro-debonding initiation and insufficient residual bonding length induced debonding failure, respectively. The initial and final stages covered about the first and last

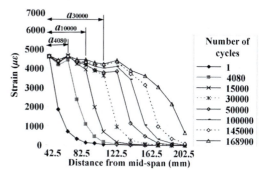

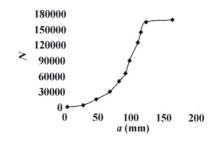

Figure 4. CFRP sheet strain distributions of specimen A-3.

Figure 5. Crack length a versus the number of cycles N for A-3.

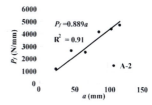

Figure 6. The typical relationship between the friction force per unit width P_f and crack length a.

10 per cent of its life, respectively. During the second stage covering most of the middle 80 per cent of its life, the cracks grew stably.

3.2.3 Interfacial friction

The friction force still existed between the CFRP sheet and the concrete in the cracked zone until a debonding failure occurred. According to test results, the typical relationship between the total frictional force per unit width P_f and the crack length a was shown in Figure 6. Through the linear regression, the relationship between P_f (in N/mm) and a (in mm) was described as:

$$P_f = (\varepsilon_0 - \varepsilon_a) \times E_f t_f = ka \quad (9)$$

where ε_0 and ε_a are the strain at mid-span and crack front respectively; E_f and t_f are the modulus of elasticity and thickness of CFRP sheet respectively. The values of k were 0.889, 0.840, 0.803, and 0.797 for specimens A-1, A-2, A-3, and A-4 respectively, and their correlation coefficients were 0.91, 0.91, 0.93 and

0.82 respectively. Then the value of k was taken to be 0.832, the average of the above four values.

4 INTERFACIAL CRACK GROWTH RATE MODEL

4.1 Calculation of the energy release rate

According to the test specimen design and loading program (Fig. 2), we have:

$$Q_T(x) = P/2 \tag{10a}$$

$$M_T(x) = \begin{cases} P(L-L_1)/2 & \text{if } x \leq L_1 \\ P(L-x)/2 & \text{if } x > L_1 \end{cases} \tag{10b}$$

$$N_c(x) = -N_f(x) = N_0 - b_f P_f, \quad N_0 = P(L-L_1)/(2d_0) \tag{10c}$$

$$h_1 + h_2 = (t_c + t_f)/2 \tag{10d}$$

where b_f is the width of CFRP sheet; and N_0 is tension force applied on CFRP sheet at mid-span. The interfacial crack energy release rate G can be calculated by submitting Equations 1, 2, 9 and 10 into Equation 8.

4.2 Proposed interfacial crack growth rate model

The interface crack growth rate, da/dN, is chosen to be related to the variable, G/G_c, determined by a form of Paris law (Paris & Erdogan 1963):

$$da/dN = C(\Delta G/G_c)^m \tag{11a}$$

$$G_c = \begin{cases} G_f & \text{if } L \geq L_e \\ G_f \left[(2 - L/L_e)L/L_e \right]^2 & \text{if } L < L_e \end{cases} \tag{11b}$$

$$L_e = \sqrt{E_f t_f / \sqrt{f_c}} \tag{11c}$$

where C and m are material constants; G is the amplitude of the energy release rate; G_c is the fracture energy; G_f is the fracture energy when the bond length L is greater than the effective bond length L_e; f_c is the cylinder concrete compressive strength and is found by using the relationship $f_c = 0.67 f_{cu}$ according to the Chinese Standard Code for Design of Concrete Structures GB50010-2010. The expressions of G_c and L_e are given referring to the literature by Chen & Teng (2001). G_f of each specimen equals the strain of the energy release rate of A-0 when debonding occurred. G and G_f can be calculated using the method detailed above.

Figure 7 illustrates the relationship between the crack length growth rate, da/dN, and the G/G_c for different specimens. The relationship is almost linear and can be expressed by the Paris law. By a linear regression it fits the experiment's results, while the parameters C and m of equation 11a are 3.55 and 6.39, respectively, with the correlation coefficient being R^2 0.94.

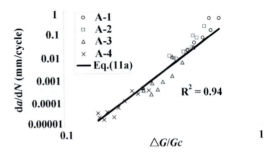

Figure 7. Logarithmic da/dN versus logarithmic G/G_c.

5 CONCLUSIONS

In this paper, the fatigue crack propagation behaviour between CFRP and the concrete was investigated experimentally and theoretically. The following conclusions can be drawn, based on the results:

1) The energy release rate can be used to describe the FRP-concrete interfacial crack propagation behaviour. The theoretical method was proposed for calculating the energy release rate of an interfacial crack in FRP reinforced concrete flexural beams.
2) Evolution of interface fatigue cracks can be divided into three stages: namely, crack initiation and rapid propagation, steady propagation and instability propagation. The first stage covers about the first 10 per cent of the concrete's life, the second, about the following 80 per cent, and the third, roughly the last 10 per cent of its life.
3) The crack growth rate da/dN can be described in a form of Paris law with the variable G/G_c. Based on test results, a semi-empirical formula was developed for the growth rate of an interfacial crack caused by flexural cracks in CFRP strengthened beams.

ACKNOWLEDGMENT

The authors would like to thank the Major State Basic Research Development Program of China (973 Program) (No. 2012CB026200), for the financial support for this project.

REFERENCES

Anderson, T.L. 1995. Fracture Mechanics: Fundamentals and Applications. Boca Raton, Florida: CRC Press.
Chen, J.F. & Teng, J.G. 2001. Anchorage strength Models for FRP and steel plates bonded to concrete. J. Struct. Eng. 127(10): 784–791.
Diab, H.M., Wu, Z. & Iwashita, K. 2009. Theoretical solution for fatigue debonding growth and fatigue life prediction of FRP-concrete interfaces. Adv. Struct. Eng. 12(9): 781–792.

Ferrier, E., Bigaud, D. & Hamelin, P., et al. 2005. Fatigue of CFRPs externally bonded to concrete [J]. *Mater. Struct.*, 38(1): 39–46.

Gheorghiu, C., Labossière, P. & Proulx, J. 2006. Fatigue and monotonic strength of RC beams strengthened with CFRPs. *Compos. Part A* 37(11): 1111–1118.

Heffernan, P.J. & Erki, M.A. 2004. Fatigue behavior of reinforced concrete beams strengthened with carbon fiber reinforced plastic laminates. *J. Compos. Constr.* 8(2): 132–140.

Kim, Y.J. & Heffernan, P.J. 2008. Fatigue behavior of externally strengthened concrete beams with fiber-reinforced polymer: state of the art. *J. Compos. Constr.* 12(6): 246–255.

Nigro, E., Di Ludovico, M., & Bilotta, A. 2011. Experimental investigation of FRP-concrete debonding under cyclic actions. *J Mater. Civil Eng.* 23(4): 360–371.

Paris, P.C. & Erdogan, F. 1963. A critical analysis of crack propagation laws. *J. Fluids Eng.* 85(7): 528–533.

Williams, J.G. 1988. On the calculation of energy release rates for cracked laminates. *Int. J. Fracture* 36(2): 101–119.

Advances in Civil Engineering and Building Materials IV – Chang et al (eds)
© 2015 Taylor & Francis Group, London, ISBN: 978-1-138-00088-9

Full-scale investigations of civil engineering structures using GPS

T. Chmielewski & P. Górski
Opole University of Technology, Opole, Poland

ABSTRACT: The static and dynamic deformation monitoring of engineering structures has been a matter of concern for engineers for many years. This paper provides a review of research and development activities from 1993 in the field of bridges, tall buildings, and tower health monitoring using GPS. The first pioneering applications of GPS to measure the structural vibrations of these structures and the assessment of the measurement accuracy of GPS are first briefly described. Then, the progress on monitoring the displacements and dynamic characteristics of bridges and tall structures, caused by traffic loads, wind, and the combined influence of solar radiation and daily air temperature variations, is presented.

Keywords: Global Positioning System, civil engineering structures, displacement monitoring, measurement errors

1 INTRODUCTION

The static and dynamic deformation monitoring of engineering structures has been a matter of concern for engineers for many years. Examples of deformation behaviour in engineering structures are historical full-scale experiments on the Eiffel Tower, on the Empire State Building, on the Stuttgart TV Tower and several others. These measurements have not only provided important clues for theoretical calculations but also provided the verification needed before a theory could be announced successful.

This paper provides an overview of the applications of the Global Positioning System (GPS) in monitoring engineering structures during the last twenty years. These applications have become a useful tool for measuring and monitoring the static, quasi-static, and the dynamic responses in civil engineering structures, exposed to thermal expansions, the action of wind, earthquakes, traffic loads, or even people's excitation on footbridges.

2 PRELIMINARY RESEARCH

The first pioneering applications of GPS to measure the structural vibrations of tall structures and large cable-stayed bridges took place in the years 1993–1996. Lovse et al. (1993) installed two GPS receivers on the Calgary Tower (completed in 1968). Two GPS receivers were used on the tower only to provide backup data in case one receiver would fail. These receivers on the tower were mounted on tripods near the base of the main communication antenna above the observation deck (Figure 1). The reference receivers

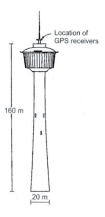

Figure 1. The Calgary Tower and location of GPS receivers (Lovse et al., 1995).

were located on a tripod on the roof of a low-rise (three-story) apartment building situated approximately 1 km north of the Tower. Data was collected at 10 Hz for about 15 min in the morning of 19 November 1993. The analysis of the measured data gave the following results:

1) the Calgary Tower, under wind loading, vibrated with a natural frequency of about 0.3 Hz;
2) the N-S amplitude of displacement was approximately ±15 mm, and the E-W was ±5 mm.

The measured motion of the Calgary Tower using GPS receivers, in both east-west and north-south directions, is shown in Figure 2 for a representative interval of 1 min.

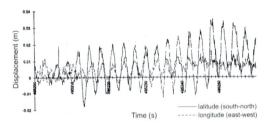

Figure 2. Displacements of Calgary Tower (Lovse et al., 1995).

Leach & Hyzak (1994) reported their results as periodic movements on a large cable-stayed bridge, measured by GPS. Because the vertical movements, which had amplitudes in the order of ±30 mm, were caused by temperature changes rather than wind loading, a very slow sampling rate of one measurement every 10 s was sufficient to recover the movements.

Another example of preliminary research on the application of GPS in monitoring large-scale structures was done by Ashkenazi & Roberts (1996). Their paper describes the way in which the UK Humber Bridge was monitored using a "kinematic GPS". A rover receiver was placed on the west side rail of the bridge deck, at the mid span. The reference receiver was placed on the top of the bridge control tower. Two measured tests were carried out on these days: the first test on 7 March 1996, and the second on 7 May 1996. During the first test, the wind was fairly low and generally from the north-east, along the length of the bridge. The authors have shown the results of the longitudinal, vertical, and lateral movements of the bridge deck, obtained from a selected period of time lasting 15 min. This was shown only for illustrative purposes. These tests were carried out only as a feasibility study of the kinematic GPS techniques for the in situ monitoring the movements of the bridge, and not as a full-scale experiment of a structural deflection analysis.

3 ASSESSMENT OF MEASUREMENT ACCURACY OF GPS

After the first feasibility studies, the GPS technology became an emerging tool for measuring and monitoring both the static and dynamic responses on a full-scale of long-period structures to the ambient loadings. As a new method, the GPS performance had to be thoroughly validated before its application as a full-scale test of structural displacement analysis. Many researchers (Tamura et al. 2002, Chan et al. 2006, Nickitopoulou et al. 2006, and Breuer et al. 2008) carried out feasibility studies to investigate the following items: (1) what is the accuracy of measurements of horizontal and vertical displacements? (2) what is the range and level of natural frequencies and mode shapes determined by a GPS? (3) does this accuracy depend on the frequency of recorded vibrations? (4) can some of the first natural frequencies be detected from a single displacement record?

In order to answer these questions many calibration tests were carried out by applying various kinds of equipment such as: an earthquake shake-simulator track, shaking platforms, slender structure simulating equipment, experimental apparatus consisting of one or three degree of freedom oscillators and a rotating arm equipment.

The major outcome of these studies showed that GPS, advanced to record 10 samples per second (sps or Hz), could record vibrations with frequencies of 0.1 to 4 Hz with an accuracy ±0.5 cm horizontally and ±1 cm vertically.

The most recent paper which deals with the error properties of ultra-high-rate GPS data was published by Moschas & Stiros (2014). In this paper, the impact of the Phase-Locked Loop (PLL) bandwidth on the noise and correlation of GPS measurements, sampled at 100 Hz, was investigated using short and long baselines, and stationary or moving GPS rovers recording vibrations with known characteristics relative to 'true' reference values. Data were collected under various satellite constellations using various values of PLL bandwidth, ie 25, 50, 100 Hz and were processed in differential mode using different software packages. The authors suggested that optimal results can be obtained using either a pre-set 50 Hz PLL bandwidth or a 100 Hz PLL bandwidth combined with a posteriori band-pass filtering of the coordinates. Such optimal results permitted an accurate recording of millimetres and indicated that 100 Hz data are useful for monitoring high-frequency structural vibrations, and also strong earthquake and high-frequency movements from vehicles.

4 MONITORING VIBRATIONS IN STRUCTURES SUCH AS BRIDGES, TALL BUILDINGS AND TOWERS

Nakamura (2000) described a field test conducted for about three weeks in March 1998 on a Japanese suspension bridge (with a main span of 720 m and two side spans of 330 m each) to measure the girder displacements induced by a strong wind using GPS. Two sets of GPS receivers were placed at the midspan of the girder. The reference receiver was installed on a land office about 1 km away from the bridge's centre. The rover and reference receivers recorded displacements at a rate of 1 sample per second (sps = 1 Hz). Accelerometers were also set at the same position as the GPS rover receivers and acceleration data in three directions was collected at the same time. The main results of the described test are as follows: 1) the natural frequencies obtained by GPS data in the lateral (first), vertical (first), longitudinal (first and second) directions and those obtained by accelerometers matched very well. In addition, these values agreed well with the values obtained analytically by the FEM and in the forced vibration tests. 2) the vertical displacement measured by GPS showed 24 h periodic movements because of the temperature changes in the cable. 3) it maybe

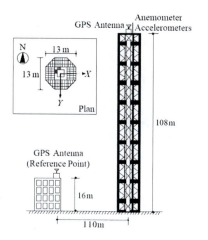

Figure 3. A 108 m tall steel tower for full-scale measurements (Tamura et al., 2002).

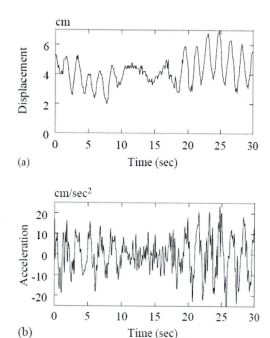

(a)

(b)

Figure 4. Example of temporal variations of wind-induced responses of actual steel tower during a typhoon: (a) RTK-GPS (Y-dir.) and (b) accelerometer (Y-dir.) (Tamura et al., 2002).

concluded that GPS technology is reliable and useful in measuring the static and gust response behaviours of long span bridges.

Meng et al. (2007) applied GPS array units and triaxial accelerometers in order to carry out a field test to record the response of the Wilford Bridge, a suspension footbridge over the River Trent in Nottingham, to a forced vibration excited by more than thirty people with a total mass of 2353 kg, as well as subsequent decayed free vibration and an ambient vibration caused by casual pedestrian traffic and a light wind loading. Designing a proper digital filter for the extraction of structural dynamics parameters was an important aspect of the structural deformation analysis of this footbridge. Measurement time series in the format of coordinates or accelerations were filtered either to reduce the noise level or to split the measurements so that only the real signals to an allotted frequency band were the output for further analysis.

Tamura et al. (2002) made two experiments to demonstrate the feasibility of RTK-GPS for measurements of the wind-induced response and its efficiency in measuring the displacement of a full-scale tower (108 m tall). In the first experiment, the accuracy of RTK-GPS in measurements of the sinusoidal displacements was examined, using an electronic exciter. The efficiency of RTK-GPS was then demonstrated in the full-scale measurement of the actual steel tower. The sketch of the steel tower and some results of the measurements are shown in Figures 3, 4 and 5.

GPS offers great potential for the structural health monitoring (SHM) of tall buildings. Several tests have been carried out until now. Çelebi (2000) made two preliminary tests to prove the technical feasibility of the application of GPS to monitoring tall buildings. The first test was done with a standard stack steel bar to simulate a thirty to forty story flexible building. In the second test they measured an ambient vibration (exposed to the wind and traffic noise) of a forty-four story building with a GPS unit temporarily deployed on its roof. The reference GPS unit was located within

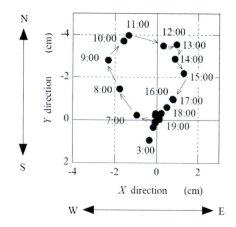

Figure 5. Deformation of the tower caused by solar radiation and daily temperature variations (Tamura et al., 2002).

500 m from the building. The signals were very noisy and the amplitudes very small (<1 cm); therefore the most common method to identify structural characteristics did not work. It was possible to identify fundamental frequency of the building at 0.23 Hz.

Ogaja et al. (2000) installed two Trimble 4700 units on the top of the Republic Plaza (280 m tall) in Singapore sampling at the rate of 1 Hz to measure the vibrations of the building due to the wind. Ogaja et al. (2001, 2003) used a pair of Leica GPS units installed on the Republic Plaza building again to generate a time

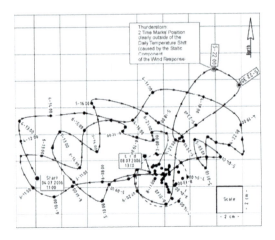

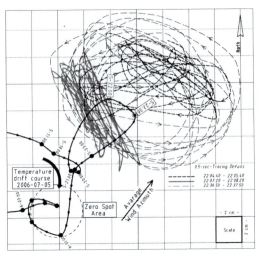

Figure 6. The Stuttgart TV Tower, daily drift of the top due to solar radiation and daily temperature variation with hourly positions and time marks (4th–8th July 2006). In the north-east of the ground plan two outliers are shown, caused by a thunderstorm with wind from South-West at a peak velocity of about 17 m/s (Breuer et al., 2008)

Figure 7. The Stuttgart TV Tower, static and dynamic components of wind response during a thunderstorm producing an outline of the daily temperature drift (see Figure 6). The wind from South-West with a peak velocity of about 17 m/s causes a north-east displacement (static component) of about 6 cm. The main direction of vibration is at a right angle to the wind direction with a displacement of 9–14 cm 5th July 2006 (Breuer et al., 2008)

series of the receiver positions. Chen et al. (2001) conducted a field test employing two NovAtel Outrider DL RT2 dual frequency GPS units to measure the vibrations of the 384 m tall Di Wang building in Shenzen, China, under relatively strong wind.

Kijewski-Correa et al. (2006, 2007) established in 2001 a 'Chicago Full-Scale Monitoring Program' to characterize the in-situ response of three tall buildings in order to verify design practices.

Breuer et al. (2008) examined field tests conducted on the Stuttgart TV tower to measure the displacements at the tower top caused by the wind and the combined influence of solar radiation and daily air temperature variations during different weather seasons and conditions. This paper presents the daily drift of the top of the Stuttgart TV Tower caused by solar radiation and daily air temperature variation during three GPS campaigns during different seasons. The daily elliptical path of the shift extended mainly in a west-east direction during summertime and in a northern direction during winter (Figure 6). This paper presents also the static and dynamic components (in the plane of the east-west and south-north directions) of the wind response during a thunderstorm (Figure 7). For the wind response and for a sample rate equal to 2 sps, GPS was able to measure only the first natural frequency (0.191 Hz) of the Tower.

Moschas & Stiros (2011) presented a new technique to reconstruct displacement records of relatively rigid structures (modal frequency: $1 < f < 5$ Hz). This technique was based on the double digital filtering of high-frequency (10 Hz) GPS recordings of an oscillation, constrained and assessed by independent accelerometer data. This technique was applied to the process of noisy GPS measurements of vibrations of a 40 m long steel footbridge excited by the coordinated jumps of a group of people.

5 CONCLUSIONS

1) In the last twenty years the GPS with 10 Hz sampling rates has become a useful tool for measuring and monitoring static, quasi-static and dynamic responses in civil engineering structures exposed to settlement, gust-winds, earthquakes, thermal expansions traffic loads, and even some unforeseen events. These measurements allow the examination of both displacement levels and vibration characteristics such as natural frequencies, mode shapes, and damping ratios. These values are the key parameters when assessing the safety of flexible engineering structures. One should remember that the sampling rate of 10 Hz mostly permits us to identify the first natural frequency of the majority of civil engineering structures (below 5 Hz).
2) GPS can measure the total wind response, ie its dynamic fluctuating component and also a static component which is directly measured.
3) Ten years ago, the GPS sampling rate of 10 Hz was considered 'high'. Recently during the last five years, rates of 20, 50 or even 100 Hz have been used for some different kinematic applications. Further studies of the error properties of ultra-high-rate GPS data are needed in the near future.

REFERENCES

Ashkenazi, V. & Roberts, G.W. 1997. Experimental monitoring of the Humber Bridge using GPS; *Proc. – Institution of Civil Engineers* 120: 177–182.

Breuer, P., Chmielewski, T., Górski, P., Konopka, E. & Tarczynski, L. 2008. The Stuttgart TV Tower – displacement of the top caused by the effects of sun and wind. *Engineering Structures* 30(10): 2771–2781.

Çelebi, M. 2000. GPS in dynamic monitoring of long-period structures. *Soil Dynamics Earthquake Engineering* 20: 477–483.

Chan, W.S., Xu, Y.L. & Ding, X.L. et al. 2006. Assessment of dynamic measurement accuracy of GPS in three directions. *Journal of Surveying Engineering ASCE* 132(3): 108–117.

Chen, Y.Q., Huang, D.F., Ding, X.L., Xu, Y.L. & Ko, J.M. 2001. Measurement of vibrations of tall buildings with GPS: a case study; *Proc. intern. symp. on NDE for Health Monitoring and Diagnostics*, Newport Beach, California: USA.

Kijewski-Correa, T., Kareem, A. & Kochly, M. 2006. Experimental verification and full-scale deployment of Global Positioning Systems to monitor the dynamic response of tall buildings. *Journal of Structural Engineering ASCE* 132(8): 1242–1253.

Kijewski-correa, T. & Pirnia, J.D. 2007. Dynamic behavior of tall buildings under wind: insights from full-scale monitoring. *The structural design of tall and special buildings* 16(4): 471–486.

Leach, M.P. & Hyzak, M.D. 1994. GPS structural monitoring as applied to a cable-stayed suspension bridge; *Proc. intern. fed. of surveyors (FIG), 20th Congr.*, Melbourne: Australia, 606.2/1-606.2/12.

Lovse, J., Teskey, W., Lachapelle, G. & Cannon, M. 1995. Dynamic deformation monitoring of tall structure using GPS technology. *Journal of Surveying Engineering* 121(1): 35–40.

Meng, X., Dodson, A.H. & Roberts, G.W. 2007. Detecting bridge dynamics with GPS and triaxial accelerometers. *Engineering Structures* 29: 3178–3184.

Moschas, F. & Stiros, S. 2011. Measurement of the dynamic displacements and of the modal frequencies of a short-span pedestrian bridge using GPS and an accelerometer. *Engineering Structures* 33(1): 10–17.

Moschas, F. & Stiros, S. 2014. PLL bandwidth and noise in 100 Hz GPS measurements. *GPS Solutions* – in press (DOI 10.1007/s10291-014-0378-4).

Nakamura, S. 2000. GPS measurement of wind-induced suspension bridge girder displacements, *Journal of Structural Engineering ASCE* 126(12): 1413–1419.

Nickitopoulou, A., Protopsalti, K. & Stiros, S. 2006. Monitoring dynamic and quasi-static deformations of large flexible engineering structures with GPS: Accuracy, limitations and promises. *Engineering Structures* 28(10): 1471–1482.

Ogaja, C., Rizos, C. & Han, S.W. 2000. Is GPS good enough for monitoring the dynamics of high-rise buildings?; *Proc. 2nd Trans Tasman Surveyors Congress*, Queenstown: New Zeland.

Ogaja, C., Rizos, C., Wang, J.L. &, Brownjohn, J. 2001. GPS and building monitoring case study: Republic Plaza Building, Singapore; *Proc. intern. symp. on Kinematic Systems in Geodesy, Geomatics and Navigation,* Banff: Canada.

Ogaja, C., Wang, J.J. & Rizos, C. 2003. Detection of wind-induced response by wavelet transformed GPS solutions. *Journal of Surveying Engineering* 129(3): 99–104.

Tamura, Y., Matsui, M., Pagnini, L.-C., Ishibashi, R. & Yoshida, A. 2002. Measurement of wind-induced response of buildings using RTK-GPS. *Journal of Wind Engineering and Industrial Aerodynamics* 90: 1783–1793.

Advances in Civil Engineering and Building Materials IV – Chang et al (eds)
© 2015 Taylor & Francis Group, London, ISBN: 978-1-138-00088-9

A study of the static performance of a single-layer hyperbolic latticed shell structure

Xingtao Wang
Department of Geotechnical Engineering, Tongji University, Shanghai, China

Z.L. Wang
Key Laboratory of Geotechnical & Underground Engineering of Ministry of Education,
Department of Geotechnical Engineering, Tongji University, Shanghai, China

ABSTRACT: A long span steel roof structure is widely applied in public buildings. The requirements for the long span steel structure are very strict. To investigate the static performance of a reticulated shell structure, in this paper, we use the large-scale general finite element analysis software ANSYS to deal with the analysis of static performance and the study of the static characteristics of the structure under six kinds of different combinations of loads in order to understand the internal forces of the structure's elements and the structure's displacement. Based on the static performance analysis, we can get the type of forces in the structure's components and the maximum value distribution of the components' forces, and we also get the sensitivity of the reticulated shell structure's vertical displacement under an undistributed loads. The present study may provide a theoretical basis for the analysis of the reticulated shell structure's static performance. We can also put forward useful suggestions for the reticulated shell structure in the engineering practice.

Keywords: hyperbolic latticed reticulated shell; numerical analysis; static performance; axial stress; vertical displacement

1   INTRODUCTION

A reticulated shell structure which is similar to a grid structure is a kind of spatial truss structure, and it has the properties of both a shell and a truss structure [1]. Having a reasonable load and great span of covering, it attracts popular attention at home and abroad and it has broad prospects for development in the spatial structure. It not only has the series of advantages of a reticulated shell structure, but also can provide a variety of beautiful shapes [2]. Throughout the history of the development of building structures, the three-dimensional spatial reticulated shell structures that can best meet the design requirements became the first choice of many construction programs [3, 4].

The spatial reticulated shell structure is a symbol for measuring a country's scientific and technological level of architecture [5], with the wide application of reticulated shell structures, its mechanical properties receive attention on a daily basis. Therefore, we should carry out this research on the static performance of the reticulated shell structure and understand the working its performance, which provides a theoretical basis for the research on the static performance of reticulated shell structures.

2   DESIGN OF RETICULATED SHELL STRUCTURE

In this research, we choose the single-layer reticulated shell structure to be researched. The single-layer reticulated shell structure has no subsidiary structure and the bottom of the concrete structure, and its structure selection is the single-layer hyperbolic latticed shell structure. The parabolic equation of the reticulated shell structure is $z = (x/a)^2 - (y/b)^2 + 22.7$ ($a = 18.6339747, b = 15.5319719$, the unit is m). Roof plane whose east-west direction contour dimension is 76.242 m and whose north-south direction contour dimension is 70.72 m is approximately elliptical. The structural steel uses Q345B grade steel, and the rod uses circular steel tubes Φ 121 × 4.5, Φ 133 × 4.5, Φ 146 × 4.5, Φ 159 × 6, Φ 180 × 6, Φ 219×6, Φ 219 × 10, Φ 219 × 16, Φ 245 × 16, Φ 273 × 16, Φ 299 × 16, and support nodes are in the form of full-constraints.

3   MATERIAL PROPERTIES CONSTITUTIVE MODELS

Calculations use the constitutive model of linear elastic material, which is characterized as follows: loading

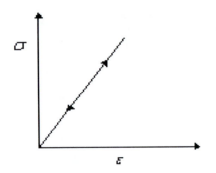

Figure 1. Linear elastic model.

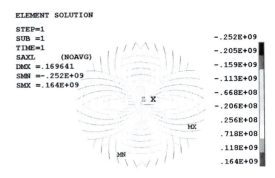

ELEMENT SOLUTION
STEP=1
SUB =1
TIME=1
SAXL   (NOAVG)
DMX =.169641
SMN =-.252E+09
SMX =.164E+09

Z X

MX

MN

-.252E+09
-.205E+09
-.159E+09
-.113E+09
-.668E+08
-.206E+08
.256E+08
.718E+08
.118E+09
.164E+09

Figure 4. Axial stress of this structure in first load condition.

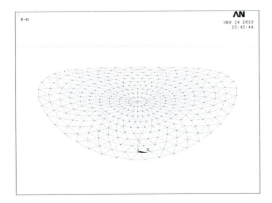

Figure 2. Finite element model of single-layer reticulated shell.

and unloading curves completely overlap and the stress-strain relationship is always linear, and the curvature of the curve is the elastic modulus of the material, and the stress-strain curve is shown in Figure 1 [6]. The selection of shell components is Q345B steel. The elastic modulus of the steel is 2.06 GPa, and its Poisson's ratio μ is 0.3.

4 FINITE ELEMENT MODEL AND ANALYSIS OF RETICULATED SHELL STRUCTURE

4.1 *The establishment of a finite element model*

The finite element model of reticulated shell structure is shown in Figure 2. BEAM188 unit is adopted in this model, and support nodes are in the form of full-constraints. In ANSYS modelling process, BEAM188 unit has many advantages [7–11]: (1) it considers the effects of shear deformation; it is suitable for the analysis of slender to medium stubby beams; it is a three-dimensional linear finite stress beam; it is also a Timoshenko beam element, and it uses independent interpolations of the cross-section's corner and the deflection. (2) It represents a three-dimensional linear finite element of the stress beam. The influence of shear deformation is considered on the basis of the classical beam element, and it considers the independent interpolations of the cross-section's corner and the deflection. (3) The BEAM188 unit can be used in

SSECDATA, SECREAD, ECTYPE, SECWRITE, and SECOFFSET to define the cross section.

4.2 *Load combinations of the finite element model*

Here are the two main types of loading:

(1) Type of permanent load

Welded hollow spherical joints are calculated at 25 per cent of the components' weight, and 0.45 kN/m² is used to calculate the purlins and the roof panels.

(2) Type of variable load

The roof live load: 0.3 kN/m². Taking some characteristics of reticulated shell into account, the load imposed on the reticulated shell structure is equivalent to the force applied to the structure's node, and it is considered in the live load. According to the characteristics of this structure, we will consider six combinations of load:

Combination of 1: 1.0 × dead load + 1.0 × live load (full cross);
Combination of 2: 1.0 × dead load + 1.0 × live load (X > 0, half of a cross);
Combination of 3: 1.0 × dead load + 1.0 × live load (Y > 0, half of a cross);
Combination of 4: 1.0 × dead load + 1.0 × live load (X < 0, Y > 0; 1/4 of a cross);
Combination of 5: 1.0 × dead load + 1.0 × live load (X > 0, Y < 0; 1/4 of a cross);
Combination of 6: 1.0 × dead load + 1.0 × live load (X > 0, Y > 0; 1/4 of a cross).

4.3 *Analysis of the results of finite element calculation*

(1) According to the finite element calculation and the analysis of the structure, under the six combinations of load, the axial stress distribution is shown in Figures 4 to Figure 9, and the detailed analysis is as follows:

a) Under the six combinations of load, the maximum and minimum stresses of the reticulated shell structure are shown in Table 1 (tensile stress is positive, and pressure stress is negative). From the following table and figures, we can see that the maximum

Table 1. The axial stress peaks of this structure in different load conditions (units: N/m²).

Load combination	1	2	3	4	5	6
Maximum (tensile stress)	164	146	266	238	238	237
Minimum (pressure stress)	−252	−275	−224	−276	−276	−275

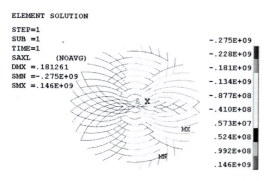

Figure 5. Axial stress of this structure in second load condition.

Figure 8. Axial stress of this structure in fifth load condition.

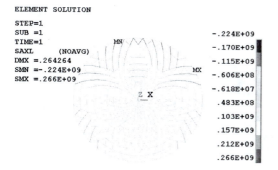

Figure 6. Axial stress of this structure in third load condition.

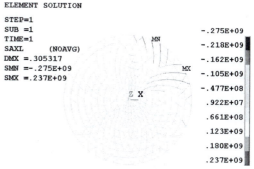

Figure 9. Axial stress of this structure in sixth load condition.

Figure 7. Axial stress of this structure in fourth load condition.

tensile stress and the maximum pressure stress occur respectively in third and fourth (or fifth) load conditions, and they are the most unfavourable combinations.

b) The components' (including radial upper chords along the direction of the cable and upper chords of the ring along the direction of the arch) axial stress along the arch direction in this structure bear the largest compressive stress; the components' (including radial upper chords along the direction of the arch and upper chords of the ring along the direction of the cable) axial stress along the cable direction in this structure bear smallest compressive or tensile stress.

c) The maximum axial tensile stress of this structure's components appears at radial upper chords near the supports of the higher corner along the cable direction, and the minimum axial compressive stress of this structure's components appears at radial upper chords near the supports of the lower corner along the arch direction.

d) The radial chords of this structure transfer an inner force in the radial direction; circle components transfer the inner force in the direction of the ring.

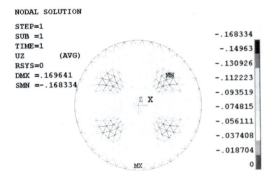

Figure 10. Vertical displacement of this structure in the first load condition.

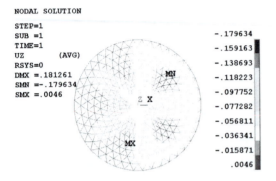

Figure 11. Vertical displacement of this structure in the second load condition.

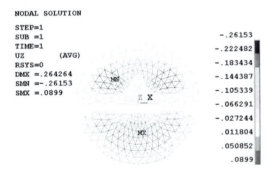

Figure 12. Vertical displacement of this structure in the third load condition.

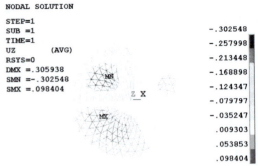

Figure 13. Vertical displacement of this structure in the fourth load condition.

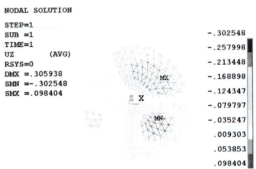

Figure 14. Vertical displacement of this structure in the fifth load condition.

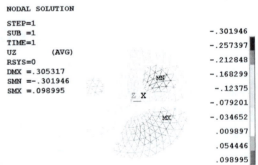

Figure 15. Vertical displacement of this structure in the sixth load condition.

(2) Through the calculation and analysis with software ANSYS, vertical displacements of this structure in Figures 10–15 under six combinations of load, and a detailed analysis of vertical displacements is as follows:

a) The maximum and minimum of vertical displacements in the reticulated shell structure are shown in the following table (we shall be as follows: the vertical displacement along the negative direction of the Z-axis generates the positive displacement, and the vertical displacement along the positive direction of the Z-axis generates the negative displacement.)

b) The combinations of a load can largely influence the distribution of the nodes' vertical displacements, and the position of maximum vertical displacements is not always apparent in the centre node of the structure.

c) From the figures and the tables, we can see that the vertical displacements of the structure develop in the direction of 45 degrees between the trend line and the X-axis; at the same radius, in the direction of 45 degrees between the trend line and the X-axis, the nodes' vertical displacements are at a maximum.

180

Table 2. The vertical displacement peaks of this structure in different load conditions (units: m).

Load combination	1	2	3	4	5	6
Maximum displacement	0.168	0.180	0.262	0.302	0.303	0.302
Minimum displacement	0.019	−0.005	−0.09	−0.099	−0.098	−0.099

d) This structure is more sensitive to an uneven distributed load; under the uneven distributed load, the distribution of the nodes' vertical displacements is not uniform, and under these conditions the nodes have larger vertical displacements, while the nodes in load combinations 4 and 6 even generate 0.099 m's of vertical displacement along the Z-axis positive direction, and this is very adverse to the overall stability of the structure.

5 CONCLUSION

This paper carries out a numerical simulation analysis on the structure of the large span steel roof. We mainly draw the following conclusions:

(1) The upper chords of the single-layer reticulated shell structure mainly suffer pressure; the components' axial stress along the arch direction in this structure bear compressive stress; the components' axial stress along the cable direction in this structure bear tensile stress or smaller compressive stress. In addition, the structure deformation is mainly composed of vertical displacements.
(2) The maximum axial tensile stress of this structure's components appears in radial upper chords near the supports of the higher corner along the cable's direction, and the minimum axial compressive stress of this structure's components appears at radial upper chords near the supports of the lower corner along the arch's direction. Therefore, in engineering practice, local reinforcement should be done on the parts with larger stress.
(3) This structure is more sensitive to the uneven distributed load; under the uneven distributed load the distribution of the nodes' vertical displacements is not uniform, and under these conditions the nodes have larger vertical displacements; and this is very adverse to the overall stability of the structure. Therefore, in the structural design, the role of the uneven distributed load cannot be ignored.

ACKNOWLEDGEMENTS

This study was supported by the Natural Science Foundation of China (No. 51174145, 61179062), the Specialized Research Fund for the Doctoral Program of Higher Education of China (No. 20120072110024) and the Fundamental Research Funds for the Central Universities.

REFERENCES

[1] Deren Yin. 1996. Design of Reticulated Shells [M]. Beijing: Chinese Architecture Industry Press.
[2] Yongyan Ma & Xiumao Yang. 2001. Development and Application of Reticulated Shell Structures [J]. Tianjin: Tianjin Construction, 2: 3–5.
[3] Th. Von. Karmen & A. S. Tsien. 1939. The Buckling of Spherical Shells by External Pressure [J], J. Aero. Sci., 7: 43–50.
[4] D. T. Wright. 1965. Membrane Forces and Buckling in Reticulated Shells [J], Journal of Structural Div., ASCE, 91(1): 325–342.
[5] Fu Li Yong &Yang Congjuan. 2002. Large Span Space Structure-History and Development of the Reticulated Shell Structures [J], Beijing: Construction & Design, 5: 3–5.
[6] Jinze Bai. 2005. LS-DYNA3D Theoretical Foundation and Case Analysis [M]. Beijing: Science Press.
[7] ANSYS Inc. 2000. ANSYS Modeling and Meshing Guide [Z]. Chengdu, China.
[8] ANSYS Inc. 2000. ANSYS Basic Procedures Manual [Z]. Chengdu, China.
[9] ANSYS Inc. 2000. ANSYS Analysis Guide of Structural Dynamics [Z]. Chengdu, China.
[10] ANSYS Inc. 2000. ANSYS Guide of Advanced Technical Analysis [Z]. Chengdu, China.
[11] Shuguang Gong & Guilan Xie. 2004. The ANSYS Operation Command and Parametric Programming [M]. Beijing: Machinery Industry Press.

Advances in Civil Engineering and Building Materials IV – Chang et al (eds)
© 2015 Taylor & Francis Group, London, ISBN: 978-1-138-00088-9

The benefits of Ubiquitous Sensor Network (USN) application in monitoring temporary construction structures

Sungwoo Moon, Byongsoo Yang & Eungi Choi
Department of Civil and Environmental Engineering, Pusan National University, Busan, Republic of Korea

ABSTRACT: Ubiquitous Sensor Network (USN) technology has a capability of providing information anywhere and anytime. The technology is being widely accepted as a tool that assists data acquisition in a real-time fashion. The USN technology can provide a local wireless network environment that can be used for remotely monitoring construction operations in a wide natural area. This paper discusses the potential application of the wireless networking technology for monitoring temporary structures during a construction operation. A USN (Ubiquitous Sensor Network)-based monitoring system that is being experimented at the Pusan National University supported by the National Research Foundation (NRF) of Korea. The test-bed has been built to test the applicability of system components of USN technology and to measure the effectiveness of data acquisition. Since construction relies heavily on temporary structures, the USN could bring a significant impact on improving construction technology.

Keywords: ubiquitous sensor network, temporary construction, remote monitoring, data acquisition

1 INTRODUCTION

In construction, temporary structures are used to build main bodies of slabs, columns, beams, etc. These structures function as a mold for concrete placement, or support the frame until permanent concrete bodies are erected. Since these temporary structures are used during the construction operation, they are exposed to accidents unless care is taken. Any collapse will result in fatal consequences. In order to prevent accidents, the operation should be carefully monitored during the operation [1], [2].

Construction sites are, however, usually located away from the staging area. This geographical constraint makes it hard to monitor any changes in the conditions of temporary structures during or after the operation. Moreover, if construction personnel are assigned to monitor the details of the operation, this will be costly. However, any remote monitoring device would be very effective in understanding the behaviour of temporary structures and take actions to prevent serious structural damage.

In this study, the USN technology has been tested for its effectiveness as a remote monitoring system that can be applied in the construction operation. The USN technology is being widely accepted as a technology that assists data acquisition in the construction industry. This capability of remote data access can be instantly installed and utilized to collect field data and provide important information on the status of any structural safety. Any sign of failure in the construction should be detected as soon as possible in order for remedial action to be taken. Any accident could result in a serious accident to the workers in the construction process, and significantly delay the job progress.

2 USN TECHNOLOGY FOR DATA ACQUISITION

The USN technology is often used to develop a ubiquitous environment. The technology provides a wireless network in a temporary local area and transmits various sensory data instantly. Usually the USN-based system uses sensor boards which have a capability of transmitting and receiving data using radio frequency signals. Since the board communicated data within a limited distance, multiple nodes are used to transmit and receive data over longer distances between the transmitter and the receiver (Figure 1).

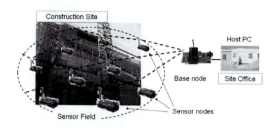

Figure 1. The concept of USN application in construction.

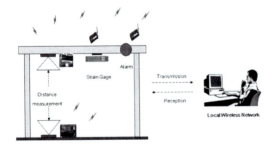

Figure 2.   Transmission and reception using USN boards.

Ultrasonic sensor (Transmitter)	Ultrasonic sensor (Receiver)	Strain gage
Weight device	Inclinometer	Alarm device

Figure 3.   Sensors for wireless monitoring of temporary construction structures.

The networking in the USN system is based on the TinyOS, which is a local wireless networking environment. The environment is an embedded operating system written as a set of cooperating tasks and processes in accordance with the ZigBee specification [3]. The TinyOS is install in the host PC to provide a local area networking.

The USN plays a key role in developing a ubiquitous environment. Working with various sensor devices, the USN board can transmit data on temperature, pressure, pollution, etc. and is applied in many areas such as environmental protection, disaster prevention, fire prevention, and logistics management, etc. [4]. One of the construction manager's main task is to overcome the constraints in the construction operation.

In construction, the USN technology can be used in monitoring the behaviour of temporary structures. Any sign of abnormal behaviour indicates the danger of construction failure. The behaviour should be constantly checked to keep the temporary construction safe. Otherwise, the construction process can fail and cause severe damage to the structures under construction and casualties to the construction workers.

Considering that the construction operation is executed in a wide natural area, the USN technology can provide easy-to-use applications for remote monitoring. The construction manager can adopt the technology to prevent costly collapses of temporary structures and to provide safe conditions for construction workers.

3   CONFIGURATION OF WIRELESS NETWORK

Although the USN technology has pros and cons, the ubiquitous capability provides benefits in data communication that overcomes the constraints of having construction sites situated over a wide area. (Figure 2). For example, a wireless local network can be established within a reasonable budget. In this case, the USN boards themselves can relay data in a multiple hopping system to form a local network in a ZigBee environment [5]. When the construction site is large in area and away from the staging office, the temporary structure can be continuously monitored through the data relay functions of the USN.

This capability makes the application of the USN technology very feasible in the construction operation. Various sensors can be connected to USN boards, and the boards in turn transmit the sensory data to the host PC in the staging area. The construction personnel in the site office can continuously monitor the progress of the construction operation, and takes action when necessary. This ubiquitous environment reduces the physical distance between the actual construction site and the management office.

4   PHOTOGRAPHS AND FIGURES

An indoor test is important to test whether the USN technology can be applied in monitoring the structural behaviour of temporary structures. Based on the indoor test, the system module can be given the functions in transmitting and receiving desired signals. A test-bed has been built to test the applicability of the USN technology in monitoring temporary structures during the construction operation.

The test-bed is 2 m in height, and 4 m in width, and made up of plywood, steel support, H-beam, etc. The simulation is designed to test the capability of the USN technology in a laboratory environment. However, the result can be used again for the application in the actual site environment. The outputs both from the indoor test-bed and the actual construction operation can be compared together to measure the reliability of the monitoring capability.

Application of the USN technology to the temporary structures requires an integration of various system components. These components include: 1) ultrasonic sensors for transmission, 2) ultrasonic sensors for receive, 3) strain gages, 4) weight devices, 5) inclinometers, and 6) alarm device. These devices are used to understand the structural behaviour of temporary structures (Figure 3).

Ultrasonic sensors measure the change in distance between two measurement points. In this study, these ultrasonic sensors were divided into a component for transmission and a component for reception. Considering that the construction site has usually many objectives, the functional separation can increase the reliability of the distance measurement. The functional

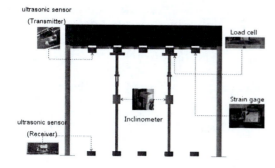

Figure 4. Test-bed for experimental application for wireless monitoring of temporary construction structures.

separation allowed for the avoidance of interference from neighbouring objects in the test-bed (Figure 4). Strain gages detect the stress on the components of the structure. Weight devices measure the loads on structural components. Inclinometers check whether the structural component is maintaining its upright position.

These sensors were calibrated to understand the linear relations of the voltage output from the sensor nodes to distances, strains, and inclinations. One of the main goals of the indoor test is to collect data for calibration from the sensors used in this study. The alarm-device signals any danger of structural failure if the sensors detect any abnormal behaviour in the temporary structures. The alarming function can be used to protect workers when construction failure occurs during the construction operation.

A multi-hop algorithm has been programmed to relay the data transmission and reception between multiple USN nodes. Since the construction site office is usually located away from the actual construction area, it is important to provide a function of remote monitoring.

In this research, it is assumed that the site is located 100 metres away from a site office at the staging area. The data from the USN boards are displayed in a web system so that these data can be utilized wherever they are on the construction sites.

The USN boards were powered by 9 volt batteries, which is made up of lithium. The lithium battery, which is rechargeable, can be reused after recharge. Since the devices are used during the construction operation, the batteries provided sufficient electric power during the operation. Since the USN boards are used remotely, the power should be supplied in a reasonable and economic way.

A test-bed of a temporary structure has been set up in a laboratory at the Pusan National University. The output will provide important knowledge and experience in implementing the system for actual construction operations. An experimental test is being conducted to acquire data, and use them for evaluating the safety conditions of a construction operation. The test result will provide useful information on the

design of a USN environment for increasing safety in the construction industry.

5 STRENGTHS AND WEAKNESSES

Being different from other industries, the construction operation has unique characteristics. The construction projects are usually executed in a natural environment. The constructed product is usually large in size and magnitude. The construction operation should be monitored to avoid construction failure. The USN technology can assist very effectively for data collection during the construction operation.

Considering these unique characteristics, however, application of the USN to construction operation requires customized modification on the systems' components. The application of USN technology should be competitive compared with other optional technologies. That is, the system should be technically and economically feasible in real application at the construction site.

Therefore, the following strengths and weaknesses should be carefully considered when it is applied to the construction operation.

Strengths:

- A wireless local network can be setup without the constraints of space and distance that are often encountered on the construction site.
- Sensor nodes can be placed anywhere they are needed in a simple manner for measuring temperature, precipitation, loads, noise, etc.
- As the construction progresses, the USN nodes can be detached and installed again in other points of the construction project.
- The construction manager can monitor the safety of the temporary structures in a staging area away from construction sites.

Weaknesses:

- Sensors should be able to interface with the USN boards for data communication because the sensors could have different types of sensory data.
- The ad-hoc local network should be reliable enough to cover the wide area of data transmission and reception on the construction site.
- Construction materials such as steel beams, concrete walls and reinforcement can cause noise on the data communication of the USN boards with the hosting computer.
- The sensors and USN boards should be protected against dust. These devices, therefore, must be covered to prevent damages during monitoring.
- The USN technology is an economically feasible alternative to traditional monitoring devices.

6 APPLICATION TO SAFETY MONITORING

Safety management is very important in construction operations in order to avoid any accidents. Any accidents on the temporary structures could result in fatal consequences. This paper introduces an effort to monitor the behaviour of temporary structures. The real-time data acquisition from the construction site can help the contractor come up with more effective ways to increase the safety of the construction operation.

Application of the USN technology to safety monitoring involves many different disciplines of hardware and software. The electronic devices of the USN should be modified to transfer and receive sensory data. These data need to be displayed on a computer screen, preferably in a web environment. In addition, the output should be compared with the structural analysis when the technology is applied to temporary structures in the construction operation. Despite the difficulties in implementing the USN technology in a construction operation, the effort will help improve the construction technology to a great extent.

In the construction operation, the USN technology can help monitor the operation of temporary construction and contribute to construction safety in that: 1) construction data can be collected and monitored in real time during the construction operation; 2) the collected data can be used to understand the degree of stress that is loaded on the structural members; 3) the collected data can be utilized to define the status of structural safety; and 4) based on the USN signal,

the contractor can take actions to avoid any possible accidents that might occur during the operation.

ACKNOWLEDGEMENT

This research was supported by the Basic Science Research Program through the National Research Foundation of Korea (NRF) funded by the Ministry of Education (2012R1A1B3003393).

REFERENCES

A. Rong and M. Cuffari, 2004, Structural Health Monitoring of a Steel Bridge Using Wireless Strain Gauges, *Structural Materials Technology VI, NDE/NDT for Highways & Bridges*, 327–330, Buffalo, NY.

J. Gehrke, and L. Liu, Sensor-network applications, 2006, *IEEE Internet Computing*, vol. 10(2), 16–17.

J. Lynch, K. Law, A. Kiremidjian, T. Kenny, E. Carryer, and A. Partridge, 2001, The design of a wireless sensing unit for structural health monitoring, *Proc. of 3rd International Workshop on Structural Health Monitoring*.

P. Brown, 2008, Use ZigBee for cost-effective WPAN sensing and control solutions, *Electronic Design*, vol. 56(21), 57–63.

TinyOS, 2009, *Installing TinyOS* <http://docs.tinyos.net/index.php/Getting started>.

Advances in Civil Engineering and Building Materials IV – Chang et al (eds)
© 2015 Taylor & Francis Group, London, ISBN: 978-1-138-00088-9

Numerical study of blind-bolted moment connections of hollow sections

P.K. Kosmidou, Ch.N. Kalfas & D.T. Pachoumis
Department of Civil Engineering, Democritus University of Thrace, Xanthi, Greece

ABSTRACT: Hollow sections, despite their many advantages, are not widely employed in Steel Construction, because of the difficulty of designing and implementing bolted connections to them. The main problem is the proper clamping of the bolts, since the use of the nuts is impractical in view of the closed geometry and the use of threads is insufficient due to the small wall thickness. In the present study, a new alternative method is proposed, whereby the bolts are clamped in threads engraved in an additional plate welded at the flange of the column, which essentially replaces the nut. Two three-dimensional Finite Element (*FE*) models, with different steel quality used for the column's plate, were designed, in order to simulate the proposed connection. The performed analysis provides information on the inherent behavior of the joint concerning the strength and the failure mode of the connection.

Keywords: blind-bolted connection, additional extended endplate with threads, hollow sections, moment connection, finite element (FE) analysis, M-φ diagrams

1 INTRODUCTION

Tubular sections have many clear advantages over open ones on mechanical as well as aesthetic grounds. In particular, the rigidity of hollow sections in weak-axis bending and in torsion combined with their very low propensity for warping make them highly resistant to flexural-torsional and purely torsional buckling. However, the design of moment connections for hollow sections by means of standard structural bolts alone is problematic, mainly because the use of nuts is impractical, and this implement prevents general application of hollow sections in Steel Construction.

Several studies have been carried out to implement the design of bolted connection on hollow sections. One of the most common types of blind-bolted moment connections is the extended endplate connection. Several researchers have investigated the behavior of such a connection with different types of blind bolts. Mourad & Globarah (1994) were among the first to study the behavior of blind-bolted connections with extended endplate and they proposed a special bolting system using high strength bolts (High Strength Blind Bolt system, HSBB). According to this system, the clamping of the bolts is carried out with nut on the outer side of the connection (Mourad et al. 1994). The main difference between HSBB and a standard bolt is in the oversized bolt hole, through which the blind bolt enters the hollow section.

Similar blind-bolted systems with the HSBB system (Mourad et al. 1994) have been proposed and studied by several researchers to simplify and standardize connections involving hollow sections. The Lindapter

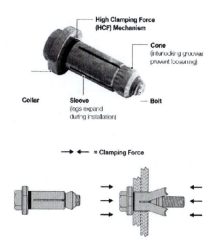

Figure 1. Installation procedure of Lindapter HolloBolt® system (Elghazouli et al. 2009).

HolloBolt® (Elghazouli et al. 2009) and the Ajax ONESIDE blind bolt® (Lee et al. 2011) systems have been adopted at international level. The difference between these two methods is in the installation procedure, which are illustrated in Figures 1–2. According to the manufacture's specification and requirements, both systems require oversized bolt holes.

An alternative method for the design of bolted connection on hollow sections is the Flowdrill® method. France et al. (1999) have conducted an extensive investigation program in order to study the moment capacity and rotational stiffness of joints bolted to tubular

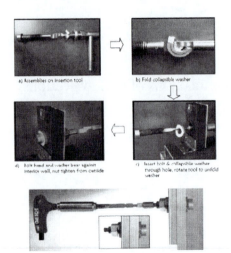

Figure 2. Installation procedure of Ajax ONESIDE® blind bolt system (Lee et al. 2011).

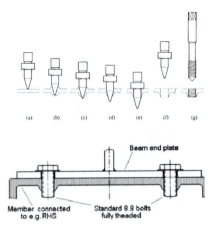

Beam end plate

Member connected to e.g. RHS   Standard 8.8 bolts fully theaded

Figure 3. Schematic of the Flowdrill® process and the resulting endplate connection (Kurobane et al. 2004).

columns using flowdrill connectors. The Flowdrill system® is a thermal drilling technique for the extrusion of holes in the front wall of the hollow section using a four lobed tungsten-carbide friction drill. According to this system the bolts are clamped in threads made in the wall of the closed section (Kurobane et al. 2004). In this system only standard fully-threaded bolts are used and standard beam and column bolt holes can be made, according to the requirements of EC3, Part 1-8. The Flowdrill® process and the resulting endplate connection are illustrated in Figure 3.

In this paper an alternative method is proposed for the design of bolted beam-to-column connections between hollow sections. The bolts are clamped in threads made in an additional plate welded at the flange of the column, which essentially replaces the nuts. The connection can be implemented with conventional standard bolts. The paper aims to investigate the behavior of this proposed connection. For this reason, two three-dimensional Finite Element (FE) models were designed using the ABAQUS® finite element software. The difference between the two analytical models was the steel quality used for the column's plate. Several iterations took place in order to find the best simulation for the threads of the bolts and the plate as well as the weld of the column's plate. The aim of this numerical study is to design a model that will be reflective to the performance of the real system, as this research is part of the experimental study carried out at the Steel Structure Laboratory of the Democritus University of Thrace in Greece.

2 FINITE ELEMENT (FE) ANALYSIS

A cross-shaped beam-to-column bolted connection with two extended endplates prepared for future experimental testing at our Lab (Fig. 4) was simulated in 3D continuum mechanics using Finite Elements (FE) with geometric and material nonlinearity in ABAQUS®. Two analytical models were designed, noted as C220S275 and C220S355 according to the steel quality used for the plate welded at the tube face. Each one consisted of one square cross-section SHS 220 × 220 × 10 mm column and two rectangular cross-section RHS 200 × 100 × 8 mm beams. M20, 8.8 grade standard bolts were used to connect the plates to the column face and were placed according to the requirements of EC3, Part 1-8. Finally, two plates were used to implement the connection:

- The first one, noted as column's plate, was welded at the flange of the column and replaced the nuts for the clamping of the bolts, as the threads were engraved in this. Its thickness was 20 mm.
- The second one represented the front plate, where the beam was welded. It had the same thickness with the column's plate.

The nonlinear material properties was obtained from the EC3, Part 1-1, as had been used the nominal values for each material. For the column, the beam and the front plate was used steel S275 and for the column's plate steel S275 and S355. At Table 1 the mechanical properties are provided for both steel qualities whose values were imported to define the material's identity in the finite element analysis.

The static system was defined according to the experimental set-up (Fig. 4). The beams constituted the horizontal components and the column the vertical. Each beam was fixed at one end at the column and in a distance of 1000 mm from the face of the column the beam was fastened by a roller support, as shown in Figure 4. The roller prevented any vertical displacement and thus constant shear and linear moment were developed in the beam. The analysis was conducted by applying monotonic displacement at the top of the column. The total displacement was obtained equal to 150 mm, which is much bigger than that which appeared at the yielding, as the main purpose is to gather information on the behavior of the

| S355 | 355 | 510 | 210000 | 0.3 |

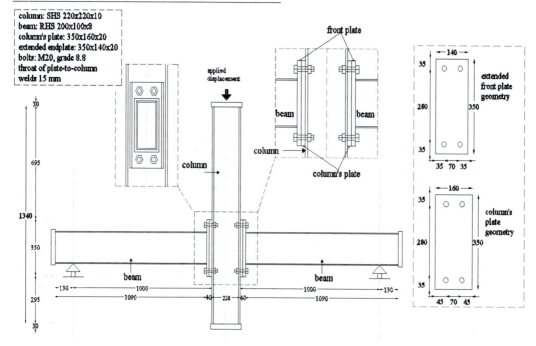

Figure 4. The experimental set-up for the analytical model.

Table 1. Nominal values for Steel quality of S275 & S355.

Steel quality	f_y MPa	f_u MPa	Modulus of Elasticity E MPa	Poisson's ratio ν
S275	275	430	210000	0.3
S355	355	510	210000	0.3

joint, in both the elastic and plastic area, based on the mechanical properties of the material in question.

Tie constraint and surface-to-surface contact elements with friction behavior were employed in the *FE* models. A tie constraint was chosen for the simulation of the welds, both of the column's plate and the beam at the front plate, and to incorporate the threads into the column's plate and the bolts. The threads were simulated as independent elements in the shape of ring, with thickness 1.25 mm, and not as a single spiral shaped element, as illustrated in Figures 5–6. Surface-to-surface contact elements with friction behavior was defined between the following elements: the flange of the column and the plate welded at its face, the column's plate and the front plate, the front plate and the bolts, the threads of the bolts and the column's plate. Taking advantages of symmetric conditions along the longitudinal and transverse planes, only a quarter of the specimen in Figure 4 was modelled to reduce the computational time, with the appropriate boundary conditions imposed (Fig. 5). For the meshing of the

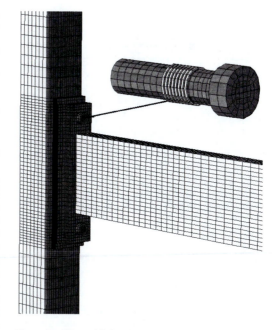

Figure 5. View of finite element mesh of the connection.

model two types of finite elements (Hibbitt et al. 1997) were used:

– Shell elements with rectangular shape, four nodes and six degrees of freedom at each node (type S4R) for the column's web and the beam.

Figure 6. The simulation of the column's plate and the bolts.

– Continuum (solid) elements with eight nodes and three degrees of freedom at each node (C3D8R) for the column's flange, the plates, the bolts and the threads.

As observed in Figure 5 a denser grid was applied at the joint's area, while in the regions with less importance of research a sparser grid was preferred.

3 RESULTS FROM THE ANALYSIS

The results of the two analytical models focused on the evaluation of the strength of the connection and the identification of the failure modes that may appear in the joint's area. The moment was calculated at the face of the column:

$$M = P * L \qquad (1)$$

where L is the distance between the roller support of the beam and the column face axis (Fig. 4).

In Figures 7–8 the moment versus rotation diagram for each analytical model is indicated. In each diagram were defined the limits for the classification of the joint according to its rotational stiffness (EC3, Part 1-8). Both joints were classified as semi-rigid joints, according to the M-φ diagrams (Figs. 7–8). The moment rotation was calculated from the ratio of the column's vertical displacement to the distance of the roller support of the beam from the front plate.

The two models presented similar performance during the analysis and the same ultimate strength. The failure was caused by the shearing off of the bolts in the tensile area, due to combined tension and shear at the shank of the bolts which were in contact with the front plate of the connection. The model C220S355

M-φ for the C220S355 model

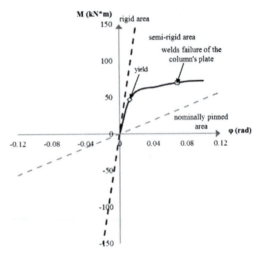

Figure 7. M-φ diagram of model C220S355.

M-φ for the C220S275 model

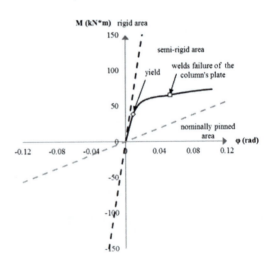

Figure 8. M-φ diagram of model C220S275.

demonstrated greater strength at the yielding than the model C220S275, due to the higher quality of the steel used for the column's plate.

At the model C220S355 initial yielding was observed at the vicinity of the threads of the column's plate, where the bolts were clamped in the tension load area, at $M_y = 47$ kN*m and at vertical displacement of column $u_y = 11$ mm (Fig. 9). Progressing through the applied displacement, the front plate suffered strong deformation at the height of the tensile area, removing the contact from the column's plate, due to the prying forces developed between the fasteners, as shown in Figure 11. The plates are in contact with each other and connected only with the bolts, while the column's plate

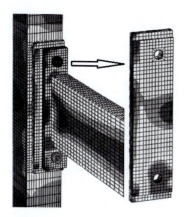

Figure 9. Connection area of specimen C220S355 at yielding.

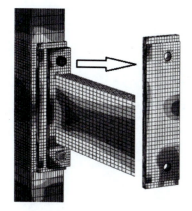

Figure 10. Connection area of specimen C220S355 at yielding.

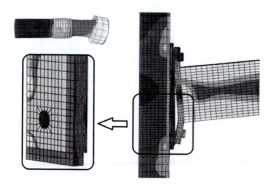

Figure 11. The failure modes appeared in the connection area of C220S355 model.

was welded at its flange. The moment versus rotation curve presented change of its slope at M = 64 kN*m and at column's vertical displacement u = 41 mm, as shown in Figure 7. Until the bolts' failure, the connection suffered only deformation without a significant increase in its strength. The first failure mode was observed in the vicinity of welds of the column's

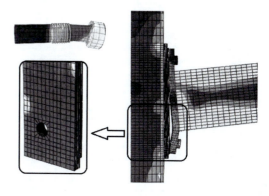

Figure 12. The failure modes appeared in the connection area of C220S275 model.

plate at $M_u = 69$ kN*m and at column's vertical displacement $u_u = 65$ mm, while the ultimate failure was caused by the shearing off of the bolts, which were clamped in the tensile area, at $M_{u,ult} = 72$ kN*m and at vertical displacement of column $u_{u,ult} = 95$ mm, as it is illustrated in Figure 11.

The specimen C220S275 had similar performance with the C220S355 model presenting the same failure mode. During the analysis, yielding started to propagate at the vicinity of threads of the plate in the tensile area. At $M_u = 65$ kN*m and at column's vertical displacement $u_u = 50$ mm, failure in the vicinity of welds of the column's plate was observed, while at $M_{u,ult} = 72$ kN*m and at vertical displacement of column $u_{u,ult} = 97$ mm, bolt failure occurred. Images from the analysis are shown in Figures 10, 12.

4 CONCLUSIONS

The analysis of the two-three dimensional finite element models has provided the following findings:

- The results show that the extended endplate connection using blind bolts and an additional extended plate welded at the column's face classified as semi-rigid joint according to the EC3 specifications, depending on the rotational stiffness of the joint.
- The model C220S355 compared to the model C220S275 demonstrated better performance, due to the higher steel quality used for the column's plate.
- Both models provided similar performance after yielding and the same ultimate strength. The failure was caused by the shearing off of the bolts in the tensile area, due to combined tension and shear at the shank of the bolts which is in contact with the front plate of the connection.
- The analytical work presented in this paper provides a basis for further study, as it is necessary to confirm the results via experiments and to investigate all the parameters that affect the performance of the connection.

REFERENCES

Elghazouli, A.Y., Málaga-Chuquitaype, C., Castro, J.M. & Orton, A.H. 2009. Experimental monotonic and cyclic behavior of Blind-Bolted angle connections. *Engineering Structures* 31:2540–2553.

European Committee for Standardisation (CEN) 2005. EN 1993-1-1, Eurocode 3: Design provisions of steel structures, Part 1-1: General rules and rules for buildings.

European Committee for Standardisation (CEN) 2005. EN 1993-1-8, Eurocode 3: Design provisions of steel structures, Part 1-8: Design of joints.

France, J.E., Davidson, J.B. & Kirby, P.A. 1999. Strength and rotational stiffness of simple connections to tubular columns using flowdrill connectors. *Journal of Constructional Steel Research* 50:15–34.

Hibbitt, D., Karlsson, B. & Sorensen P. 1997. ABAQUS/ manual. *Hibbitt, Karlsson & Sorensen Inc.*

Kurobane, Y., Packer, J.A., Wardenier, J. & Yeomans, N. 2004. Design Guide 9 for structural hollow section column connections. CIDECT (ed.), *TÜV Verlag*. Cologne: Germany.

Lee, J., Goldsworthy, H.M. & Gad, E.F. 2011. Blind Bolted T-stub connections to unfilled hollow section columns using extended T-stub with back face support. *Engineering Structures* 33:1710–1722.

Lindapter International Ltd. 1995. Type HB HolloBolt for blind connection to structural steel sections and structural tubes. Bradford, England.

Mourad, Sh. 1994. Behavior of Blind Bolted Moment Connections for Square HSS columns. *Ph.D. thesis.* McMaster University, Hamilton, Ontario, Canada.

Detection of building materials

Advances in Civil Engineering and Building Materials IV – Chang et al (eds)
© 2015 Taylor & Francis Group, London, ISBN: 978-1-138-00088-9

Tests and simplified behavioural model for steel fibrous concrete under compression

C.E. Chalioris & F.A. Liotoglou
Department of Civil Engineering, Democritus University of Thrace, Xanthi, Greece

ABSTRACT: This work includes uniaxial compression tests of 27 plain and fibrous concrete cylinders with various concrete strengths, types and amount of steel fibres. Full stress-strain behavioural curves are obtained. A simple model for the estimation of the steel fibrous concrete compressive behaviour is also developed based on 125 stress-strain curve and 257 strength data from compression tests of normal and high strength concrete. Comparisons between test data curves and analytically predicted ones showed a good agreement.

Keywords: steel fibrous concrete, compression, test, model, high strength concrete

1 INTRODUCTION

1.1 Steel Fibrous Concrete (SFC)

The use of discontinuous steel fibres in concrete as non-conventional mass reinforcement has long been recognized that ameliorates the post-cracking characteristics of concrete, provides for crack propagation control and increases energy absorption capability in the tensile regime (Karayannis 2000a). The ability of SFC to overcome to some extent the limitations of concrete and to convert its brittle behaviour to a ductile one has well been established. SFC beams under flexure (Ashour et al. 1993, Chalioris 2013a), shear (Kwak et al. 2002, Chalioris et al. 2011, Chalioris 2013b) and torsion (Chalioris et al. 2009, Karayannis 2000b) have been tested to assess the favourable influence of the used steel fibres.

1.2 Compressive behaviour of SFC

It is known that the compressive behaviour of SFC is influenced by the properties of its constituent materials and especially by the volume fraction, the aspect ratio and the bond characteristics of the steel fibres added (Nataraja et al. 1999, Marar et al. 2011). Although for small amounts of steel fibres the compressive strength is not significantly increased, the post-peak compressive behaviour, or else, the post-cracking ductility seems to be improved.

In the literature there are reported many analytical models developed to represent the stress-strain curves for plain and fibrous concrete under compression (Oliveira Júnior et al. 2010). Most of them are based on test data regression analysis in order to estimate a factor that considers the influence of fibres on the curve form, depend on many parameters, are valid

for fibrous concretes mixtures with specific properties (short-ranged concrete strength and amount or type of fibres) and predict the entire pre-peak and post-peak compressive stress-strain response. In this work a simplified and easy to apply behavioural model for SFC under compression is proposed. This model utilizes empirical formulas derived from test results of a broad range of parametrical studies from the literature and is applicable to the analysis and design of SFC elements.

An experimental program of 27 compressive tests that enrich the existing database providing the entire stress-strain behaviour of SFC mixtures with various concrete strengths, types and volume fractions of steel fibres is also included herein.

2 EXPERIMENTAL PROGRAM

2.1 Materials

Three different concrete mixtures (A, B and C) were used with Greek type pozzolan cement (CEM II 32.5 N) containing 10% fly ash. Crushed and natural river sand with a high fineness modulus and crushed stone aggregates with a maximum size of 16 mm were used. Further, 2.5 lt, 1.0 lt and 0.5 lt of Retarder (Pozzolith 134 CF) also added in mixtures A, B and C, respectively. Mix proportions for casting one cubic meter of concrete are summarized in Table 1. Cement, sand and stone aggregates were first dry-mixed.

The steel fibres added in fibrous concrete mixtures were hooked-ended type with a length to diameter ratio (aspect ratio) equal to $\ell_f/d_f = 25\,\text{mm}/0.80\,\text{mm} = 31.25$ in group A concrete mixtures, $\ell_f/d_f = 44\,\text{mm}/1.0\,\text{mm} = 44.00$ in group B and $\ell_f/d_f = 60\,\text{mm}/1.0\,\text{mm} = 60.00$ in group C. Two different steel fibre

Table 1. Mix proportions of plain concrete mixtures.

	Cement	Water	Sand Crushed	Natural	Stone aggregate* Crushed
A	1.00	0.50	1.47	1.43	2.29
B	1.00	0.55	1.77	1.85	2.67
C	1.00	0.62	1.83	1.91	2.76

*maximum size: 16 mm.

volume fractions, V_f, were used in each concrete mixtures; 1% that corresponds to a dosage of 80 kg per 1 m³ concrete and 3 % or 240 kg per 1 m³ concrete.

The steel fibres were first dispersed clump-free by hand, gradually and added slowly in small amounts in order to avoid fibre balling, while mixing continued. Water was added and mixed gradually to ensure that the produced SFC mixtures would obtain uniform material consistency, adequate workability and homogeneous fibre distribution. The prepared fresh SFC mixtures were placed in the cylinders and adequately vibrated. During mixing and casting of the fresh mixtures no steel fibre gravitation was observed.

2.2 Compressive tests

Typical cylinders with dimensions of diameter/height = 150/300 mm were cast from the plain concrete batches to be used as reference specimens and from the batches of each SFC mixture. The specimens were tested 28 days after casting under uniaxial compression using a universal testing machine with a maximum capacity of 3000 kN and under displacement control mode at a constant rate of stain (about 2 mm per min). Before performing the test, a layer of special cement was applied on the top and bottom surface of the specimens, to ensure the flatness of their surface.

Axial platen-to-platen deformations of the cylinders were recorded with the measurement of linear electronic strain gauges (LVDT), with total range of 50 mm and precision of 0.01 mm. In order to avoid the disturbance of the extensometer's gauge reading by the cracks formed on the concrete surface, "O" rings at the middle third of the specimen were also used. In the SFC specimens the "O" rings were removed near before cracking, so that a higher force after the maximum compressive strength was applied. In this way, more measurements were obtained for the post-cracking stress-strain curve. The measurements of the applied load and the corresponding axial displacement converted to compressive stress and strains.

2.3 Experimental results

Test results of the experimental program are presented in Figure 1 in terms of compressive stress versus strain curves.

Each examined concrete mixture group (A, B and C) includes three plain concrete reference specimens,

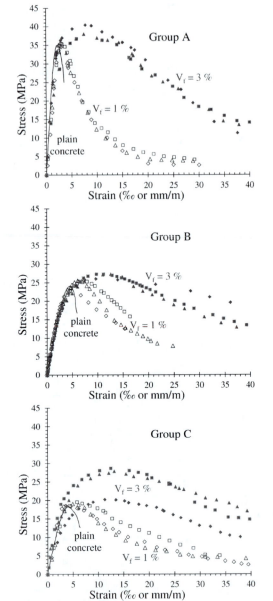

Figure 1. Experimental compressive stress versus strain curves.

three SFC specimens with $V_f = 1\%$ and three SFC specimens with $V_f = 3\%$. The average plain concrete curve from each group and all the SFC curves are displayed and compared in Figure 1.

3 BEHAVIOURAL MODEL

3.1 Parameters of the model and used database

Based on the test data of this study, Figure 2 displays the typical behavioural curves of a plain concrete

196

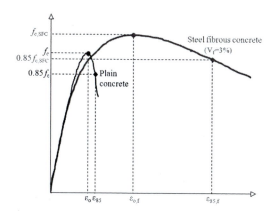

Figure 2. Typical stress versus strain compressive behaviour.

and a SFC specimen under axial compressive loading. In most of the examined cases and especially in fibrous concrete specimens with high amount of steel fibres ($V_f = 3\%$), SFC exhibits increased compressive strength, higher value of the strain at ultimate strength and significantly improved post-cracking response with respect to the corresponding plain concrete specimen.

This way, the parameters of interest which are used in the developed behavioural model are (a) the compressive strength (f_c and $f_{c,SFC}$), (b) the corresponding strain at ultimate strength (ε_o, $\varepsilon_{o,f}$) and (c) the ductility parameter which is represented with the strain ratio $\mu_{85} = \varepsilon_{85}/\varepsilon_o$ for plain concrete and $\varepsilon_{85,f}/\varepsilon_{o,f}$ for SFC, where ε_{85} and $\varepsilon_{85,f}$ are the strains that correspond at stress equal to 85% of the compressive strength for the case of plain concrete ($0.85f_c$) and SFC ($0.85f_{c,SFC}$), respectively, at the descending stress-strain curve, after the peak stress.

In order to determine the increase of the aforementioned parameters due to the addition of steel fibres, test results from experimental works around the world are used as database (Tan et al. 1993, Ashour et al. 1993, 2000, Gao et al. 1997, Barros et al. 1999, Nataraja et al. 1999, Padmarajaiah et al. 2001, Daniel et al. 2002, Kwak et al. 2002, Song et al. 2004, Duzgun et al. 2005, Yazici et al. 2007, Köksal et al. 2008, Unal et al. 2007, Mohammadi et al. 2008, Oliveira Júnior et al. 2010, Marar et al. 2011, Pawade et al. 2011, Nili et al. 2012, Chalioris 2013b). These parameters are expressed in formulas using regression analysis of test data and are related by the known fibre factor, F, that was first proposed by Narayanan and Darwish (1987) and given by:

$$F = \beta \cdot V_f \cdot \ell_f / d_f \qquad (1)$$

where $\beta =$ the fibre shape/bond factor that accounts for the different bond characteristics of the steel fibre and can be taken as 0.75 for deformed fibres, 0.50 for round fibres and 1.0 for indented fibres; $V_f =$ the volume fraction of the steel fibres; and ℓ_f, $d_f =$ the length and the diameter of the fibre, respectively.

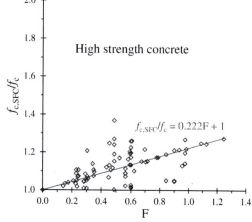
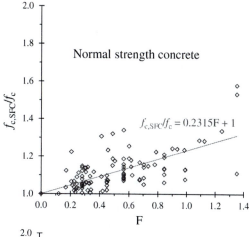

Figure 3. Relationship between the compressive strengths ratio and the fibre factor, F, for normal and high strength concrete.

3.2 Compressive strength

The relationship between the increase of the compressive strength due to the steel fibres and the fibre factor, F, for normal strength (≤ 50 MPa) and high strength concrete (> 50 MPa) is shown in Figure 3.

3.3 Compressive strains

Figure 4 presents the influence of the fibre factor, F, to the increase of the compressive strain at ultimate strength due to the presence of steel fibres. Based on the examined test results (data points of Figure 4) the formula shown in Figure 4 calculates the ratio of the SFC to the reference plain concrete strain at ultimate strength using the value of the fibre factor.

3.4 Ductility parameter

The relationship between the ductility parameter μ_{85} and the fibre factor, F, is estimated based on the results of test data curves, as displayed in Figure 5.

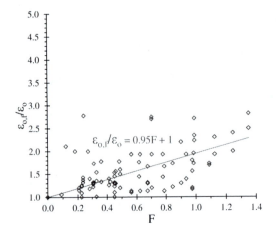

Figure 4. Relationship between the compressive strains ratio and the fibre factor, F.

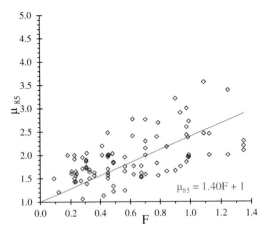

Figure 5. Relationship between the ductility parameter μ_{85} and the fibre factor, F.

The ascending part of the SFC compressive stress-strain behaviour till the point of $[\varepsilon_{o,f}, f_{c,SFC}]$ can be described using the well-known Hognestad parabola. Afterwards, the post-peak response is assumed linear till the point of $[(\mu_{85} \times \varepsilon_{o,f}), 0.85 f_{c,SFC}]$.

4 COMPARISONS

Comparisons between experimentally obtained full stress-strain curves and analytically calculated ones are presented in Figures 6 and 7 for normal concrete strength (tests of this research) and high concrete strength (Marar et al. 2011), respectively.

5 CONCLUDING REMARKS

Compression tests of plain concrete and SFC cylinders showed that the fibre factor, F, significantly influences the compressive strength and especially

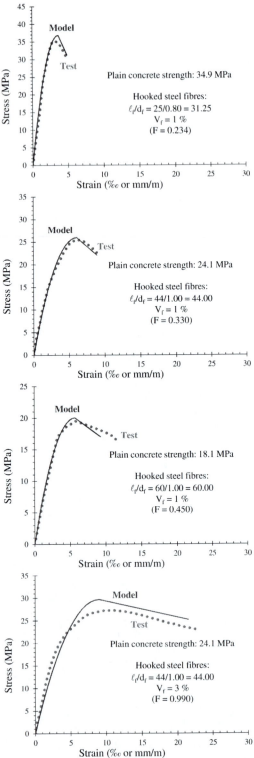

Figure 6. Comparisons between tests and model predictions.

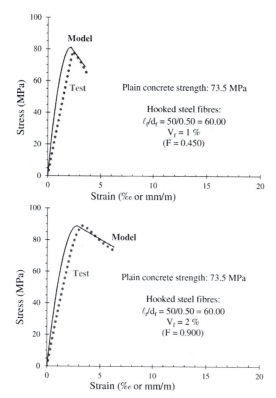

Figure 7. Comparisons between tests and model predictions.

the post-cracking behaviour. SFC mixtures with high values of F exhibit ductile response. A simplified and easy to apply model for the calculation of the SFC compressive behaviour is also developed based on the experimental data derived from this study and 20 works of the literature. A reasonable agreement between tests and predictions is achieved.

REFERENCES

Ashour, S.A. & Wafa, F.F. 1993. Flexural Behavior of High-Strength Fiber Reinforced Concrete Beams. *ACI Structural Journal* 90(3): 279–287.

Ashour, S.A., Wafa, F.F. & Kamal, M.I. 2000. Effect of the concrete compressive strength and tensile reinforcement ratio on the flexural behavior of fibrous concrete beams. *Engineering Structures* 22: 1145–1158.

Barros, J.A.O. & Figueiras, J.A. 1999. Flexural behavior of SFRC: Testing and Modeling. *Journal of Materials in Civil Engineering, ASCE* 11(4): 331–339.

Chalioris, C.E. & Karayannis, C.G. 2009. Effectiveness of the use of steel fibres on the torsional behaviour of flanged concrete beams. *Cement and Concrete Composites* 31(5): 331–341.

Chalioris, C.E. & Sfiri, E.F. 2011. Shear performance of steel fibrous concrete beams. *Procedia Engineering* 14: 2064–2068.

Chalioris, C.E. 2013a. Analytical approach for the evaluation of minimum fibre factor required for steel fibrous concrete

beams under combined shear and flexure. *Construction and Building Materials* 43: 317–336.

Chalioris, C.E. 2013b. Steel fibrous RC beams subjected to cyclic deformations under predominant. *Engineering Structures* 49: 104–118.

Daniel L. & Loukili A. 2002. Behavior of High-Strength Fiber-Reinforced Concrete Beams under Cyclic Loading. *ACI Structural Journal* 99(3): 248–256.

Duzgun, O.A., Gul, R. & Aydin, A.C. 2005. Effect of steel fibers on the mechanical properties of natural lightweight aggregate concrete. *Materials Letters* 59: 3357–3363.

Gao, J., Suqa, W. & Morino, K., 1997. Mechanical Properties of Steel Fiber-reinforced, High-strength, Lightweight Concrete. *Cement and Concrete Composites* 19: 307–313.

Karayannis, C.G. 2000a. Analysis and experimental study for steel fibre pullout from cementitious matrices. *Advanced Composites Letters* 9(4): 243–255.

Karayannis, C.G. 2000b. Nonlinear analysis and tests of steel-fiber concrete beams in torsion. *Structural Engineering and Mechanics* 9(4): 323–38.

Köksal, F., Altun, F., Yiğit, I. & Şahin, Y. 2008. Combined effect of silica fume and steel fiber on the mechanical properties of high strength concretes. *Construction and Building Materials* 22(8): 1874–1880.

Kwak, Y.K., Eberhard, M.O., Kim, W.S. & Kim, J. 2002. Shear Strength of Steel Fiber-Reinforced Concrete Beams without Stirrups. *ACI Structural Journal* 99(4): 530–538.

Marar, K., Eren, O. & Yitmen, I. 2011. Compression Specific Toughness of Normal Strength Steel Fiber Reinforced Concrete (NSSFRC) and High Strength Steel Fiber Reinforced Concrete (HSSFRC). *Materials Research* 14(2): 239–247.

Mohammadi, Y., Singh, S.P. & Kaushik, S.K. 2008. Properties of steel fibrous concrete containing mixed fibres in fresh and hardened state. *Construction and Building Materials* 22: 956–965.

Narayanan, R. & Darwish, I.Y.S. 1987. Use of steel fibers as shear reinforcement. *ACI Structural Journal* 84(3): 216–227.

Nataraja, M.C., Dhang, N. & Gupta, A.P. 1999. Stress-strain curves for steel-fiber reinforced concrete under compression. *Cement & Concrete Composites* 21: 383–390.

Oliveira Júnior, L.A., Borges, V.E.S, Danin, A.R., Machado, D.V.R., Araujo, D.L., Debs, M.K.E. & Rodrigues, P.F. 2010. Stress-strain curves for steel fiber-reinforced concrete in compression. *Revista Materia* 15(2): 60–266.

Padmarajaiah, S.K. & Ramaswamy, A. 2001. Behavior of Fiber-Reinforced Prestressed and Reinforced High-Strength Concrete Beams Subjected to Shear. *ACI Structural Journal* 98(5): 752–761.

Pawade, P.Y., Nagarnaik, P.B. & Pande, A.M. 2011. Performance of steel fiber on standard strength concrete in compression. *International Journal of Civil and Structural Engineering* 2(2): 483–492.

Song, P.S. & Hwang, S., 2004. Mechanical properties of high-strength steel fiber-reinforced concrete. *Construction and Building Materials* 18: 669–673.

Tan, K.H., Murugappan, K. & Paramasivam, P. 1993. Shear Behavior of Steel Fiber Reinforced Concrete Beams. *ACI Structural Journal* 90(1): 3–11.

Unal, O., Demir, F. & Uygunoglu, T. 2007. Fuzzy logic approach to predict stress–strain curves of steel fiber-reinforced concretes in compression. *Building and Environment* 42: 3589–3595.

Yazici, S., Inan, G. & Tabak, V. 2007. Effect of aspect ratio and volume fraction of steel fiber on the mechanical properties of SFRC. *Construction and Building Materials* 21: 1250–1253.

Advances in Civil Engineering and Building Materials IV – Chang et al (eds)
© 2015 Taylor & Francis Group, London, ISBN: 978-1-138-00088-9

Use of the maturity method to predict compressive and flexural strengths in high strength concrete in Kuwait

F. Al-Fahad, J. Chakkamalayath & A. Al-Aibani
Construction and Building Materials Program, Energy and Building Research Centre,
Kuwait Institute for Scientific Research, Kuwait

ABSTRACT: The hot marine environment in Kuwait makes it impractical to adapt experiences and empirical equations developed in other countries. The maturity concept accounts for the combined effect of time and temperature on concrete strength. The utilization of the maturity index method will be beneficial in fast track constructions as it will help contractors and engineers determine the in-place concrete strength at any time. This paper presents the development of compressive and flexural strength maturity curves for a high strength concrete mix of 60 MPa. The concrete mix was cured by two different curing methods and the temperature-time factor for developing the maturity curve was obtained using maturity meter.

Keywords: maturity curves, maturity index method, compressive strength, flexural strength, temperature-time factor, datum temperature

1 INTRODUCTION

The environment in Kuwait is characterized as a hot marine environment. Normal temperatures in summer exceed 50°C. These unique harsh conditions make it impractical to adapt empirical equations developed in other countries. The strength of a given concrete mix is a function of its age and temperature history. The maturity concept accounts for the combined effect of time and temperature on concrete strength development. The concept is based on the fact that temperature is a critical factor in the progress of cement hydration and hence the development of its strength. The maturity of concrete is determined by multiplying an interval of time by the internal temperature of the concrete. This product is summed over time and the maturity of this concrete is equal to the sum of these time-temperature products, or:

$$M(t) = \sum (T_a - T_o) \Delta t$$

where:

M(t) = the temperature time factor at age t
Δt = a time interval
T_a = average concrete temperature during time interval, and
T_o = datum temperature

This is known as the Nurse-Saul Temperature – Time Factor (TTF) maturity and it accounts for the temperature effects on the strength development of concrete. This is illustrated in the following Figure 1 which shows that two concrete samples with the same

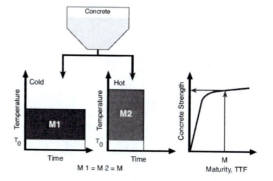

Figure 1. Nurse-Saul's maturity rule using Temperature-Time Factor (Kehl., Constantino and Carrasquillo 1998).

maturity will have the same strength, even though each may have been exposed to different curing conditions.

The evaluation of the maturity age and the right time to remove the formworks under different curing and environmental conditions always represent a challenge in Kuwait as early estimates may lead to cracks and low quality concrete, while later estimates affect the overall project cost. The utilization of the maturity index method, therefore, will be beneficial as it will help contactors and engineers determine and assess the in-place concrete strength at any time regardless of what curing system is used and its condition (Kehl, 1998; Vázquez-Herrero., 2012). The dependence on the in-place strength will also help in decision-making, as it will eliminate the doubts and confusions resulting from the discrepancies between the behaviour of laboratory-stored samples and the actual behaviour of the concrete on site.

In this research study, one high strength concrete mix of 60 MPa was selected and utilized to evaluate and establish its corresponding TTF maturity curves. The ConReg System™ was utilized to evaluate the tendency curves and the rate of maturity. Two concrete cylinders were used to produce a compressive strength maturity curve and two beams were utilized to develop the flexure maturity curve. Other cylinders and beams were used to modify the produced curves by evaluating compressive and flexure strength through direct compression and flexure after 1, 3, 7, 14, 21 and 28 days. The produced curves represent the actual behaviour of the concrete mixes and the actual relation between the measured temperature and the expected age of maturity. Concrete specimens were cast and subjected to different exposure conditions and different curing regimes, including water curing and the external coating of concrete specimens by a spraying method.

2 METHODOLOGY

Ordinary Portland cement (OPC Type 1) satisfying the requirements of Kuwaiti and International Standard specifications was selected as the binder material. A compatible superplasticizer from the polycarboxylate family was selected to increase the workability of high strength mix as a low w/c ratio of 0.31 was adopted. Imported lime stone aggregate of maximum size of 20 mm (3/4″), washed sand and crushed sand conforming to the standard specifications were selected.

The sieve analysis for the three different aggregates used in the concrete mixes was conducted and it showed that the percentage of materials finer than a 75-μm (No. 200) sieve by washing for the 20 mm aggregates was found to be 0.72%, while it was 17.26% for the crushed sand and 5.1% for the washed sand. Sieve analysis of coarse aggregate showed that the average size of the aggregate was less that 19 mm. A well-graded particle size distribution was observed for crushed stone. A concrete mix of 60 MPa with detailed mix proportions was designed.

The mix was cast twice and subjected to two curing methods: water curing and surface spray coating curing surface after 24 hours of mixing. An acrylic resin based non-degrading curing compound was selected. The dosage of curing compound was selected based on the manufacturer's recommendation, which was 4.5 to 5.5 m²/litre of sample. This curing method was used to help retain the humidity in the concrete specimens as shown in Plate 1. Table 1 (Below) represents the mix designs for the mixes. The quantity of concrete was calculated to cast 21 cylinders (10 * 20 cm) and 21 beams (50 * 10 * 10 cm).

Concrete specimens were connected to the maturity meter for 28 days as shown in Plate 2 to measure the Temperature Time Factor (TTF) values for each

Table 1. Mix design for 1 m³ concrete.

Materials	Mix
Concrete Mix design strength, MPa	60
Cement type	OPC
Cement (kg)	500
Microsilica (kg)	40
Water (lts)	155
Water absorption of total agg. by weight, %	0.5
20 mm aggregates, kg	630
10 mm aggregates, kg	470
Crushed sand, kg	250
Washed sand (dry), kg	400
Target slump, mm	100

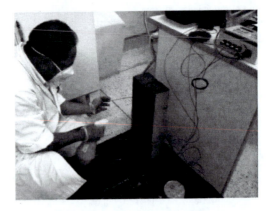

Plate 1. Application of spray coating curing.

mix. The datum temperature was selected as −10°C for the maturity meter and direct values of TTF were obtained from the maturity meter. This was conducted in parallel with measuring the compressive and flexure strengths for each of the concrete mixes. Cylinder specimens were tested for their compressive strength at different ages, and beams specimens were tested for their flexure strength at the same specified ages. Three specimens were used for each testing and the average strength was recorded.

3 RESULTS AND DISCUSSION

Tables 2 and 3 represent the compressive and flexural strength values for the selected mixes, Tables 4 and 5 represent its corresponding TTF values, while Figures 2 and 3 represent the developed maturity curve for the same mix.

The mixes had the required workability and the mixes did not show any segregation and bleeding during casting. The test results show that the designed mixes attained the target compressive and flexural strength within twenty-eight days. The maturity function depended on the temperature development in the initial days. In the case of compressive strength, the obtained results showed that the surface spray curing

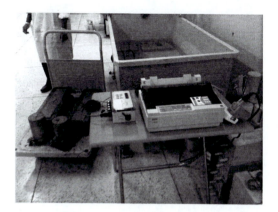

Plate 2. Connecting specimens to maturity meter.

Table 2. Average compressive strength values for concrete mixes under different curing conditions.

Age	Avg. Compressive Strength (MPa)	
	Water cured	Spray cured
Day 1	17.49	27.80
Day 3	42.02	46.47
Day 7	53.05	53.05
Day 14	62.69	60.48
Day 21	62.81	64.94
Day 28	67.69	65.91

Table 3. Average flexural strength values for concrete mixes under different curing conditions.

Age	Avg. Flexural Strength (MPa)	
	Water cured	Spray cured
Day 1	3.22	2.75
Day 3	5.28	3.13
Day 7	5.38	3.28
Day 14	6.6	3.58
Day 21	7.17	3.63
Day 28	7.86	3.69

Table 4. TTF values for 60 MPa mix under water curing conditions.

Hrs	TTF (°C-hrs)	Comp. strength (MPa)	TTF (°C-hrs)	Flexural strength (MPa)
24	707	17.49	698.5	3.22
72	2090.5	42.02	2054.5	5.28
168	4852.5	53.05	4776.5	5.38
336	9632	62.69	9475.5	6.6
504	14164.5	62.81	13939.5	7.17
672	18908	67.69	18590	7.86

Table 5. TTF values for 60 MPa mix under spray.

Hrs	TTF (°C-hrs)	Comp. strength (MPa)	TTF (°C-hrs)	Flexural strength (MPa)
24	786.5	27.80	786	2.75
72	2406.5	46.47	2386	3.13
168	5513.5	53.05	5462.5	3.28
336	10484.5	60.48	10727.5	3.58
504	16169	64.94	15984	3.63
672	21489	65.91	21233	3.69

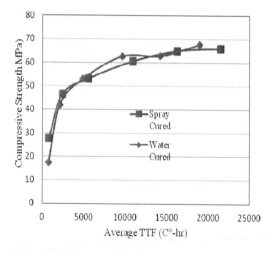

Figure 2. Compressive strength-maturity curve.

method increased the specimens' initial compressive strengths by the third day, but both curing methods led to the same target designed strength after twenty-eight days. Fig. 2 shows that, even though both mixes are designed for achieving a target strength of 60 MPa, the maturity curve for flexural strengths is different for both mixes. It is observed that the strength-maturity relationship depends on the curing conditions and the effect is more predominant in the case of flexural strength-maturity relationships. This shows that the maturity curve has to be developed separately for different properties, and also for different curing conditions for the same mix. A minimum flexural strength of 3.8 MPa is required for structural concrete before being subjected to flexural loading as in the case of opening pavements for traffic. The test results show that the selected mix under water curing had attained this strength within three days, whereas mixes, with surface spray of curing compounds, could not achieve this strength even after twenty-eight days.

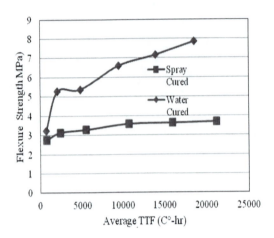

Figure 3. Flexure strength maturity curve.

4 CONCLUSIONS

- The maturity function for a concrete mix is directly dependant on the temperature development in the initial days.
- The surface spray curing method can be used for structural elements used in compression, as it increased the specimens' initial compressive strengths up to the third day.

- The strength-maturity relationships depend on the type of mix and curing conditions, thus the maturity curve has to be developed separately for different properties and for different curing conditions for the same mix.
- The water curing method is recommended for structural elements used in flexure, as it gave better results when compared with the surface spray curing method.

REFERENCES

ASTM C918, 2007. Standard Test Method for Measuring Early-Age Compressive Strength and Projecting Later-Age Strength, *ASTM Standards*.

ASTM C1074, 2004. "Standard Practice for Estimating Concrete Strength by the Maturity Method, *ASTM Standards*.

Kehl, R., Constantino, C. and Carrasquillo, R., 1998. Investigation of Match-Cure Technology and the Maturity Concept as New Quality Control. Quality Assurance Measures for Concrete, *Report 1714-s, Center for Transportation Research, University of Texas*.

Vázquez-Herrero, C., Martínez-Lage, I. & Sánchez-Tembleque, F. 2012. A new procedure to ensure structural safety based on the maturity method and limit state theory. *Construction and Building Materials, 35*, pp. 393–398.

Advances in Civil Engineering and Building Materials IV – Chang et al (eds)
© *2015 Taylor & Francis Group, London, ISBN: 978-1-138-00088-9*

An electrochemical impedance spectroscopy monitor for concrete with coated rebars and admixtures

A. Husain, S. Abdulsalam & S. Al-Bahar
Construction and Building Materials programme, Energy and Building Research Centre,
Kuwait Institute for Scientific Research, Kuwait

ABSTRACT: The efficiency of several types of admixtures and epoxy coated steel in stopping or reducing corrosion has been investigated. The monitoring of concrete structures in aggressive service conditions requires a quantitative assessment of the properties and corrosion rates of structural steel surrounded by concrete. This paper deals with predicting the short and long-term degradation of reinforced concrete structures in the Kuwait environment. For this reason, a rapid evaluation method has been developed to monitor polarization resistance, impedance, and the capacitance of admixture materials, using Electrochemical Impedance Spectroscopy (EIS). EIS tests for corrosion rate tests were carried out respectively on reinforced concrete specimens made of plain, silica fume, calcium nitrite, and on epoxy coated steel in the concrete. It was found, using EIS and the DC polarization method in the laboratory, that calculated polarization resistance was in agreement with that which was measured in the field. Also, quantitative and non-destructive evaluation of the corrosion rate evaluation is possible in the field by this method.

Keywords: corrosion of reinforced concrete structure, electrochemical impedance spectroscopy (EIS), silica fume, calcium nitrite, and epoxy coated steel

1   INTRODUCTION

A 1992 survey found that the general cost of corrosion to the construction industry in Kuwait was approximately KD 26 million. This amount is not insignificant, being approximately 0.5 per cent of the gross domestic product (Al-Matrouk et al., 1996). Thus, it is clear that corrosion-induced deterioration of reinforced concrete structures has economic consequences. The Arabian Gulf region provides an extremely aggressive environment, which is characterized by high ambient temperature and humidity conditions, and severe ground and ambient salinity with high levels of chlorides and sulphates in the soil and groundwater. Structures exposed to the marine environment, groundwater conditions, and industrial pollution have suffered the most. In Kuwait, concrete structures, designed for a typical service life of fifty years or more, have required extensive and costly repairs as a result of reinforcement corrosion after only twenty to twenty-five years (Al-Bahar and Attiogbe, 1995; Al-Qazweeni et al., 1992; Attiogbe et al., 1994a, b; Hamid, 1995; Matta; Shalaby and Daoud, 1990; and Rasheeduzzafar et al., 1984).

Corrosion protection systems reduce the risk of corrosion in reinforced concrete structures in many parts of the world. The use of protection systems, such as corrosion-inhibiting chemical admixtures and supplementary cementing materials, increases the cost of construction, but the cost increase is marginal compared with the costs for repair of the structures as a result of premature deterioration. It is estimated that when these protection systems are used, the increase in construction cost is typically only 10 per cent of the cost for the repair of the premature corrosion-induced deterioration that occurs in the absence of any protection systems (Berke et al., 1988). In general, corrosion protection systems are rarely used in normal concrete construction practices in Kuwait. This is largely due to the local construction industry being relatively unfamiliar with the use of these systems as well as the lack of comprehensive data on the long-term effectiveness of the systems in the environmental conditions prevailing in the region.

In this paper, the use of EIS from the point view of the concrete scientist will be presented, with an emphasis on its application simultaneously with accelerated exposure. EIS is used by laboratory concrete researchers for several purposes, among them:

- The detection of changes due to exposure
- The prediction of lifetime corrosion protection,
- The identification of corrosion processes that lead to failure
- The ranking of concrete materials
- The measurement of water uptake by admixtures
- The development of models for cementitious material during the incubation period of curing

The objectives of this research project are to identify corrosion protection systems that are most suitable for application in reinforced concrete structures in Kuwait. Moreover, to establish performance data on the selected corrosion protection systems under the typical local service conditions of Kuwait.

2 EXPERIMENTAL PROCEDURES

The test programme was designed to satisfy standard and popular laboratory test methods and outdoor experimental test evaluations. Table 1 presents a summary of the test programme that shows the types of tests run, specimen sizes used, and measurement parameters tested.

The laboratory test programme is supplemented by ASTM G 109: This test method evaluates the corrosivity of admixtures in a chloride environment on embedded steel reinforcements in concrete exposed to such an environment. The specimen dimensions are given in Table 1. Each specimen is cyclically ponded (two weeks wet and two weeks dry) with a 3% NaCl solution. The test is completed when the average macrocell current is greater than or equal to 10 μA, and the test is continued to ensure the presence of sufficient corrosion for visual evaluation.

2.1 Time-to-corrosion initiation

Measurement proceeds after 96 h of saltwater ponding, followed by a vacuum removal of the saltwater and an immediate freshwater rinse and a vacuum removal again. This is followed by 72 h of air-drying. The weekly measurements involve readings of half-cell potential, corrosion rate, and concrete resistivity, which are recorded with respect to a copper-copper sulphate reference electrode (CSE) of −350 mV.

2.2 Outdoor exposure studies

These evaluations are carried out within the tidal zone, submerged zone, and onshore, exposed to coastal humidity and overall weather conditions. A visual examination was supported by the removal of the steel reinforcement from the specimens to determine the mass lost. The chloride permeability was evaluated by extracting the concrete powder from the concrete specimens in order to examine the chloride ingress profile.

2.3 Electrochemical Impedance Spectroscopy (EIS)

EIS was utilized to characterize the performance of the various corrosion protection systems in an interactive media around the steel reinforcements during the hydration process and after the final setting of the concrete. Very small excitation amplitudes, in the range of 5 to 10 mV peak-to-peak, minimally disturb the sample and the attached corrosion products or the absorbed species during the testing. This technique was applied over a frequency spectrum from 100 KHz to 0.01 Hz on two sets of specimens, specimens involved in the corrosion rate measurement test (lollipop specimens), and a set of specimens specially designed to examine the role of internally added chlorides on the corrosion behaviour of steel reinforcement protected by different concrete systems. Table 2 shows the chemical compositions of the concrete specimens used for the impedance testing.

Concrete lollipop test specimens of 380 × 200 × 76 mm in size were used for this study. Each specimen

Table 1. Test programme summary.

Test	Specimen size	Measurement
ASTM G-109	Beams: 279 × 152 × 114 mm	Macrocell current Half-cell potential
Corrosion Rate (Lollipop) Time-to-Corrosion	Prisms: 380 × 200 × 76 mm Blocks: 300 × 300 × 200 mm	Macrocell current Half-cell potential Corrosion rate Macrocell current Half-cell potential
ASTM C-1202 (AASHTO T 277-86)	Cylinders: 100 × 200 mm	Total charge (conductivity)
AASHTO T 259-80	Slabs: 300 × 300 × 75 mm	Chloride profile
Chloride Diffusivity	Cylinders: 100 × 50 mm	Chloride content
Outdoor Exposure	Beams: 120 × 120 × 350 mm With 2 bars With 1 bar Cylinders: 100 × 200 mm	Visual examination Half-cell potential Steel mass lost Chloride content

Table 2. Concrete specimens used for Electrochemical Impedance Spectroscopy (EIS) testing

Specimen type	Specimen designation no.	Concrete specifications*
Plain Concrete	Pl-G2(2,3) Pl-G2(2,6)	Cement type I (0.5)
Epoxy-Coated Concrete	PlC-G2(1,3) PlC-G2(1,6)	Cement type I (0.5) Coated rebar
Epoxy with Damaged-Coated Concrete	Pld-G2(1,1) Pld-G2(1,2)	Cement type I (0.5) Coated rebar with pinholes
CN Corrosion-Inhibitor with Concrete	CN-G3(2,1) CN-G3(2,2) CN-3(2,1)2 CN-G3(2,2)2	Cement type I (0.5) with Calcium-Nitrite inhibitor
Silica-Fume Concrete	SF-G4(2,5)2 SF-G4(2,5) SF-G4(2,6)2	Cement type I with (0.5) with silica fume
Concrete with silica fume and corrosion inhibitor	G4(2,1)2 G4(2,2)	Cement type I (0.5) with silica fume and calcium-nitrite inhibitor.

*Water-to-cement ratios are given in parentheses.

had a single steel reinforcement bar (14 mm in diameter) embedded in concrete, centred and positioned 5 cm from the bottom. The samples were immersed in a 3% saline solution approximately 15 cm deep. Graphite rods were used as counter electrodes. Data acquisition was controlled using a Solarton Model 1287 Potentiostat-Galvanostat interfaced with a Solarton Model 1260 Frequency Response Analyzer (FRA) to provide sweep-frequency measurement. An equivalent circuit and data analysis software package known as the Solarton Z-plot was used for analysis of the resulting spectra.

3 RESULTS AND DISCUSSION

Table 3 summarizes the performance of the different concrete systems, based on the corrosion activities, concluded after time-to-corrosion testing. The results for half-cell potential, corrosion current density, and corrosion penetration rate indicated that the performance of silica-fume concrete exceeded that of calcium-nitrite concrete, and both performed better than ordinary concrete. Nevertheless, the corrosion morphology of steel rebars, pulled out of concrete specimens at the end of the test, illustrated the poor performance of ordinary concrete. Also, it exposed the hidden damage that took place in the epoxy coating layer above the steel rebar, in spite of acceptable readings of half-cell potential, indicating the vulnerability of the epoxy-coated rebars in a chloride-laden environment.

3.1 Corrosion rate measurement (Lollipop test)

This test simulates the wicking of chlorides by concrete in seawater to demonstrate the effectiveness of corrosion protection systems exposed to the marine environment. The results show the corrosion potentials of embedded, uncoated steel bars in different systems. Concrete with calcium-nitrite inhibitor and in combination with silica fume performed effectively in protecting the embedded steel bars, compared with concrete without these systems or with silica fume alone.

The behaviour of OUSF50U, as the value of the corrosion potential crossed the threshold limit to regions of uncertain corrosion, a 90 per cent probability of

corrosion took place on the surface of the steel reinforcements. In the case of epoxy-coated bar, protected by silica fume (OUSF50C), the corrosion potential values lingered around zero, showing the effectiveness of a multilevel protection system. In the cases in which epoxy-coated bars were damaged, the corrosion potential of OUSF50D approached the threshold value, demonstrating that silica fume alone is not capable of effective protection in this environment.

Visual examination of the field test specimens of each group revealed evidence of corrosion-induced deterioration, such as cracking, rust stain, scaling, spalling, and salt deposits. Splash and partially submerged zone specimens exhibited more severe symptoms of deterioration than those onshore. Concrete with silica fume sustained lower diffusion rates than ordinary concrete of the same water-to-cement ratios, which is simply ascribed to the variations between the pore structures of the materials.

3.2 Electrochemical Impedance Spectroscopy (EIS) analysis

The impedance results, shown in Figures 1 and 2 are in the form of Nyquist and Bode plots. The plots are for plain concrete, epoxy-coated concrete, damaged epoxy-coated concrete, calcium-nitrite inhibited concrete, silica-fume concrete (SF) and a mixture of silica-fume and calcium-nitrite concrete. Qualitative analysis of the Nyquist impedance spectra indicated

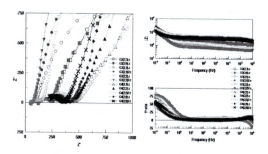

Figure 1. Nyquist and bode plot for plain concrete (PL), Silica fume (SF), Calcium nitrite (CN) & Si-Fume + Ca-nitrite.

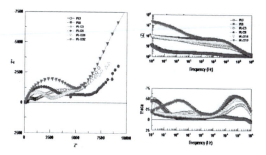

Figure 2. Nyquist and Bode plot for plain concrete (PL), epoxy-coated rebar (PL-C), and damaged epoxy-coated rebar (PL-CD).

Table 3. Corrosion activities of concrete systems after time-to-corrosion testing

Parameter	Half-cell potential (mV-CSE)	Corrosion current density ($m\mu/cm^2$)	Corrosion current density ($m\mu/cm^2$)
OUSF50U	−125	0.07	8.1 E-4
OUCN50U	−185	0.10	11.6 E-4
OU-50U	−220	0.14	16.3 E-4

that the corrosion performances of the different concrete mixtures are in agreement with the ASTM test results for corrosion rate and potential values obtained in the previous sections. The impedance results showed that the mixed silica-fume and calcium-nitrite concrete structure exhibited the highest impedance and capacitance values among the specimens.

This is shown clearly in Figure 1 after one year of exposure. As it is expected from the protective nature of the silica-fume concrete with calcium nitrite inhibitor, the diffusivity of the chloride species was slowed down in reaching the steel substrate.

In other words, the presence of a Warburg diffusion response indicated that the slow diffusion processes of the corrosive species through pores in the material and the corrosion product is the rate-controlling factor for the mechanism of concrete protection for this particular sample. In contrast to the concrete with calcium-nitrite inhibitor, the plain concrete, and the silica-fume concrete samples showed the lowest impedance and capacitance behaviour, respectively. This was due to the high rate of permeability and adsorption of the chloride corrosive species into the steel substrate surface. In the case of the inhibited concrete sample, the attempt to study such an inhibitor with concrete showed that improper concentrations of the inhibitor and/or specimen solution submersion may cause corrosion to the protection surface. It also tends to become active when the molar ratio of chloride to hydroxyl and nitrite ions in the pore solution reaches a critical value. Finally, the mixed concrete design of silica fume with calcium nitrite exhibited a very high impedance and capacitance values that exceeded the performance of the other concrete samples.

In the case of the epoxy-coated fusion-bonded concrete (Fig. 2), the epoxy-coated rebar had two time constants and a Warburg diffusion tail, while the damaged coating indicated polarization behaviour with a diffusion tail and was compared to the plain concrete specimen. It is usually expected that the EIS spectrum of a coated steel reinforcement in concrete will be in the form of a capacitive response if the coating is solid. The EIS spectra in Figure 2 did not show such responses. This may be attributed to weakness in the interfacial adhesion property of the coated film with respect to the cement cover.

4 CONCLUSION

- The experimental testing programme adopted in this study provides technical data that will add to the available knowledge worldwide on the performance of the selected corrosion protection systems in Kuwait
- The results demonstrated the effectiveness of epoxy-coated steel bars in resisting chloride attack, especially in multilevel protection setups, where they are used in combination with silica fume or calcium nitrite.
- For the very first time, EIS has reported clearly evidence of the detection of the localized corrosion damaged epoxy-coated rebar in a chloride-laden environment, in spite of the DC readings of half-cell potential.
- The results of the outdoor testing programme and long-term chloride ponding showed a consolidation of all of the conclusions above.

REFERENCES

Al-Bahar, S., & E. Attiogbe. 1995. Corrosion-induced deterioration of reinforced concrete structures in Kuwait. In concrete under severe condition: Environment and Loading Proceedings CONSEC'95, *International Conference, Sapporo, Japan. Edited by K. Sakai, N. Banthia and O.E Gjorv. Vol. 1: 564–573.*

Al-Qazweeni, O. Daoud, M. Al-Fadhala & K. Hijab. 1992. Causes of deterioration and repair measures in concrete foundations at al-Ahmadi Refinery. *Kuwait Institute for Scientific Research, Report No. KISR 4139, Kuwait.*

Attiogbe, E.K, S. al-Bahara, and H. Kamal.1994. Evaluation of a concrete Floor system in Safir International Hotel. *Kuwait Institute for Scientific Research, Report No. KISR 4235, Kuwait.*

Hamid, A.A 1995. Improving structural concrete durability in the Arabian gulf. *Concrete International 17(7): 32–35.*

Rasheeduzzafar, F.H, Dakhil, & A.S. Al-Gahtami. 1984. Deterioration of concrete structures in the environment of the Middle East. *ACI journal 81(1): 13–20.*

Shalaby, H.M. & O.K. Daoud. 1990. Case studies of deterioration of coastal concrete structures in two oil refineries in the Arabian Gulf region. *Cement and Concerete Research 20: 975–985.*

Advances in Civil Engineering and Building Materials IV – Chang et al (eds)
© 2015 Taylor & Francis Group, London, ISBN: 978-1-138-00088-9

Reliability of visual inspection, an example of timber structure in Valencia, Spain

M. Diodato, C. Mileto & F. Vegas López-Manzanares
Instituto de Restauración del Patrimonio, Universitat Politècnica de València, Valencia, Spain

ABSTRACT: In the process of surveying a building, the detailed assessment of historic timber structures is fundamental in order not to encounter unexpected issues during project and construction phases. Due to the general lack of knowledge about historic timber structures, this assessment it is often underestimated or not done, leading to the complete elimination of significant heritage. This article deals with the reliability of the assessment of wood degradation through visual inspections and the use of wood awls. The case study is a timber roof in Valencia, Spain, yet this method is easily applicable to many other cases. The comparison between the assessment and the real state of conservation will show how the prediction is more reliable the more accessible the analyzed elements are. Finally, even when the accessibility is reduced, the assessment can anyway show to the designer where further studies are needed and where conservation issues could emerge during the restoration works.

Keywords: timber, visual inspection, assessment, wood degradation, roof structure, Valencia

1 INTRODUCTION

In the Valencian region the presence and use of wood in traditional architecture is limited to roof and floor structures. The simple configuration of these structures in the common civil architecture produces a generalized lack of interest about their conservation. During the restoration of these historic buildings, the timber structures are not usually properly analyzed and therefore the probability of their preservation is significantly reduced.

Even without the use of sophisticated instruments it is possible to perform a visual inspection on the timber elements with the help of a wood awl in order to identify the material decay and consequent reduction of the cross section caused both by wood boring insects and rot. This type of assessment can be easily done in the majority of timber structures.

The goal of the inspection is to get near enough to the wood so that all the faces of each element can be observed from very close and through piercing the material with a wood awl is possible to identify the degradation under the surface. The depth that can reach the wood awl with the manual strength and the resistance of the wood are clear indicators of the health of the material.

It is clear that, without the help of instruments that can reach the hidden parts of the timber elements, the assessment has a degree of uncertainty. In order to evaluate this uncertainty a case study has been considered. A timber roof in Valencia has been visually inspected

Figure 1. Mid-19th century building in the historic center of Valencia, Spain. Case study.

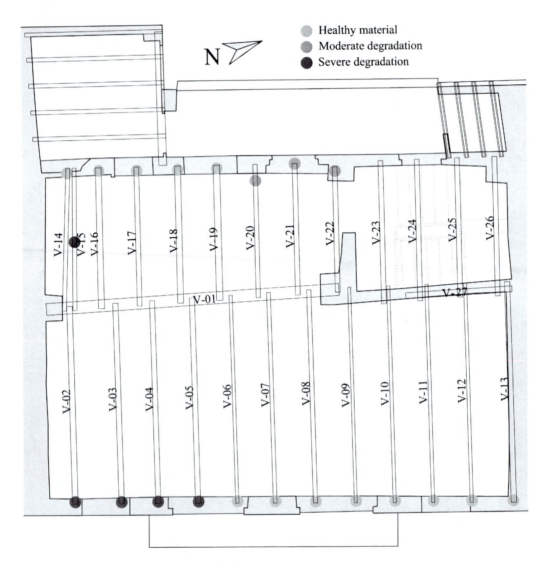

Figure 2.   Plan of the roof with the codes of the timber elements studied.

in order to assess its degradation and afterwards this assessment has been compared with the actual state of conservation that became visible once the layers that covered the roof structure were removed. This comparison finally allowed the estimation of the reliability of this method.

2   DESCRIPTION OF THE STRUCTURE

The case study is the roof structure of a mid-19th century building in the historic center of Valencia, Spain.

The building has a gable roof; the structure has a main beam with a square section (22 × 22 cm) that shapes the roof's ridge. This beam rests on top of two

pillars and two timber brackets which are supposed to be contemporary to the construction of the structure since they are embedded in the masonry of the pillars and the nails used in the union with the main beam seem hand made.

The main beam supports two series of rafters with different slopes profiling the two fields of roof: 12 rafters supported by the façade wall, facing east, and 13 rafters supported by the rear wall, facing west, four of these are not analyzed because hidden behind the false ceiling above the staircase (V-23 to V-26), three of them are supported by a plate inside the wall, V-27. The sections of these rafters are variables because some of them are replacements: the original ones, V-02 to V-05, V-13 and V-16 to V-22, have a cross section of approximately 10 × 20 cm while the

Figure 3. Roof structure.

substitutions, V-06 to V-12, have a cross section of approximately 6 × 21 cm; finally V-14 and V-15 have completely different and smaller sections. The timber structure was covered, on the west field of roof, with a layer of flat bricks on top of a series of battens perpendicular to the rafters, finally protected by a layer of tiles; on the other hand, the east field of roof was built with fiber cement corrugated sheets with almost no secondary structure. As the rafters, also the battens have very different sizes: 8 × 3.5 cm, 6.5 × 4 cm or 4.5 × 7.5 cm.

The two rooms added west to the main building use to be the kitchen and the bathroom, the roof structure of the first has four beams (7.5 × 11 cm) which hold another layer of flat bricks, while the second is covered by four beams (8 × 8.5 cm) decorated with two three-quarter molding. These two structures are new additions from the mid-20th century and will remain outside the present analysis although, as the rest of the structure, have been preserved and restored during the conservation works.

The microscopic identification of the wood species revealed that the original elements of timber belonged to the *Pinus nigra-sylvestris* group of species. Considering the population of trees in the Valencian inland and supply area, it is very probable that the wood is *Pinus nigra*.

3 ASSESSMENT

3.1 *Visual inspection*

In the effort to understand the structure, it is important to systematize the process of knowledge to every single supporting element so each structural timber piece has to be designated with a univocal code. The fundamental issue with timber structures is that it is impossible to extrapolate the presence of decay from one rafter to another.

In this case, the two fields of roof have a very different degree of accessibility. The east field of roof

Figure 4. Rafters: V-02, V-03, V-13, V-20 and V-22.

with its light and almost provisional structure leaved a small space that could be used to access with a camera to the rafters' end while in the west field of roof the rafters' ends were fully embedded within the masonry.

So, even if the most part of the rafters' surface was visible, the inspection concentrated on the usual problematic points: the ends. The difficult access and the direct contact with the masonry can results in the unsuspected entrance of both water and termites.

As a general fact it should be noted that the attack of wood-boring insects from the family of *Anobiidae* from the *Coleoptera* order, however not active, had been widespread. Nonetheless the damage caused concerned only the superficial centimeters and it was not a risk for the stability of the structure.

The fundamental agents that can cause a deep damage and a significant reduction of the rafters' section are two: the rot caused by the proliferation of fungal attacks on humid wood due to infiltrations and the termites of the order *Isoptera*, and especially the species *Kalotermes flavicollis* and *Retuculitermes lucifugus*, endemic in the Valencian region.

As can be seen in Figure 2 the degradation seemed to be concentrated in two specific areas.

Concerning the eastern rafters, the ends of V-02 to V-05 could be photographed and the damage made by the termites was clearly revealed (Fig. 4) however no insects were detected. Similarly, rafter V-13 was visibly affected by the infiltration of water due to the poor condition and execution of the roof. In all these cases the supporting area was greatly reduced.

Regarding the western rafters the interpretations were more uncertain. Even if it was not possible to assess in details the state of conservation, the ends of rafters V-21 and V-22 seemed damaged, while in rafter V-20, near the end, some infiltration water had been channeled inside the element that was partially rotted; this decay could be prolonged also to the material inside the wall.

Finally rafter V-15 had a diffuse attack of insects of the family of *Cerambycidae* or *Curculionida*, which had significantly decreased its section. This element, together with the V-14, appears to be a last-minute remedy possibly caused by the lack of bigger substitution element.

3.2 *Real state of conservation*

Once the protection of the roof structure was dismantled it was possible to see the real state of conservation of the elements and their ends and to compare the results with the visual inspection.

In the case of rafters V-02 to V-05 and V-13 the visual assessment had a 100% of correspondence to the real damage. The rafters' ends had completely disappeared due to termites even if the attack was not active. It also seemed that the problem was amplified by the humidity caused by the integrated gutter that historically passed near the rafters' ends and that maybe caused also the degradation and former substitution of the rest of rafters in the same field of roof.

Figure 5. Rafters: V-19, V-20, V-21 and V-22.

On the other hand, in the western rafters where the inspection had been more difficult because of the reduced accessibility to the ends inside the masonry, the reliability was partially reduced.

Rafters V-20 to V-22 were considered with the same degree of degradation while, once revealed, they have different extent of decay (Fig. 5).

V-22 had a great damage similar to the eastern rafters that could just be glimpsed from outside. V-21 was in good health as the rafters V-16 to V-18. Finally V-20 had a moderate extent of damage that is similar to the one found in V-19 which however was not detected in the inspection.

4 CONCLUSIONS

From the described example, can be inferred that the reliability of the assessment method through visual inspection and testing with wood awl is good.

While in the case of having visual access, at least with a camera, to the analyzed elements ensures the absolute reliability of the assessment, this is slightly reduced in the other cases. In the circumstance where is not possible to fully examine the rafters' ends because they are inside the masonry, the prediction given by the use of wood awls can give an indication of where further problems may be found. Although this part of the results will not give a 100% of correspondence with the reality, the indications are very helpful during the design and mainly during the construction works so the designer and the builder are prepared and alert to manage issues in specific areas of the building that they may need a detailed intervention.

With this simple method it is then possible to adjust the contingencies that always rise up the prize of the restoration work.

ACKNOWLEDGMENT

We would like to thank Salvador Tomas Márquez, the building engineer of the conservation works of the building for granting access to it and for providing the last picture of V-22.

REFERENCES

Diodato, M. 2015. Variaciones constructivas y formales en forjados y cubiertas. In Centro histórico de Valencia. Ocho siglos de arquitectura residencial. Edited by F. Vegas and C. Mileto. Valencia: Generalitat Valenciana.

Kasal, B. and Tannert T. 2010. In situ assessment of structural timber, State of the art report of the RILEM TC 215-AST. Dordrecht: Springer.

Macchioni N. 1998. Inspection techniques for ancient wooden structures. In: Structural analysis of historical constructions II: possibilities of numerical and experimental techniques. Edited by P. Roca, J.L. Gonzalez, E. Onate, P.B. Lourenco. Barcelona: CIMNE.

Nuere Matauco, E. 2000. La Carpintería de Armar Española. Madrid: Editorial Munilla Lería.

Vegas López-Manzanares F. and Mileto C. 2011. Aprendiendo a restaurar. Un manual de restauración de la arquitectura tradicional de la Comunidad Valenciana. Valencia: Colegio Oficial de Arquitectos de la Comunidad Valenciana.

Advances in Civil Engineering and Building Materials IV – Chang et al (eds)
© 2015 Taylor & Francis Group, London, ISBN: 978-1-138-00088-9

The systematic application of non-destructive testing techniques for ancient wood buildings

Wei Qian, Jian Dai, Xin Li & LiHong Chang
College of Architecture and Urban Planning, Beijing University of Technology, Beijing, China

ABSTRACT: In China, timber building is a main part of ancient buildings, accounting for more than 70%. Obviously, wood materials are perishable, especially in the natural environment. So ancient buildings are often faced with the unforeseen danger of collapse in the natural environment. This characteristic of wood material can cause ancient buildings to frequently face the unpredictable risk of collapse. Moreover, in recent years mass improper remedies by man meant that these ancient buildings' wood components, which have been disassembled and replaced without effective detection and scientific evaluation, has caused specimens of timber buildings with a large amount of historical original information to rapidly disappear. Therefore, we need do our best to obtain information about the damage status of timber structures of ancient buildings authentically and accurately in our country. In this study, we will break through the technology difficulty of accurate testing, non-destructive or micro-damage testing and suitable testing etc. and provide effective basic-research support for comprehensive security detection and health assessment of ancient timber buildings.

Keywords: Ancient buildings; Wood structure; Wood; Non-destructive testing

1 INTRODUCTION

In China, wood buildings are a main part of ancient buildings, accounting for more than 70% thereof. Obviously, wood materials are perishable, especially in the natural environment. So ancient buildings are often faced with the unforeseen danger of collapse in the natural environment. This characteristic of wood material can cause ancient buildings to frequently face the unpredictable risk of collapse. Moreover, in recent years mass improper remedies by man meant that these ancient buildings' wood components, which have been disassembled and replaced without effective detection and scientific evaluation, has caused specimens of timber buildings with a large amount of historical original information to rapidly disappear. Therefore, we need do our best to obtain information about the damage status of timber structures of ancient buildings authentically and accurately in our country. Non-destructive testing techniques have begun to be used in the protection of cultural heritage by virtue of security, portability, and quantitative data etc.

In Japan, timberwork accounted for ninety percent of cultural relic buildings, the non-destructive testing research of wood structure buildings can be traced back to the 60's. Application of non-destructive testing technology is earlier and more mature in Europe and the United States but most of them tend to be applications or research on the masonry of historic buildings. In recent years, some scholars have studied the non-destructive testing technology of ancient architectures in our country, and achieved certain results. However, both domestically and abroad the research has been on the application of individual non-destructive testing instrument for research. Due to the fact that the wood structures of ancient architecture is very complicated; now there are still lack of integrity, the systematic research on non-destructive testing techniques for the component part of wood structure buildings. In this study, we try to explore, through integrated and optimized technology methods, a variety of testing equipment and data acquisition instruments, to provide non-destructive testing for all the parts of the wood buildings. Eventually they will serve safety evaluations for ancient wood buildings.

2 THE COMPOSITION OF WOOD STRUCTURE BUILDINGS AND THE SUITABILITY OF NON-DESTRUCTIVE TESTING TECHNIQUES

Ancient Chinese wood building monomer can generally be divided into four main parts including bearing structure, the building envelope, foundation and the affiliated cultural relics. The bearing structure is generally a timber frame, including all kinds of traditional wood building units such as beams, pillars, rafters, purlines, and bucket arches etc. The building envelope is composed of the brick walls, roof, windows, and doors. The affiliated cultural relic is composed of

Table 1. Non-destructive testing technology for ancient wood building.

Parts	Material	Testing techniques
Bearing structure	Wood	Stress wave detection, Impedance test
Building envelope	Brick stone Wood	Infrared thermal imaging instrument, non-metal ultrasonic instrument
Foundation	Brick stone	Ground penetrating radar
Affiliated cultural relic	Stone, Painting	X-ray fluorometric analyzer

murals, inscriptions, painting, etc. The different functions of every structure and different materials need different non-destructive testing equipment and methods. (Table 1). This is different from the traditional detection method, using looking and drumming experience to determine the damage type and the damage degree of ancient building.

Firstly, we use stress wave detection to detect the damage of wooden parts of ancient buildings in combination with an impedance test. Secondly, the robustness of the brick and stone walls of an ancient building can be detected by an infrared thermal imaging instrument, and a non-metal ultrasonic instrument. Thirdly, we may use ground-penetrating radar to find the defects such as cracks etc. Finally, for the affiliated cultural relics, the efflorescence of stone tablets and the material composition of the coloured pattern need detection with an X-ray fluorometric analyser. We try to grasp the maximum damage information of ancient buildings through different means, and practice at the scene of the ancient building repair. The ultimate aim is to provide basic data for the safety evaluation of the ancient wood building structure.

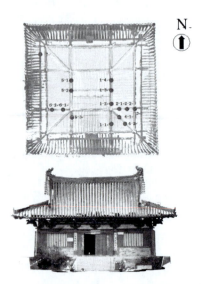

Figure 1. Test point plan of DaYun Yuan in ShanXi.

3 FIELD EXPERIMENT

3.1 Non-Destructive Testing for the bearing structure

We have been establishing a non-destructive testing system for Chinese ancient wood buildings since "The compass project- A detailed and precise survey for ancient architecture" funded by the State administration of cultural heritage which began in 2008. The field experiment of non-destructive testing was conducted for early Chinese wooden buildings in southern Shanxi. We collected a lot of useful data about the damage information of the main components and maintenance structure through the non-destructive testing. DaYun Yuan, located in the northwest of Ping Shun country, was built in the Earlier Song dynasty and belongs to the state protected historic sites. First, we detected the representative beams and pillars, rafters, purlines of Dafuo Hall, the most important building in DaYun Yuan by stress wave equipment (Figure 1). Based on the different sizes of the wood components, 6–10 probes were placed around them. Test results can display the propagation velocity of the stress wave in the wood, and show the damage situation by images (Figure 2). Through the different colours in the image,

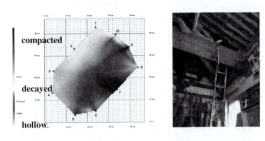

Figure 2. Stress wave detection for bearing structure.

we judge the damage type and damage degree of the wood components. Green can explain that the internal wood is compacted Dark green means more compacted than light green. Yellow to red show decayed wood components. Blue means that the internal wood is hollow.

From the NDT testing result of all tested components, we know that most of them are healthy except for the No.5 beam and the No.4 bucket arches (Figure 3). A large amount of data show that the surface cracks but internal may be very dense and the surface looks like a good condition but internal may be decayed. Therefore, people cannot just judge internal damage from seeing the wooden parts.

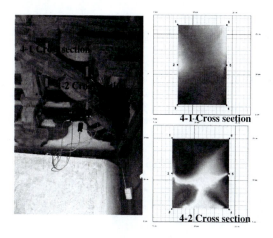

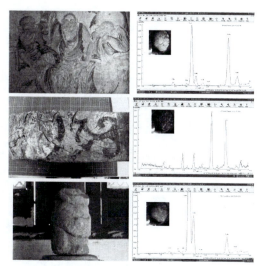

Figure 3. The decayed situation of the No.4 bucket arches.

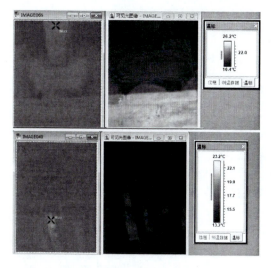

Figure 4. Roof testing of building envelope.

3.2 Non-Destructive Testing for the building envelope

The safety of building roof and the structure of building are equally important for ancient buildings. Infrared thermal imaging instruments can be used for testing. The infrared energy of the tested objects will display different colours from different thermal conductivity of each several part by the imaging (Figure 4). We observed that the colour of the damaged part of the roof was different from the surrounding objects. It is quicker and more accurate finding the damaged parts of the roof by infrared thermal imaging instrument.

3.3 Non-Destructive Testing for the affiliated cultural relic

Affiliated cultural relics of ancient buildings mainly contain stone materials and pigments. An X-ray fluorometric analyser is adapted to find their composition.

Figure 5. NDT of affiliated cultural relics.

Not only Da Yun Yuan, but also other ancient buildings in southern ShanXi have undergone NDT of their affiliated cultural relics by us (Figure 5).

After we tested the stone pillars, stone tablets, and stone carvings of Da Yun Yuan and analysed their composition, we found a high proportion of calcium in all stones, and almost never saw Silicon Dioxide (SiO_2) Therefore, these stone materials must have been taken from the local limestone. There is also a phenomenon of significance that the content of sulfur in these stone material surfaces was very high. Then we detected a stone fragment as a sample and broke it to analyse its composition We found that the sulfur content is very low, so we can conclude that the sulfur is due to external contamination. Limestone hardness is very high, but it is easy to be damaged by acid. It may be destroyed by acid rain and the pollution of sulfur dust are very serious around it. Therefore, this phenomenon is to be paid more attention, and corresponding protection work needs to be done. In the surrounding environment, we can find the sources of sulfur, mainly around the coal mines, and coking plants, etc.

We tested the composition of the wood units' painting in LongMen temple. The result showed us that all colours are drawn by inorganic mineral pigment. The main composition of white intonaco was plumbum (Pb), so it may be identified as cerusite (plumbous carbonate $PbCO_3$) with traditional Chinese intonaco material. A large amount of plumbum (Pb), copper (Cu), and ferrum (Fe) were detected in the black painted on the intonaco. The black identified is not from ink but the mixture of the bottom of the pot black (ferriferrous oxide Fe_3O_4) and malachite [(basic cupric carbonate $Cu_2(OH)_2CO_3$)]. On the other hand, the black from the bottom of the pot is cheaper than ink. From the analysis of the spectra, the composition of the yellow contains not only Fe but also aurum (Au). This shows that the artisans are using ferric oxide as yellow pigment, alternatively using gold foil to map the

Figure 6.   NDT for foundation of Prince Gong's Mansion.

glittering effect. At last, Cu was detected in the blue and green of the painting. This suggests that azurite [$Cu_3(CO_3)_2(OH)_2$] as traditional blue pigment, and malachite as traditional green pigment were used. Test results show that all of the paintings are traditional inorganic pigment without signs of modern synthetic pigments to repair. In ancient buildings, both expensive gold and cheap pigment such as black from the bottom of the pot were used widely, it reflected the wisdom of the ancient working people.

We also detected the wall painting in DaYun Yuan's MiTuo Hall established in A.D.940, we found different results between the front of the hall and at the back of the hall. In the front of MiTuo Hall, the wall painting intonaco was made from cerusite (plumbous carbonate $PbCO_3$) In addition, deep red is from cinnabar, orange red is from red lead, green is from malachite and yellow is from yellow rust or aurum(Au). In a nutshell, the wall painting in the front of MiTuo Hall was made with traditional inorganic pigment and used a traditional process. It is likely to be made in Tang Dynasty or Five Dynasty. In the meantime, the detected result of the wall painting at the back of MiTuo Hall shows us that intonaco was made with gypsum, and other colours such as red, green, blue, etc..., and are almost the same inorganic components. It proved that these pigments were made with synthetic organic pigments in the late Qing Dynasty or the Republic of China.

3.4 Non-Destructive Testing for foundation defects

The foundation defects of ancient buildings may be done by Ground Penetrating Radar. Ground penetrating radar can find the not compacted local strata, underground cracks of buildings from uneven settlement, and underground cavities and pipelines. Ground Penetrating Radar has been used for detecting roads and bridges. It is very useful for detecting ancient buildings' foundation defects. We have been using Ground Penetrating Radar in actual testing projects. In Prince Gong's Mansion, the uneven settlement was detected and the building was in dangerous. Through the NDT of the foundation, we obtained the data and knew where underground cracks and uncompact layers appeared. Then, the ancient building repair personnel can reference our results to conduct the repair work.

3.5 Precision correction by different Non-Destructive Testing devices

In the actual detection work, not only one tool is used for ancient buildings' local detection. Because we need to verify each other to obtain more accurate data through a variety of detection methods and every NDT equipment is not applicable to all testing environment. People need to have better choices to decide on the equipment. For example, we detected the damage of wood pillars by stress wave detection and an impedance test (Figure 7). According to the stress wave detection and impedance test results, we obtained the corresponding data to different damage types such as moth-eaten, split, decayed, etc. Comparing the stress wave velocity and the change of the resistance value, the damaged area inside wooden components can be corrected. It ensures the accuracy of the field detection.

4 CONCLUSIONS

In this paper, we propose that a system of non-destructive testing for ancient wood building needs to be established to evaluate the safety of ancient wood buildings scientifically. By virtue of different

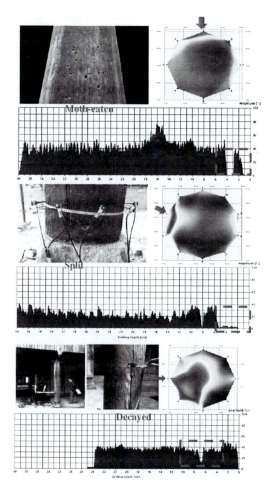

Figure 7. Precision correction by two NDT equipment.

Of course, non-destructive testing is not universal to ancient buildings. Because the structure and composition of Chinese ancient wood buildings are very complicated, single NDT technique and device is not enough to solve all problems. This is a key problem for us how to choose the suitable techniques to each part of ancient buildings and analyse their data. We need lots of experiments and practice and deeper study to achieve systematic NDT of Chinese ancient wood buildings.

ACKNOWLEDGEMENTS

Research subject of state science and technology (Grant No.2013BAK01B03); National Natural Science Foundation of China (NSFC) (Grant No. 51278003); The Natural Science Foundation of Beijing (Grant No.8132009)

REFERENCES

[1] Xinfang Duan, Non-destructive detection of decay and in sect attacked ancient wood members in Tibet by stress wave methods, J, China building industry press, Beijing, 2008.
[2] Duan Xin-fang, Li Yu-dong, Wang Ping, Review of NDE technology as applied to wood preservation, J, China wood industry, 2002.
[3] Shanxi Chong De ancient architecture planning and design limited company, The protection and repair plan of Cheng Tang King Temple, R, Zhang Zi county, Shanxi Province, 2012.
[4] Xuan li, Evaluation of technologies and restoration of modern historic buildings in Beijing based on deterioration mechanism of microbiology, D. Beijing University of Technology, 2013.
[5] Yuan An et al., Inspection of decay distribution in wood column by stress wave and exploration techniques, J. Journal of building mater lals, 2008.

suitable equipment, we respectively carried out non-destructive detection to four parts: bearing structure, building envelope, foundation, affiliated cultural relic. From the results we conclude that NDT is more convenient, accurate, and visual than traditional methods. The data obtained can be used for recording the situation (damaged type, material composition, etc.) of ancient wood buildings. This way we can maximize the protection of ancient buildings.

Engineering management

Advances in Civil Engineering and Building Materials IV – Chang et al (eds)
© 2015 Taylor & Francis Group, London, ISBN: 978-1-138-00088-9

Risk of megaprojects in transport infrastructure

V. Hromádka, J. Korytárová, L. Kozumplíková, P. Adlofová, D. Bártů & M. Špiroch
Brno University of Technology, Faculty of Civil Engineering, Brno, Czech Republic

ABSTRACT: The main objective of the paper is to describe present approaches to the risk management of large infrastructure projects carried out mainly in the area of transport – road and highway – infrastructure and to suggest new possible approaches and to test them in a case study. This paper works with the standard methods of risk analysis, such as sensitivity analysis, expert valuation, and simulation, which are described in the theoretical part of the paper. The case study carried out within this paper is based on a real megaproject comprising the modernization of the most frequently used highway in the Czech Republic – D1. The case study concludes with the present approaches to risk management and the authors suggest the next possible approaches to risk analysis and use them as data for the project.

Keywords: risk, efficiency, transport infrastructure, megaprojects, simulation

1 INTRODUCTION

The social impact of public projects could be very important; the bigger the project, the bigger the impact it is possible to expect. Therefore, it is not enough to define a precise objective of the project which has to be achieved, even if this is the most important element of the project management. It is necessary to take care of needed resources, and the efficiency of their utilization, in order to reach defined objectives. There are many methods for the assessment of efficiency, so it is not so big a problem to assess, if the project is efficient enough. However, each project (and especially big projects or megaprojects) is connected with a specific number of factors, which can threaten the achievement of project objectives and cause the project uncertainty. The paper deals with the risk management and risk assessment of big projects and megaprojects in the Czech Republic.

2 PRESENT STATE REFERENCES

Gellert and Lynch (2003) categorize megaprojects into four types: infrastructure (e.g., dams, ports, and railroads); extraction (e.g., minerals, oil, and gas); production (e.g., massive military hardware such as fighter aircrafts, chemical plants, and manufacturing parks); and consumption (e.g., tourist installations, malls, and theme parks). This paper is focused on megaprojects in transport, especially roads and highways, and infrastructure.

It is very important to take care of the total impact of these projects on society and their efficiency, because the impacts of these projects (technical, financial, economic, legal and environmental)are huge. High levels of public attention and/or political interests are attracted by these projects because of their substantial cost, and the direct and indirect impacts these projects have on the community, the environment, and on budgets (Capka, 2004)

Megaprojects are characterized by complexity, uncertainty, ambiguity, dynamic interfaces, significant political or external influences, and time periods reaching a decade or more (Floriceland Miller, 2001). At least three features associated with megaprojects are notable: large sum of resources, high human, social and environmental impact, and extreme complexity (Capka, 2004; Flyvbjerg, Bruzelius, and Rothengatter, 2003).

The complexity of megaprojects is brought about by a number of contributing factors such as tasks, components, personnel, and funding, as well as numerous sources of uncertainty and their interactions (Mihm, Loch, and Huchzermeier, 2003; Sommerand Loch, 2004).

All mentioned information should lead to the accurate risk management, because the risk connected with uncertainty of expected results is one of the most important issues within these kind of projects.

Miller and Lessard (2007) define risk as a possibility that events will turn out differently than anticipated.

The risk can be considered as an element of uncertainty, which more or less influences expected results of the human work and projects as well. The risk is on the one hand connected with expectation to achieve extra good economic results; on the other hand it could be threatened by the danger of the failure leading to losses, which can significantly affect the financial stability of the project and can lead to its downfall (Smejkal, 2005).

The risk can exist in a lot of different forms and it can have very different impact even if the attention is paid just to the construction sector and construction projects. Furthermore, many researchers analyse risks from different perspectives. E.g. Shen et al (Shen, 2007) suggest safety risks, health risks and ecological risks as an integral part of construction project sustainability performance checklist. Other authors focus on natural risks, e.g., in order to enhance risk-based decision making in case of floods (Jim Hall and Dimitri Solomatine, 2008).

Regarding the high number of possible risks, which can be met during the planning and the realization of investment projects, it is necessary to manage particular risks (Sanderson, 2012). The main objective of the project's risk management is to increase the probability of the success of the project and to minimize the danger of its failure.

The steps of the project risk management are as follows:

- determination of risk factors of the project,
- assessment of the importance of risk factors,
- project risk assessment,
- valuation of the risk of the project and the suggestion and the acceptance of operations for its decreasing,
- preparation of the plan of correction operations (Fotr, 2006).

Generally, the risk management has to be carried out efficiently. It must permeate all areas, functions and processes of the project (Schieg, 2010).The experiences from the past provide a big number of examples of risk that have influenced a lot of European megaprojects (Berlin Airport, Stuttgart 21, and Tunnel Blanka in Prague). In addition to cost overruns, Flyvbjerg (2003) concludes that more than 86% of megaprojects have not been finished on time.

For the risk management of projects and megaprojects it is crucial to define and choose an appropriate method or methods. The most important methods, according to the opinion of the authors, are described in the next chapter.

3 METHODOLOGY AND METHODS

3.1 *Determination of risk factors*

The determination of risk factors and the evaluation of their importance are basic steps for the company risk management (Edwards, 1999). The company risk factor can be characterized as a variable, whose possible future development can,in a positive or negative way,influence its economic situation. The identification of risk factors is not in principle a difficult issue, but it poses heavy requirements on experiences and professional knowledge of the evaluator. Considering factors those have been taken as sure yet could be the next step of the risk factor identification.There is

only a small number of inputs, which the development can considered as sure. It is necessary to dispute values of factors, which have not yet been considered as dangerous variables for the company (Minasowicz, 2009).

In the frame of the risk management system it is possible to create and use various tools, which makes the risk factors identification easy (e.g., help-list, checklist, interviews with experts or group discussions. The result of the factors identification is a written summary of all relevant risk factors.

3.2 *Assessment of the importance of risk factors*

The importance of risk factors assessment is the next very important step of the risk analysis. The importance of the specific risk factor provides information, whether it is necessary to carry out the next detailed analysis which assesses the total number of risks, or it is only the residual risk, which the subject is willing to accept and which will not be then analysed. For the risk factor importance assessment there are distinguished expert evaluations and the sensitivity analysis. (Hromádka, 2010)

The essence of the expert evaluation consists of the determination of the probability of the occurrence of the risk factors and the intensity of the negative impact. The probabilities of the occurrence of the risk factors and their intensities can reach five degrees, from very low to extra high.Factors, whose probabilities of occurrence and intensity of the negative influence are reaching at least the middle degree, or whose probabilities of occurrence are low, but the intensity of negative impact is extra high, are considered as important.The sensitivity analysis consists of the assessment of the sensitivity of certain economic criteria(Net Present Value, profit, and costs) on factors, which influences these criteria (demand for production and the production capacity utilization, sale prices, prices of raw materials, investment costs, bank rates, tax rates and the others).

3.3 *Project risk assessment*

It is necessary to quantify important risks in a suitable way. The risk can be assessed in a numeral form or indirectly using specific managerial characteristics. This paper is focused mainly on risk assessment in the numeric form, which consists of the calculation of statistical characteristics (mean, variance, standard deviation, and coefficient of variability), which in the financial management express the rate of the risk. The starting point is the determination of the probability distribution of the evaluation criterion (NPV, IRR, profit, etc.). For the determination of the risk in the numeral form, it is possible to use simpler approach in the form of probability trees. More difficult approach consists of the definition of the simulation model, which describes the probability distribution of the selected valuation criteria characterizing the risk for the company. For example, Monte

Carlo method enables effective assessment of the mean of the random variable and its next probability characteristics. The basic input necessary for the Monte Carlo method utilization is the probability distribution of the random variable. In the case of continuous random variables, it is possible to describe the variable with the distribution function or the probability density (Korytárová, 2011). The use of Monte Carlo method should be supported by use of quality random number generator.

4 CASE STUDY AND RESULTS

The investor in projects of public transport (road) infrastructure in the Czech Republic is usually the Road and Motorway Directorate of the Czech Republic, whilst the provider of financial resources is usually the State Fund of Transport Infrastructure. The usage of public resources must be transparent and efficient, so there is the necessity, which is defined by national directives, to approve of the efficiency of prepared project in road infrastructure using the feasibility study. The efficiency is within this study declared with Net Present Value (NPV), Benefit-Cost Ratio (BCR,) and Internal Rate of Return (IRR). NPV and IRR are common indexes used in investment valuation. BCR is defined as Net Present Value related to the discounted investment costs.

The cash-flows used within the analyses are in the form of investment and operation costs for infrastructure, and socio-economic costs and benefits connected with the projects. These are taken into account following the cost categories:

- costs of investor (investment, operation, and maintenance),
- operation costs of infrastructure users,
- costs for travel time consumption,
- costs connected with car accidents, and
- external costs connected with air pollution and noise.

Benefits are understood as the positive difference between costs of zero option and costs of investment option.

However, the objective of the case study is to demonstrate the risk analysis used in the frame of the big projects in road infrastructure used in the Czech Republic, and to suggest alternative approaches for risk analysis respecting the importance of these types of projects.

The case study is based on one of the most important infrastructure projects in the Czech Republic – reconstruction and modernization of the highway D1, the busiest motorway in the Czech Republic. The total duration of the investment phase of the project is seven years, expected investment costs are 860 million EUR and the length of the modernized part of the highway is 160.8 km.

The results of the economic efficiency calculated within the carried out feasibility study in the structure

mentioned above are as follows:

- NPV = 880 million EUR,
- IRR = 13,5%,
- BCR = 2,231.

4.1 Risk analysis of the project

Basic risk analysis is included nearly in each feasibility study intended for valuation of projects financed by State Fund of Transport Infrastructure. The essential approach to projection of the risk into the economic valuation is the usage pro the risk database. The expected price of investment is increased by the risk premium.The final risk premium includes the rates for particular parts or for the construction as a whole in following risk categories:

- Risks arising from the location of the construction;
- Risks arising from technological developments;
- Environmental risks;
- External risks;
- Legislative and legal risks;
- Economic risks.

The next simple approach is represented by the sensitivity analysis. The sensitivity analysis tests the project resistance against changes of inputs influencing the final efficiency of the project. The projects of road infrastructure are usually tested against changes of investment costs and socio-economic benefits. There is tested relative change of NPV by e.g., ten percent change of values of input parameters.

The sensitivity of the efficiency to changes of investment costs and socio-economic benefit is tested in the case study.Ten percent change of investment costs evokes 7.7 percent change of NPV. Ten percent change of socio-economic benefits makes seventeen percent change of NPV. The sensitivity test provides the information that the total efficiency is quite a lot sensitive on changes of socio-economic benefits and not so much sensitive on changes of investment costs.

The risk analysis also includes the identification of risk factors and the assessment of their importance. However, identified risk factors are very general and it is very difficult to derive any conclusion leading to the risk avoidance or mitigation of the risk. Identified risk factors are as follows:

- Project will not be finished in time and international funding will be threatened;
- Changes in constructions;
- Increasing of investment costs;
- Poor project documentation;
- Failure of technological discipline;
- Changes in legislation;
- Changes (decreasing) of the traffic intensity.

4.2 Next possible steps of the risk analysis

Megaprojects are very difficult to finance, and in many cases they are financed using public financial

Table 1. Probability characteristics in mil. EUR.

Characteristic	Value
Base Case	22 874.65
Mean	22 871.30
Median	22 795.11
Standard Deviation	9 709.87
Minimum	−11 589.38
Maximum	60 079.60
Range Width	71 668.98

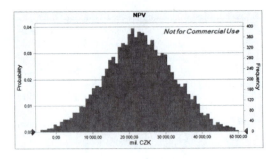

Picture 1. Probability distribution of NPV (results displayed in CZK).

resources. Therefore, it is very important to monitor not only expected efficiency of their utilization, but also the risks and risk factors that could threaten the achievement of expected results. It is possible to recommend proceeding according to the five steps defined in Chapter 2.

The first step includes detailed identification of risk factors using methods described in Chapter 3. The Highway and Motorway Directorate of the Czech Republic dispose of the detailed list of risks, which could help a lot in their identification.

The second step consists of the risk factors importance assessment. It is possible to divide risk factors into two groups First group includes risk factors, for which it is possible to determine mathematical relationship between the risk factor and the valuation criterion (NPV, IRR and BCR). Their importance is assessed using sensitivity analyses. Other risk factors must be evaluated by expert valuation comparing the probability of their appearance, and intensity of their impacts.

The third step consists of the risk assessment using scenario analysis or simulation. Both methods are described in Chapter 3; this case study presents elementary results of the simulation. Variables investment costs, and socio-economic benefits are considered as random variables (normal probability distribution); other inputs into the NPV calculation are considered as constants. The objective of the simulation is to assess the expected value of the NPV and the next probability characteristics expressing the risk to the project. The essential outputs of the risk analysis are displayed in the following Table 1.

The mean value as a basic probability characteristic confirms very good result of the project and its high efficiency. However, other results of simulation display quite a big range of the achieved values, including also the negative values. The standard deviation is quite big as well, which informs about big uncertainty of results. However, crucial information is displayed in Picture 1.

From the probability distribution displayed in Picture 1, it is evident that the probability of positive NPV is very big, more than 97%. Therefore, from this aspect it is possible to conclude that even though the uncertainty of the project is quite big, it is reasonably certain that the project will be successful and efficient from the economic aspect.

In the next step it is necessary to analyse results of the simulation and to suggest appropriate operations leading the risk mitigation, if it results from the analysis.

It is also necessary to focus on risk factors that do not influence the final values of NPV or IRR (there it is not possible to define exact mathematical relation), but could have important impact on the project success. It is possible to come out from results of the expert valuation and risk factors with high level of importance to analyse and to find appropriate solution for their avoidance or mitigation, e.g., in the form of CAR/EAR insurance (Hanák & Rudy, 2010).

As an example of practice protests and objections of civic organizations can be mentioned. These organizations can be focused on protection of laws of specific groups (residents in relevant locations) or on protection of general laws of the society (mostly with the environmental protection orientation). The second group of civic organization can discover a lot of mistakes in the project preparation, and their activities could lead to the delays of the project, and appearance of next expenditures intended to eliminate mistakes. These organizations have very erudite teams of lawyers and specialists, so they must be considered as strong opponents. The protection consists of the precise preparation of the projects, mostly during the proceedings leading to the territorial and building permissions.

5 CONCLUSIONS AND RECOMMENDATIONS

In the case of projects of big road and highway infrastructure the economic efficiency is assessed using the CBA analysis and the software HDM-4 Model. Economic efficiency assessed with this model is objective and correct. However, this tool does not include (instead of sensitivity analysis) any approach to the risk assessment and consequent management. Therefore, it is possible to recommend the utilization of more detailed risk analysis (a detailed list of risk factors, expert valuation, scenario analysis or simulation) and follow-up definition of steps intended for the risk avoidance or mitigation. The simulation carried out within this paper is processed using software

Crystal Ball, but it is possible to find another tool, for the MonteCarlo simulation realization.

ACKNOWLEDGEMENTS

This paper has been written with the support of The Ministry of Education, Youth and Sports Program COST CZ project LD14113 Effectiveness of Megaproject in the Czech Republic.

REFERENCES

Capka, J. R., (2004). Megaprojects – They are a different breed. Public Roads Magazine, 68, http://www.fhwa.dot.gov/publications/publicroads/04jul/01.cfm (accessed on 24.05.11).

Damodaran, A. (2002). Investment Valuation – Tools and Techniques for Determining the Value of Any Asset. John Willey and Sons, New York.

Edwards, P. (1999). Risk and risk management in construction projects: concepts, terms and risk categories re-defined. Journal of Construction Procurement. Vol. 5, No. 1, pp. 42–56.

Floricel, S., & Miller, R. (2001). Strategizing for anticipated risks and turbulence in large-scale engineering projects. International Journal of Project Management, 19, 445–455.

Flyvbjerg, B., Bruzelius, N., & Rothengatter, W. (2003). Megaprojects and risk: An anatomy of ambition. Cambridge, UK: Cambridge University Press.

Fotr, J. & Souèek, I. (2006). Podnikatelský zámìr a investièní rozhodování. Grada Publishing, a.s., Praha. ISBN 80-247-0939-2

Gellert, P. K., & Lynch, B. D. (2003). Mega-projects as displacements. International Social Science Journal, 55, 15–25.

Hall, J.& Solomatine, D. (2008). A framework for uncertainty analysis in flood risk management decisions, International Journal of River Basin Management, 6:2, 85–98.

Hanák, T. & Rudy, V. (2010). Stavební a montážní pojištìní (Construction/Erection All Risk Insurance),Stavebnictví, 4:11–12, 55–59.

Hromádka, V. & Kindermann, T. (2010). Analysis of costs for the realization of the building object. In People, Buildings and Environment 2010. Brno, Akademické nakl.CERM. P. 124–130. ISSN: 1805–6784.

Korytárová, J. &al. (2011). Management rizik souvisejících s dodávkou stavebního díla.(Management of Risks Related with Delivery of Building Object) Brno, Akademické nakladatelství CERM, s.r.o. 148 p. ISBN 978-80-7204-725-3.

Mihm, J., Loch, C., & Huchzermeier, A. (2003). Problem-solving oscillations in complex engineering projects. Management Science, 49, 733–750.

Miller, R., & Lessard, D. (2007). Evolving strategy: Risk management and the shaping of large engineering projects. MIT Sloan School of Management. Working Paper, 4639–4607.

Minasowicz, A. (2009). Feasibility study of construction investment projects assessment with regard to risk and probability of NPV reaching. OTMC Journal. Vol. 1, No. 1, pp. 10–14. ISSN 1847-6228.

Sanderson, J. (2012). Risk, uncertainty and governance in megaprojects: A critical discussion of alternative explanations. International Journal of project management 30(4): 432–443.

Shen, L.; Hao, J., I.; Tam, V., W.; Yao, H. (2007) A checklist for assessing sustainability performance of construction projects, Journal of Civil Engineering and Management, 13:4, 273–281.

Schieg, M. (2006) Risk management in construction project management, Journal of Business Economics and Management, 7:2, 77–83.

Smejkal, V.; Rais, K. (2005) Øízení rizik ve firmách a jiných organizacích. Grada Publishing, a.s., Praha. ISBN 80-247-1667-4

Sommer, S. C., & Loch, C. H. (2004). Selectionism and learning in projects with complexity and unforeseeable uncertainty. Management Science, 50, 1334–1347.

Advances in Civil Engineering and Building Materials IV – Chang et al (eds)
© 2015 Taylor & Francis Group, London, ISBN: 978-1-138-00088-9

Airfield pavement maintenance and rehabilitation management: Case study of Shanghai Hongqiao International Airport

Jian-Ming Ling, Zeng-Ming Du, Jie Yuan & Long Tang
Key Laboratory of Road and Traffic Engineering of the Ministry of Education, Tongji University, Shanghai, China

ABSTRACT: A decision tree approach for the Maintenance and Rehabilitation (M&R) applied in Shanghai International Airport is introduced. Because of the lack of historical data, the approach is established by historical experience. In order to obtain a more accurate evaluation of pavement conditions, four indexes calculated by monitored survey results are defined along with the Pavement Condition Index (PCI). Then the M&R treatments and intervention levels are detailed. For various pavement conditions, recommended and minimum M&R measures are proposed considering the decisions' uncertainties. Finally, with the establishment of a decision tree, the whole decision progress could be conducted and followed by a case study M&R decisions on aprons in Shanghai Hongqiao International Airport.

Keywords: decision tree, airfield pavement, maintenance and rehabilitation, pavement distress

1 INTRODUCTION

One of the major requirements of Airport Pavement Management System (APMS) is the ability to develop a pavement M&R program for the airport agency. Generally, the optimal pavement M&R strategy has been obtained through: (i) questionnaire surveys, (ii) historical practice, (iii) mathematical optimization, or (iv) any combination of these three. Expert opinion of the pavement practitioners could be incorporated into the decision process by questionnaire surveys, but it is subjective and inconsistent (Walls & Smith, 1998; Lamptey et. al.2008). Historical practice reflects continuance of past practices and would be easier to implement in the APMS. Unfortunately, due to the vulnerability of political forces or funding uncertainties, cost-effectiveness of historical practice is questionable (Lamptey et al. 2008). During the past two decades, numerous studies have established cost-effective maintenance and/or rehabilitation strategies using mathematical optimization thanks to higher computational capabilities, and the increasing availability of data (on cost and effectiveness) of pavement management databases of most agencies. Dialectically, the development of mathematical optimization would also be hindered due to the complex calculation progress and incomplete monitoring data.

In most of the practices reviewed, historical practice is still a more popular decision approach (Denehey & Edward J, 1997; Hicks, Seeds & Peshkin, 2000; Nasir Gharaibeh et al. 2010). Usually, the decision tree would be adopted to express the empirical decision

progress. Here, the decision tree framework applied in Shanghai Hongqiao International Airport is detailed and followed by a case study.

2 METHODOLOGY

In Shanghai Hongqiao International Airport, Portland Cement Concrete (PCC) Pavement plays a dominant role, hence the introductions of airfield pavement M&R management on PCC slab below.

2.1 Distress manifestation of Jointed Concrete Pavement (JCP)

In the most of APMSs, including PAVER, IAPMS and AIRPAVE, PCI is adopted to evaluate the functional performance (Barling, J.M. & Fleming, P. 2005). As a composite distress index, PCI could indirectly reflect the roughness, skid, and a structural integrity (not capacity) and be regarded as a criterion to make M&R strategies. However, it would be inappropriate to make decisions on detail maintenance measures through PCI since PCI could not reflect the certain condition of pavement. Table 1 briefly describes the four JCP distresses used by the decision tree in conjunction with the PCI to recommend treatments.

2.2 Treatments and intervention levels

Annual maintenance decision making uses the results of annual condition surveys to recommend one of the

Table 1. JCP Distress Manifestations.

JCP Distress	Brief Description	Calculation*
FJC	Spalled and/or unsealed joints and transverse cracks that can transfer load	(Corner Spalls + Joint Spalls + Unsealed Joints + Cracks) × k**
FL	Distresses resulting in structure failure: punchouts, faulted joints or cracks, failed concrete patches, D-cracking, wide or large spalls, etc.	(Faulted FJC + Punchouts + D-Crackings) × k
SS	Shattered slab or Faulted slab***	(Shattered Slabs + Faulted Slabs) × k
CP	Any concrete patch	(Concrete Patches) × k

*The calculation uses the number of distresses monitored in actual survey.
**k represents the scaling factor due to various areas of sections.
***Faulted Slab is the slab which is with five or more failures or with failure area over $10\,m^2$.

Table 2. JCP Treatments and Intervention Levels.

Level	JCP M&R Treatment	Index
NN	Routine Monitoring	M01
PM	Grouting	M02
	Grooving and Grinding	M03
	Joint Sealing	M04
	Repair of Spalled Cracks or Joints	M05
	Partial Depth Repair (PDR)	M06
LR	Partial Depth Repair (PDR)	M07
	Full Depth Repair (FDR)	M08
	Precast Slab	M09
MR	Full Depth Repair (FDR)	M10
	Precast Slab	M11
	Asphalt Concrete Overlay	M12
	FDR & Asphalt Concrete Overlay	M13
HR	PCC Overlay	M14
	Reconstruction	M15

M&R levels listed below. The JCP treatments are categorized into five levels: need nothing (NN), preventive maintenance (PM), light rehabilitation (LR), medium rehabilitation (MR), and heavy rehabilitation (HR). Furthermore, the correspondence between intervention levels and actual interventions commonly used to treat JCP could be identified as shown in Table 2.

2.3 Recommended and minimum treatments for various pavement conditions

The survey assessment was performed following the guidelines provided in the Pavement Condition Rating Manual (CAAC, 2009). It could be of significance to make M&R decisions based on the various deterioration conditions. Table 3 displayed the recommended and minimum treatments for various pavement conditions in Shanghai Hongqiao International Airport.

2.4 Decision Tree of M&R

Designing the decision tree is the most important part during the M&R decision making. Based on the historical practice, the decision tree adopted by Shanghai Hongqiao International Airport is depicted in Figure 1.

3 CASE STUDY

3.1 General information

The study airport, Shanghai Hongqiao International Airport, located in Eastern China. In 2013, the agency decided to take M&R measures on apron AE, AF and AH. The survey results are listed in Table 4 and shown in Figure 2.

As can be seen from the Table 4, scaling, small, and large patches play dominant roles in the distress survey results. PCI values of most units are over 85. It could be deduced form engineering practice that preventive maintenance could satisfy the M&R requirements of most surveyed units.

3.2 Results and discussion

According to the M&R decision approach proposed above, the annual recommended M&R plan could be given as shown in Figure 3, and Figure 4. The legends of the figures are displayed in Table 5.

As shown in Figure 3, and Figure 4, most pavement units require PM or NN which is in coincidence with the engineering judgment. Then, based on the M&R levels given by decision tree, some of the M&R measures are listed in Table 6 considering the distress survey results per unit.

4 SUMMARY AND CONCLUSION

A decision tree framework for M&R decision making in Shanghai International Airport is detailed in this

Table 3. Recommended and Minimum Treatments.

Code	Brief Description	Recommended Level	Recommended Treatments	Minimum Level	Minimum Treatment
C01	Large area of Severe Structure Distresses	HR	M15	MR	M12
C02	Severe deteriorated surface layer with some structure distresses	HR	M14	MR	M12
C03	Severe deteriorated surface layer with many large patches	HR	M14	MR	M12
C04	Asphalt overlay should be an option	MR	M12	LR	M09
C05	Structure distresses should be treated timely	MR	M13	LR	M08
C06	Many distresses are identified with some structure distresses	MR	M13	LR	M08
C07	Quite many patches exist and a number of new distresses are identified	MR	M10	LR	M07,M08
C08	Deterioration may accelerate in the near future	LR	M07–M09	PM	
C09	Some patches exist and few structure distresses are identified	LR	M07	PM	
C10	Many patches exist with no structure distress	PM	M02–M06		
C11	Some faulted slabs exist	LR	M08/M09	PM	
C12	No failures exist with some patches	PM	M02–M06		
C13	Few distresses and patches exist, but monitored structure distresses should be treated	LR	M08,M09		
C14	No structure distresses exist	PM	M02-M06		
C15	Few distresses exist	PM	M02-M06		
C16	Perfect condition	NN	M01		

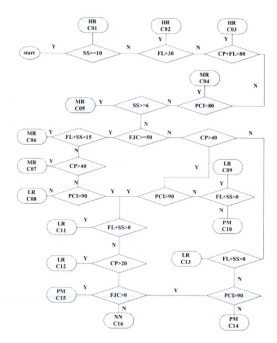

Figure 1. M&R Decision Tree.

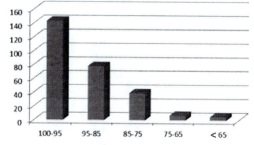

Figure 2. PCI Monitoring Results.

Table 4. Distress Survey Results*.

Distress Type	Number of Slabs with Distress	Ratio (%)
Transverse cracks	11	1.3
Corner break	4	0.5
Joint seal damage	93	10.9
Joint spalling	106	12.4
Popouts	4	1.8
Scaling	276	32.4
Corner spalling	64	7.5
Small patches	111	13.0
Large patches	173	20.3

*Only the distresses identified are listed in the table.

paper. For lack of historical data, the decision tree framework is established by experience. However, it could be concluded from the case study, that the decision to use the tree framework proposed here could give quite satisfying results.

As is known to us, the establishment of the decision tree approach hinders its application. Luckily, with the development of data mining and knowledge discovery, the decision tree could be constructed by the historical

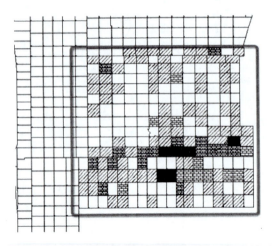

Figure 3. Spatial Distribution of M&R measures for AE and AF.

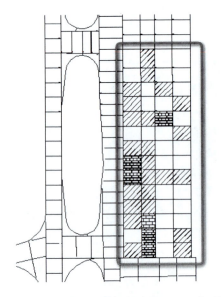

Figure 4. Spatial Distribution of M&R measures for AH.

Table 5. M&R Levels Depicted in Figures.

Legends	□	▨	▦	▨	■
Recommended M&R level	NN	PM	LR	MR	HR

data (Zhou, G. Q., et al.). Therefore the future work should focus on the establishment of the decision tree approach.

Table 6. M&R measures of some units in AE, AF and AH.

Unit Name	PCI	Recommended M&R treatment
AE1-1	97.5	M01
AE1-11	91.2	M05
AE1-17	76.4	M08/M09
AE1-44	87.8	M07
AE2-1	93.8	M04/M06
AE2-10	62.7	M15
AE2-13	84.7	M07
AE2-14	100	M01
AE2-26	64.8	M15
AH1-1	89.3	M06
AH1-13	95.7	M01
AH1-15	79	M08/M09
AH1-17	88.8	M07

ACKNOWLEDGMENT

The authors would like to thank the National Natural Science Foundation of China, Grant No. 51278364, for the financial support for this project.

REFERENCES

Barling, J.M. & Fleming, P. 2005. Safe Management of Airfield Pavements. Proceedings of the Institution of Civil Engineers: Transport, 158, Issue TR2, 2005, pp. 97–106.
Civil Aviation Administration of China. 2009. MH/T 5024-2009, Specifications on Pavement Evaluation and Management for Civil Airports. Beijing (in Chinese).
Denehey & Edward J. 1997. Implementing New York State DOT's Pavement Preventive Maintenance Program. In Transportation Research Record 1597, Transportation Research Board.
Hicks, Seeds & Peshkin, 2000. Selecting a Preventive Maintenance Treatment for Flexible Pavements. Foundation for Pavement Preservation, Washington, DC.
Lamptey, G., Labi, S. & Li, Z., 2008. Decision support for optimal scheduling of highway pavement preventive maintenance within resurfacing cycle. Decision Support Systems, 46 (1), 376–387.
Nasir Gharaibeh et.al. 2010. Evaluation and Development of Pavement Scores, Performance Models and Needs Estimates for the TXDOT Pavement Management Information System-Final Report. Texas A&M Transportation Institute.
Walls III, J. & Smith, M., 1998. Life cycle cost analysis in pavement design – interim technical bulletin. Washington, DC: Federal Highway Administration.
Zhou, G. Q., et al. 2010. Integration of GIS and Data Mining Technology to Enhance the Pavement Management Decision Making. Journal of Transportation Engineering-Asce 136(4): 332–341.

Advances in Civil Engineering and Building Materials IV – Chang et al (eds)
© *2015 Taylor & Francis Group, London, ISBN: 978-1-138-00088-9*

A fuzzy evaluation of investment risk in China's green building project

Jin Liu

School of Engineering, Fujian Jiangxia University, Fuzhou, China

ABSTRACT: In this paper, an evaluation index system of investment risk in China's green building project has been established, a model based on fuzzy synthetic evaluation for evaluating the investment risk of China's green building projects was put forward. Finally, a case study was presented to verify the feasibility and validity of the model which provides important reference for the owner.

Keywords: green building project, investment, risk, fuzzy evaluation

1 INTRODUCTION

China is in the process of industrialization and urbanization. In the new situation of constructing a harmonious saving oriented society, developing a green building project is the future development direction of China's construction industry [1], which can help to achieve strategic transformation and upgrading of the construction industry. However, in the green building project, compared with the traditional building, there are stricter requirements from the target cost construction time investment and implementation, and other aspects, therefore the risk in green building project is more prominent than in traditional building, especially as green building project in our country is still in a primary stage, the participating units lacking experience and the management not being mature, which exacerbates the risk of green building project. In order to ensure the smooth development of green building project in China, it is of great significance to study the risk of the green building project. In this paper, an evaluation index system of investment risk in China's green building project has been established, a model based on fuzzy synthetic evaluation for evaluating the investment risk in China's green building project was put forward. A case study was presented to verify the feasibility and validity of the model which is helpful to the owner to manage, and control investment risks in the green building project.

2 ESTABLISHMENT OF MODEL

2.1 *Establishing evaluation index system of investment risk in China's green building project*

According to a large number of actual survey data, experience of relevant experts and scholars on green building project, and the literature about identification of risk factors domestic and foreign, the international engineering project investment risk was divided into the following categories:

2.1.1 *Political risk*
Political risk is the risk brought by the war, the international situation, regime change, and policy change [2]. Political risk includes four risk factors, that is the war and civil strife, the low efficiency of the government sector, too much approval procedures, unreasonable interference on the project form government, imperfection of legality and regulations, and the change policy and standard.

2.1.2 *Economic risk*
Economic risk is the risk brought by the country's economic policy or other uncertainty factors, such as exchange rate, interest rate, and inflation, which are important factors affecting the investment success.

2.1.3 *Natural risk*
Natural risk is the risk caused by natural environmental factors [3], which include geography, hydrology, the meteorology of Location of the project's location, flood, fire, earthquake, typhoon, lightning and irresistible force.

2.1.4 *Contractor risk*
Contractor Risk is the risk caused by contractors, which includes poor ability of contractor, lack of personnel who hasan experience and technology of green building project, cannot an experience and technology of green building, cannot fully understand the green requirements from the owners' point of view, lack of information, documents, and insurance of green building project.

Table 1. Investment risk index of green building project.

Total risk	Primary risk	Secondary risk
Investment risk of China's green building project	Political risk R_1	war and civil strife R_{11}, the low efficiency of government sector and too much approval procedures R_{12}, unreasonable interference on the project form government R_{13}, imperfect of legal, regulations, the change policy and standard R_{14}
	Economic risk R_2	exchange rate R_{21}, interest rate R_{22}, and inflation R_{23}
	Natural risk R_3	geography R_{31}, hydrology R_{32}, and meteorology R_{33}
	Contractor risk R_4	poor ability of contractor R_{41}, lack of personnel who has the experience and technology for green building project R_{42}, cannot fully understand the green requirements from the owners'f point of view R_{43}, lack of information and document R_{14}, and relevant insurance products of green building project R_{45}.
	Design risk R_5	lack of integrated design experience R_{51}, under or incorrect estimation of design cost R_{52}, incorrect site survey R_{53}
	Technical risk R_6	project management level of the owner R_{61}, decision-making ability of the owner R_{62}, and the experience of the owner R_{63}

2.1.5 Design risk

Design risk is a risk caused by the design. Engineering design is very important to engineering quality, the design mistakes can lead to rework, repair, and great loss. Design risk includes lack of integrated design experience, under or incorrect estimation of design costs and incorrect site survey.

2.1.6 Technical risk

Technical risk is a risk caused by lack of experience and management ability of owner [4]. The risk of green building project are very complex, therefore, the experience, level of project management and decision-making ability of the owner are confronted with severe tests.

The author has summarized the various risk factors of green building project: an evaluation index system of investment risk in China's green building project has been established (see Table 1). The investment risk in the green building project is called total risk, the next level of risk is called primary risk, that includes political risk, economic risk, natural risk, design risk, and technology risk. The index shows the sources of risk and the levels of risk which refer to the specific risk factors.

2.2 Establishing evaluation set

Due to the need for risk analysis, evaluation set was divided into five levels, which includes "very large, large, medium, small, and very small". Y is used to express evaluation set, that is Y = {very large, large, medium, small, and very small}.

2.3 Establishing evaluation matrix

$$R = \begin{bmatrix} R_{11} & R_{12} & ... & R_{1l} \\ R_{21} & R_{22} & ... & R_{2l} \\ ... & ... & ... & ... \\ R_{m1} & R_{m2} & ... & R_{ml} \end{bmatrix} \qquad (1)$$

Amongst them, R_{ij} $(i = 1, 2, \ldots, m; j = 1, 2, \ldots, l)$ is membership degree of the i evaluation index to the j evaluation level, which reflects the fuzzy relation expressed by membership representation between evaluation index and evaluation level; m is the number of evaluation index, and l is the number of evaluation level.

2.4 Determining weight

In this paper, Analytic Hierarchy Process was used to determine the weight. Analytic Hierarchy Process is a method of system analysis which combines quantitative analysis and qualitative analysis and is often used in risk assessment [5]. Its basic steps are as follows: (1) Building the hierarchical structure. Each evaluation index will be decomposed into several levels from top to bottom according to different attributes, which form the hierarchical structure [6]. (2) Establishing a judgement matrix. After the establishment of the hierarchical structure, and multiple comparisons amongst each evaluation index on the same level according to a criterion, the relative important degree of evaluation index can be determined according to the evaluation criterion, and then judgement matrix can be established. (3) Determine the weight of evaluation index through the judgment matrix, and the weight will do the consistency test [7]. (4) Sorting out the weights.

2.5 Fuzzy synthetic evaluation

Fuzzy algorithm was used to calculate, calculation formula is as follows [8]:

$$B = A \times R \qquad (2)$$

Amongst them, A is weight set of evaluation index, and R is evaluation matrix [9].

First, one-level fuzzy synthetic evaluation should be calculated. The results of one-level fuzzy comprehensive evaluation is $B_i = A_i \times R_i$, then two-level fuzzy synthetic evaluation should be calculated; two-level evaluation matrix is expressed as follows: $R = [B_1, B_2, \ldots, B_i, \ldots, B_m]^T$, m is the number of one-level

Table 2. Weight and evaluation matrix of investment risk in green building project.

Primary risk	Weight	Secondary risk	Weight	Very big	Big	Model	Small	Very small
Political risk R_1	0.1297	War and civil strife R_{11}	0.1214	0.2	0.4	0.1	0.2	0.1
		Low efficiency of government sector and too long approval procedures R_{12}	0.3526	0.4	0.1	0.3	0.1	0.1
		Unreasonable interference on the project form government R_{13}	0.2731	0.3	0.1	0.3	0.2	0.1
		Imperfect of legal, regulations the change policy and standard R_{14}	0.2529	0.1	0.2	0.5	0.1	0.1
Economic risk R_2	0.1131	Exchange rate R_{21}	0.2892	0.6	0.1	0.1	0.1	0.1
		Interest rate R_{22}	0.3364	0.4	0.2	0.2	0.1	0.1
		Inflation R_{23}	0.3744	0.2	0.3	0.3	0.2	0
Natural risk R_3	0.0845	Geography R_{31}	0.1096	0.7	0.2	0.1	0	0
		Hydrology R_{32}	0.3428	0.1	0.1	0.2	0.5	0.1
		Meteorology R_{33}	0.5476	0.4	0.1	0.2	0.1	0.2
Contractor risk R_4	0.3563	Poor ability of contractor R_{41}	0.1163	0.1	0	0.2	0.6	0.1
		Lack of personnel who owns experience and technology of green building project R_{42}	0.2130	0.5	0.3	0.1	0.1	0
		Cannot fully understand the green requirements from owners R_{43}	0.1782	0.3	0.2	0.3	0	0.2
		Lack of information and document R_{44}	0.2561	0.2	0.1	0.4	0	0.3
		Relevant insurance products of green building project R_{45}	0.2364	0.1	0.5	0.2	0.1	0.1
Design risk R_5	0.2125	Lack of integrated design experience R_{51}	0.3825	0.4	0.2	0	0.2	0.2
		Under or incorrect estimation of design costs R_{52}	0.3706	0.2	0.3	0.1	0.2	0.2
		Incorrect site survey R_{53}	0.2469	0.3	0.3	0.1	0.2	0.1
Technical risk R_6	0.1069	Project management level of the owner R_{61}	0.2785	0.5	0.1	0.2	0	0.2
		Decision-making ability of the owner R_{62}	0.4162	0.2	0.2	0.3	0.1	0.2
		Experience of the owner R_{63}	0.3053	0.4	0.1	0.2	0.2	0.1

risk, the total evaluation vector is represented as $B = A \times R$[10]. The results of comprehensive evaluation were manipulated by weighted average method.

3 APPLICATION EXAMPLES

A number of overseas experts were invited to evaluate the investment risk of a green building project.

3.1 Determining one-level evaluation matrix

Risk degree of two-level was evaluated by a number of experts, then one level evaluation matrix could be determined (see Table 2).

3.2 Determining weight

Analytic Hierarchy Process was used to determine the weight. Results as shown in Table 2.

3.3 Establishing evaluation matrix

Risk degree of two-levels was evaluated by a number of experts, then one -level evaluation matrix could be determined (see Table 2).

3.4 Fuzzy synthetic evaluation

$B_1 = A_1 \times R_1 = [0.1214, 0.3526, 0.2731, 0.2529]$

$$\times \begin{bmatrix} 0.2 & 0.4 & 0.1 & 0.2 & 0.1 \\ 0.4 & 0.1 & 0.3 & 0.1 & 0.1 \\ 0.3 & 0.1 & 0.3 & 0.2 & 0.1 \\ 0.1 & 0.2 & 0.5 & 0.1 & 0.1 \end{bmatrix} = [0.2725, 0.1617, 0.3263,$$

0.1395, 0.1000].

Similarly, $B_2 = [0.3830, 0.2085, 0.2085, 0.1374, 0.0626]$, $B_3 = [0.3300, 0.1110, 0.1890, 0.2262, 0.1438]$, $B_4 = [0.2465, 0.2433, 0.2477, 0.1147, 0.1477]$, $B_5 = [0.3012, 0.2618, 0.0618, 0.2000, 0.1753]$, $B_6 = [0.3446, 0.1416, 0.2416, 0.1027, 0.1695]$.

235

Then, the two-level fuzzy synthetic evaluation can be calculated.

$$B=A \times R=[0.1297, 0.1131, 0.0845, 0.3563, 0.2125, 0.1069] \times$$

$$\begin{bmatrix} 0.2725 & 0.1617 & 0.3263 & 0.1395 & 0.1000 \\ 0.3830 & 0.2085 & 0.2085 & 0.1374 & 0.0626 \\ 0.3300 & 0.1110 & 0.1890 & 0.2262 & 0.1438 \\ 0.2465 & 0.2433 & 0.2477 & 0.1147 & 0.1477 \\ 0.3012 & 0.2618 & 0.0618 & 0.2000 & 0.1735 \\ 0.3446 & 0.1416 & 0.2416 & 0.1027 & 0.1695 \end{bmatrix}$$

$$=[0.2952, 0.2114, 0.2091, 0.1471, 0.1402].$$

Weighted average method was used to manipu late the results. $0.2952 \times 90 + 0.2114 \times \times 70 + 0.2091 \times 50 + 0.1471 \times 30 + 0.1402 \times 10 = 26.568 + 14.798 + 10.455 + 4.413 + 1.402 = 57.636$, synthetic score of this green building project is 57.636, risk rank is medium-large.

4 CONCLUSION

In this paper, an evaluation index system of investment risk in China's green building project has been established. An Analytic Hierarchy Process was used to determine the weight, a model based on fuzzy synthetic evaluation to evaluate the investment risk in China's green building project was put forward, an example, showing the feasibility and validity of the model which provided important reference for the owner.

REFERENCES

Fei Xianhui. The Institutional Analysis of Chinese Green Building Policy [D]. Beijing: Beijing Forestry University, 2011.
Guo Lijuan. Promoting the Study of Key Problems of Green Construction Management [D]. Jinan: Shandong University, 2011.
Li Bo. The Development and application of Green Building [J]. Value Engineering. 2014, (1):108–109.
Liu Jin, and Gao Xuanneng. Improved Fuzzy Synthetic Evaluation of Investment Risk in International Engineering Project [J]. Journal of Huaqiao University, 2011, 32 (6): 689–693.
Liu Jin, and Gao Xuanneng. Fire Risk Assessment of Hospital Building Based on Improved Risk Matrix Method [J]. Mathematics in Practice and Theory, 2012, 42(1): 115–121.
U.S. Marsh Green Building Team. Green Building Assessing the Risks, 2009.
Ding Guli, and Chen Jin. Study on Green Construction Risk Evaluation Based on Grey System Theory [J]. Journal of Engineering Management, 2009, 24(2): 182–185.
Wu Wenjing. Study on Green Residential Incremental Investment Risk Assessment and Coping Strategies from the Perspective of Developers [D]. Hangzhou: Zhejiang University, 2013.
Hu Xingfu. An Empirical Analysis of the Construction Project Risk Source and Risk Factors [J]. Scientific Research in Sichuan Building, 2011, (04): 53–59.
Zhao Hong. Study on Evaluation of Green Building [D]. Shijiazhuang: Hebei University of Science and Technology: 2012.

Advances in Civil Engineering and Building Materials IV – Chang et al (eds)
© 2015 Taylor & Francis Group, London, ISBN: 978-1-138-00088-9

The efficiency and risks of megaprojects in the Czech Republic

V. Hromádka, T. Hanák, J. Korytárová & E. Vítková
Brno University of Technology, Faculty of Civil Engineering, Brno, Czech Republic

ABSTRACT: Current profound insufficient funding of road construction infrastructure results in an enormous pressure on the economy and the efficiency of road construction. The topic of this paper reflects a methodological insight into streamlining the process of economic evaluation of transport infrastructure projects with respect to their Whole Life Costs (WLC). The WLC development and the combined methods of Life Cycle Costs (LCC), Life Cycle Assessment (LCA), and Cost Benefit Analysis (CBA), which work with the analysis of the studied road construction itself, in this case, of a section of a road or a highway. The WLC expands the existing practice in the assessment of the impact of road construction on society and the environment by its cost and the benefits that arise in the vicinity of the construction, and which may affect or be affected by other entities that are not directly related to the construction and its usage.

Keywords: risk, efficiency, traffic infrastructure, megaprojects, cost, benefits

1 INTRODUCTION

The paper deals with the issue of the valuation of the economic efficiency and risks of large-scale projects and megaprojects in the area of traffic, mostly road and highway, infrastructure in the Czech Republic. These types of projects have a huge social importance and socio-economic impacts including the consumption of a large amount of public resources. From this perspective, it is necessary to take care of the process of decision making about the project realization, its transparency and economic and social benefits for the society as a whole. The paper presents the overview of published information about examined issue, then it introduces essential methods and approaches for the economic and risk valuation, and the situation in the Czech Republic is demonstrated in the case study of ongoing projects of the D1 highway modernization.

2 PRESENT STATE REFERENCES

The economic valuation of large-scale projects and megaprojects in the area of traffic, mostly road and highway, infrastructure in the Czech Republic, is based on the essential principles of the CBA (Cost-Benefit Analysis). The CBA approach is in detail described in the European Commission Methodology (Florio, 2002).

For the real expression of the utility of the project for the society it is very important to describe in detail, and to quantify the benefits for the society and the impacts caused by realization and the operation of these projects.

Ongoing improvement and more realistic social cost-benefit ratios are still key areas of research (Priemus, 2010). In the Czech Republic, HDM-4 model (which is explained in Section 3) is used for economic valuation of important projects of the road and highway infrastructure.

Together with the economic efficiency, it is also crucial to evaluate risks of megaprojects during the preparation phase. Miller and Lessard (2007) define risk as the possibility that events will turn out differently than anticipated. Bruzelius, Flyvbjerg, & Rothengatter (2002) identify four main groups of risks (cost risk, demand risk, financial market risk, and political risk).

The cost risk consists mostly of the uncertainty of achievement of maximum supposed costs for the construction delivery, and arising of costs overruns. The demand risk is connected with the uncertainty of the interest of the planned infrastructure mostly by future users. The financial risk is connected with the threat of instability of financial resources provided by investors, or other providers of finances. The risk with quite a big possible impact is the political risk. The political risk is connected mostly with megaprojects having the potential of a big impact on a huge number of beneficiaries, and could be in conflict with interests of some stakeholders or civic organizations. Risk management has to be carried out systematically; five steps include (Kardes et al., 2013) the following tasks:

– Define risks: both exogenous and endogenous threats and opportunities must be identified.
– Assess and quantify risks: once threats and opportunities are identified, they should be assessed and prioritized. (Cavusgil & Deligonul, 2011).

- Determine risk response strategies: management decides on a set of actions to respond to each risk (risk management strategies) – elimination or avoidance, reduction, transfer, and retention.
- Implement: risk responses are implemented as planned.
- Monitor and update: the outcomes are monitored and necessary modifications are made.

In the risk reduction strategy, financial hedging and defining pre-specified project content, such as fixed price, time, quantity, etc. are the most common techniques to prevent possible risk (Miller, 1992). The risk transfer is usually connected with appropriate insurance and in the case of the risk retention, the investor accepts the risk without next measure.

The main challenge in the risk management of megaprojects consists of the complexity of megaprojects. The complexity of megaprojects is brought by a number of contributing factors such as tasks, components, personnel, and funding, as well as numerous sources of uncertainty and their interactions (Mihm, Loch & Huchzermeier, 2003; Sommer & Loch, 2004).

3 METHODOLOGY AND METHODS

The infrastructure projects and megaprojects are very important for the development of national economy. The WLC expands the existing practice in the assessment of the impact of road construction on the society and environment by costs and benefits that arise in the vicinity of the construction, and may affect or be affected by other entities that are not directly related to the construction and its usage. Many of these aspects can be concentrated in a Whole Life Costs (WLC) methodological approach. The WLC develop and combine methods of Life Cycle Costs (LCC), Life Cycle Assessment (LCA), and Cost Benefit Analysis (CBA), which work with the analysis of the studied road construction itself, in this case, the section of a road or a highway. Appropriateness of this combination is confirmed by further studies. Life Cycle Assessment (LCA) (Finnveden et al., 2009) and Life Cycle Costing (LCC) (Ciroth et al., 2008) are sustainability tools that take into account the full Life Cycle (LC) of a product, from raw material extraction, and production to use, and final disposal.

A very complex approach to evaluation of investment projects in the transport infrastructure on the base of CBA, is the HDM-4 model. The model had been developed during 1993–2000 at University of Birmingham with the financial support of World Bank. HDM-4 model (Highway Development and Management) is focused on evaluation of investment projects in the area of transport infrastructure, mostly highway projects. It concerns about complex software tool for analyses of roads and highways and investment decision making in this area. The methodological guideline of the Road and Motorway Directorate of the Czech Republic prescribes the obligation to use the HDM-4 model for the valuation of the economic efficiency of projects in road and highway infrastructure. This methodological guideline is in the Czech Republic gradually updated by next costs and benefits connected with valuated infrastructure. Currently, following items are valuated regarding the cash flow: cost of the transport route (cost of construction and reconstruction, cost of maintenance, and repairs of transport routes); costs of users (fuels, lubricants, tire wear, repair and maintenance of vehicles, and other expenses of trucks - crew wages, insurance, depreciation, overheads, etc.); other costs (valuation of time passengers, time value of the carriage of goods, loss from traffic accidents); other external costs (loss from traffic noise; losses from motor vehicle exhaust gases).

The net cash flows (NCF) for calculation of criteria (Net Present Value, Internal Rate of Return, and Benefit Cost Ratio) are based on information about the present state and future development of traffic for zero and investment variant. Zero variant represents the variant "without project", and investment variant supposes realization of the evaluated project. For both variants of cash flows it is possible to know information about the traffic flow, its intensity, its structure, and changes due to the project realization. The cash flows for valuation criteria NPV, IRR and BCR are calculated as economic costs, e.g., without taxes (especially VAT or excise duty).

The CBA analysis is accompanied by basic risk analysis using the sensitivity analysis, where investment costs and expected value of benefits are tested. The standard valuation period is currently 30 years and time value for money used as a discount rate is 5.5%.

Other important approaches for the investment valuation are methods based on impact of projects on their surroundings. These are the Life Cycle Costs (LCC) method, Life Cycle Assessment (LCA) method and Whole Life Costs (WLC) method (Guidance Paper F, 2004).

4 CASE STUDY AND RESULTS

The case study consists of the economic and risk valuation of the modernization of the highway D1, the most important highway in the Czech Republic.

4.1 Description of the case study

The highway D1 was finished in 1980, but several sections are in the operation since the seventies. Nowadays, this most important road in the Czech Republic is in need of modernization. Nearly the whole part of D1 between two biggest Czech cities – Prague and Brno is considered for modernization. The end for the whole road modernization is planned for 2018, the total length of the modernized section is 160.8 km, and the total expected cost is 861 mil. €. The whole modernization is divided into 21 sections.

The project of modernization consists of complete repair to the surface and extensions of the hard

Table 1. Cashflow analysis.

Variable	Zero-investment	option
Costs of investor	−679.2 mil. €	(cost)
Operational costs	984.5 mil. €	(benefit)
Costs of travel time	547.0 mil. €	(benefit)
Costs of car accidents	29.8 mil. €	(benefit)

shoulder (lay-by lanes) of 2 × 0.75 m. New road construction will have cement concrete surface using recycled materials.

4.2 The economic analysis

The economic analysis is based on the cash flow analysis respecting the amount of savings between costs within the investment option and zero option. The investment option includes the project realization and the zero option is an option without project. The net cash flows intended for the valuation are calculated as a difference between costs of the investment option and costs of zero option. Valuated costs are defined by User Guidance for the Czech Roads Valuation and are as follows:

Investment costs and costs for maintenance, and residual value of investment (costs of investor);

- Operational costs of vehicles (fuels, and tires);
- Costs for travel time consumption;
- Car accidents.

The main criteria for the economic efficiency assessment are Net Present Value (NPV), and Internal Rate of Return (IRR). Basic inputs into analysis, which is carried out within the software HDM-4, are information about present and future state of related roads and highways (type, and quality), information about investment costs and residual value of investments, and the development of traffic flows in related area in the frame of valuated period.

The results of the calculation of indexes are as follows:

Net Present Value: 861.151 mil. €.

Internal rate of Return: 13.5%.

It is possible to analyse the difference of cash-flows (costs) of zero and investment option. The net cash-flow, as a difference between zero and investment option is, according to the monitored variables, displayed in Table 1.

Table 1 displays the difference between zero and investment option for each of valuated socio-economic cashflows. Net cashflow from investment and maintenance can be considered as a cost of project; other net cash flows are in the form of benefits. In total benefits the savings of operational costs contribute 63%, savings of costs for travel time contribute 35%, and savings of costs for car accidents contribute 2%.

According to the results it is possible to state that the valuated investment into modernization of the highway D1 is efficientfromthe economic point of view.

4.3 The risk analysis

The risk analysis is carried out as a part of the Study of Economic Efficiency in quite a simple form. The risk analysis consists of:

- Identification of critical variables using sensitivity analysis;
- Identification of risks using expert valuation.

4.3.1 The sensitivity analysis

The sensitivity analysis provides the information about the total impact on change of selected critical variables on the Net Present Value and the Internal Rate of Return. As the critical variables there are considered investment costs and benefits of the project. It is generally possible to state that decreasing of costs and increasing of benefits lead to the better results, whilst the opposite situation leads to the decreasing of the efficiency. Calculations are usually made for increasing (decreasing) of critical variables of 10%, 20%, and 30%. The valuated project is from this aspect very resistant; neither the decreasing of benefits of 30%, nor the increasing of investment costs of 30% cause negative values of the valuation criteria (NPV, and IRR).

4.3.2 Expert valuation

Expert valuation in the risk analysis allows to take into account also risk factors, which are not suitable for the sensitivity analysis, because it is not possible to find appropriate mathematical relation amongst the risk factors and valuation criteria. It is very important to consider all possible risk factors with nonzero impact on the project and nonzero probability of occurrence. According to the investor experiences, the risk catalogue involving expected and common risks usually connected with these kinds of projects might be used. According to the risk catalogue of investor (Road and Motorway Directorate of the Czech Republic) there were identified the following risks:

- Project will not be finished in time and international fundingwill be threatened;
- Changes in constructions;
- Increasing of investment costs;
- Poor project documentation;
- Failure of technological discipline;
- Changes in legislation;
- Changes (decreasing) of the traffic intensity.

The expert valuation of risk provides information about risk factors that can eventually negatively influence the investment project in the realization and operation phase. It is important to conclude that mentioned risk factors are suitable for all investment projects and good project preparation (financing, design, building permission etc.) can avoid or reduce quite a lot of mentioned risks. For the risk factors connected with economic results it is possible to recommend detailed risk analysis including scenario analysis or the MonteCarlo simulation.

4.4 Results of the case study

The project solved within the case study is considered as efficient with acceptable rate of risk. According to the analysis of results it is possible to recommend the project for realization.

5 CONCLUSIONS AND RECOMMENDATIONS

The economic efficiency analysis and the risk analysis belong to the most important documents of pre-investment phase of the project life cycle. These documents should prove the project feasibility from theeconomic and risk aspect. The case study introduced the results of economic efficiency analysis of one of the most important projects of transport infrastructure in the Czech Republic. The analysis is carried out respecting the Czech legislation ordering the use of HDM-4 model. The results of the economic analysis are correct and respect wide scale of impacts (positive and negative) of project on the society. The Study of the Economic Efficiency also includes the part of risk valuation. There is elaborated the sensitivity analyses proving resistance of the projecttochanges of critical variables (investment costs, and benefits). Risk analysis also includes the identification of risks according to the risk catalogue, but regarding the importance and the size of the project it is possible to recommend more detailed risk analysis involving scenario analysis, or MonteCarlo simulation. However, it is not onlyimportant to describe, to identify, and to assess the risk, it is also very important the make suggestion to avoid or to decrease the risk, or to make the risk acceptable foran investor.

ACKNOWLEDGEMENTS

This paper has been written with the support of The Ministry of Education, Youth and Sports Program COST CZ project LD14113 Effectiveness of Megaproject in the Czech Republic.

REFERENCES

Bruzelius, N., Flyvbjerg, B., & Rothengatter, W. (2002). Big decisions, big risks. Improving accountability in mega projects. Transport Policy, 9(2), 143–154.

Cavusgil, S. T., & Deligonul, S. (2011). Exogenous risk analysis in global supplier networks: Conceptualization and field research findings. Information Knowledge Systems Management, 10, 1–19.

Ciroth, A., Huppes, G., Klöpffer, W., Rüdenauer, I., Steen, B., Swarr, T., 2008. Environmental Life Cycle Costing, first ed. CRC Press, Publishing House Taylor and Francis, SETAC Press, Pensacola, FL.

Finnveden, G. (1997). Valuation methods within LCA – where are the values? Int. J. Life Cycle Assess. 2, 163–169.

Florio, M. (2002) Guide to Cost-Benefit Analysis of Investment Projects.[PDF document].European Commission. Available from: http://www.strukturalni-fondy.cz.

Guidance Paper F concerning the Council Directive 89/106/EEC (2004) relating to construction products: Durability and the construction products directive CONSTRUCT 99/367, CONSTRUCT 04/655 Rev.1, European Commission.

Kardes, I, Ozturk, A., Cavusgil, S. T., Cavusgil, E. (2013). Managing global megaprojects: Complexity and risk management. International Business Review 22 (2013) 905–917.

Mihm, J., Loch, C., & Huchzermeier, A. (2003). Problem-solving oscillations in complex engineering projects. Management Science, 49, 733–750.

Miller, K. D. (1992). A framework for integrated risk management in international business. Journal of International Business Studies, 23, 311–331.

Miller, R., & Lessard, D. (January 2007). Evolving strategy: Risk management and the shaping of large engineering projects. MIT Sloan School of Management. Working Paper, 4639–4607.

Priemus, H. (2010). Mega-projects: Dealing with Pitfalls, European Planning Studies Vol. 18, No. 7, July 2010.

Sommer, S. C., & Loch, C. H. (2004). Selectionism and learning in projects with complexity and unforeseeable uncertainty. Management Science, 50, 1334–1347.

Geotechnical engineering

Advances in Civil Engineering and Building Materials IV – Chang et al (eds)
© 2015 Taylor & Francis Group, London, ISBN: 978-1-138-00088-9

Research on the prediction method of Gompertz model for high subgrade settlement

Xiangxing Kong
The First Highway Survey and Design Institute of China Communications Construction Company Ltd., Xi'an, China

ABSTRACT: According to the development law of high embankment settlements, the Gompertz curve model is established. Combined with specific engineering examples, the Gompertz curve well reflects the change process of high embankment settlement, and the prediction result of the model with the measured settlement value is very accurate and reliable.

Keywords: Gompertz model, high subgrade, settlement, prediction method

1 INTRODUCTION

The construction of expressways in the west, generally occur with high filling and excavation of roadbed structures [1–2]. After completion of construction of high fill embankments, subgrade settlements often appear and even cause the destruction of pavement structures and engineering accidents [3]. In order to ensure the safety of the normal operation of highways, according to the law of development of high filling subgrade settlement, observation data and combining site, Gompertz curve model is applied to predict the settlement of the roadbed.

2 MECHANISM OF HIGH FILLING SUBGRADE SETTLEMENT

In the early period of construction, just after loading, the soil is in elastic status. The soil pore water do not be exhausted [4–5]. Due to the soil having instantaneous shear de-formation, with the increase of load, the settlement has approximately linear increase. In the later period of construction, with increasing load, the load imposed on the high embankment becomes bigger and bigger. The pore water in the foundation soil is gradually discharged, meanwhile the excess pore water pressure gradually decreases. So compaction causes more de-formation, and the soil in the embankment will be in elastic-plastic state. With the continuous development of the plastic zone, settlement rate increases quickly, until the load no longer increases.

After construction, due to the pore water pressure being close to disappearing completely, the consolidation process is not complete; settlement will continue with the passage of time, but significantly decreases the sedimentation rate. If the time of settlement is up to

infinity, the settlement will be at the ultimate state. And its rate is zero, so the settlement at that time is really the final. In fact, fifteen years is generally considered as the enough long time.

3 THE FORECAST MODEL OF GOMPERTZ CURVE

At present, there are two categories for calculating the method of the settlement of the embankment. First, according to the constitutive model of consolidation theory combined with various soils, it is the settlement calculation of the application of numerical methods, such as visco-elastoplastic finite element model method [6–7]. The method of the calculation of the finite element method is very high, generally only used to calculate the key end of large engineering. In addition, there is a large gap between with some constitutive model and engineering practice, and needs a large number of soil tests to determine the soil parameters, so it is difficult to use in general engineering designs. Second, based on the measured data, is a calculated settlement prediction method of quantity and time relation, such as hyperbola and exponential curve method. This method is concise and practical, and can satisfy the engineering precision requirement. Method for forecasting subgrade settlement of this selection is one of the Gompertz models.

Gompertz method is proposed by statistician and mathematician, B.Gompertz, and is also a growth curve, which is widely used in economics, management, and statistics. Gompertz forecasting model is a growth curve model, and can be used to express and describe the process of growth and development. The practice indicates a prediction model that can be used as a reliable, effective method to forecast the settlement prediction of roadbed.

4 CALCULATION OF PARAMETERS

The three parameters in the model can be obtained using the three estimations. First, it is made into a modified exponential form, then the data is divided into three sections (if not divisible by three, it can increase and decrease the individual data which is a multiple of three), each containing a number of segments, i.e., seek and obtain.

$$y_t = e^{(k+ab^t)} \tag{1}$$

$$y' = k + ab^t \tag{2}$$

$$y_1 = k + ab^1 \tag{3}$$

$$y_2 = k + ab^2 \tag{4}$$

$$y_3 = k + ab^3 \tag{5}$$

$$y_t = k + ab^T \tag{6}$$

$$\sum\nolimits_1 y_t = \sum_{t=1}^{n} y_t = nk + ab(b^0 + b^1 + b^2 + \cdots + b^{n-1}) \tag{7}$$

$$\sum\nolimits_2 y_t = \sum_{t=n+1}^{2n} y_t = nk + ab^{n+1}(b^0 + b^1 + b^2 + \cdots + b^{n-1}) \tag{8}$$

$$\sum\nolimits_3 y_t = \sum_{t=2n+1}^{3n} y_t = nk + ab^{2n+1}(b^0 + b^1 + b^2 + \cdots + b^{n-1}) \tag{9}$$

$$(b^0 + b^1 + b^2 + \cdots + b^{n-1}) = \frac{b^n - 1}{b - 1} \tag{10}$$

$$\sum\nolimits_1 y_t = nk + ab\frac{b^n - 1}{b - 1} \tag{11}$$

$$\sum\nolimits_2 y_t = nk + ab^{n+1}\frac{b^n - 1}{b - 1} \tag{12}$$

$$\sum\nolimits_3 y_t = nK + ab^{2n+1}\frac{b^n - 1}{b - 1} \tag{13}$$

$$b = \sqrt[n]{\frac{\sum_3 y_t - \sum_2 y_t}{\sum_2 y_t - \sum_1 y_t}} \tag{14}$$

$$a = \frac{b - 1}{(b^n - 1)^2 b}(\sum\nolimits_2 y_t - \sum\nolimits_1 y_t) \tag{15}$$

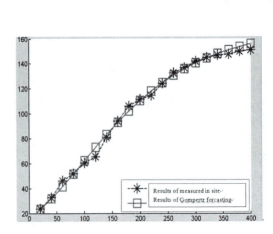

Figure 1. Results of Gompertz forecasting and measured in K123 + 560.

$$k = \frac{1}{n}(\sum\nolimits_1 y_t - ab\frac{b^n - 1}{b - 1}) \tag{16}$$

$$k = \frac{1}{n}\left[\frac{\sum_1 y_t \sum_3 y_t - (\sum_2 y_t)^2}{\sum_1 y_t + \sum_3 y_t - 2\sum_2 y_t}\right] \tag{17}$$

By using the method of analogy, Gompertz prediction parameters can be calculated as below:

$$a = \frac{b - 1}{(b^n - 1)^2 b}(\sum\nolimits_2 \ln y_t - \sum\nolimits_1 \ln y_t) \tag{18}$$

$$b = \sqrt[n]{\frac{\sum_3 \ln y_t - \sum_2 \ln y_t}{\sum_2 \ln y_t - \sum_1 \ln y_t}} \tag{19}$$

$$k = \frac{1}{n}\left[\frac{\sum_1 \ln y_t \sum_3 \ln y_t - (\sum_2 \ln y_t)^2}{\sum_1 \ln y_t + \sum_3 \ln y_t - 2\sum_2 \ln y_t}\right] \tag{20}$$

5 ENGINEERING EXAMPLES

Combined with the subgrade settlement observation data in Hunan province, the Changde-Zhangjiajie expressway, the K103 + 380, K123 + 560, are chosen as the case study. The high fill subgrade filler is typical, where it is filled with a low liquid limit clay and shale, an embankment height is about 13m.

In Figure 2, the time-scattered settlement is "S" shape. The first 12 points (20–240 days) data are selected as samples, Matlab 6.5 is used to calculate related parameters of a Gompertz curve model (as shown in Table 1), then the results of Gompertz curve model for settlement prediction of roadbed (as shown in Table 2) are obtained, as well as the sum squared error (prediction precision) 168.13.

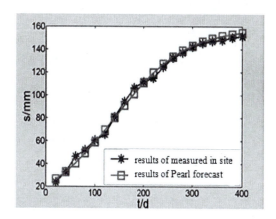

Figure 2. results of Pearl and forecasting and measured in K103 + 380.

Table 1. results of Gompertz and forecasting and measured in K123 + 560.

Tine/d	Gompertz forecasting/mm	Measured in site/mm	Absolute error/mm
20	23.6187	23.9	0.3
40	32.1777	32.9	0.7
60	41.7554	46.3	4.6
80	52.0064	52.1	0.1
100	62.5736	60.6	2.0
120	73.1273	65.7	7.5
140	83.3896	80.9	2.5
160	93.1464	94.4	1.3
180	102.2482	106.3	4.1
200	110.6040	111.6	1.0
220	118.1724	114.9	3.2
240	124.9498	124.6	0.3
260	130.9612	132.7	1.7
280	136.2502	137.0	0.7
300	140.8721	142.3	1.4
320	144.8878	145.2	0.3
340	148.3601	147.1	1.3
360	151.3503	148.0	3.3
380	153.9166	150.2	3.7
400	156.1125	151.4	4.8

Table 2. results of Pearl and forecasting and measured in K103 + 380.

Tine/d	Pearl forecasting/mm	Measured in site/mm	Absolute error/mm
20	25.32749	22.944	2.383488
40	31.62893	31.584	0.044928
60	39.0169	44.448	5.4311
80	47.4527	50.016	2.5633
100	56.79898	58.176	1.37702
120	66.81466	63.072	3.742656
140	77.17133	77.664	0.49267
160	87.49296	90.624	3.13104
180	97.40851	102.048	4.63949
200	106.6032	107.136	0.5328
220	114.8543	110.304	4.550304
240	122.0436	119.616	2.427648
260	128.1486	127.392	0.756576
280	133.2204	131.52	1.700352
300	137.3577	136.608	0.749664
320	140.6829	139.392	1.290912
340	143.3236	141.216	2.107584
360	145.4007	142.08	3.320736
380	147.0224	144.192	2.830368
400	148.281	145.344	2.937024

Table 3. Parameters of Gompertz method.

K	a	b
5.1264	−2.3314	0.8426

(3) In order to improve the prediction's accuracy, measured samples should be taken more frequently and compared over a period of time with the original samples.

6 CONCLUSION

In the forecasting of road foundation settlement, different methods can often provide valuable information.

(1) The settlement of a high embankment has its own law of development. Gompertz model can express S shape relationships between settlement and the time of the whole process of subgrade. The settlement not only is shown very well, but also can be accurately applied in the prediction of a high embankment.

(2) Using the Gompertz model to forecast the settlement of a roadbed, in order to obtain good prediction effect, enough measured data is the basis. The measured settlement value should be checked, and obvious error eliminated.

REFERENCES

[1] Highway traffic division. The highway engineering quality problems prevent guide [M]. Beijing: China Communications Press, 2002.
[2] Xu Jiayu, Cheng Ka Kui. Road engineering [M]. Shanghai: Tongji University press, 1995.
[3] Zai Jin min, Mei Kunio. Settlement of the whole process of prediction method [J]. Rock and soil mechanics, 2000, 21 (4): 322–325.
[4] Gu Xiaolu, Qian Hong Jin, et al. The foundation and foundation [M]. Beijing: China Architecture Industry Press, 1993.
[5] PI Dao Ying, Sun Youxian. An algorithm of [J]. Control and decision of multi model adaptive control, 1996, 11 (1): 77–80.
[6] Zhao Minghua, Liu Yu, et al. The soft soil roadbed settlement development law and its forecasting [J]. Journal of Central South University, 2004, 35 (1): 157–161.
[7] Bates J Granger C W J. Th examination of forecasts [J]. Operations Research Quarterly, 1969 (12): 319–325.

Advances in Civil Engineering and Building Materials IV – Chang et al (eds)
© 2015 Taylor & Francis Group, London, ISBN: 978-1-138-00088-9

An analysis of tunnel monitoring results based on modulus maxima method of wavelet transform

Xiangxing Kong

The First Highway Survey and Design Institute of China Communications Construction Company Ltd., Xi'an, China

ABSTRACT: In recent years, the theory and method of wavelet analysis is widely used in signal processing, pattern recognition, data compression, image processing, and quantum physics. Wavelet packet decomposition and coefficient shrinkage de-noising method of wavelet transform is compared with modulus maxima. Their advantages and disadvantages are analysed and summarized, and their respective scopes obtained. The Noissin, chosen as the original signal with noise, is analysed and de-noised by the modulus maxima method of wavelet transform. Meanwhile the usage conditions and key computing parameters are also obtained. Finally, the modulus maxima method of wavelet transform is successfully adopted to de-noise the monitoring results of a shield tunnel. The data revised are reliably provided for healthy tunnel diagnosis.

Keywords: tunnel monitoring, results analysis, modulus maxima method, wavelet transform

1   INTRODUCTION

In 1910, Hear proposed the theory of wavelet orthonormal basis, which is the earliest wavelet theory [1]. In 1936, Littlewoods and Paley established a binary frequency component grouping theory for Fu Lie series, which carried on the division to the frequency. It was the earliest theory in the source of multi-scale analysis [2]. In 1981, according to the group theory, the French physicist, Morley, first proposed the concept of wavelet analysis in the geological analysis of geological data [3–5]. As a signal analysis expert, Mallet proposed the concept of a multi-resolution analysis, and gave a general method for constructing an orthogonal wavelet basis [6]. The most commonly used multi resolution analysis technique has two kinds: one is the analysis of time limited resolution, and the other is spline multi-resolution analysis.

2   COMPARATIVE ANALYSIS OF DE-NOISING METHOD

There are many signal de-noising methods based on wavelet transform, which can be summarized in the following three categories.

(1) The modulus maxima de-noising method, based on the wavelet transform, is mainly suitable for a signal mixed with white noise, and contains more singular points. The theory of signal singularity detection and the wavelet transform modulus is proposed based on the maximum principle by Mallet. It effectively

uses propagation characteristics of different multi-scale spaces in de-noising signals and white noise. In de-noising and singularity information, this method effectively retains a signal and no excess concussion, and is a very good estimate.

However, during reconstruction, using an alternating projection method, in order to ensure the accuracy of a reconstructed signal and improve the signal-to-noise ratio, the modulus maxima de-noising method is usually iterated dozens of times, so the calculation speed is slow. In addition, it is important to choose the wavelet coefficients of the wavelet decomposition scale, and a small scale is easily affected by noise, and a large scale could make signal loss an important local singularity.

(2) The wavelet packet decomposition de-noising method, based on a coefficient of shrinkage, is mainly suitable for the case where the signal is mixed with white noise. The method, based on Donoho and Johnstone, put forward discrete wavelet transform. Wavelet packet decomposition can provide a more precise signal. According to the analysis of the characteristics of the signal, it is adaptive selection of frequency band. Compared with wavelet transform, wavelet packet decomposition can obtain richer time-frequency localization information, and is more suitable for the analysis of nonstationary signal processing.

However, the de-noising effect depends on the ratio of the signal to the noise, and is especially suitable for a high ratio of the signal to the noise. In addition, in some cases, such as the existence of discontinuous point signal, de-noising will produce the Vo Gibbs phenomenon. In addition, the selection of threshold plays an important effect on the de-noising effect.

(3) Wavelet transform de-noising method based on translation invariant is mainly suitable for the signal mixed with white noise and contains a plurality of discontinuous point. This method is a kind of de-noising method based on Donoho's threshold method to improve the de-noising threshold method, in order to solve the possible pseudo Gibbs phenomenon occurred in the signal the discontinuous point. The basic idea is cyclic shifts of the original signal with noise in translation within a certain range. In order to obtain a new signal phase difference in the time domain and the original signal, the soft threshold function or hard threshold function of wavelet coefficients are shrunk, which can get the signal de-noising by the inverse wavelet transform.

The signal is contrary to the cyclic shift, which gets a de-noised signal and the original signal with the same phase. With the translation changed, this process is repeated and on average the results obtained estimate the signal noise signal after de-noising.

However, the process of de-noising is convenient, and the speed of calculation is very slow.

3 MODULUS MAXIMUM METHOD BASED ON WAVELET TRANSFORM

Noissin is selected as the original signal with noise, and the wavelet decomposition with 4series, the wavelet function being chosen as db3. The original signal is shown in Figure 3, after wavelet transform coefficient levels were similar to Figure 4. The detail of the coefficient is shown in Figure 5. The wavelet transform modulus maxima de-noising, after all levels remained modulus maxima, is shown in Figure 6, using wavelet modulus maxima de-noising method of reconstruction signal as shown in Figure 7. From the figure, the noise of original signal is mainly concentrated in the 1, 2 level, through the two levels after de-noising, and then through the decomposition of the 3, 4 level, signal to remove the noise coefficient of the vast majority, to retain the integrity of the signal coefficients, and finally through the reconstruction on the restoration of the signal after de-noising. The wavelet transform modulus maximum de-noising method not only can effectively suppress noise, but also can well preserve the high frequency details of the original signal. The de-noising signal distortion is small, and the

de-noising effect is very good. At the same time, the method of the de-noising signal with no excess concussion is the original signal and a very good estimate, and has a good surface quality. It is important to choose the wavelet decomposition scale, which should be a suitable decomposition scale wavelet transform.

It does not yet represent a unified standard of perfection in the application of theoretical research and the methods of wavelet de-noising. There are various methods applicable to conditions and limitations in practice. Overall, because signal in the discontinuous point can produce pseudo Gibbs phenomenon, decomposition coefficients of wavelet packet often have some difficulties in the process of noise based

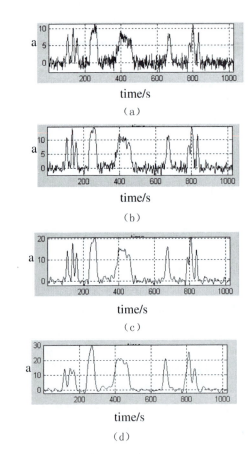

(a)

(b)

(c)

(d)

Figure 2. Approximate coefficient.

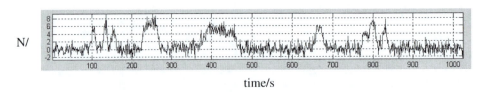

time/s

Figure 1. Noissin chosen as the original signal with noise.

on wavelet transform. Translation invariant de-noising methods can improve the above shortcomings, but in practice there are slow computational speeds and poor convenience.

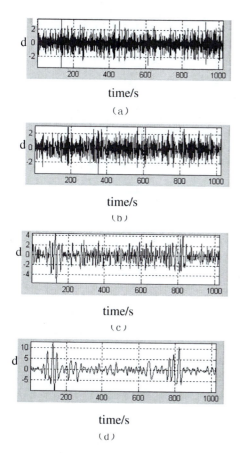

Figure 3.   Detail coefficients.

4   ANALYSIS AND APPLICATION

Health monitoring data is an important scientific basis for shield tunnel health diagnosis, and its reliability and accuracy directly affect the health status of the shield tunnel evaluation and analysis of results [7, 8]. In fact, due to a variety of external objective conditions, such as engineering environment, climate conditions, and measuring instruments, errors caused by tunnel monitoring data often occur. It is difficult to accurately determine the non-stationary signal, which is either a self-mutation expression, or the effects caused by the mutation phenomenon, or it is the comprehensive embodiment of the two combined. Because the effects of these mutations are in the signal curve and show spike and mutation, they are all in the high frequency part of the signal. To accurately analyse this kind of signal, the wavelet transform modulus must be suitable for the characteristics of the maxima method of de-noising, which can effectively remove the noise interference, and can well preserve the original signal's high frequency detail.

Based on the above principle, the wavelet modulus maxima de-noising algorithm can be summarized as follows: (1) the signals contain noise of wavelet, which transforms into general 4~5 scales, and calculate each wavelet coefficient modulus maxima. (2) With the biggest scale as the start and the threshold selected, if the absolute value is less than the corresponding value of extreme point, then the extreme point will be removed, or be retained, which can get new modulus maxima.

5   CONCLUSION

Compared with modulus maxima, wavelet packet decomposition and coefficient shrinkage de-noising methods of wavelet transform are analysed and the

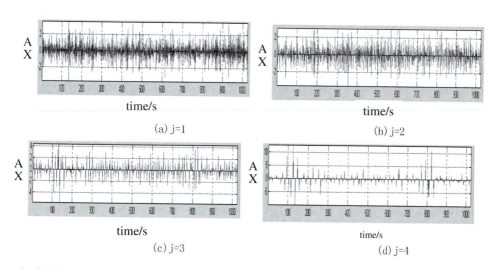

Figure 4.   Modulus maximum of wavelet transform after signal de-noising.

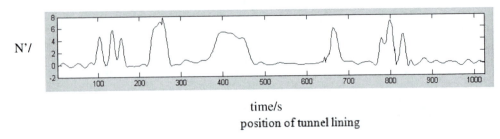

N'/

time/s
position of tunnel lining

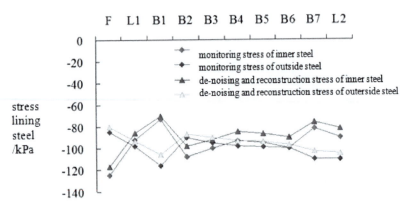

Figure 5.   A comparison between health monitoring results and de-noising and reconstruction of tunnel lining stress.

advantages and disadvantages are summarized and the respective scopes obtained.

(1) Noissin chosen as the original signal with noise is analysed and de-noised by the modulus maxima method of wavelet transform Meanwhile the usage conditions and key computing parameters are also obtained.

(2) The modulus maxima method of wavelet transform is successfully adopted to de-noise the monitoring results of a shield tunnel The data revised are reliably provided for healthy tunnel diagnosis.

REFERENCES

[1] Jiang Peng. Application Research of Wavelet Theory in Noise and Data Compression [D]. Hangzhou, Zhejiang University, 2004.
[2] Gao Zhi, Yu schallsea. Principle and Application of Wavelet Analysis in Matlab Toolbox [M]. Beijing, National Defence Industry Press, 2004.
[3] Ge Zhexue, Sha Wei. Wavelet Analysis Theory and Matlab R2007 Realization [M]. Beijing, Publishing House of Electronics Industry, 2007.
[4] Guo Jian. Method of Bridge Health Monitoring Based on Wavelet Analysis [J]. Engineering Mechanics, 2006, 23 (12): 129–135.
[5] Lin Dachao, Shi Huiji, Bai Chunhua. The blasting Vibration Wavelet Transform Based on Time-Frequency Characteristic Analysis [J]. Chinese Journal of Rock Mechanics and Engineering, 2004, 23 (1): 101–106.
[6] Panou K D, Sofianos A I. A Fuzzy Multi-criteria Evaluation System for the Assessment of Tunnels Surface Roads: Theoretical Aspects-Part I. Tunnelling and Underground Space Technology, 2002, 17 (2): 209–219.
[7] Liu Bo, Tao Longguang, Ding Chenggang, et al. Prediction for Ground Subsidence Induce by Subway Tube Tunnelling [J]. Journal of China University of Mining & Technology, 2006, 35 (3): 356–361.
[8] Kong Xiangxing. Analysis of Lining Deterioration Effect on Underwater Shield Tunnel Deformation [J]. Highway Engineering, 2012, 1 (37): 26–31.

Advances in Civil Engineering and Building Materials IV – Chang et al (eds)
© 2015 Taylor & Francis Group, London, ISBN: 978-1-138-00088-9

Fuzzy modelling of runoff and outflow of permeable pavements

R. Ramirez, H. Kim, H. Jeong & J. Ahn
School of Civil and Environmental Engineering, Pusan National University, Busan, Korea

ABSTRACT: Infiltration to permeable pavements is a complex phenomenon, and often requires attention for modeling to predict the surface runoff and outflow volumes. Several models are available for this infiltration process from sophisticated nonlinear finite element models to simple empirical ones which can estimate the surface runoff and/or outflow volumes. In the present study, an effort has been made to apply a fuzzy model on the results of field experiments using rainfall depth, peak intensity and antecedent dry period as inputs and runoff and outflow as outputs. A set of data in literatures, reported by a research team at North Carolina State University from their test beds, were utilized for fuzzy hydrologic modeling. It seems that the application of a fuzzy model to predict the performance of permeable pavements is promising when appropriate sets of input and output data are selected. The model may be verified further based on the data collected from different types of pavement under different weather conditions.

Keywords: fuzzy model, permeable pavement, hydrologic data, surface runoff volume, outflow volume

1 INTRODUCTION

The infiltration of rainwater to permeable pavements is a complex hydrological process. The transformation from rainfall to pavement runoff and outflow involves many hydrological components that are believed to be highly nonlinear, time varying, spatially distributed and not easily described by simple models. Rainfall-runoff models have been extensively applied in water resources management and have important role in operation and design of hydraulic systems. The main objective in the development of those models is the attempt to simulate the hydrological process in input-output system model structure that generates the process of rainfall metamorphosis into runoff.

In recent years, artificial neutral network (ANN) and fuzzy logic have been widely applied in engineering problems. ANN has been accepted as an efficient alternative tool for modeling complex hydrologic systems and widely used for prediction (i.e. Sajikumar & Thandaveswara 1999). Fuzzy logic method was first developed to explain the human thinking and decision system by Zadeh (1965). Several studies have been carried out employing fuzzy logic-based hydrology and water resources planning (i.e. Mahabir *et al.* 2003).

Fuzzy inference systems are nonlinear models that describe the relationship between the inputs and the output of a system using a set of fuzzy IF-THEN rules. The functioning of fuzzy systems is based on fuzzy sets theory (Zadeh, 1965) and they were first proposed by Mamdani (1974) as a modeling tool for the control of a dynamic plant. The major strength of fuzzy systems resides in their ability to interpret the behavior of complex systems purely from data, and still provide some insight about their internal operation. Fuzzy systems are very flexible modeling tools as their architecture and the inference mechanisms can be adapted to the given modeling problem.

In this study, an effort has been made by employing a fuzzy model to predict surface runoff and outflow volumes of permeable pavements based on rainfall depth, peak intensity and antecedent dry period data from the eastern North Carolina parking lot from the period of June 2006 to July 2007.

2 FUZZY MODELING

The transformation of rainfall into surface runoff and/or outflow volume is dynamic and nonlinear process which is affected by a number of factors which are often inter-related. A fuzzy model developed and intended to predict the holding capacity of suction caisson anchors is utilized in this study and details can be found in Kim & Ahn (2014).

2.1 *Fuzzy sets*

Fuzzy sets are an extension of classical set theory and are used in fuzzy logic. In classical set theory the membership of elements in a set is assessed in binary terms according to a crisp condition – an element either belongs or does not belong to the set. By contrast, fuzzy set theory permits the gradual assessment of the membership of elements in relation to a set; this is described with the aid of a membership function $\mu = [0,1]$. Fuzzy sets are formed from membership functions which may act as an indicator function mapping all elements to either 1 or 0, as in the classical notion.

2.2 Membership function

Every element in the universe of discourse is a member of a fuzzy set to some grade, maybe even zero. The set of elements that have a non-zero membership is called the support of the fuzzy set. The function that ties a number to each element of the universe is called the membership function.

2.3 Selection of input and output variables

Input and output variables are selected for the model on the basis of the study objectives. Only three input variables i.e. rainfall depth, peak intensity, and antecedent dry period are applied for hydrologic modeling of pavements. The surface runoff and outflow volumes are taken separately as output variables. Flow diagram of fuzzy model is shown in Figure 1.

2.4 Fuzzy rules

The rules may use several variables both in the condition and the conclusion of the rules. The fuzzy rule base was formed based on the historical data and intuition. Fuzzy rule-based contains fuzzy rules that include all possible fuzzy relations between inputs and output. These rules are expressed in the IF-THEN format. Fuzzy IF-THEN rules are conditional statements that describe the dependence of one linguistic variable on another.

2.5 Performance evaluation of the model

For better appreciation of the model, the predictive effectiveness of fuzzy rule-based model is judged on the basis of performance indicators. To judge the predictive capability of the model for runoff and outflow volumes of permeable pavements, root mean squared error (RMSE) and correlation coefficient (CC) are employed. The root mean squared error and correlation coefficient are calculated as follows:

Root mean squared error (RMSE)
$$J_2 = \sqrt{\frac{\sum |\hat{y} - \tilde{y}|^2}{N_t}} \qquad (1)$$

Correlation coefficient (CC)
$$J_4 = \frac{\sum_{N_t} (\hat{y} - \bar{y})^2}{\sum_{N_t} (\tilde{y} - \bar{y})^2} \qquad (2)$$

where $\hat{y}$ is the forecasting value, $\tilde{y}$ is the data measured in the field, $\bar{y}$ is the average of the measurement, and N_t is the number of data points.

3 APPLICATION

3.1 Permeable parking lot

Collins et al. (2008) monitored a permeable parking lot in eastern North Carolina from June 2006 to July 2007. The parking lot consists of four types of permeable pavement and standard asphalt. The four permeable sections were pervious concrete (PC), two types of permeable interlocking concrete pavement (PICP) with small-sized aggregate in the joints, and concrete grid pavers (CGP) filled with sand as depicted in Figure 2. The effects of different parameters such as the rainfall depth, rainfall intensity, antecedent dry period, season, and lot age were evaluated for all pavement types.

Hydrologic differences among pavement sections were evaluated for surface runoff volume, total outflow volume, peak flow rate, and peak delays. All permeable pavements significantly and substantially reduced surface runoff volumes and peak flow rates when compared to standard asphalt (AS). The study revealed that permeable pavement surface runoff and peak flow reductions were most dependent on rainfall intensity, whereas outflow volume and peak flow reductions were most dependent on rainfall depth.

3.2 Hydrological data

To evaluate the performance of the fuzzy hydrologic model, a split sample experiment was performed by splitting the total collected dataset from east North

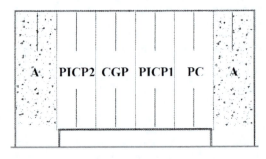

Figure 2. Permeable pavement parking lot design (plan view) in eastern North Carolina (Collins et al. 2008).

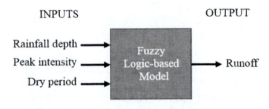

Figure 1. Schematic of a fuzzy inference system.

Carolina parking lot into training and test subsets. The first 43 data were used for training the model and the remaining 8 were used for testing the model's performance. The statistical characteristics of the output vectors (runoff and outflow) for each data set are given in Table 1.

The statistical characteristics of the test set suggest that it contains extreme maximum and minimum values of runoff and outflow with a low degree of variability (standard deviation). The model was fitted to the training data set and this model was then employed for simulation.

3.3 Results

The selection of proper number of inputs and output possesses the prime importance and needs to be selected carefully. The effects of rainfall depth, rainfall intensity, antecedent dry period, season, and lot age on the performance of the fuzzy model were first evaluated for all pavement types. Visual inspection of simple plots (hydrographs) can provide significant information about how close the predictions are to the observations for different surface runoff and outflow volumes. Hence, rainfall depth, rainfall intensity and antecedent dry period data from eastern North Carolina permeable parking lot from the period of June 2006 to July 2007 were used as input variables to describe the daily time series and development of the fuzzy hydrologic model.

In the current study, the performance of the fuzzy model was evaluated qualitatively and quantitatively by visual observation and employing statistical indices such as the root mean squared error (RMSE) and

correlation coefficient (CC). The optimal learning parameters were determined by trial and error for membership function. All input and output variables were separately divided into subsets i.e. 60–40%, 70–30% and 80–20% that means the 60% datasets were used for calibrating the model and the remaining 40% dataset were taken for its testing purpose. More subsets are considered to increase the accuracy of prediction. The fuzzy model was constructed for the study under fuzzy logic toolbox in soft computing program MATLAB R2012a.

The qualitative evaluation of the model is based on the visual comparison i.e. overall shape of the observed and predicted hydrographs. The qualitative assessment of the fuzzy model was made by regenerating permeable pavement runoff and outflow volumes, and by comparing these with observed ones. Using fuzzy logic rule-based model, sample plots of observed and predicted values for the Eastern North Carolina permeable parking lot from June 2006 to July 2007 are depicted in Figures 3-6. The plots show fair agreement between the observed and predicted values.

Figures 3–4 show the observed and estimated hydrographs i.e. surface runoff and outflow volumes of the fuzzy hydrologic model for the CGP permeable pavement from June 2006 to May 2007 calibration period. Similarly, Figures 5–6 show the observed and

Table 1. Statistical characteristics of hydrological data*.

Type	Statistic	Runoff (mm)		Outflow (mm)	
		Train	Test	Train	Test
AS	Mean	13.44	16.88	13.44	16.88
	STDEV	14.40	12.78	14.40	12.78
	Max.	76.14	39.89	76.14	39.89
	Min	0.62	0.66	0.62	0.66
PC	Mean	0.03	0.22	11.92	13.85
	STDEV	0.07	0.27	12.03	11.06
	Max.	0.36	0.7	57.57	33.92
	Min	0.00	0	0.30	1.89
PICP1	Mean	0.09	0.28	8.27	11.2
	STDEV	0.13	0.30	10.21	10.19
	Max.	0.60	0.70	48.09	29.81
	Min	0.00	0.00	0.00	0.41
CGP	Mean	0.34	2.21	8.55	9.25
	STDEV	0.44	4.23	11.79	7.45
	Max.	1.78	12.60	61.38	17.98
	Min	0.00	0.00	0.00	0.00
PICP2	Mean	0.10	0.22	13.23	13.15
	STDEV	0.17	0.26	12.56	10.41
	Max.	0.98	0.70	57.21	32.25
	Min	0.00	0.00	0.61	1.51

*Collins *et al.* 2008

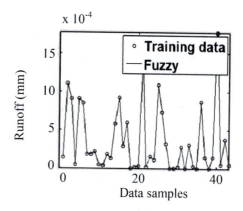

Figure 3. Model training for surface runoff of CGP pavement.

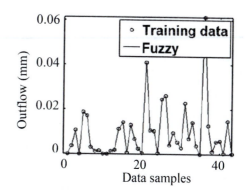

Figure 4. Model training for outflow of CGP pavement.

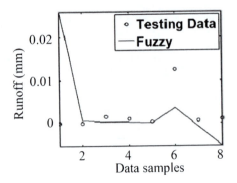

Figure 5. Model testing for surface runoff of CGP pavement.

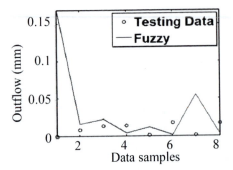

Figure 6. Model testing for outflow of CGP pavement.

Table 2. Performance evaluation of the fuzzy model.

Type	Index	Runoff (mm)		Outflow (mm)	
		Train	Test	Train	Test
AS	RMSE	0.02	7.51	0.02	7.51
	CC	1.00	0.42	1.00	0.42
PC	RMSE	0.00	1.33	0.02	7.74
	CC	1.00	−0.37	1.00	0.45
PICP1	RMSE	0.00	1.27	0.02	7.58
	CC	1.00	0.39	1.00	0.44
CGP	RMSE	0.00	1.51	0.02	8.92
	CC	1.00	0.71	1.00	−0.24
PICP2	RMSE	0.00	1.57	0.02	9.76
	CC	1.00	−0.28	1.00	0.53

estimated hydrographs i.e. surface runoff and outflow volumes for the CGP permeable pavement from June 2007 to July 2007 verification period.

The quantitative performance of developed model was also evaluated by applying statistical indices such as the root mean squared error (RMSE) and correlation coefficient (CC). The values of these indices were determined using Equations 1 and 2 and are presented in Table 2.

The RMSE statistic is a measure of residual variance and is indicative of the model's ability to predict high flows. Considering the magnitude of the surface runoff and outflow volumes during the period of study, the fuzzy model is able to forecast the flows with reasonable accuracy, as can be evidenced from the low RMSE values. The correlation statistics evaluate the linear correlation between the observed and the computed surface runoff and outflow volumes, which is consistent during calibration period but not during validation period. The study revealed that the CGP and PICP2 yielded the acceptable and highest value of correlation coefficient for runoff and outflow volumes, respectively.

4 CONCLUSION

The applicability of a fuzzy logic-based approach to predict surface runoff and outflow volumes of permeable pavements using the data from eastern North Carolina parking lot from June 2006 to July 2007 has been illustrated. In the study, rainfall depth, peak intensity and antecedent dry period data were optimal inputs which presented appropriate surface runoff and outflow simulations independently.

Overall, the results of this study indicate that the developed fuzzy logic-based model is a suitable tool for modelling the nonlinear relationship between interrelated hydrologic data. The fuzzy hydrologic model may be verified further based on the data collected from different types of permeable pavement under different weather conditions. In addition, further work will be carried out to model the performance in water quality improvement based on fuzzy theory

ACKNOWLEDGEMENT

This research was supported by a grant from Advanced Water Management Research Program funded by Ministry of Land, Infrastructure and Transport of Korean government.

REFERENCES

Abebe, A.J., Solomatine, D.P. & Venneker, R.G.W. 2000. Application of adaptive fuzzy rule-based models for reconstruction of missing precipitation events. Hydrological Sciences-Journal-des Sciences Hydrologiques 45(3): 425–436.

Altunkaynak, A. 2010. A predictive model for well loss using fuzzy logic approach. Hydrol. Processes 24(17): 2400–2404.

Arliansyah, J., Maruyama, T. & Takahashi, O. 2003. A development of fuzzy pavement condition assessment. Proceedings of Japan Society of Civil Engineers 746: 275–285.

Bardossy, A. 1996. The use of fuzzy rules for the description of the elements of the hydrological cycle. Ecological Modelling 85: 59–65.

Chang, F.J., Chen, Y.C. & Liang, J.M. 2002. Fuzzy clustering neural network as flood forecasting model. Nordic Hydrology 22(4): 275–290.

Chu, H.J. & Chang, L.C. 2008. Application of the optimal control and fuzzy theory for dynamic groundwater remediation design. *Water Resources Management* 20(4): 647–660.

Collins, K.A., Hunt, W.F. & Hathaway, J.M. 2008. Hydrologic and water quality comparison of four different types of permeable pavement and standard asphalt in Eastern North Carolina. *11th International Conference on Urban Drainage*, Edinburgh, Scotland, UK.

Fwa, T.F. & Shanmugam, R. 1998. Fuzzy logic technique for pavement condition rating and maintenance-needs assessment. *4th International Conference on Managing Pavements*, Durban, South Africa: 465–476.

Gautam, D.K. & Holz, K.P. 2001. Rainfall-runoff modelling using adaptive neuro-fuzzy systems. *Journal of Hydroinformatics* 3: 3–10.

Kim, Y. & Ahn. J. 2014. Modelling the capacity of suction caisson anchors based on fuzzy theory. *The 24th International Ocean and Polar Engineering Conference,* Busan, Korea.

Lohani, A.K., Goel, N.K. & Bhatia, K.K.S. 2011. Comparative study of neural network, fuzzy logic and linear transfer function techniques in daily rainfall-runoff modeling under different input domains. *Hydrol. Processes* 25: 175–193.

Mahabir, C., Hicks, F.E. & Robinsons, F.A. 2003. Application of fuzzy logic to forecast seasonal runoff. *Hydrol. Processes* 17: 3749–3762.

Mamdani, E.H. 1974. Application of fuzzy algorithms for control of simple dynamic plant. *Proceedings IEE* 121(12): 1585–1588.

Mamdani, E.H. & Assilian, S. 1975. An experiment in linguistic synthesis with a fuzzy logic controller. *International J. Man-Machine Studies* 7: 1–3.

Maskey, S., Guinot, V. & Price, R.K. 2004. Treatment of precipitation uncertainty in rainfall-runoff modelling: a fuzzy set approach. *Advances in Water Res.* 27: 889–898.

MATLAB and Statistics Toolbox Release 2012a, The Mathworks, Inc., Natick, Massachusetts, United States.

Nayak, P.C., Sudheer, K.P. & Ramasastri, K.S. 2004. Fuzzy computing based rainfall-runoff model for real time flood forecasting. *Hydrol. Processes* 17: 3749–3762.

O'Loughlin, G., Huber, W. & Chocat, B. 1996. Rainfall-runoff processes and modelling. *J. Hyd. Res.* 34(6): 733–751.

Ozger, M. 2009. Comparison of fuzzy inference systems for stream flow prediction. *Hydrol. Sci. J.* 54(2): 261–273.

Sajikumar, N. & Thandaveswara, B.S. 1999. A nonlinear rainfall-runoff model using artificial neural networks. *J. Hydrol.* 216: 32–55.

Sen, Z. & Altunkaynak, A. 2006. A comparative fuzzy logic approach to runoff coefficient and runoff estimation. *Hydrol. Processes* 20: 1993–2009.

Sun, L. & Gu, W. 2011. Pavement condition assessment using fuzzy logic theory and analytic hierarchy process. *Journal of Transportation Engineering* 137: 648–655.

Taheri Shahraiyni, H., Bagheri Shouraki, S., Fell, F., Schaale, M., Fischer, J., Tavakoli, A., Preusker, R., Tajrishy, M., Vatandoust, M. & Khodaparast, H. 2009. Application of the active learning method to the retrieval of pigment from spectral remote sensing reflectance data. *International J. Remote Sensing* 30: 1045–1065.

Takagi, T. & Sugeno, M. 1985. Fuzzy identification of systems and its application to modelling and control. *IEEE Transactions on System, Man and Cybernetics* 15(1): 116–132.

Tayfur, G. & Singh, V.P. 2006. ANN and fuzzy logic models for simulating event-based rainfall-runoff. *Journal of Hydraulic Engineering* 132: 1321–1330.

Zadeh, L.A. 1965. Fuzzy sets. *Information and Control* 8: 338–353.

Advances in Civil Engineering and Building Materials IV – Chang et al (eds)
© *2015 Taylor & Francis Group, London, ISBN: 978-1-138-00088-9*

A review of test beds and performance criteria for permeable pavements

H. Jeong, H. Kim, B. Teodosio, R. Ramirez & J. Ahn
School of Civil and Environmental Engineering, Pusan National University, Busan, Korea

ABSTRACT: Permeable pavement systems have been successfully constructed in limited traffic load applications such as parking lots, sidewalks, and roadways. The benefits of these systems include a reduction in urban flooding, water quality deterioration, and the heat island effect phenomenon. In addition, the benefits of these systems extend to reflection suppression, noise reduction, and improved skid and splash resistance. Recently, Low Impact Development (LID) has been introduced and adopted by professionals. However, there is still further research to conduct in order to produce standards for permeable pavement system design criteria. This paper presents different literature as reference to propose a design standard for permeable pavement systems. The suggested criteria elaborate on the design of pervious concrete, porous asphalt, and porous blocks. In addition, several research facilities which monitor the performance of permeable pavements in literature are introduced and summarized. The literature will serve as a basis for the permeable pavement test beds that the authors plan.

Keywords: permeable pavements, test beds, design criteria, low impact development

1 INTRODUCTION

The available design of road pavements globally is based on impervious surfaces but lately, environmental problems, including climate change, have prompted designers and contractors to reconsider stormwater management, specifically in relation to permeable pavements. The benefits of permeable pavement systems include hydroplaning reduction, reflection diffusion, noise reduction, improved skid resistance, and splash prevention. However, the design criteria for a permeable pavement are not yet well established based on the integrated performance in terms of structural, hydrological, and pollution reduction aspects. Most applications of permeable pavements are for low-traffic roadways, parking lots, bicycle roads, and sidewalks.

In Japan, since 1973, permeable pavements have been installed as sidewalks in main cities. In 1993, Japanese professionals attempted to apply permeable pavements to low-traffic roads. In the past years, flooding has been experienced in most developed areas. Hence, the Japanese government encouraged researchers and practitioners to explore the possible applications of stormwater management. A design guide was published in 2005 entitled, "*Road surface stormwater treatment manual*" (PARI, 2011).

In the United States, the consequences of urbanization during the 1970s prompted the use of permeable pavements. Because of recent climate change effects, research on permeable pavement systems has become active. Evaluation systems and certifications have been conducted and awarded to sustainable projects

since this could attenuate or prevent further environmental deterioration (Thelen and Howe, 1978; Hunt, 2009).

Around late 1990s in South Korea, an eco-friendly permeable pavement was introduced and adopted (Ministry of Land, Transport, and Maritime Affairs, 2011).

A typical permeable pavement design includes the surface pavement, the base course and the sub-base course. The surface can be constructed using asphalt, concrete and block pavements. The pavement surface must have the ability needed for a specific structural design traffic load. Furthermore, the hydrologic capacity should be sufficient to let the stormwater infiltrate through its system controlling surface runoff. Presented in the succeeding sections are the reviews of permeable pavement designs according to the guides of: (1) the Seoul Metropolitan Government (Seoul Metropolitan Government, 2013), and the Ministry of Land, Infrastructure and Transport of Korea (Ministry of Land, Infrastructure and Transport, 2011); (2) the Japan Road Association (Japan Road Association, 2007); and (3) the National Asphalt Pavement Association (NAPA) (National Asphalt Pavement Association, 2003), the American Concrete Institution (ACI) (American Concrete Institution, 2010) of the United States and the University of New Hampshire Stormwater Center (UNHSC) (University of New Hampshire Stormwater Center, 2009).

The test bed monitoring the performance of permeable pavements is introduced by the North Carolina State University (Collins et al, 2010), the Riverside County Flood Control and Water Conservation

(Williams, 2011), the University of New Hampshire (The UNH Stormwater Center, 2008), and Washington State University (Herrera Environmental Consultants, 2013).

2 PERMEABLE PAVEMENTS

2.1 Porous asphalt

A typical design value for porosity of porous asphalt is 20 ~ 30%. In the case of sidewalk design porosity in Japan, the value used is commonly lower than 12%.

To conduct KS A 5101-1, a Korean standard for draindown ratio, a 4.75 mm sieve and pan are used. An asphalt mixture with a mass of 1,000 ± 5 g is flatly spread on the test sieve with a mixing temperature. After an hour, the amount of asphalt that flows through the test sieves and accumulates in the pan should be quantified. The draindown ratio value is suggested to be below 0.3.

A Cantabro abrasion ratio is given by KS F 2397, which measures the abrasion loss of the porous asphalt pavement. Korea and the US perform temperature and aged/unaged tests, respectively.

The Tensile Strength Ratio (TSR) is specified according to a guide from South Korea, and one from the University of New Hampshire Storm Center (UNHSC), using one-cycle freezing and thawing experimentations. If the TSR (retained tensile strength) values fall below 80% when tested per NAPA IS 131 (with a single freeze-thaw cycle rather than 5), then in Step 4, the contractor shall employ an anti-strip additive, such as hydrated lime (ASTM C977) or a fatty amine, to raise the TSR value above 80% according to the guide of the UNHSC. In the quantification of dynamic stability, the method used in both the US and Korea is similar. Furthermore, in the determination of

hydraulic conductivity, Korea and Japan used the same suggested procedures. Stability and moisture-induced damages are only considered by Seoul, Korea, for low noise permeable asphalt pavement specifications.

In Korea, in order to satisfy the required compactive effort with respect to the design criteria, a gyratory compactor and Marshall automatic compactor are suggested for specimens with similar field compaction conditions. Japan performed density tests. In addition, Marshall Stability and flow index values are deduced to determine the Optimum Binder Content (OBM) and to measure the resistance to creep and plastic flow, accordingly.

2.2 Pervious concrete

The Japanese standard includes a design with flexural rigidity over 2.5 MPa, permeability over 0.01 cm/s, and porosity from 15% to 25%. In addition, the Japanese

Table 2. Porous asphalt mix design criteria (University of New Hampshire Stormwater Center, 2009).

Sieve Size (inch/mm)	Percent Passing (%)
0.75/19	100
0.50/12.5	85–100
1.375/9.5	55–75
No.4/4.75	10–25
No.8/2.36	5–10
No.200/0.075	2–4
Binder Content	6–6.5%
Fibre Content by Total Mixture	0.3 cellulose or
Mass	0.4 mineral*

*Cellulose of mineral fibres may be used to reduce drain-down.

Table 1. Performance criteria for porous asphalt.

| Contents | Korea | | Japan C*** | | U.S. UNHSC |
	A*	B**	Roadway	Sidewalk	
Draindown ratio (%)	≤0.3				≤0.3
Porosity (%)	20 ± 0.3 20 ± 2	20 ± 1	≥20	≥12	16 ~ 22
Cantabro abrasion ratio (%)	≤20 (20°C) ≤30 (−20°)	≤20			≤20 (unaged) ≤30 (aged)
Tensile strength ratio (TSR)	≥0.85				≥0.80
Hydraulic conductivity (cm/sec)	≥0.01	≥0.01	≥0.01	≥0.01	
Dynamic stability (time/mm)	≥3000	≥3000	≥3000		
Moisture induced damage		≥0.7			
Marshall stability (kN)		≥5.0	≥3.43	≥3.0	
Flow index value (1/100 cm)				20–40	
Density (g/cm)				≥1.95	

*Ministry of Land, Infrastructure, and Transport, 2011. *Porous Asphalt Mixture Production & Construction Provisional Guidelines.*
**Seoul Metropolitan Government, 2013. *Low-noise Porous Asphalt Pavement Specifications.*
***Japan Road Association, 2007. *Porous Pavement Guidebook.*

standard distinguishes the pervious concrete usage of roadways, parking lots, and sidewalks.

2.3 Permeable block

South Korea divides the usage of permeable blocks into roadways and sidewalks, while Japan uses

Table 3. Typical* ranges of material proportions in pervious concrete (University of New Hampshire Stormwater Center. 2009. *UNHSC Design Specifications for Porous Asphalt Pavement and Infiltration Beds*).

	Proportions, kg/m^3
Cementitious materials	270–415
Aggregate	1190–1480
w/cm	0.27–0.34
Aggregate: cement ratio	4–4.5:1
Fine: coarse aggregate ratio	0–1:1

*These proportions are given for information only. Successful mixture design will depend on properties of the particular materials used.

Table 4. Typical Performance criteria for Pervious Concrete.

Contents	Korea*	Japan**		U.S.***
		Roadway	Sidewalk	
Flexural rigidity (MPa)	≥2.5	≥4.5	≥2.5	
Permeability (cm/sec)	≥0.01	≥0.01	≥0.1	
Porosity (%)	15–25	15–20	20–25	15–25

*Seoul Metropolitan Government. 2013. *Permeable Blocks Pavement Design, Construction and Maintenance Criteria.*
**Japan Road Association. 2007. *Porous Pavement Guidebook.*
***American Concrete Institute (ACI). 2010. *Report on pervious concrete.*

Table 5. Performance criteria for Permeable block.

	Korea* (Permeable block)		Japan** (Interlocking block)
	Roadway	Sidewalk	
Flexural rigidity (MPa)	≥5	≥4	≥3
Compressive strength (MPa)	≥20	≥16	≥20
Permeability (cm/sec)	≥0.1		≥0.01

*Seoul Metropolitan Government 2013. *Permeable Blocks Pavement Design, Construction and Maintenance Criteria.*
** Japan Road Association. 2007. *Porous Pavement Guidebook.*

permeable blocks for low limited traffic roadways and parking lots. Furthermore, Korea divides permeable blocks into two types, the typical permeable block, and the open-jointed paving block.

Japan, on the other hand, has distinguished two types of permeable blocks. These are interlocking blocks, and pervious concrete plates. Japan follows a suggested design criteria that includes a permeable block with a bending strength of more than 3 to 5 MPa, a compressive strength of over 16 to 20 MPa, and a hydraulic conductivity equivalent or greater than 0.01 cm/s.

3 TEST BED

As a result of development that transformed spacious regions to impermeable surfaces, many researchers are investigating the feasibility of Low Impact Development (LID) to manage stormwater runoff. Eight permeable pavement prototype test beds are presented. The permeable pavement systems were constructed by the North Carolina State University (Collins et al., 2010), the Riverside County Flood Control and Water Conservation (Williams, 2011), the University of New Hampshire (The UNH Stormwater Center, 2008) shown in Figure 1, and Washington State University (Herrera Environmental Consultants, 2013).

The structures constructed by each research institution exemplify different sections and the exfiltration processes used to manage stormwater. Table 6 tabulates the research facilities of the North Carolina State University (NCSU), the Riverside County Flood Control (RCFC), the University of New Hampshire (UNH), and Washington State University (WSU).

4 CONCLUSION

The main objective of the reviews conducted by the proponents of this study is to identify the present practices in designing permeable pavement systems for stormwater management. The review focused on three types of paving materials which include porous asphalt, pervious concrete, and permeable blocks. The

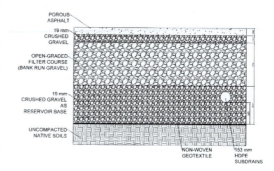

Figure 1. Porous asphalt of the University of New Hampshire – Department of Environmental Services.

Table 6. Permeable pavement classification for NCSU, RCFC, UNH and WSU.

Permeable Pavement System	NCSU*		RCFC**		UNH***		WSU****	
Permeable pavement classification	Pervious concrete (200 mm)	Concrete grid pavers with sand fill (150 mm)	Pervious concrete (216 mm)	Porous asphalt (127 mm)	Pervious concrete (150 mm)	Porous asphalt (100 mm)	Pervious concrete (200 mm)	Porous asphalt (75 mm)
Dimension (sq. m)	500	620	780.4		Not Specified		613	613
Usage	Parking lot		Parking lot and driveway		Research facility		Parking lot	Parking lot
Base/ Sub-base type	Washed #57 aggregate (200 mm)	Sand $k = 5.6 \times 10^{-3}$ cm/s (230 mm)	#57 aggregate (AASHTO #8) (~230 mm)	#57 aggregate (AASHTO #8) and Class 2	38-mm and 9-mm gravel (150 mm each) and reservoir base (350 mm)	19-mm gravel (100 mm and 530 mm) and reservoir base (610 mm)	Washed aggregate (450 mm)	Washed aggregated (450 mm)
Natural soil	Loamy sand	Loamy sand	Not Specified	Not Specified	Uncompacted ($k = 3.53 \times 10^{-4}$ cm/s)	Uncompacted natural soil	Not Specified $k = 2.22 \times 10^{6}$ cm/s	
Impermeable liner	None	None	Installed	Installed	None	None	None	Installed
Perforated drain	None	None	Installed	Installed (150 mm) 100 mm from grade line	HDPE 300 mm from geotextile	HDPE (150 mm) PVC (under and after the permeable pavement)	Two 100 mm PVC (under and after the permeable pavement)	Two 100 mm
Monitoring well	None	Installed	None	None	None	None	6 Installed	None
System infiltration maintained (cm/s)	1.11	2.39×10^{-3}	Not Specified	Not Specified	Not Specified	Not Specified	0.44 to 0.50	0.20 to 0.22
System infiltration clogged (cm/s)	3.61×10^{-3}	1.36×10^{-3}	Not Specified	Not Specified	Not Specified	Not Specified	Measured but not Specified	Measured but not Specified
Remarks	Water storage for rain-fall is 20 m^3; Design void = 20%	None	None	From surface, sandwich base course installed (#57-class 2-#57)	From surface, base course installed (38 mm-reservoir base-9 mm)	Non-woven geotextile installed (19 mm-reservoir base-19 mm)	Temperature sensors (51 mm below pavement and every 150 mm)	None

*Collins, K., Hunt, W. and Hathaway, J. 2010. *Side-by-side Comparison of Nitrogen Species Removal for Four Types of Permeable Pavement and Standard Asphalt in Eastern North Carolina.*
**Williams, W. D. 2011. *Riverside County Flood Control and Water Conservation District Standard Manual.*
***The UNH Stormwater Center. 2008. *Pervious Concrete Pavement for Stormwater Management.*
****Herrera Environmental Consultants, Inc., and Washington Stormwater Center. 2013. *Guidance Document for Western Washington Low Impact Development (LID) Operation and Maintenance (O&M).*

design criteria of South Korea, Japan, and the United States were compared and evaluated to acquire knowledge regarding practical designs of the said structures. Furthermore, research facilities with prototype test beds of permeable pavements were reviewed from the NCSU, the RCFC, the UNH, and WSU. The reviewed literature will be used to propose practical design criteria for the planned LID research facility test beds in Pusan National University (PNU), Yangsan Campus.

ACKNOWLEDGEMENT

This research was supported by a grant from the Advanced Water Management Research Program, funded by the Ministry of Land, Infrastructure, and Transport of the Korean government.

REFERENCES

AASHTO. 1993. *Guide for Design of Pavement Structures*, Washington, DC: 640.

American Concrete Institute (ACI). 2010. *Report on pervious concrete.*

Ashley, E. 2008, *Using pervious concrete to achieve SEED™ points*. National Ready Mixed Concrete Association.

Atlanta Regional Commission. 2001. *Georgia Stormwater Management Manual*: 3.3–33 and 3.3-40.

Baas, W. P. 2006. *Pervious concrete pavement surface durability in a freeze-thaw environment where rain, snow and ice storms are common occurrences*. Ohio Ready Mixed Concrete Association: 4.

Collins, K., Hunt, W. & Hathaway, J. 2010. *Side-by-side Comparison of Nitrogen Species Removal for Four Types of Permeable Pavement and Standard Asphalt in Eastern North Carolina*. Journal of Hydrologic Engineering, June 2010, pp. 512–521.

Cruoch, L.K., Cates, M.A., Dotson, V.J., Honeycutt, K.R. & Badoe, D.A. 2003. *Measuring the effective air void content of Portland cement pervious pavements*. Cement, Concrete and Aggregates B. 25, No. 1: 16–20.

Debo, T.N. & Reese, A.J. 2002. *Municipal Storm Water Management* (2nd ed.): 976. CRC Press.

Florida Concrete and Products Association (FCPA). 1990. *Pervious Pavement Manual*, Orlando, Florida: 57.

Ferguson, B.K. 1998. *Introduction to Stormwater: Concept, Purpose, Design*: 272. Wiley.

Haselbach, L.M. & Freeman, R.M. 2006. Vertical porosity distributions in pervious concrete pavement. *AVI Materials Journal* 103(6): 452–458.

Herrera Environmental Consultants, Inc., and Washington Stormwater Center. 2013. *Guidance Document for Western Washington Low Impact Development (LID) Operation and Maintenance (O&M).*

Hunt, W. & Collins, K. 2008. *Urban Waterways: Permeable Pavement Research Update and Design Implications.* North Carolina State University.

Japan Road Association. 2007. *Porous Pavement Guidebook.*

Mahboub, K., Canler, J., Rathbone, R., Robl, T. & Davis, B. 2008. The effects of compaction and aggregate gradation on pervious concrete. *Proceedings of the NRMCA Concrete Technology Forum*, Denver, CO.

Marolf, A., Neithalath, N., Sell, E., Wegner, K., Weiss, J. & Olek, J. 2004. Influence of aggregate gradation on the acoustic absorption of enhanced porosity concrete. *ACI Materials Journal* 101(1): 82–91.

Ministry of Land, Infrastructure and Transport. 2011. *Porous Asphalt Mixture Production & Construction Provisional Guidelines.*

National Asphalt Pavement Association. 2003. *National Asphalt Pavement Association Journal.*

National Asphalt Pavement Association. 2002. *Design, Construction and Maintenance of Open-graded Asphalt Friction Courses Journal.*

Seoul Metropolitan Government. 2013. *Low-noise Porous Asphalt Pavement Specifications.*

Seoul Metropolitan Government. 2013. *Permeable Blocks Pavement Design, Construction and Maintenance Criteria.*

The UNH Stormwater Center. 2008. *Pervious Concrete Pavement for Stormwater Management*. University of New Hampshire.

University of New Hampshire Stormwater Center. 2009. *UNHSC Design Specifications for Porous Asphalt Pavement and Infiltration Beds.*

Williams, W. D. 2011. *Riverside County Flood Control and Water Conservation District Standard Manual*. Riverside County, USA.

Advances in Civil Engineering and Building Materials IV – Chang et al (eds)
© 2015 Taylor & Francis Group, London, ISBN: 978-1-138-00088-9

Analysis of the influential factors of organic matter in soil

Baotong Shi & Xiangxing Kong
The First Highway Survey and Design Institute of China Communications Construction Company Ltd.,
Xi'an, China

ABSTRACT: This paper makes a summary of the method for determination of the organic matter in soil, the influential factors of the occurrence of organic matter, and the chemical structure. The factors determining soil organic C levels are climate, soil mineral parent materials, vegetation and organisms, water environment, and land management practices. The chemical structure of organic matter is changeable and its composition is complex, with benzene as the main body structure, and with hydroxyl, carboxyl, carbonyl, and alkyl complex connected. The hydrogen in the structure is often replaced by other perssad such as halogens, and amino acids.

Keywords: influential factors, organic matter, chemical structure, physical structure.

1 INTRODUCTION

Organic matter has an important influence on the physical and mechanical properties of soil. Usually, it is difficult to handle for the organic matter because of its high liquid limit and low strength Therefore, we need the characteristics of organic matter in soil and the conditions for their occurrence.

2 METHOD FOR DETERMINATION OF THE ORGANIC MATTER CONTENT

To directly determine the content of organic matter composition is difficult, as most studies are after the determination of total content of C multiplied by a conversion factor 1.72 to 2.0. they mostly include wet calcination, dry burning, and dichromate oxidation method. After calcination method refers to the organic soil drying in 440–5500c high temperature calcination to quality unchanged, ignition loss as the organic matter content of quantitative indicators. Dichromate oxidation method refers to the use of dichromate oxidation characteristics, which will be organic matter composition oxidation with using dichromate oxidation in the process of consumption that organic matter content of salt. Dichromate sutra by Cl^- in the soil and influenced by Fe^{2+} and MnO_2, causes the high content of organic matter. Generally, calcination the method can make the organic whole oxidation causing low organic matter content determination, which must be multiplied by a coefficient of between 1.0 to 2.86, and 1.3 is widely used as a mean value. Dichromate method

does not need multiplied by the coefficient of all generally think, and this method can make the organic composition oxidation. However, J. Skjemstad put forward its oxidation degree, which related to organic matter particle size (Baldock, 2000).

3 THE DECIDING FACTOR OF ORGANIC MATTER CONTENT

3.1 Climate

Climate, such as temperature, humidity, solar radiation and etc., will affect plants growth and the mineralization of organic C. However, this has no effect on the chemical structure of organic C. Temperature (of more than 15 degrees) of the organic C content degradation speed increases faster than the generation of organic C, and results in the decrease of organic C content.

3.2 Parent rock mineral composition and biological products

According to the different soil chemical properties, organic matter can be formed in the mineral surface adsorption. Soil particles are surrounded by role and limit the creatures of the erosion of the organic matter. When soil is active with calcium carbonate, Al and Fe content of high organic matter content is relatively high. Especially calcium carbonate content in lower soil, organic C stability depends on the formation of the complex organic matter Ca. Sørensen's (1972, 1975) studies have shown that montmorillonite's influence on the organic matter content is higher than kaolinite's (Fan, 2004).

3.3 *Vegetation and soil organisms*

Vegetation type, quantity, distribution, and biodegradable properties of soil organic C content and organic matter structure has an important influence. Vegetation can be broken down into simple sugars, amino acids, organic acids, alkyl compounds such as lipids, oil and wax, cellulose, lignin, and protein. Polysaccharide, cellulose, and hemicellulose are carbohydrates. 80–95% of organic C mineralization is due to microbial action (chemical oxidation); soil organisms can also play an important role of organic residue crushing, and migration. The number and type of bacteria and fungi in soils has an important influence on the chemical structure of organic matter also, and the complexity of the organic molecules (N content and C:N ratio, concentration of lignin/polyphenol, lignin/N ratio, and acid soluble carbohydrates), which also affect the microbial mineralization rate and mineralization.

4 THE CHEMICAL STRUCTURE OF THE SOIL ORGANIC MATTER

Humus is first used in the process of the extraction of PH for the extraction of 10 to 13 of alkaline solution, then be extracted with acid solution with PH 2 segregation (Oades, 1988). However, studies have shown that when the soil PH is 13 in the solution of 4 hours, aromatic aldehyde is isolated from lignin. Alkaline solution can also induce polymerization of acidification process. So soil organic components of extract and the process of chemistry itself exists in the organic matter composition is not completely consistent. Spectroscopy and modern technology commonly use mass spectrometry to determine the structure of organic matter, and this method, which can determine the structure of organic matter in situ, has higher accuracy (Maccath 1990).

Humus in soil cannot get rid of the aqueous solution and organic fraction of amorphous organic matter, including humic substances, and the humus biological molecules (Stevesen, 1990). Types and concentrations of different extracts can lead to extracts of organic matter composition, which show different chemical properties (Trembelm, 2002). The humic acid has a strong adsorption of calcium, whilst humic acid calcium is poorly soluble in water.

5 CONCLUSION

In this paper, source of organic matter mainly comes from organic plant residues, and organic matter. There are two major arguments: (1) the mechanism of biodegradation theory: there is some biological polymer degradation, but the wholeness of biopolymers is not damaged just after the modification, and the modification of biological polymer is to form an organic skeleton; (2) The abiotic poly condensation theory: biodegradable products are produced by monomer polymerization to form an organic skeleton again.

Light decomposition under the action of unsaturated fatty acids in the spontaneous oxidation and cross-connection role is regarded as the bottom of the formation mechanism of the organic matter.

REFERENCES

Baldock, J.A., and Nelson, P.N. soil organic matter. [M]. Florida USA: Florida CRC press, boca raton, 2000.

Fan Zhaoping. The influence of organic matter in sludge solidification mechanism and counter measures research [D]. Nan Jing: Hehai University, 2004.

M. Oades J. The retention of organic matter in soils [J]. Biogeochemistry, 1988, 5 (1): 35–70.

MacCarthy, P., R.L. Malcolm, C.E. Clapp, and P.R. Bloom. An introduction to soil humic substances. [M]. Madison: Soil Science Society of America, 1990.

Stevenson, F.J. Humus Chemistry. Genesis, Composition, Reactions. [M]. New York: 1994.

Tremblay, Hélène, Josée Duchesne, Jacques Locat. Influence of the nature of organic compounds on fine soil stabilization with cement [J]. Canadian Geotechnical Journal, 2002, 39 (3): 535–546.

Advances in Civil Engineering and Building Materials IV – Chang et al (eds)
© 2015 Taylor & Francis Group, London, ISBN: 978-1-138-00088-9

Analysis and review of consolidation test of bentonite

Baotong Shi & Xiangxing Kong
The First Highway Survey and Design Institute of China Communications Construction Company Ltd.,
Xi'an, China

ABSTRACT: This paper makes a summary about structure characteristics, experimental methods of consolidation, and compression features of bentonite. Montmorillonite, the main component of bentonite, is 2:1 type cell, which has two layers of silicon sandwiching a layer of aluminium. There is only the van der Waals force of oxygen atoms, and no hydrogen bond between the cells, so the key force is very weak and water molecules can easily enter the cell. This gives bentonite the properties of water absorption expansion, and desiccation shrinkage. For the consolidation experiment of saturated bentonite, we can apply bentonite mud samples, or bentonite samples, after a compaction test and vacuum saturation.

Keywords: consolidation, test, bentonite, compression properties, analysis and review.

1 INTRODUCTION

Compacted bentonite and its mixture with sand have a performance such as high expansibility, low permeability, and fine nuclide adsorption, therefore, they are chosen by the world as buffer materials and backfill materials in multiple engineering barrier systems. Bentonite, and its mixture with sand in deep geological disposal, are under confining stress condition, which will be out of shape under complicated stress such as load and suction, thus producing force to the surrounding rock and soil mass. the research is about the deformation characteristics of bentonite and its mixture with sand, which are under confining stress condition,; the resulting lateral stress change rule is extremely important.

2 STRUCTURAL CHARACTERISTICS OF BENTONITE

Bentonite is a type of clay mineral, in which the main composition is montmorillonite. The crystal structure and mineral composition of montmorillonite determine the main properties of bentonite. The basic unit of montmorillonite is a connection of oxygen atoms which are interconnected by the interaction between the molecules called van der Waals force. The link force is very weak, so the water is easy to enter the molecules between the cells and make the cells' distance increase.

Therefore, montmorillonite lattice is active, and it will be expansive after absorbing water. The volume will be several times. At the same time, it

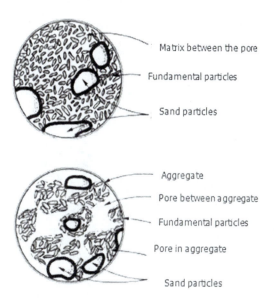

Figure 1. The structure of bentonite.

can be a significant contraction after dehydration, accompanied by microcracks (Liu, 2002).

From a micro point of view, bentonite fabric is composed of elementary particles, the particle conglomerate, and pore space, and it can be roughly divided into two types, which are dispersion and concentration body as shown in Figure 1.

The characteristics of bentonite at micro level can affect the macroscopic deformation, strength, and hydraulic characteristics. Usually, the clay is studied by using a scanning electron microscope to observe the microstructure. Before flooding, the pore is very clear; in the process of soaking the bentonite will swell, and the pore is populated gradually. After completion of suction, the sample is saturated, and the pore is completely filled with bentonite.

3 COMPRESSION PROPERTIES OF BENTONITE

3.1 Compression feature of mud bentonite

Mud bentonite samples come from bentonite powder. Put the sample in the consolidation box and fill the gap of the consolidation box with water in order to make sure the saturated state of the bentonite is achieved (Qin, 2008).

Many experimental results show that, when the pressure is low, the e – log curve conforms to the only conclusion to the isotropic compression curve for normal consolidation of clay in the Cambridge model. When the vertical pressure is high, the compression curve deviates from the original line, which tends to be more levelled. This is because the Cambridge model is mostly suitable for clays such as kaolin. When the vertical stress is larger, increasing pressure causes the decrease of pore in the body.

In addition, the study found that the lateral stress produced in the process of the compression of bentonite is higher than in ordinary clay.

3.2 Compression properties of vacuum saturation bentonite

First of all, spray the deionized water evenly on the dry bentonite powder, to achieve soil samples in target water content. Soil samples are mixed fully in the process of modulation, avoiding to appear caking phenomenon. Store the soil sample for twenty-four hours to keep the water content of the soil sample homogeneous. The soil samples are separated with the circular knives, each layer of soil samples have roughly the same weight, the compaction times of each layer is also the same, whilst a number of compactions depend on the initial void ratio required in the test. Finally, cut off a part of the soil sample above the cutting ring with a knife, level it, make a cylinder specimen whose cross-sectional area is S, and height is H. Use a fold type saturator to clamp the compacted sample with filter papers put on both sides, put it into the glass container filled with deionized water, and use vacuum pump for saturation. Afterwards, push saturated samples into the DO meter for consolidation experiments.

The study found that, when the vertical stress is high, the compression curve of the compacted specimen and the mud specimen is consistent.

3.3 Compression feature of unsaturated bentonite

Deformation of unsaturated soil in addition to being affected by the stress, is also associated with moisture in the soil. In unsaturated soil, the degree of saturation or water content, and suction, can both reflect the influence of deformation of soil, resulting from the change of moisture content. The engineers are more likely to accept with moisture content, and the practical application is also very convenient (Southeast University, 2008). Experiments such as bentonite vacuum saturation can be used as the method for sample preparation, do not need to be carried on vacuum saturation at last (Sun, 2009). During experiments cover both up and down surfaces of the specimen with a breathable waterproof membrane, so that we can ensure the moisture content of the sample is constant, then carry on the normal loading process (Wang, 2000). Unsaturated bentonite samples in the compression process make the transition from the elastic state to the elastic-plastic state, slope of e – log p curve after the fold point. For the sample with the same initial water content and different initial dry density, the yield stress is different, which is larger when the dry density is larger. The ompression curve in the high stress range has little to do with the initial density of the sample, and it mostly depends on the water content of the sample. Because unsaturated samples of compacted bentonite under the same water content have similar saturation, the suction value tends to be equal, the corresponding σ curve also tends to be consistent.

4 COMPRESSION FEATURE OF THE MIXTURE OF BENTONITE AND SAND

Sand particles in the mixture compaction state, according to the content of bentonite, can be roughly.

The sample preparation process can be consistent with that of the bentonite mud sample. Many experiments show that at the last several load levels of the compression test, curve slope gradually gets smaller. The reasons for this phenomenon is probably that the soil sample under. When the content of bentonite is high, sand particles are surrounded by bentonite. When the content of bentonite is low, the sample is compressed in the corresponding vertical stress, when volume shrunk to a certain degree, and the sand particles contact each other to form sand grain skeleton. At this time the external load is mostly undertaken by the sand grain skeleton. Therefore, e – log p is the void ratio curves of skeleton inside soil samples.

5 CONCLUSION

Most of the experiments show that for saturated samples of the mixture in different initial void ratios, the compression curves are nearly equal in the large vertical load. In large vertical loads, the change trend of the e – log p curve dips. As the initial dry density of the compacted mixture increases gradually, the compression curves of saturated samples tends to be resilient.

REFERENCES

Liu Quansheng, Wang Zhijian. Sand mixture of bentonite expansion force influencing factors research [J]. Journal of rock mechanics and engineering, 2002, 21 (7): 1054–1058

Qin Bing, Cen Henghan, Liu Yuemiao, Wang Ju. Gao Miaozi swell-shrink deformation properties of bentonite and its influencing factors of the study [J] Chinese Journal of Geotechnical Engineering, 2008, 30(7): 1005–1010

Southeast university. Soil mechanics. [M]. Beijing. China building industry press. 2008

Sun Wenjing. Swelling the hydraulic and mechanical properties of unsaturated soil and elastoplastic constitutive model [D]. Master's degree thesis. Shanghai. Tongji University. 2009

Wang Zhijian quan-sheng liu. Sand compacted bentonite mixture experimental study on the expansion properties. Rock and soil mechanics [J]. Chinese Journal of Geotechnical Engineering, 2000, 21(4):331–334

Advances in Civil Engineering and Building Materials IV – Chang et al (eds)
© *2015 Taylor & Francis Group, London, ISBN: 978-1-138-00088-9*

Analysis and review of the rheological theory of soft soil

Baotong Shi & Xiangxing Kong
The First Highway Survey and Design Institute of China Communications Construction Company Ltd.,
Xi'an, China

ABSTRACT: The research about soil creep mostly has four aspects, including macroscopic, mesoscopic, nano aspect, and the onlookers. From a macro aspect, we see soil as an homogeneous continuum, and from a micro aspect, soil is considered as a multiphase mixture that consists of liquid phase, gas phase, and solid phase. From a nano aspect, we consider that soil is composed of an accumulation of particles made of molecules and atoms. Macroscopic theory mostly includes linear viscoelastic deformation theory, genetic creep theory, and technology theory. From a mesoscopic aspect, a rheological model based on the mesoscopic level has not appeared. Various microstructures of soil are put forward, which reflect the development of the microstructure.

Keywords: rheological theory, soft soil, macroscopic, mesoscopic.

1 INTRODUCTION

Soil creep refers to the deformation process with the development of time, even under constant load. A lot of damage about the retaining wall's displacement and stability, caused by foundation creep, results in long-term settlement and the tilt of structures. The landslide caused by soil creep poses a great loss to the national economy. Consequently, the research on soil creep is significant. The study of soft soil rheological is carried on, mainly for macroscopic, nano aspect and the onlookers (Hu, 2013).

2 MACROSCOPIC SCALE

On the macroscopic scale, soil is considered as a homogeneous continuum, regardless of the physical foundation of soil rheology and the mesoscopic structure of soil. In constructing the soil rheological model by using irreversible thermodynamics and continuum mechanics, and by fitting the rheological test results, quantitatively estimate the rheological deformation of soil. The characteristics of this king of research is: for the simulation of soil rheological macroscopic phenomena and revealing the physical essence, the level of study pays more attention to the former.

Basic theory includes linear viscoelastic deformation theory, genetic creep theory and technology theory, etc.

2.1 Linear viscoelastic deformation theory

$$\alpha_0 + \alpha_1\tau + \alpha_2 i = \beta_1\gamma + \beta_2\dot{\gamma} \qquad (1)$$

In the formula, the ultimate shear stress: $\alpha_0 = \tau_T$; relaxation time: $\alpha_1 = 1$; $\alpha_2 = \alpha T_r = \eta/G_0$; viscous coefficient: $\beta_2 = \eta$; $\beta_1 = G_\infty$ G_0 and G_∞ are instantaneous and long-term limit value of shear modulus (Hu, 2008).

The elasticity of the object is simulated by an elastic element model, namely a spring. The spring obeys Hooke's law $\tau = G_\gamma$ representing the spring by H. Viscous characteristics of objects are simulated by drilling a pot model. A glue pot with a perforated piston is filled with gas. After the load application, the sinking velocity of the piston in the glue pot is described by Newton's Law.

Glue pot is represented by N. The plasticity of the object is simulated by friction components. Friction components obey Saint – Venant's Law, whilst the components are represented by SV. All three elements can have a variety of combinations to form different models reflecting different characteristics of viscoelastic soil. They are mostly Maxwell, Kelvin, and Merchant.

2.2 Genetic creep theory

The deformation caused by the past sometime of loading is equal to the sum of the deformation caused by each unrelated load to the moment. In other words, a deformation at some time ago is not only related to the stress of this moment, but also is associated with the history of the deformation. If there is a stress $\tau(t)$ that changes over time in a moment, then the total deformation for the moment when $t > v$:

$$\gamma = \frac{\tau(t)}{G_0} + \frac{1}{G_0}\tau(v)K(t-v)\Delta v \qquad (2)$$

The creep equation can be obtained, and relaxation equation can be obtained by solving creep equation.

2.3 Technology theory

The equation of state deformation can be the relation equation between stress and time, which can also be a relation equation between the deformation rate, stress and time, and can also be a relation between the strain rate, stress and deformation. The first equation of state describes theory of ageing, the second the flow theory, and the third, the hardening theory.

3 MESOSCOPIC SCALE

Although the rheological model based on macroscopic level has not appeared at present, a large number of domestic and foreign studies have shown that combined water is an important factor affecting the rheological properties of soft soil.

Studies have shown that as the clay minerals, organic matter, and combined water content in soil is increasing, the soft soil rheological phenomenon is more obvious (Vyalov, 1987). Furthermore, soft soil can be considered as a medium composed of the solid phase substance, gaseous substances, liquid substances, and rheological phase materials.

Research has also demonstrated that the viscous shear resistance caused by relative motion between membrane of soil, combined water, and the particle contact points, lead to the rheological state. Moreover, some studies show that the water content has a significant influence on the rheology of soil sample, and soil samples with high water content are compared with soil samples with low moisture content; the rheological phenomenon is more obvious. Under the condition of drainage, a sample of the initial water content is higher, the creep deformation value of the sample is greater; under the undrained conditions, the initial water content of the sample with low stress level does not have obvious influence on creep deformation. However, on the high stress level, the creep deformation value of samples with the lower initial water content is higher than the value of samples with the high initial moisture content. Initial shear force and plastic viscosity, respectively, present the approximate linear relationship between water content, and the soil pressure, and also have certain influence on the rheological state, it is thought that the change of the pressure in the soil induced by combined water film thickness change, leads to a change in viscosity.

In organic matter with higher content, the viscosity coefficient is decreasing, whilst the soil rheological deformation resistance decreases with the increase of organic matter content, and the shear deformation rate increases. The creep stress threshold of soil also increases with the increase of organic matter content, and presents the approximate linear decreasing relationship; the rheological phenomenon is more obvious.

When a shear stress is applied, due to rheological phase material lubrication, leading to larger cementation bonding between soil particles, the apparent mechanical friction is smaller. In this state, the soil particles are more prone to sliding and rotation.

4 MICRO AND NANO LEVELS

Compared to the macro-level, the research of meso-level has more help to reveal the microstructure and rheology relationships. On the nano level, soil is seen as particles formed by the accumulation and studying the effects of micro-force between the particles, and doing research including particle aggregation, ion effect, chemical adsorption studies and so on, by means of quantum statistical mechanics, molecular dynamics and so on.

Creep deformation of soft soil includes reversible deformation and irreversible deformation, such as sticky elastic-plastic deformation, etc. Reversible deformation of soft soil is mostly caused by changes of rubber particle size, volume deformation of soil skeleton, and reversible volume changes of enclosing small bubbles. When the force increases, the particles move closer, and thickness of the water film decreases. By removing the loading, repulsion pushes particles, which results in partial volume deformation recovery. Irreversible deformation of soft soil is the result of reversible displacement of particles and their rearrangement. As the load increases, the structural strength values on the rise does not exceed the maximum value, although the clay-structural defects are always moving in the direction of stabilising the development of creep. When the stress exceeds the structure yielding stress, because cementation and the water film thickness in the soil particles are different.

5 CONCLUSION

The research about soil creep mostly has four aspects, including macroscopic, mesoscopic, nano aspect and the onlookers. On the macro scale, soil is seen as homogeneous continuum, and on the micro scale, soil is considered as multiphase mixture that consists of liquid phase, gas phase and solid phase. On the nano aspect, we consider that soil is composed of accumulation of particles by molecules and atoms. Macroscopic theory mostly includes linear viscoelastic deformation theory, genetic creep theory and technology theory, etc. On the mesoscopic level, the rheological model based on mesoscopic level has not appeared. Various microstructures of soil are put forward, which reflects the development of microstructure.

REFERENCES

Hu Guijie. Experimental study on composition and distribution of soil influencing soil strength and the rheological

[D]. Guangzhou. South China University of technology, 2013.

Hu hua. Experimental study on multiple factors affect stable rheological parameters of soft soil characteristics [J]. Soil mechanics. 2008 (S1): 507–510.

Vyalov. Rheology of soil mechanics [M]. Translated by Du Yupei. Beijing science press, 1987.

Yang Aiwu, Kong lingwei. Analysis of micro-mechanism and change of structural strength in the process of dredger soft soil creep. [J]. Journal of engineering geology, 2014, 22 (2): 181–187.

Advances in Civil Engineering and Building Materials IV – Chang et al (eds)
© 2015 Taylor & Francis Group, London, ISBN: 978-1-138-00088-9

Review and analysis of the basic features of organic matter in soil

Rui Huang & Xiangxing Kong
The First Highway Survey and Design Institute of China Communications Construction Company Ltd.,
Xi'an, China

ABSTRACT: Present a summary of the classification of organic matter, the physical and chemical functions of organic matter, the factors determining soil organic C levels, and its chemical structure. Soil organic matter is divided into two main parts, living components and non-living components, and the non-living components consist of particulate organic matter, dissolved organic matter, humus, and inert organic matter. The main physical functions of organic matter are stabilization of soil aggregates and water retention. The main chemical functions of organic matter are exchange capacity and buffering capacity.

Keywords: basic features, organic matter, soil, chemical function.

1 INTRODUCTION

Organic soil existing in nature has a high moisture content, a high compressibility, and a high creep rate, with low volume density and low strength at the same time, thus becoming bad soil for engineering (Keller, 1982, Lu, 2011 & Mu, 2008). The presence of organic matter also has many negative influences on the solidification of soil (Chen, 2005). For example, in the winter of 1985, there was a reinforcement project of swamp facies peat soil in the southwest area of our country, during the construction: when using deep mixing reinforcement method to the peat soil cement only, even if the mixing ratio was up to 30%, the strength of reinforced soil is also difficult to achieve 300 kPa, thus the cement reinforced peat soil had lost economic and technical benefits.

In order to analyse the reasons for characteristics of organic soil, we must do in-depth research on the composition of the organic matter, and its basic characteristics. From the molecular structure, and physical and chemical characteristics of organic matter, the fundamental reasons for various features of organic soil are analysed.

In this paper, the comprehensive domestic and foreign research achievements were given, which gave a summary of the composition of organic matter in soil, physical and chemical characteristics, influence factors of soil organic C levels, and chemical structure.

2 THE DEFINITION AND CLASSIFICATION OF ORGANIC MATTER

It has a diverse composition and complex structure, Oades (1988), MacCarthy et al. (1990); Stevenson

Table 1. Organic matter composition table.

Organic matter Composition		Definition
The active ingredient	Quantity of plants	Plant tissue still have a cellular structure
	Microbial biomass	Microorganisms living cells
Inactive ingredients	Organic matter particles	Derived from plants and animals; fragments of the organic matter which have obvious cell structure
	Dissolved organic matter	Particle size <0.45 um water-soluble organic compounds in soil solution
	Humus	Organic matter that, except for organic matter particles and dissolved organic matter, is still retained in the soil
	Inert organic matter	Advanced carbonized organic matter including charcoal, charred plant materials, graphite and coal, etc.

(1994) has given a relatively comprehensive organic matter composition, as shown in Table 1.

Humus can be further divided into humus acid, fulvic acid, and humic acid according to its soluble characteristics in acid and alkali solutions. Humus acid is dissolved in alkaline solutions, but not dissolved in acidic solutions; fulvic acid is dissolved in alkaline solutions, and also dissolved in acidic solutions; humic acid is insoluble in alkaline solutions.

In aerobic and anaerobic conditions, animals and plants' bodies produce humification. Humus as relatively stable product of biochemical degradation is continuously accumulated; humus accounted for in organic matter content is greater than 60%. The accumulation of humus in geological time formed the oil and coal, and produced oxygen and water at the same time. Humus also plays an important role in migration to the surface of the earth and concentration changes of metal elements.

The humus in water is part from land-based sources transfer, the other part is produced by organisms in water. Internal sedimentary humus's C13/C12 is higher than terrestrial humic, and also has a more aliphatic composition, aromatic compounds, phenolic compounds, and polymers. Carbon content is relatively low, but nitrogen, sulphur, hydrogen root, hydroxyl, and carboxyl content is higher. Aliphatic products come mostly from algae humic or sapropel, and we know that a high content of fat coal and oil is formed from plant sources. Differences in sources of organic matter cause different components of internal sedimentary organic matter sources humus, and terrigenous humus.

3 THE PHYSICAL FUNCTIONS OF ORGANIC MATTER

3.1 Stabilizing effects for the soil particle aggregates

When mineral grains rely mostly on connection effects of organic matter, they can form different levels of aggregate. Oades and Waters (1991), Oades (1993) and Golchin et al. (1997 c) put forward three levels aggregates: <20 um particle group; 20–250 um micro aggregate; and >250 um aggregate.

The formation of particle groups is directly related to the chemical property of soil mineral composition; different mineral compositions combine with different organic matter compositions, which form particle groups that have a different structure and size. Formation mechanism of micro aggregate is mostly the bonding effects of the polysaccharide produced by plant roots and microbes.

Micro aggregate of organic matter and biological product stick together around the nucleus to form larger aggregate particles. The deposition of humus in water is not only related to flocculation, but also related to adsorption of clay particles, the combination of invertebrate animal waste, and blisters polymerization, as well as the hydraulic factors of influencing sedimentary.

3.2 Water holding effects

The holding water quality of organic matter composition can even reach 20 times the quality of soil granule (Stevenson, 1994). Organic matter can increase water-holding capacity directly by reducing evaporation and increasing infiltration capacity, and can also be based on the soil aggregates and the influence of pore distribution indirectly influencing the water holding capacity. Silva and Kay (1996) have given the relationship between volume moisture content ($m^3\ m^{-3}$) and matric potential (MPa), clay content (%), organic C content (%), and bulk density BD ($Mg\ m^{-3}$), (Chen 2005). As seen from Equation 1, the volume moisture content is exponentially positively related to the organic content.

4 THE CHEMICAL FUNCTIONS OF ORGANIC MATTER

4.1 Cation exchange capacity

Organic matter provides soil with 25–90% cation exchange capacity. When the soil has a low content of clay or mostly has clay minerals with a low cation exchange capacity, the importance of the organic matter is more obvious (the effect on the kaolin is stronger than the bentonite, and in sandy soil, organic matter provides the main cation exchange capacity). Organic cation exchange capacity comes mostly from carboxyl, phenol, and olefinic alcohol and imide, and thus cation exchange capacity has a great relationship with the pH value.

4.2 The influence of buffering effect of the pH values

Organic weak acid functional groups can form the base pairs, and also form a conjugate acid which makes organic matter have effective buffering effect; the diversity of organic functional groups makes buffer effect possible under the condition of a wide range of pH values. Organic matter provides soil with about two-thirds of the buffering capacity. Adding organic matter to soil makes the pH change; the change process itself is related to the soil chemical properties, organic matter and soil moisture content, the leaching conditions, and other factors.

Adding soil organic matter to acidic soil can make metal ions of the metal complex dissolve, at the same time causing the mineralization of organic N, and denitrification which increase pH. Pocknee and Sumner (1997) put forward that the rise of pH is mostly related to the ratio of N element content to the total cation, but also related to decarboxylate. Adding organic matter into the alkaline soil can accelerate the mineralization of organic P, S, the nitration of organic N, leaching of mineralized organic N, decomposition of organic ligand, and dissolution of CO_2, which then makes the pH decrease.

4.3 Complexing with inorganic cation

The existence of various functional groups of organic matter means that it can react with inorganic cations, such as simple cation exchange, and back pressure, in which complex organic ligands are involved. Humus, which can be combined with metal ions, carboxylic, and phenolic groups is the most important component of ligands. The stability of humus complexes is related to the characteristics of metal, and the binding energy of functional groups and the environment (e.g., pH, Eh value). Facts have proved that different metals in sea water and lake water can be combined with different humus, and in acidic environment, the bonding strength of the terrigenous humus and ferric iron and aluminium ion in water is stronger than bivalent ions, and when the humus are put forward from sea sediments, in the pH value of 7, the binding reaction is reverse.

If humus is rich in nitrogen element, its ability of combining with the ferrous iron is stronger than ferric iron, but obviously different composition of humus can also lead to different binding ability with metal ion. Fe, Al, and humus complex is easy to combine with phosphorus, and boron, whilst molybdenum and vanadium can combine with humus in soluble oxide forms.

The influence on soil properties of the complexation of organic matter to inorganic cations and its process mostly includes: (1) by complexation of Fe^{3+}, Al^{3+} and Ca^{2+}, which makes absorbed P replaced, and increases non-soluble P; (2) to promote the release of plants' needed nutrients; (3) to improve the content of the surface layer of soil trace element. (4) to promote the organic matter component concentration in the soil; (5) to have buffering effect on over high concentration toxic substances or heavy metals; (6) to benefit from organic matter with soil depth.

5 CONCLUSIONS

Through analysing the summary of the classification of organic matter, the physical and chemical functions determine soil organic C levels and its chemical structure. Soil organic matter is divided into two main parts: living components and non-living components, and the non-living components are consist of particulate organic matter, dissolved organic matter, humus, and inert organic matter. The main physical functions of organic matter are stabilization of soil aggregates and water retention. The main chemical functions of organic matter are cation exchange capacity, buffering capacity, and complexation of inorganic cations.

REFERENCES

Chen, HuieQing Wang. The behaviour of organic matter in the process of soft soil stabilization using cement [J]. Bulletin of Engineering Geology and the Environment, 2005, 65(4):445–448.

Keller, George H. Organic matter and the geotechnical properties of submarine sediments [J]. Geo-Marine Letters, 1982, 2(5):191–198.

Lu Yan, Nie Lei, Xu Yan. The mechanism of the physical and mechanical properties of peat soil organic matter analysis [J]. Chinese Journal of Geotechnical Engineering, 2011(4):655–660.

Mu Chunmei, Li Baifeng. Organic matter content and mechanical properties of soft soil effect analysis [J]. Hydrogeology engineering geology, 2008(3):42–46.

Advances in Civil Engineering and Building Materials IV – Chang et al (eds)
© 2015 Taylor & Francis Group, London, ISBN: 978-1-138-00088-9

Analysis of the soil organic matter influence on the physical and mechanical properties

Rui Huang & Xiangxing Kong
The First Highway Survey and Design Institute of China Communications Construction Company Ltd.,
Xi'an, China

ABSTRACT: It is found that organic matter content has an obvious effect on the physical and mechanical characteristics of soil, based on the field survey data. With the increase of organic matter content, the proportion of water content in the soil is reduced, static sounding penetration resistance is reduced, the data is linear fitting, and the fitting equation is given.

Keywords: organic matter, influence, physical property, mechanical property.

1 INTRODUCTION

According to geotechnical engineering specifications, when the organic content of soil is between 5%–10% it is called organic soil. The main components of the organic matter are humus and cellulose. Humus includes humic acid, fulvic acid, and humilic acid. Organic matter has a lot of pores, a big specific surface area, strong adsorption, and so on. Therefore, soft soil contains different types of organic matter. It usually has special engineering geological properties, such as big void ratio, high natural water content, strong compressibility, and low strength. The presence of organic matter had a great influence on the engineering properties of soil. The understanding, in general, is that with the water content increasing, the dispersion of soil increased, the water content increased, the dry density decreased, and the swell-shrink increased; in addition, the compressibility increased, the intensity decreased, and the bearing capacity decreased. The organic soft soil in the Pearl River Delta region contains a large amount of water and wide ranges of plastic soil, the larger dilatability soil, and low permeability soil. The features are not conducive to engineering.

According to the geotechnical investigation GB50021-2001, it is stipulated that we can use Loss on Ignition instead of organic content. Loss on Ignition refers to the soil sample burned at a temperature of 550°C to constant weight. The reduced proportion means less weight of soil samples and and original soil samples. At 550°C, the water of crystallization, combined water, oxide and volatile components in the soil are volatile at the same time, therefore Loss on Ignition of soil samples is often greater than the weight of the organic matter content. However, the higher the organic matter content in soil is, the closer Loss on Ignition is to the organic content. In other words, the more ignition loss there is, the greater the organic content is. Therefore, we use ignition loss instead of organic content.

2 THE INFLUENCE OF ORGANIC MATTER ON THE NATURAL SOIL MOISTURE CONTENT

Through the statistics and analysis of large numbers of investigation data, we obtain the basic physical and mechanical properties of soft soil with different organic matter contents (Mu, 2008). The chart below is the relationship between the natural moisture content and the organic matter content of the soil in Guangzhou. With the increase of organic matter content, the soil moisture content increases. Solid line is a fitting equation between water content and organic content as shown in Figure 1, $W = 4.487*OC + 60.949$. W is natural moisture content, and OC is organic matter content. The increase of organic matter content in the soft soil, on the one hand, improves the structure of the soft soil, and increases the porosity. On the other hand, it changes the distribution of soil colloid, and makes an additional enhancement to the soil. Both of these aspects are conducive to maintaining the soil's moisture, so that the moisture content increases. At the same time, the pore size distribution and arrangement situation has changed too, and the soft soil almost becomes an empty frame structure. The change of the soil's physical property indexes, and the improvement soil structure will cause changes in the mechanical properties of soft soil (Shan, 1998), inevitably affect the mechanical effect of soft soil. Building structures on this kind of soft soil ground often make sliding problems in the process of filling in the engineering.

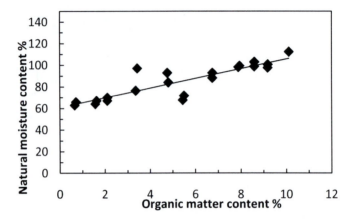

Figure 1.   The relationship between the moisture content and organic matter.

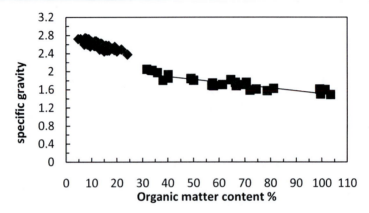

Figure 2.   The relations of gravity with organic matter content.

3   THE INFLUENCE OF ORGANIC MATTER ON THE SPECIFIC GRAVITY OF SOIL

Specific gravity refers to the soil particles burned at a temperature of 105°C to 110°C to constant weight, the proportion of the quality of constant weight soil, and the quality of the same volume of distilled water at 4°C. It is actually part of the solid density. Its size is associated with the mineral composition of soil. The larger the specific gravity of minerals, the larger the specific gravity of the soil. The specific gravity of iron minerals is large, and the specific gravity of organic matter is small. The specific gravity of humus is very small. The content of it in the soil has a large effect on the specific gravity of soil. The determination of the specific gravity of organic soil should use neutral liquids to do so. When organic soil has interaction with water, the value of its specific gravity will increase additionally. Therefore, we must use neutral liquid instead of pure water to determine it, and use a vacuum suction method to replace the boiling method so that it can eliminate the air in the soil. According to the Post-Graduate Laboratory of Nanjing Institute of Surveying[2] and Mapping and the data of

peat swamp soil in Jilin (Xie, 2003), we find that the specific gravity rapidly reduces with the increase of organic matter content when the organic matter content is less than 30%. We can get the regression equation, Gs = −0.0162*OC + 2.788. Gs is the specific gravity of soil. The decrease in specific gravity becomes slow when organic matter content is more than 30%. The regression equation is, Gs = −0.0063*OC + 2.138, as shown in Figure 2.

4   THE INFLUENCE OF ORGANIC MATTER ON THE MECHANICAL PROPERTIES OF SOIL

Organic matter content has an effect on the mechanical properties of soft soil, such as low strength, strong structure, large rheological, and creep properties. It does great harm to the buildings on soft organic soil. For example, it produces a large amount of settlement of foundation soil, the time of settlement completion is very long, the proportion of secondary consolidation settlement is large, and post-construction settlement is great. Therefore, it is difficult to accurately estimate

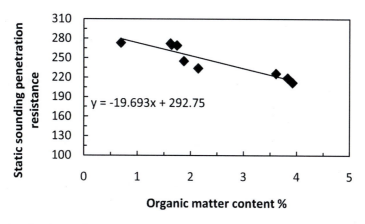

Figure 3. The relation of static sounding penetration resistance with organic matter.

the settlement development trends in the process of design and construction. It also makes for a higher cost of foundation treatment, increases engineering investment, and sometimes, a satisfactory result cannot be obtained even if we spend a lot of money. According to in situ test of soft organic soil in Guangzhou Nansha [2] development zone, we obtain in situ experimental data. Figure 3 shows static sounding penetration resistance with the changes of the organic matter content. With the increase of organic matter content, the soil mechanical property index will also change. With the increase of organic matter content, the soil mechanical property indexes are sharply reduced.

5 CONCLUSIONS

It is found that organic matter content has an obvious effect on the physical and mechanical characteristics of soil, based on the field survey data. Though tested soft

organic soil in Guangzhou Nansha, with the increase of organic matter content, the proportion of water content in the soil is reduced, static sounding penetration resistance is reduced, The data is linear fitting, and the fitting equation is given.

REFERENCES

Mu Chunmei and Li Baifeng. Organic matter content and mechanical properties of soft soil effect analysis [J]. Journal of hydrogeological engineering geology, 2008, 3:42–45.

Shan Xiuzhi, Wei Youqing, Yan Huijun, etc. The influence of soil organic matter content of soil water dynamics parameters [J]. Journal of soil, 1998, 35 (1): 1–8.

Xie Liping and Wu Fengjun. The proportion of organic soil, peat soil and organic matter content of relevance [J]. Journal of survey of city, 2003, 4:52–54.

Hydraulic engineering

Advances in Civil Engineering and Building Materials IV – Chang et al (eds)
© *2015 Taylor & Francis Group, London, ISBN: 978-1-138-00088-9*

The parameter sensitivity of the Duncan-Chang model for the simulation of a concrete faced rockfill dam

Ling Chen, Fuguo Tong, Niannian Xi & Gang Liu
*College of Hydraulic and Environment Engineering, China Three Gorges University,
Hubei Province, China*

ABSTRACT: Duncan-Chang E-B model was widely used in the numerical simulation of Concrete Faced Rockfill Dam (CFRD). Because many parameters were involved in this model, it is necessary to know which parameters play key role in the distribution of stress and deformation of rockfill dam. This paper presented the parameter sensitivity to the deformation of rockfill dam through finite element method calculation. The results show that the deformation of rockfill dam strongly related to failure ratio (R_f), the friction angle (φ), the modulus exponent (n), and bulk modulus number (K_b). On the contrary, bulk modulus exponent (m), the cohesion (c) and modulus number (K) are not sensitive to the deformation of rockfill dam.

Keywords: rockfill dam, nonlinear Finite Element Method, Duncan-Chang E-B model, parameter sensitivity.

1 INTRODUCTION

As the Concrete Faced Rockfill Dam (CFRD) has certain advantages, such as fewer supplies, practical geological conditions and seismic effects, etc., it has been widely used in hydraulic engineering for nearly 20 years. In south-western China, there are dozens of concrete faced rockfill dams that have been built or are under construction or being planned to be built. Therefore, it plays a crucial role to master the deformation of CFRD and find material properties for optimizing the design and performance evaluation of the dam (Chen, Y.F. et al. 2011).

Numerous studies show that material parameters of a dam are some of the main factors to affect stress and deformation states of a dam (Chen, H.Y. 1988, Wang, H.S. & Weng, Q.D. 1990, Cao, Y.H. et al. 2005, Yang, Q.Z. & Wen, B. 2011). The changes of parameters will inevitably lead to the change of calculation results that do not match the measured values, and sometimes make a big difference. If the parameters are in accordance with the equally important to consider, it would necessarily be less effective (Zhang, J.Z. et al. 2008). Therefore, the appropriate choice for the dam material parameters is central. Currently, Duncan – Chang model is most commonly used in finite element analysis of stress and deformation states of CFRD abroad (Xiao, H.W. 2004, Cao, Y.H. et al. 2005, Yang, Q.Z. & Wen, B. 2011, Zhang, J.Z. et al. 2008, Li, Y.L. et al. 2013). Today among many studies of the material parameters of E-B model, the calculation parameters are given, but there are still some deficiencies. Conventional triaxial tests used to determine the value of the material is very convenient, but it cannot ensure the accuracy of the parameters. Inversion analysis of the measured results can ensure more accurate calculation parameters (Zhou, Y.G. et al. 2010). However, Duncan – Chang model has many parameters. Parameter inversion analysis is more difficult to calculate and the workload is large (Li, Y.L. et al. 2013). For the study of multivariate sensitivity analysis, orthogonal design method is currently widely used; there are two methods to deal with results —range analysis and variance analysis. Range analysis method is simple and convenient, but not accurate enough (Hai, Y. et al. 2008). Variance analysis method is relatively complicated, but the accuracy of the orthogonal design parameter sensitivity analysis will be greatly reduced, thereby affecting the reliability of the inversion accurate analysis (Yan, Q. et al. 2010).

The traditional parameter sensitivity analysis is univariate analysis method, which is used to select an index value and change it, assuming the other parameters remained unchanged, by comparing the value of the benchmark curve with parameters to reflect the sensitivity of each parameter (Hou, H.G. & Wang, Y.M. 1985, Ni, H. et al. 2002, Xu, H. et al. 2012, Li, Y.L. et al. 2013). This method can be more intuitive to reflect the impact of various parameters on the basis of index values, which have more clarity to identify regularity.

In this paper, the central typical cross-section of Liu River concrete faced rockfill dam is selected as a computing model, using Duncan – Chang E-B model to analyse the seven model parameters of failure ratio (R_f), the friction angle (φ), the modulus exponent (n), bulk modulus number (K_b), bulk modulus exponent (m), the cohesion (c) and modulus number (K) using the univariate analysis to probe the relation between deformation and the stress on a dam and the parameters of Duncan – Chang E-B model. It is in order to seize the main contradiction, and thus achieve a multiplier effect in parametric testing and construction design.

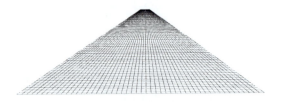

Figure 1. Finite element mesh.

2 THE CALCULATION METHODS

2.1 The Duncan – Chang E-B model

This model is a hyperbolic nonlinear elastic model as Duncan – Chang proposed. It takes the tangent modulus and bulk modulus of the soil as a computational model. The stress-dependent elastic tangent modulus and bulk modulus were respectively expressed as follows:

$$E = KP_a \left(\frac{\sigma_3}{P_a}\right)^n \left[1 - \frac{R_f(1-\sin\varphi)(\sigma_1-\sigma_3)}{2c\cos\varphi + 2\sigma_3\sin\varphi}\right] \quad (1)$$

$$B_t = K_b P_a \left(\frac{\sigma_3}{P_a}\right)^m \quad (2)$$

Whilst this model also consider that the internal friction angle φ changes with confining pressure σ_3, for coarse-grained soil and cohesion, internal friction angle φ and confining pressure σ_3 have the following relationship:

$$\varphi = \varphi_0 - \Delta\varphi \lg\left(\frac{\sigma_3}{P_a}\right) \quad (3)$$

Where: E = the tangential elastic modulus; c = the cohesion; φ = the friction angle; P_a = atmospheric pressure; R_f = the failure ratio; K = the modulus number; n = the modulus exponent; K_b = bulk modulus number; m = the bulk modulus exponent. When using Duncan – Chang model to calculate stress deformation of a rockfill, seven parameters (c, φ, R_f, K, n, K_b and m) can be identified by triaxial tests or previous engineering practice.

2.2 The Geometric Model

This paper takes a Liu pond faced rockfill dam as a calculated and analysed object, selecting the middle of the riverbed typical cross-section of the dam as a calculation model, whose sectional height is 105 m, upstream slope is 1:1.4, and dam surface elevation is 350 m. Take along the flow direction as the X-axis positive. Take straight up as the Y-axis positive. The coordinate origin is located at the surface of the dam foundation: 2500 computing grid cells, and 5202 element nodes (Figure 1).

2.3 Simulation of the construction process

Construction sequence for CFRD has a significant impact on a dam deformation, so that the calculation and analysis should realistically simulate the actual filling process of the dam. In order to make the results closer to the observations, the filling construction of each layer is calculated as an incremental load step to iterative calculation.

2.4 The values of parameters

When the Duncan model is used to calculate the deformation of CFRD, it is very important to determine seven model parameters of failure ratio (R_f), the friction angle (φ), the modulus exponent (n), bulk modulus number (K_b), bulk modulus exponent (m), the cohesion (c) and modulus number (K). Whilst selecting filler, it is soil density that is one of the key factors to consider. Whilst doing parametric analysis, it needs to consider another parameter of the soil, namely soil density. Each of the fillers has eight model parameters. Each model parameter has different degrees of impact on the sensitivity of stress and deformation, and the numerical results. In order to ensure reliable results, each of the model parameters should be considered. Thus, if too many types of filler material are considered, it will increase the complexity and difficulty of the study. Before the sensitivity analysis, it is essential to make appropriate simplification of the problem based on the actual situation of the project. A representative filler parameter is selected as a reference value. Where, $c = 8.860$, $\sin\varphi = 0.237$, $R_f = 0.865$, $K = 948$, $n = 0.535$, $K_b = 430$, $m = 0.05$, $R_d = 2.132 \text{ g/cm}^3$.

All calculated combinations have changed with this reference value. That is to take -30%, -20%, -10%, -5%, 5%, 10%, 20%, 30% of each parameter, and 88 groups calculated combinations in total.

3 ANALYSIS OF THE RESULTS

As a statistical quantity, the weighted arithmetic mean is widely used in the social and economic statistical aspects. It has an important number of indicators, reflecting the general level of the side of things for a developing change. When the calculation results are analysed, a weighted arithmetic average method reflects the trends in stress and deformation, correlated with parameters for element nodes.

$$\bar{x} = \frac{\sum\limits_{i=1}^{n} f_i x_i}{\sum\limits_{i=i}^{n} f_i} \quad (4)$$

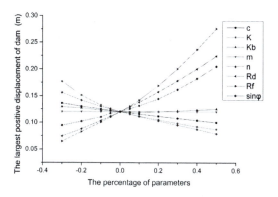

Figure 2. The correlation curve between the largest positive displacement of the dam in the horizontal direction and the parameters.

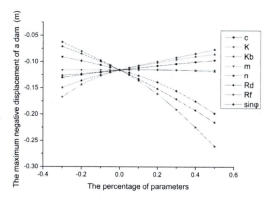

Figure 3. The correlation curve between the maximum negative displacement of the dam in the horizontal direction and the parameters.

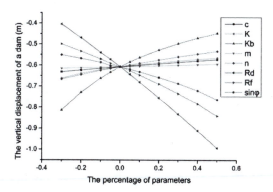

Figure 4. The correlation curve between the vertical displacement of the dam and the parameters.

3.1 The correlation between displacement of the dam and model parameters

As can be seen from the four figures:

It is equal for the absolute value of the maximum positive displacement and the maximum negative displacement of the dam along the horizontal direction.

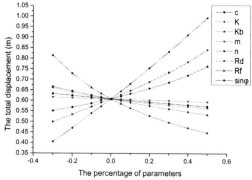

Figure 5. The correlation curve between the total displacement of the dam and the parameters.

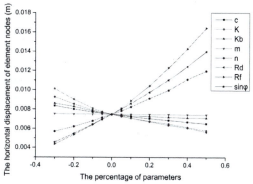

Figure 6. The correlation curve between the horizontal displacement of element nodes and parameters.

In other words, the displacement of the dam along the horizontal direction is symmetrical. Therefore, the following will only discuss the positive displacement in the horizontal direction of the dam. Dam displacement vary strongly with the three model parameters – the friction angle (φ), failure ratio (R_f) and density (R_d). However, the horizontal displacement of a dam changes with failure ratio (R_f) fastest, followed by density (R_d) or the friction angle (φ). The settlement amount and the total displacement of the dam vary with density (R_d) fastest, followed by failure ratio (R_f), and the friction angle (φ). The two model parameters – bulk modulus number (K_b) and bulk modulus exponent (m) have little effect on the horizontal displacement of a dam, and the settlement amount and the total displacement of a dam change with the cohesion (c) and modulus number (K), and bulk modulus exponent (m) slowly. Also, bulk modulus exponent (m) is not sensitive to the horizontal displacement, the settlement amount, and total displacement of a dam.

3.2 The correlation between the displacement of element nodes and model parameters

Displacement of element nodes quickly changes with the three model parameters – failure ratio (R_f), density

285

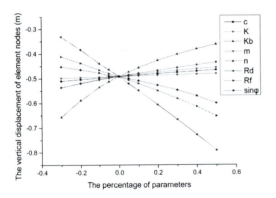

Figure 7. The correlation curve between vertical displacement of element nodes and the parameters.

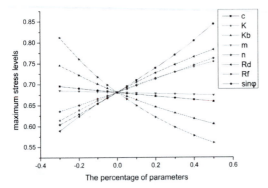

Figure 8. The correlation curve between maximum stress levels of a dam and model parameters.

(R_d), and the friction angle (φ). However, the horizontal displacement of element nodes changes with failure ratio (R_f) fastest, followed by density (R_d). The settlement amount of element nodes vary with density (R_d) fastest, followed by failure ratio (R_f), and the friction angle (φ). The three model parameters – the bulk modulus exponent (m), the cohesion (c), and the bulk modulus number (K_b) have little effect on the horizontal displacement of element nodes, and the settlement amount of element nodes change with the cohesion (c), and modulus number (K), and bulk modulus exponent (m) slowly. Also, the cohesion (c), and bulk modulus exponent (m) have least effect on displacement of the settlement of element nodes.

3.3 *The correlation between the stress level of a dam and model parameters*

The stress levels of element nodes are used for the weighted arithmetic mean.

The maximum stress levels of a dam are 0.85 or less, which is within the safe allowable range. The relation between the maximum stress levels of a dam and the four model parameters – the friction angle (φ), density (R_d), the modulus exponent (n), and modulus number

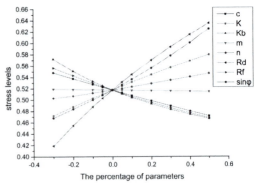

Figure 9. The correlation curve between stress levels of element nodes and model parameters.

(K) – is monotonically proportional within limits, but with decrease of the cohesion (c), bulk modulus number (K_b), bulk modulus exponent (m), and failure ratio (R_f). In addition, the maximum stress levels of a dam vary with the friction angle (φ), failure ratio (R_f), bulk modulus number (K_b), and density (R_d) quickly.

The stress levels of element nodes are 0.65 or less, which is within the safe allowable range. The relation between the stress levels of element nodes and the five model parameters – density (R_d), the friction angle (φ), modulus number (K), the modulus exponent (n), and bulk modulus exponent (m) – is monotonically proportional within limits, but with decrease of the cohesion (c), failure ratio (R_f), and bulk modulus number (K_b). In addition, the stress levels of element nodes change with density (R_d), the friction angle (φ), and modulus number (K) quickly.

4 CONCLUSION

Both the horizontal displacement and the vertical settlement of a dam increase with the parameters of the friction angle (φ), failure ratio (R_f), and density (R_d). The displacement of element nodes also has a similar principle. On the contrary, they decrease with the parameters of the cohesion (c), modulus number (K), and the modulus exponent (n). The displacement and stress change with failure ratio (R_f), density (R_d), and the friction angle (φ) quickly, but with the bulk modulus exponent (m) and the cohesion (c) the two model parameters have only small change.

Thus, the failure ratio (R_f), density (R_d), and the friction angle (φ) have a significant influence on the stress and displacement of a dam. However, the bulk modulus exponent (m) and the cohesion (c) have substantially little influence on those. Therefore, when selecting the calculation parameters, more consideration should be given to the impact on failure ratio (R_f), density (R_d) and the friction angle (φ). Within the scope of 0.55 to 0.85, the maximum stress levels of a dam are in the range of allowable safety.

ACKNOWLEDGEMENTS

This work was financially supported by the Natural Science Foundation of China (51279090), Science and Technology Research Project of China Education Ministry (211113).

REFERENCES

Cao, Y.H., Chang, X.L. & Zhou, W., 2005. Analysis on Influence of Rockfill Material upon Sensitivity of Stress & Deformation of Niuniu CFRD, Hubei Water Power 4:17–20.

Chen, H.Y. 1988. Finite Element Analysis of Embankment Dams, Nanjing: Hohai University Press.

Chen, Y.F., Hu, R., Li, D. & Zhou, C., 2011. Modeling Coupled Processes of Non-Steady Seepage Flow and Non-Linear Deformation for a Concrete-Faced Rockfill Dam, Computers and Structures 89(2011):1333–1351.

Hai, Y., Wang, R.J. & Zhu, Y.M., 2008. Sensitivity Analysis of E-B Model Parameters Effecting on Dam Slope Stability in Core Dam, Journal of China Three Gorges University (Natural Sciences) 3(5):13–17.

Hou, H.G. & Wang, Y.M. 1985. Orthogonal Test Method, Changchun: Jilin People's Publishing House.

Li, Y.L., Li, S.Y., Ding, Z.F. & Tu, X., 2013. The Sensitivity Analysis of Duncan-Chang E-B Model Parameters Based on the Orthogonal Test Method, Journal of Hydraulic Engineering 7:873–879.

Ni, H., Liu, Y.R. & Long, Z.G., 2002. Applications of Orthogonal Design to Sensitivity Analysis of Landslide, Chinese Journal of Rock Mechanics and Engineering 21(7):989–992.

Wang, H.S. & Weng, Q.D. 1990. The Thematic Section of Hydraulic Structures, Beijing: China Water Power Press.

Wang, Y.P. & Zhao R.C. 2007. An Analysis of the Sensitivity of the Factors Affecting Slope Stability Based on Orthogonal Experiment, Journal of Yancheng Institute of Techno logy Natural Science Edition 12(3):67–69.

Xiao, H.W. 2004. Effect of Duncan-Chang E-B Model Parameters on Stress and Deformation of High CFRD Dam, Journal of Yangtze River Scientific Research Institute 6:41–44.

Xu, H., Gong, S.H. & Liu, X.A., 2012. Simulation and Experimental Study on the Droplet Simulated Motion of Double-Nozzle Impact Sprinkler, Journal of Hydraulic Engineering 43(4):480–486.

Yan, Q., Wu, C.B. & Zhang, Y., 2010. Analysis of Parameter Sensitivity of Duncan E-B Model Based on Uniform Design, China Rural Water and Hydropower 7(4):82–85.

Yang, Q.Z. & Wen, B. 2011. The Adaptability Analysis of Duncan Chang E-B model for the Settlement Calculation of Medium-Altitude Dam with Concrete Panel Protection, Jilin Water Resources 12:5–7.

Zhang, J.Z., Miao, L.C. & Wang, H.J., 2008. Analysis of Parameters Sensitivity of Duncan-Chang Model and Study on the Controlling Deformation, Industrial Construction 3:75–79.

Zhou, Y.G., Chen, S.H. & Zhou, H., 2010. Back Analysis on Mechanical Parameters of Induced Joints in Arch Dams, Journal of Hydraulic Engineering 41(5):575–580.

Advances in Civil Engineering and Building Materials IV – Chang et al (eds)
© 2015 Taylor & Francis Group, London, ISBN: 978-1-138-00088-9

Effects of a crescent rib on the water flow in a bifurcated pipe

Yan Li

Department of Hydraulic Engineering, Tianjin Agricultural University, Tianjin, China

ABSTRACT: Bifurcated pipes with internal crescent ribs are widely used in the diversion system of a pumped storage power station, the water flow of which is bidirectional and complicated. In this paper, hydraulic characteristics of crescent-rib bifurcations are studied by numerical simulation method. The simulated head loss, velocity distribution, and flow pattern are analysed to determine the influences of different rib breadth ratios.

Keywords: bifurcated pipe, crescent rib, rib breadth ratio, hydraulic characteristics, numerical simulation.

1 INTRODUCTION

With the rapid development of socioeconomics and the increase of power supply demand, the construction of pumped storage power station has been paid much attention. Bifurcated pipes with internal crescent ribs are widely used in the diversion systems of pumped storage power stations. Even with the advantages of reasonable force, easy design and a reliable structure, the water flow of bifurcated pipes with internal crescent ribs are very complicated, which should be studied further (Liang & Chen 2010, Liu *et al.* 2004).

Based on a Y-shaped symmetric bifurcated pipe, the current variation is studied by using numerical simulation method when a crescent rib is embedded in this bifurcation, and the hydraulic characteristics are analysed to determine the influences of different rib breadth ratio.

2 GOVERNING EQUATION AND BOUNDARY CONDITION

2.1 Governing equations

(1) The continuity equation:

$$\frac{\partial u_i}{\partial x_i} = 0 \tag{1}$$

(2) The momentum equation:

$$\frac{\partial(\rho u_i)}{\partial t} + \frac{\partial(\rho u_i u_j)}{\partial x_j} = -\frac{\partial p}{\partial x_i} + \frac{\partial}{\partial x_j}(\mu\frac{\partial u_i}{\partial x_j} + \tau_{ij}) \tag{2}$$

(3) k equation:

$$\frac{\partial(\rho k)}{\partial t} + \frac{\partial(\rho k u_j)}{\partial x_j} = \frac{\partial}{\partial x_j}\left[\left(\mu + \frac{\mu_t}{\sigma_k}\right)\frac{\partial k}{\partial x_j}\right] + G_k - \rho\varepsilon \tag{3}$$

(4) ε equation:

$$\frac{\partial(\rho\varepsilon)}{\partial t} + \frac{\partial(\rho\varepsilon u_j)}{\partial x_j} =$$

$$\frac{\partial}{\partial x_j}\left[\left(\mu + \frac{\mu_t}{\sigma_\varepsilon}\right)\frac{\partial\varepsilon}{\partial x_j}\right] + c_1\rho S\varepsilon - c_2\rho\frac{\varepsilon^2}{k + \sqrt{v\varepsilon}} \tag{4}$$

In the above equations, u_i is the velocity component of i direction, u_j is the velocity component of j direction, ρ is the fluid density, p is the pressure, t is the time, τ_{ij} is the turbulent Reynolds stress, $\tau_{ij} = -\rho\overline{u'_i u'_j}$, k is the turbulent kinetic energy, ε is the turbulent dissipation, μ is the kinetic viscosity of fluid, μ_t is the turbulent viscosity, G_k is the production item of turbulent kinetic energy caused mean velocity gradient, $\sigma_\varepsilon = 1.2$: $\sigma_k = 1.0$: $S = (2S_{i,j}S_{i,j})^{1/2}$, $S_{i,j}$ is the time-average strain rate, $S_{i,j} = \frac{1}{2}(\frac{\partial u_i}{\partial x_j} + \frac{\partial u_j}{\partial x_i})$. The definitions and expressions of other parameters are available in the related literature (Tao 2001, Shih *et al.* 1995).

The hydrodynamic equations were solved using a finite volume method, two-order upwind. A pressure modified method is employed to decouple the velocities and pressure. The fully implicit difference method is employed in the numerical calculation.

2.2 Calculating range and boundary condition

Figure 1 shows the calculating range of the bifurcated pipe. In generating condition (the current flows from main tube to branch tube), the sectional pressure of main tube is specified for inflow boundary; the sectional average velocity of branch tube is specified for outflow boundary. In pumping condition (the current flows from branch tube to main tube), the sectional average velocity of branch tube is specified for inflow boundary; the sectional pressure of main tube is specified for outflow boundary.

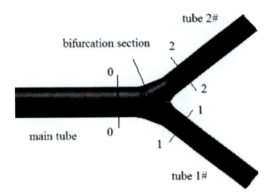

Figure 1.   Calculating range of the bifurcated pipe.

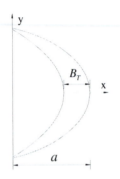

Figure 2.   Relative parameters of crescent rib.

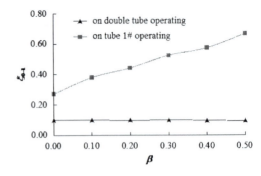

Figure 3.   Variation of head loss coefficient with rib breadth ratio in generating condition.

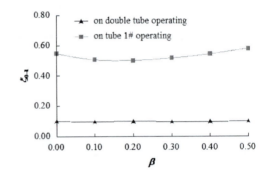

Figure 4.   Variation of head loss coefficient with rib breadth ratio in pumping condition.

3   SIMULATION RESULTS AND ANALYSIS

Hydraulic characteristics of the above bifurcated pipe are discussed by adjusting rib breadth ratio and keeping other body parameters.

Rib breadth ratio is denoted by β, which is equal to B_T divided by a (see Figure 2).

3.1   Head loss

Head loss and head loss coefficient of the bifurcated pipe are calculated using the following relationships (Li & Xu 2000):

$$h_{ij} = E_i - E_j \tag{5}$$

$$\xi_{ij} = h_{ij} / \left(v_0^2 / 2g\right) \tag{6}$$

where, E is total head loss, the subscripts i and j represent the different water-crossing section (see Figure 1). ξ is the water head loss coefficient, $v_0^2/2g$ is the velocity head of the main tube.

Because the body type of a bifurcated pipe is symmetrical, the head loss coefficient of tube 1# is fundamentally the same as tube 2#. Only the variation of the head loss coefficient of tube 1# with rib breadth ratio is described here.

Figure 3 shows that head loss coefficient of tube 1# remains unchanged on double tube operating, and is improved on tube 1# operating, with the increase of rib breadth ratio in generating condition.

Figure 4 shows that head loss coefficient of tube 1# remains unchanged on double tube operating with the increase of costal board width ratio in pumping condition. On tube 1# operating, head loss coefficient decreases first, and then increases with the increase of costal board width ratio; the lowest value is at costal board width ratio of 0.2. However, in general, the change of head loss coefficient is not very evident.

3.2   Flow pattern

In generating condition, flow patterns of the bifurcation section as rib breadth ratio changes are shown in Figure 5 and Figure 6. On double tube operating, the flow pattern is smooth and symmetrically distributed. Because it is located on the symmetrical plane of the water flow and parallel with the flow direction in main tube, the crescent rib does not withhold the flowing water. Consequently, the variation in head loss coefficient by rib breadth ratio is much smaller.

On single tube operating, the flow pattern is complex and asymmetrically distributed. As the rib breadth ratio increases, the water flow is gradually shrunk through the crescent rib. In the meantime, a low speed zone exists behind the crescent rib in the operational pipe and the intensity of vortex in the non-operational pipe is strengthened.

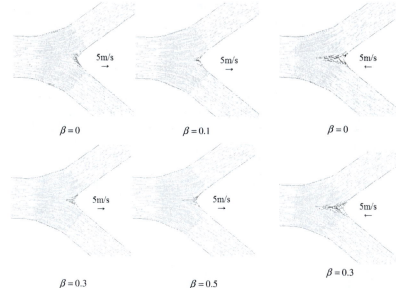

β = 0   β = 0.1

β = 0.3   β = 0.5

Figure 5. Velocity distribution of the bifurcation section on double tube operating in generating condition.

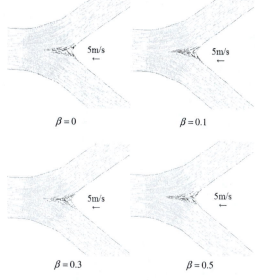

β = 0   β = 0.1

β = 0.3   β = 0.5

Figure 7. Velocity distribution of the bifurcation section on double tube operating in pumping condition.

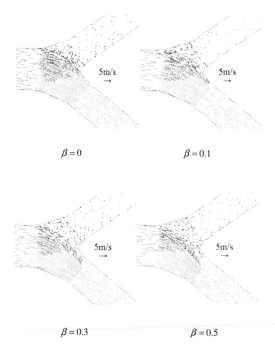

β = 0   β = 0.1

β = 0.3   β = 0.5

Figure 6. Velocity distribution of the bifurcation section on tube 1# operating in generating condition.

In pumping condition, the stream from tube 1# join the other stream from tube 2# to form a jet of water that goes on to the main tube on double tube operating. Although the crescent rib is at the confluence of two streams, it is also on the symmetrical plane of the water flow; therefore, the flow is less affected by the crescent rib (see Figure 7).

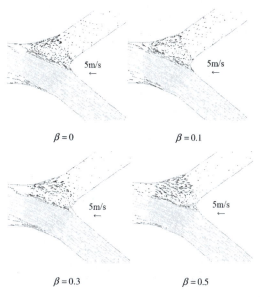

β = 0   β = 0.1

β = 0.3   β = 0.5

Figure 8. Velocity distribution of the bifurcation section on tube 1# operating in pumping condition.

On a single tube operating, the curvature of the streamline is very evident when the stream from the operational pipe flows into the main tube. In the meantime, a vortex is formed in the bifurcation section and consequently, the head loss is high. As the rib breadth ratio increases, the crescent rib plays a guiding role in the flow; the head loss is decreased. With the further increase of rib breadth ratio, flow area is rather changeable, the crescent rib affects the flowing direction, and the head loss increases (see Figure 8).

4 CONCLUSION

In this paper, the hydraulic characteristics of crescent-rib bifurcations are studied, and the primary achievements are as follows:

(1) On double tube operating, the head loss coefficient of the bifurcated pipe remains unchanged with the increase of rib breadth ratio both in generating condition and pumping condition.
(2) On single tube operating, the head loss coefficient is improved in generating condition, which decreases first and then increases with the increase of rib breadth ratio in pumping condition.
(3) On double tube operating, regardless of in generating condition or in pumping condition, the flow pattern is symmetrically distributed. Because the crescent rib is on the symmetrical plane of the water flow, it has low impact on the water flow of the bifurcated pipe.
(4) On single tube operating, regardless of in generating condition or in pumping condition, the flow pattern is asymmetrically distributed. The flow pattern changes complexly with the increase of rib breadth ratio.

REFERENCES

Liang, C.G. & Cheng, Y.X., 2010. Hydraulic Optimization of Pipe Bifurcation of Pumped-storage Power Station by CFD Method. Journal of Hydroelectric Engineering 29 (3), 84–91.
Liu, P.Q., Qu, Q.L. & Wang, Z.G., 2004. Numerical Simulation on Hydrodynamic Characteristics of Bifurcation Pipe with Internal Crescent Rib. Journal of Hydraulic Engineering (3), 42–46.
Tao, W.S., 2001. Numerical Heat Transfer. Xi'an: Xi'an Jiaotong University Press.
Shih, T.H., Liou, W.W. & Shabbir, A. et al., 1995. A new k-ε eddy viscosity model for high Reynolds number turbulent flows. Computers and Fluids 24(3), 227–238.
Li, W. & Xu, X., 2000. Hydraulics. Wu Han: Wuhan University of Hydraulic and Electrical Engineering Press.

Advances in Civil Engineering and Building Materials IV – Chang et al (eds)
© 2015 Taylor & Francis Group, London, ISBN: 978-1-138-00088-9

Analysis of the dam foundation with vacuum preloading

Baotong Shi & Xiangxing Kong
*The First Highway Survey and Design Institute of China Communications Construction Company Ltd.,
Xi'an, China*

ABSTRACT: In view of the traditional vacuum preloading, the vacuum pressure cannot transfer to the deep foundation, so the deep foundation reinforcement effect is poor. This paper presents a case study on the application of vacuum preloading on the deep dam foundation of the Huaihe River. In order to overcome the disadvantage of the traditional building, it uses the pipeline system instead of sand cushion layer for the horizontal drainage channel. Furthermore, the vacuum pressure under membrane, ground settlement and pore water pressure were monitored, and the change of moisture content and void ratio before and after reinforcement were analysed. The test results show that the vacuum preloading is effective in improving the deep foundation.

Keywords: dam foundation, vacuum preloading, pore pressure, settlement ratio.

1 INTRODUCTION

Vacuum preloading is an economic and effective soft ground treatment method, which was first proposed by Professor W.K. Jellman in 1952 (Chai, 2003). Since the vacuum equipment, sealing materials, and drains were restricted, this technique has not been used in large-scale applications. With the solving of the above problems, the vacuum preloading technology was promoted. A large number of engineering practices show that vacuum preloading has the advantage of a disposable loading, both environmental and economic (Liu, 2012). The vacuum pressure cannot transfer to the deep foundation, so the reinforcement of deep foundation is poor. A method using a pipeline system to replace sand cushion was proposed, which can effectively avoid the loss of vacuum pressure (Shang, 1998). This paper presents a case study on the application of vacuum preloading on the deep dam foundation of the Huaihe River, and a brief analysis of the vacuum degree, excess pore pressure, and settlement.

2 SOIL CONDITION

The project was conducted using vacuum preloading, located in the Huaihe River, Jiangsu Province, China. It was covered under deep muddy soil, of large thickness, which has high moisture content, low strength, and poor bearing capacity, causing body subsidence that has been unstable. The soil profile included four layers that required improvement. The first layer was crust soil with located 0–1 m deep and with wide distribution on the site. The second layer is the silt, which

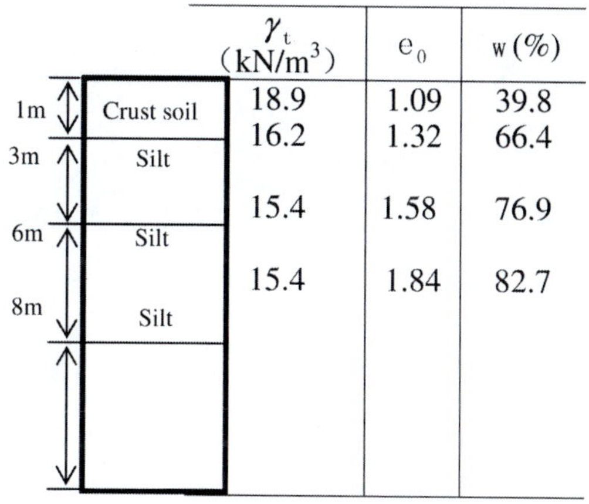

		γ_t (kN/m^3)	e_0	$w\,(\%)$
1m	Crust soil	18.9	1.09	39.8
3m	Silt	16.2	1.32	66.4
6m	Silt	15.4	1.58	76.9
8m	Silt	15.4	1.84	82.7

Figure 1. The soil condition.

is saturated plastic flow state, with high compressibility, and low mechanical strength. The next layer is the silt, which has high humus content, water content, high sensitivity, low strength and large compressibility. The organic matter content is $3.08 \sim 3.88\%$. The last layer is silt, it is saturated plastic flow state, with compressibility and low mechanical strength, as shown in Figure 1.

The project covers an area of 10000 m^2; the site layout is shown in Figure 2. In order to overcome the disadvantage of the traditional building, it uses pipeline system instead of sand cushion layer for horizontal drainage channel, as shown in Figure 3. Drain spacing distance is 1.0 m, the depth is 15 m.

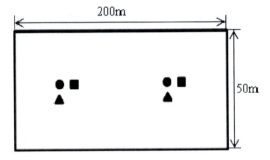

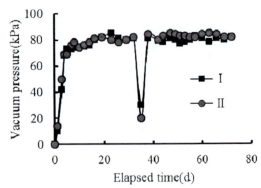

● Vacuum under membrance

■ Pore water pressure transducer

▲ Settlement

Figure 2. The site layout.

Figure 3. The pipeline system.

Figure 4. Measured vacuum pressure under membrane the settlement.

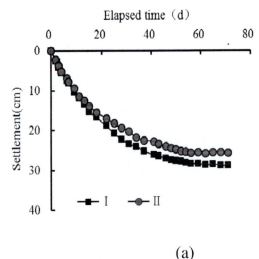

(a)

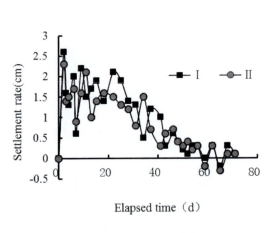

(b)

Figure 5. Settlement time curve of zone (a), and settlement rate with time (b).

3 RESULTS AND ANALYSIS

3.1 *The vacuum pressure under membrane*

As shown in the Figure 4, in the early stage when the vacuum preloading was applied, the vacuum pressure under membrane is quickly increased to 80 kPa, and it is stable. The dropped points resulted from power failure.

As Figure 5 shows, it is settlement rate. When the vacuum is applied, the settlement is produced and the maximum settlement rate reaches 2.5 cm/d. The settlement is stable in the later period and the maximum settlement is 30 cm.

3.2 *The excess pore pressure*

The Figure 6 shows the excess pore pressure versus time curve. After the vacuum is applied, the pore pressure dissipates quickly, and then becomes stable. The maximum excess pore pressure is approximately the vacuum under membrane, and the pore pressure dissipation gradually decreases with depth.

294

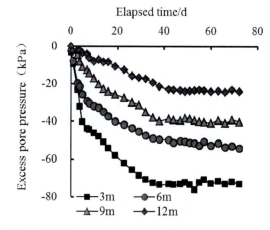

Figure 6. Excess pore pressure at different depths.

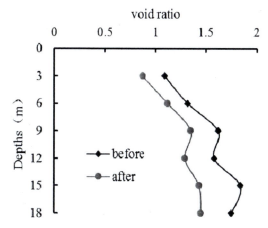

Figure 7. The physical and mechanical properties of the soft soil before and after vacuum preloading.

3.3 Soft soil properties

The variations of soil water content and the void ratio with depth are presented respectively in Figure 7. After treatment, the decrease of soil water content and void ratio were clearly shown.

4 CONCLUSION

In this paper, the vacuum and excess pore pressures were analysed. The following conclusions are drawn from the study:

(1) In the early stage, the pore pressures dissipate rapidly and decrease when the depth increases.
(2) The settlement ratio is fast when the vacuum preloading is applied, and the maximum settlement is 30 cm.
(3) After the treatment, the physical properties of the soil have been changed, and the moisture content and void ratio reduced.

REFERENCES

Chai J, Miura N, Bergado D T. Preloading clayey deposit by vacuum pressure with cap-drain: analyses versus performance [J]. Geotextiles and Geomembranes, 2008, 26(3): 220–230.

Liu Song-yu, Han Wenjun, Zhang Ding-wen, etc. Field pilot tests on combined method of vacuum preloading and pneumatic fracturing for soft ground improvement [J] Journal of Geotechnical Engineering, 2012, 34 (4): 591–599.

Shang J Q, Tang M, Miao Z. Vacuum preloading consolidation of reclaimed land: a case study [J]. Canadian Geotechnical Journal, 1998, 35(5): 740–749.

Others

Advances in Civil Engineering and Building Materials IV – Chang et al (eds)
© 2015 Taylor & Francis Group, London, ISBN: 978-1-138-00088-9

Development and use of a finite element with twenty four degrees of freedom

G.I. Sidorenko
St. Petersburg State Polytechnical University, St. Petersburg, Russia

ABSTRACT: A compound rectangular finite element with 24 degrees of freedom, degenerating into triangular finite element at the border has been developed. A rectangular finite element is created on the basis of four triangular finite elements with 18 degrees of freedom. The interpolation function is taken as the complete fifth-order polynomial in Cartesian coordinates. The normal slope along an edge of the triangle is constrained to vary cubically. The twenty one constants are expressed explicitly in terms of eighteen degrees of freedom for the triangular element. Consider application of the developed compound rectangular finite element to solve the task of bending plates. The exact formulas for the coefficient of stiffness matrix and vectors of the right part for four triangular and compound rectangular finite element with 24 degrees of freedom are deduced. It is shown that the developed finite element gives very high accuracy.

Keywords: finite element method, compound rectangular finite element, Cartesian coordinates, interpolation function, twenty four degrees of freedom, analytical formulas, blending plate, stiffness matrix.

1   INTRODUCTION

The application of Finite Elements (FE) is required to solve certain tasks in computational mechanics. FE node parameters are a value of the target function and its derivatives (Babuska *et al.* 2004, Bathe 2006, Chaskalovic 2008, Cowper *et al.* 1969, Zienkiewicz 1971, Zienkiewicz *et al.* 2005, Gallagher 1973, Strang & Fix 1973, Reddy 2005, Rozin 1971, Sidorenko 2008, Solin *et al.* 2003).

We will consider a regular division of the calculation area into finite elements. In this case, the calculation area, which looks like poligon, would consist only of rectangular and triangular FEs. The triangular FE in each strip could be the only boundary. At the border, a rectangular element degenerates into a triangular FE. In such a FE, the following nodal parameters are accepted:

$$\psi, \quad \frac{\partial \psi}{\partial \xi}, \quad \frac{\partial \psi}{\partial \eta}, \quad \frac{\partial^2 \psi}{\partial \xi^2}, \quad \frac{\partial^2 \psi}{\partial \xi \partial \eta}, \quad \frac{\partial^2 \psi}{\partial \eta^2}. \tag{1}$$

2   FINITE ELEMENT WITH TWENTY FOUR DEGREES OF FREEDOM

2.1   *Compound rectangular finite element*

A rectangular FE with 24 degrees of freedom is created on the basis of four triangular FE. In Figure 1 the circuit of creation of a rectangular FE with 24 degrees of freedom is shown. Each rectangular FE is divided into

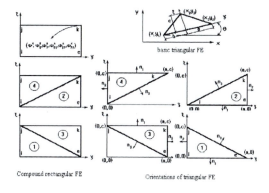

Figure 1.   Creation of rectangular FE with 24 degrees of freedom on basic four triangular FE with 18 degrees of freedom.

two triangular FEs and "stiffness matrix", and components of a vector of the right part of a rectangular FE are created by association of the appropriate matrixes and coefficients of triangular FE vectors. As the diagonal in a rectangular FE can be led in two different ways, for each rectangular element we would receive two components of a vector. Finally, their meanings are accepted as an average arithmetic means for two variants of splitting.

Consider a triangular FE. We enter a local system of coordinates (ξ, η) as shown in Figure 1. The axes of the local system of coordinates are parallel to the axes of the global system of coordinates. We accept a change of function $\psi^e(\xi, \eta)$ within the limits of a triangular FE as the complete polynomial of the fifth degree (Cowper *et al.* 1969, Rozin 1971, Sidorenko 2008):

$$\psi^e(\varsigma,\eta)=\sum_{i=1}^{21}\alpha_i\varsigma^{m_i}\eta^{n_i},\qquad(2)$$

where m_i, n_i = the appropriate degrees. The vector of the nodal parameters for triangular FE e^1 looks like:

$$(q)^1 = \begin{pmatrix}(q)_i^1\\(q)_l^1\\(q)_j^1\end{pmatrix},$$

$$(q)_i^1 = \left(\psi^1\,|\,\psi_\varsigma^1\,|\,\psi_\eta^1\,|\,\psi_{\varsigma\varsigma}^1\,|\,\psi_{\varsigma\eta}^1\,|\,\psi_{\eta\eta}^1\right)^T$$

and contains 18 components.

To find the unknown factors α_i ($i = 1, 2, 3, \ldots, 21$) it is necessary to impose 21 conditions on function $\psi^e(\xi,\eta)$. Equating function ψ in node, and also its first and second derivatives to the appropriate nodal parameters, we shall receive 18 conditions. For the construction of the closed system of the equations in regard to α_i it is necessary to add three conditions. The three additional equations have been obtained based on the conditions that the normal slop $\partial\psi/\partial n$ along each side of triangular finite element had a cubic variation.

We will consider the polynomial of the fifth degree, as polynomials of lower degrees meet the established conditions automatically. Polynomial can be written as:

$$\psi_5^e(\varsigma,\eta)=(P)^T\cdot(\alpha)^e,\qquad(3)$$

where

$$(\overline{\alpha})^e = \left(\alpha_{16}^e\,|\,\alpha_{17}^e\,|\,\alpha_{18}^e\,|\,\alpha_{19}^e\,|\,\alpha_{20}^e\,|\,\alpha_{21}^e\right)^T,$$

$$(\overline{P})^T = \left(\varsigma^5\,|\,\varsigma^4\eta\,|\,\varsigma^3\eta^2\,|\,\varsigma^2\eta^3\,|\,\varsigma\eta^4\,|\,\eta^5\right).$$

We will write down a derivative on perpendicular from $\psi_5^e(\xi,\eta)$ as

$$\partial\psi_5^e/\partial n = ((P)^T_\varsigma\cdot\varsigma_n+(P)^T_\eta\cdot\eta_n)\cdot(\alpha)^e,$$

$$(\overline{P})^T_\varsigma = \left(5\varsigma^4\,|\,4\varsigma^3\eta\,|\,3\varsigma^2\eta^2\,|\,2\varsigma\eta^3\,|\,\eta^4\,|\,0\right),\qquad(4)$$

$$(\overline{P})^T_\eta = \left(0\,|\,\varsigma^4\,|\,2\varsigma^3\eta\,|\,3\varsigma^2\eta^2\,|\,4\varsigma\eta^3\,|\,5\eta^4\right),$$

$$\varsigma_n = \cos(n,\varsigma),\quad\eta_n = \cos(n,\eta).$$

We will consider four orientations of a triangular FE as shown in Figure 1. The condition of derivative change on the perpendicular from function ψ as the polynomial of the third degree on the line 1 and the line 2 for all considered elements is reduced to

equality zero of unknown coefficients α_{17} and α_{20}. The additional condition on line 3 for considered elements looks as follows:

At orientation FE 1,3:

$$5a^3\cdot\alpha_{16}+\left(3ac^2-2a^3\right)\cdot\alpha_{18}+\left(3a^2c-2c^3\right)\cdot\alpha_{19}+5c^3\cdot\alpha_{21}=0.\qquad(5)$$

At orientation FE 2,4:

$$-5a^3\cdot\alpha_{16}-\left(3ac^2-2a^3\right)\cdot\alpha_{18}+\left(3a^2c-2c^3\right)\cdot\alpha_{19}+5c^3\cdot\alpha_{21}=0.\qquad(6)$$

$$[V]^j\cdot(\alpha)^j=(q)^j,\qquad(7)$$

Taking these additional conditions into consideration the system of the permitted equations concerning a vector $(\alpha)^j$ (j = number of triangular FE) will be written as:

$$[V]^j\cdot(\alpha)^j=(q)^j,\qquad(7)$$

where matrixes $[V]^j$ are received by simple substitution of nodes coordinates (for example for FE 1: nodes i, l, j) into polynomial (2); $(\alpha)^j$ = vector of unknown polynomial coefficients (2) of element e^j; $(q)^j$ = vector of nodal parameters of element e^j. Vector of unknown coefficients $(\alpha)^j$ of element e^j is determined unequivocally from the equation (7):

$$(\alpha)^j = \left[[V]^j\right]^{-1}\cdot(q)^j.\qquad(8)$$

The inverse matrixes $[[V]^j]^{-1}$ are received by the author analytically. Function on element e^j is defined as ($j = 1, 2, 3, 4$):

$$\psi^j = \left(1\,|\,\varsigma\,|\,\eta\,|\,\varsigma^2\,|\,\varsigma\eta\,|\,\eta^2\,|\ldots|\,\eta^5\right)\left[[V]^j\right]^{-1}\cdot(q)^j,\qquad(9)$$

where in vector $(1\,|\xi|\eta|\xi^2\,|\xi\eta|\eta^2|\ldots|\eta^5)$ the elements $\xi^4\eta$ and $\xi\eta^4$ are excluded.

The expression (3) is represented through form functions, which are received by the author in an obvious kind for four orientations of triangular FE ($j = 1, 2, 3, 4$).

$$\psi^j(\xi,\eta)=\left(\varphi_{j1}(\xi,\eta)\varphi_{j2}(\xi,\eta)\ldots\varphi_{jk}(\xi,\eta)\ldots\varphi_{j18}(\xi,\eta)\right)\cdot(q)^j.\qquad(10)$$

Thus, function $\psi^e(\xi,\eta)$ within the limits of an FE with 24 degrees of freedom with number e is represented by the formula

$$\psi^e(\xi,\eta)=\Psi_e^T\cdot\Phi^e(\xi,\eta),\qquad(11)$$

where Ψ_e and $\Phi^e(\xi, \eta)$ = matrixes of the size $N_e \times 1$, which elements are nodal parameters and functions $\phi_n(\xi, \eta)$, $n = 1, .., N_e$, FE accordingly, N_e = general number of nodal parameters FE.

2.2 Application

Consider application of the developed FE with 24 degrees of freedom to solve the task of bending plates. In this case, the solution is taken from minimization (12):

$$J[\psi] = \sum_{e=1}^{E} J_e(\psi) = \sum_{e=1}^{E} \iint_{\Omega^e} (\nabla^2 \psi)^2 d\Omega. \quad (12)$$

Taking into account formula (11) we shall receive:

$$\frac{\partial^2 \psi}{\partial \xi^2} = \Psi_e^T \cdot B_\xi^e, \quad \frac{\partial^2 \psi}{\partial \eta^2} = \Psi_e^T \cdot B_\eta^e,$$

$$(\nabla^2 \psi)^2 = \Psi_e^T \cdot B_e \cdot (\Psi_e^T \cdot B_e)^T,$$

where B_ξ^e = matrixes of the size $N_e \times 1$ with elements

$$\frac{\partial^2 \varphi_n(\xi, \eta)}{\partial \xi^2},$$

B_η^e = matrixes of the size $N_e \times 1$ with elements

$$\frac{\partial^2 \varphi_n(\xi, \eta)}{\partial \eta^2},$$

and B_e = block matrix of the size $N_e \times 2$, which blocks are the matrixes B_ξ^e and B_η^e.

In (12) $J_e[\psi]$ designates contribution FE with number $e = 1, E$ and area Ω^e. The contribution of FE with number $e = 1, E$ in $J[\psi]$ is estimated by the formula:

$$J_e[\psi] = \frac{1}{2} \int_{\Omega^e} \Psi_e^T \cdot B_e \cdot (\Psi_e^T \cdot B_e)^T d\Omega - \int_{\Omega^e} f \cdot \Psi_e^T \cdot \Phi_e \cdot d\Omega =$$

$$= \frac{1}{2} \Psi_e^T \cdot K^e \cdot \Psi_e - \Psi_e^T \cdot Q^e, \quad (13)$$

$$J[\psi] \approx \sum_{e=1}^{E} J_e[\psi],$$

where K^e = symmetric stiffness matrix of the size N_e with elements

$$k_{ln}^e = \int_{\Omega^e} \left(\frac{\partial^2 \varphi_l(\xi, \eta)}{\partial \xi^2} \cdot \frac{\partial^2 \varphi_n(\xi, \eta)}{\partial \xi^2} + \frac{\partial^2 \varphi_l(\xi, \eta)}{\partial \eta^2} \cdot \frac{\partial^2 \varphi_n(\xi, \eta)}{\partial \eta^2} \right) d\Omega$$

$$l, n = 1, 2, ..., N_e \quad (14)$$

and Q^e = matrix of the size $N_e \times 1$ with elements

$$Q_l^e = \int_{\Omega_e} f(\xi, \eta) \cdot \varphi_l(\xi, \eta) d\Omega. \quad (15)$$

"Stiffness matrix" and vector of the right part are received by the author in an obvious kind. For associations FE 1,3 and 2,4 the symmetric stiffness matrixes are shown below:

$$\begin{bmatrix} [K_{ii}]^{1,3} & [K_{il}]^{1,3} & [0] & [K_{ij}]^{1,3} \\ & [K_{ll}]^{1,3} & [K_{lk}]^{1,3} & [K_{lj}]^{1,3} \\ & & [K_{kk}]^{1,3} & [K_{kj}]^{1,3} \\ & & & [K_{jj}]^{1,3} \end{bmatrix} \quad (16)$$

$$\begin{bmatrix} [K_{ii}]^{2,4} & [K_{il}]^{2,4} & [K_{ik}]^{2,4} & [K_{ij}]^{2,4} \\ & [K_{ll}]^{2,4} & [K_{lk}]^{2,4} & [0] \\ & & [K_{kk}]^{2,4} & [K_{kj}]^{2,4} \\ & & & [K_{jj}]^{2,4} \end{bmatrix} \quad (17)$$

For example, matrix $[K_{ii}]^{1,3}$ looks like

$$[K_{ii}]^{1,3} = \begin{bmatrix} P_1 & I_9 & I_9 & I_{13} & P_4 & I_{13} \\ & I_1 & P_3 & I_{10} & I_{14} & I_{11} \\ & & I_1 & I_{11} & I_{14} & I_{10} \\ & & & I_2 & I_{12} & P_5 \\ & & & & P_2 & I_{12} \\ & & & & & I_2 \end{bmatrix}. \quad (18)$$

All submatrixes in (16), (17) have dimensions 6×6. The author managed to allocate 11 symmetric ($P_1 \ldots P_{11}$) and 80 asymmetrical ($I_1 \ldots I_{80}$) basic functions, in which all stiffness matrix coefficients are expressed. Below the functions ($P_1 \ldots P_4$) and ($I_1 \ldots I_{10}$) are shown as an example:

$$P_1 = \frac{120a^4 + 130a^2c^2 + 120c^4}{7a^3c^3},$$

$$P_2 = \frac{3a^4 + 2a^2c^2 + 3c^4}{45ac},$$

$$P_3 = \frac{148a^4 + 145a^2c^2 + 148c^4}{140 \cdot a^2c^2},$$

$$P_4 = \frac{18a^4 + 10a^2c^2 + 18c^4}{21 \cdot a^2c^2},$$

$$I_1(x, y) = \frac{77x^4 + 112x^2y^2 + 187y^4}{70xy^3},$$

$$I_2(x,y) = \frac{x \cdot (3x^4 + 4x^2y^2 + 9y^4)}{280y^3},$$

$$I_3(x,y) = \frac{y \cdot (150x^4 + 220x^2y^2 + 120y^4)}{7x^3 \cdot (x^2 + y^2)^2},$$

$$I_4(x,y) = \frac{y \cdot (357x^4 + 624x^2y^2 + 392y^4)}{70x \cdot (x^2 + y^2)^2},$$

$$I_5(x,y) = \frac{27x^8 + 51x^6y^2 + 113x^4y^4 + 41x^2y^6 + 77y^8)}{70x^3y \cdot (x^2 + y^2)^2},$$

$$I_6(x,y) = \frac{xy \cdot (279x^4 + 478x^2y^2 + 324y^4)}{2520 \cdot (x^2 + y^2)^2},$$

$$I_7(x,y) = \frac{27x^8 - 27x^6y^2 + 33x^4y^4 + 85x^2y^6 + 123y^8}{630xy \cdot (x^2 + y^2)^2},$$

$$I_8(x,y) = \frac{y \cdot (111x^8 + 163x^6y^2 + 105x^4y^4 - 45x^2y^6 + 27y^8)}{2520x^3 \cdot (x^2 + y^2)^2},$$

$$I_9(x,y) = \frac{54x^4 + 61x^2y^2 + 80y^4}{14x^2y^3},$$

$$I_{10}(x,y) = \frac{15x^4 + 20x^2y^2 + 19y^4}{140y^3}.$$

where a and c = the sizes of rectangular FE (Figure 1), but x and y = variables, which can accept value a or c. Finally, "stiffness matrix" of rectangular FE with 24 degrees of freedom is defined as:

$$[K] = \frac{1}{2}\left([K]^{1,3} + [K]^{2,4}\right). \tag{19}$$

Vector of the right part for FE with 24 degrees of freedom (Q^e) is defined by vectors components $(Q^e)^{1,3}$ and $(Q^e)^{2,4}$:

$$\left(Q^e\right) = \frac{1}{2}\left(\left(Q^e\right)^{1,3} + \left(Q^e\right)^{2,4}\right). \tag{20}$$

Vectors $(Q^e)^{1,3}$ and $(Q^e)^{2,4}$ are shown below:

$$\left(Q^e\right)^{1,3} = \begin{pmatrix} (Q)_i^{1,3} \\ (Q)_l^{1,3} \\ (Q)_k^{1,3} \\ (Q)_j^{1,3} \end{pmatrix}, \left(Q\right)^{2,4} = \begin{pmatrix} (Q)_i^{2,4} \\ (Q)_l^{2,4} \\ (Q)_k^{2,4} \\ (Q)_j^{2,4} \end{pmatrix}. \tag{21}$$

The expressions for elements of vectors of the right part have been received by the author (Sidorenko

2008). Finally, to find vector nodal parameters Ψ the system of the linear algebraic equations has been received:

$$K \cdot \Psi - Q = 0. \tag{22}$$

3 NUMERICAL COMPUTATIONS

The task about the bending of a thin plate as a square with the side equal to 1 under the concentrated force action at the centre is considered as a test task. Using this test task the comparison of results received using FE with 24 degrees of freedom and other precision FE, ensuring a continuity of the second order has been conducted.

The given task is well known in the theory of a bend of thin plates (Zienkiewicz 1971, Van-Dzi-De 1959, Raschetno-teoreticheskiy 1960). This task has been well investigated and the solution is well known (Van-Dzi-De 1959, Raschetno-teoreticheskiy 1960). If it is assumed that the concentrated force is distributed at the centre of the small radius area, the deflection of a plate is equal to (Van-Dzi-De 1959, Raschetno-teoreticheskiy 1960, Zienkiewicz 1971):

$$\psi = 0{,}0116 \cdot \frac{Pa^2}{D},$$

where D = cylindrical rigidity (Van-Dzi-De 1959).

This task is numerically solved with the use of various FEs (Melosh 1963, Bogner 1965, Deak & Pean 1967, Clough & Felippa 1968, Bell 1969, De Veubeke 1965). The comparison of numerical solutions is given in works (Zienkiewicz 1971, Gallager 1973). The author receives numerical solutions (FE with 24 degrees of freedom and FE with 16 degrees of freedom) with an increase of degrees of freedom.

The comparison of various FEs is conducted with the accuracy of deflection definition at the centre of the plate. The results are shown in a Figure 2. As we see, the FE with 24 degrees of freedom, developed by the author, gives the best results. The fastest way to achieve the exact solution is to increase of number of degrees of freedom.

4 CONCLUSIONS

The compound rectangular finite element with 24 degrees of freedom, degenerating into a triangular finite element at the border has been developed. An interpolation function at each element represents a complete polynomial of the fifth degree.

The obvious analytical expressions for form functions have been received. The exact formulas for coefficients of a stiffness matrix and vectors of the right part for four triangular and compound rectangular FE, with 24 degrees of freedom, have been deduced. Is has been shown that the developed FE gives very high accuracy.

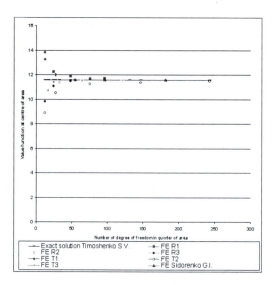

Figure 2. The comparison of numerical results using various FEs. R1 – unmatched rectangular FE with 12 degrees of freedom (Melosh 1963); R2 – rectangular FE with 16 degrees of freedom (Bogner 1965); R3 – rectangular FE with 12 degrees of freedom (Deak & Pean 1967); T1 – triangular FE with 9 degrees of freedom (Clough & Felippa 1968); T2 – triangular FE with 12 degrees of freedom (Bell 1969); T3 – triangular FE with 18 degrees of freedom (De Veubeke 1965).

REFERENCES

Bell K.A. 1969. Refined Triangular Plate Bending Efficient. Int. J. Num. Meth. Eng., 1, №1, pp.101–122.
Bogner F.K., Fox R.L., Schmidt L.A. 1965. The Generation of the Interelement Compatible Stiffness and Mass Matrices by the Use of Interpolation Formulas. Conf. on Matrix Methods in Structural Mechanics. Wright-Patterson Air Force Base, Ohio, October, 1965.
Clough R.W. and Felippa C.A. 1968. A Refined Quadrilateral Element for Analysis of Plate Bending. Proc. 2nd Conf. Matrix Methods in Struct. Mech., Air Force Inst. Of Tech., Wright-Patterson A. F. Base, Ohio, October, 1968.

Cowper G.R., Kosko E., Lindberg G.M., Olson M.D. 1969. Static and dynamic applications of high-precision triangular plate bending element, AIAA J., vol.7, № 10, pp.1957–1965.
Deak A.L. and Pian T.H. 1967. Application of the Smooth Surface Interpolation to the Finite Element Analysis. AIA Aerospace J. 187–189, January, 1967.
De Veubeke B.F. 1965. Bending and Stretching of Plates, Proc. 2nd Conf. Matrix Methods in Struct. Mech., Air Force Inst. Of Techn., Wright-Patterson A. F. Base, Ohio, October, 1965.
Gallagher Richard H. 1973. Finite element analysis. Fundamentals Prentice-Hall, Inc., Englewood Cliffs, New Jersey, 1973.
Melosh R.J. 1963. Basis of Derivation of Matrices for the Direct Stiffness Methods. AIA Aerospace J., 1, pp.1631–1637.
Sidorenko G.I. 2008. Chislennyy analiz uravneniy Navye-Stoksa. Vedomosti SPBGPU, No. 6 (70), s. 57–66. (rus)
Strang Gilbert, Fix George, 1973. An analysis of The Finite Element Method. Prentice Hall, Inc. Englewood Cliffs, N. J.
Raschetno-teoreticheskiy spravochnik proektirovshchika/Pod red. A.A.Umanskogo. 1960. M. – 1029 s. (rus)
Rozin L.A. 1971. Raschet gidrotekhnicheskikh sooruzheniy na ETSVM. L., 212 s. (rus)
Rozin L.A. 1977. Metod konechnykh elementov v primenenii k uprugim sistemam. M.: Stroyizdat, 129 s. (rus)
Van-Dzi-De 1959. Prikladnaya teoriya uprugosti, M: Izd. fiz-mat.lit., 400 s. (rus)
Zienkiewicz O.C. 1971. The finite element method in engineering science. MCGRAW-Hill, London.
Zienkiewicz O.C., Taylor, R.L, Zhu, J.Z. 2005. The Finite Element Method: Its Basis and Fundamentals (Sixth ed.). Butterworth-Heinemann.
Bathe K.J., 2006. Finite Element Procedures. Cambridge, MA: Klaus-Jürgen Bathe.
Reddy, J.N., 2005. An Introduction to the Finite Element Method (Third ed.), McGraw-Hill.
Babuska I., Banerjee, U. Osbom J.E. 2004. "Generalized Finite Element Methods: Main Ideas, Results, and Perspective". International Journal of Computational Methods 1 (1): 67–103.
Solin P., Segeth K., Dolezel I, 2003. Higher-Order Finite Element Methods. Chapman & Hall/CRC Press.
Chaskalovic J, 2008. Finite Elements Methods for Engineering Sciences. Springer Verlag.

Advances in Civil Engineering and Building Materials IV – Chang et al (eds)
© 2015 Taylor & Francis Group, London, ISBN: 978-1-138-00088-9

An energy audit of kindergartens to improve their energy efficiency

Nikolay Vatin, Olga Gamayunova & Darya Nemova
St. Petersburg State Polytechnical University, St. Petersburg, Russia

ABSTRACT: The analysis of results of power inspections and the search for possible solutions to energy saving and increasing the power efficiency of budgetary organizations (kindergartens) taking into account the features of these objects and justification of their economic efficiency, are given in this article. It compares the consumption of electricity, heat, and cold water, it examines the activities of kindergartens with the highest and lowest amount of thermal energy, it describes the basic potential for savings in the consumption of fuel and energy resources.

Keywords: audit, energy efficiency, energy saving, efficiency indices, external enclosure structures, insulation coating, building energy saving, heat conductivity, investment, payback period, kindergarten.

1 INTRODUCTION

Energy audits are relatively new to Russia, but are rather popular areas of activity, which allow identification of energy conservation reserves and ways to improve energy efficiency.

Reliability and value of the results of any audit are directly related to the impartiality of the people or company who carry out an energy audit. Typically, an energy audit by a specialized consulting firm has a sufficient technical base. As an energy audit requires serious instrumentation, these activities themselves are beginning to engage energy suppliers and also manufacturers of modern energy-intensive equipment.

The experience of many Russian companies shows the quality energy audits always lead to a decrease in the share of energy consumption. The costs of the energy audit and subsequent modernization costs are recouped in a very short period.

According to the law, all budgetary institutions (kindergartens (KG), secondary schools and other educational institutions, organizations of health care, culture and the arts, administrative institutions, etc.) should undergo mandatory energy audits. For budgetary institutions the energy audit is quite important because they are quite energy intensive.

As a rule, budgetary institutions consume electricity and heat energy. Sometimes they use different types of fuels (for example, natural gas, wood, coal, etc.). Fuel is typically used in boiler rooms, gas, for example, is used in the kitchens and cafes. Heat energy consumption in Russia is about 360 million Gcal per year: it is around 20% of the total produced in Russia. Electricity consumption is over 100 billion kWh per year. It is about 10% of the produced electric power in our country. In these conditions the costs for utilities budgetary organizations in Russia increases annually by 15–20%. Therefore, undoubtedly reduction of consumption in budgetary institutions will, undoubtedly, direct the released funds to finance more urgent goals and objectives, primarily to finance the goals of these same institutions.

2 ENERGY AUDIT AND ANALYSIS OF THE RESULTS

2.1 Methodology of the audit

It is possible only after an energy audit to determine the real potential for energy savings in each institution. It consists of several steps, including documentary, visual, and instrumental examinations, followed by analysis and preparation of a detailed report and energy performance certificate consumer of energy resources.

At the moment, there is no unified sufficiently substantiated structure to energy audits and their descriptions. Nevertheless, generally this structure can be represented as follows:

STAGE I. Preliminary:

- collecting basic information about the object
- collecting technical information

STAGE II. Instrumental survey of the object:

- processing system of products (raw materials)
- system of heating and hot water supply
- electrical systems
- water and wastewater system
- ventilation and air conditioning systems
- condition of the system of energy accounting, data of metering devices energy and kind of energy resources (error of measurement, verification, and certification)

STAGE III. Analysis of information:

- analysis of the limits of energy saving
- creating an overall fuel and energy balance
- comparing characteristics of the specific consumption with the baseline consumption for each facility and for certain types of energy
- identifying the energy efficiency of products by data of energy performance certificates
- defining inefficient objects

STAGE IV. Development of energy efficiency measures:

- calculating potential annual savings in physical and monetary terms
- defining the equipment needed to implement the recommendations, the calculation of its approximate cost, the cost of delivery, and installation and commissioning
- determining all opportunities to reduce costs
- identifying possible side effects from the implementation of the recommendations that affect real economic efficiency

STAGE V. Monitoring.

After an energy audit, the specialists produce an energy performance certificate. It is a mandatory regulatory document reflecting the balance of fuel and energy resources. It contains performance indicators of their use and an action plan to improve the use of energy resources.

Consider the methodology of an energy audit on the example of several preschools (kindergartens) in Tver.

2.2 Energy audit of public institutions (for example, kindergartens of Tver)

In order to evaluate the effectiveness of the use of energy resources, opportunities to improve this were identified and the cost were determined to implement energy saving measures, conducted through energy audits of several preschools (kindergartens) in Tver.

In all the kindergartens, there is no gas supply system, so the analysis of this system is not subject of this article.

Analysis of the cost of electricity, heat, and cold water in terms of value allowed us to determine the structure of the cost of fuel and energy resources in each of the surveyed objects (Fig. 1).

For a more complete analysis of the annual consumption of fuel and energy resources in the kindergartens of Tver, we define the annual consumption of these resources per 1 m² building area (Table 1).

The specific consumption of fuel and energy resources characterizes the energy efficiency of the building, depending on the type of building, its purpose, and application. As can be seen from Table 1, the most specific annual consumption of electricity and cold water was observed in kindergarten number 115 (63.57 kWh/m² and 7.86 m³/m², respectively). The most specific heat appeared in kindergarten number 93 (0.43 Gcal/m²).

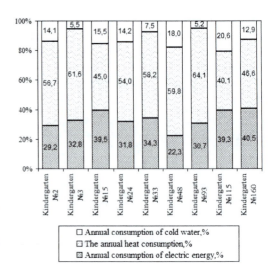

Figure 1. The cost of fuel and energy resources in each of the surveyed objects.

Table 1. The annual consumption of energy resources per 1 m² building area.

Object	The total building area, m²	Annual electricity consumption per 1 m² (kWh/m²)	Annual heating energy consumption per 1 m² (Gcal/m²)	Annual consumption of cold water to 1 m² (m³/m²)
KG № 2	2118.60	32.36	0.25	3.68
KG № 3	1101.60	41.69	0.31	1.66
KG № 15	980.15	61.80	0.28	5.72
KG № 24	1120.80	32.92	0.22	3.47
KG № 33	1117.60	46.80	0.32	2.41
KG № 48	1633.80	29.04	0.31	5.53
KG № 93	1355.75	51.78	0.43	2.08
KG № 115	1020.20	63.57	0.26	7.86
KG № 160	1082.00	47.70	0.22	3.59

Let us compare the activities of kindergartens with the lowest and highest consumption of energy resources in order to find the savings potential of these resources. For example, consider the activities of kindergartens with the highest and lowest consumption of thermal energy (Table 2).

Based on the data presented above, we can conclude that the overrun of thermal energy is mainly related to the type of window and the condition of the facades. Percent of deterioration of buildings affects on the heat energy consumption in a lesser degree.

3 THE MAIN POTENTIAL FOR SAVINGS IN THE CONSUMPTION OF FUEL AND ENERGY RESOURCES

Certainly, potential savings must be sought where the bulk of energy is consumed. Firstly, we are talking

Table 2. The main characteristics of objects with the highest and lowest consumption of heat energy.

Characteristics	KG 93	KG 24	KG 160
Specific consumption of heat energy, Gcal/m^2	0.43	0.22	0.22
The total building area, m^2	1,355.75	1,120.80	1,082.0
Year built	1987	1937	1991
Annual consumption of heat energy, Gcal/year Including:	588.15	251.697	238.808
• Heating and Ventilation (including air heaters)	470.52 (0.347)[1]	201.36 (0.179)	191.05 (0.177)
•Hot water	117.63 (0.087)	50.34 (0.045)	47.76 (0.044)
The type of windows	wood	plastic	plastic, wood
Percentage depreciation of the building, %	38	35	30
Energy efficiency class	E (very low)	D (low)	D (low)
Condition of facades	satisfactory, a lot of cracks	good, in some places there is damage and leaks	good
Building foundation	monolithic, plastered	monolithic, plastered	monolithic, interconnects, joints have been insulated

[1] indicated in brackets the heat consumption per 1 m^2.

about electrical and heat energy. For example, in the cost of fuel and energy consumption in the kindergartens of Tver, the largest consumption is of heat energy, and the smallest consumption is of cold water (Fig.1). However, we must not neglect the potential savings of electrical energy, because in some institutions there is about the same ratio in the consumption of electrical vesus thermal energy.

After analysing the data obtained during the energy audit of nine kindergartens in Tver, we can offer the following standard energy saving measures and energy efficiency:

– Warmer walling (exterior walls). According to statistics, additional insulation walling with effective insulation in the construction of exterior walls, coverings, ceilings, and partitions reduces the cost of heating the building by up to 40–50%. The existing options for building insulation differ based on design solutions and materials used in the construction. For insulation enclosing structures was found to be the most beneficial application, for example, insulating panels of mineral wool, insulating boards made from basalt rock, slabs (blocks) of foamed glass, etc.
– Replacement of windows. Windows are the most vulnerable point in walling, despite continuous improvement. In conventional wooden windows with double glazed walling, through leaks into the room, outside air can replace half of the room air for 1 hour. However, over time various slots may form in these windows, whereby there is excessive infiltration, which leads to an increase in annual losses of heat.
– Installing heat-screens behind of radiators. In the absence of a heat reflective screen possible overruns thermal energy around 5–7% of the heat transfer device. A heat reflective screen for radiator

heating completely isolates the wall from the heat, thereby reducing heat loss. A thermally reflective screen can increase the temperature inside the room by at least at 1–2°C. Energy saving is achieved by reducing of the heat needs for room heating.
– Flushing pipelines and rising heating system. It is necessary to clean the heating system of accumulated impurities that hinder the current source of energy (hot water) in the system. With the accumulation of large blockages in radiators, radiator efficiency can decline by up to twofold, it will not be energy efficient. Therefore, this event involves the rational use of thermal energy in the heating system.
– Installation of control valves with the possibility of pre-setting risers and heating elements, provide the required distribution of carrier flow through the system.
– Replacing bulbs with energy saving ones. The cost of energy-saving lamps far exceeds the cost of ordinary incandescent lamps, but over time it still pays off. The main advantage of energy-saving lamps is their high luminous efficacy, several times greater than that of incandescent lamps. Other obvious advantages of energy-saving lamps are their lifespan, the range of colours available, their light, the slight heat they give, etc.
– Personnel development on energy efficiency. It must be continuous and consist of various forms of vocational education.

4 RESUME

This article describes the main stages of the audit, the analysis of the resulting energy audits of public institutions (for example, the kindergartens of Tver), and suggests possible ways of energy conservation and energy efficiency.

We paid attention to the training and professional development in the field of energy auditing.

Thus, a well-organized audit is the basis of energy conservation and main part of it. We are unable to deal with the losses without having a precise idea of exactly where and to what extent, they occur. That is why energy audits should be done regularly.

REFERENCES

Alihodzic, R., Murgul, V., Vatin, N., Aronova, E., Nikolić, V., Tanić, M. & Stanković, D. 2014. Renewable energy sources used to supply pre-school facilities with energy in different weather conditions. *Applied Mechanics and Materials*. Vol. 624: 604–612.

Aronova, E., Radovic, G., Murgul, V. & Vatin, N. 2014. Solar Power Opportunities in Northern Cities (Case Study of Saint-Petersburg). *Applied Mechanics and Materials*, Vols. 587–589: 348–354.

Bocheninskii,V.P., Vatin, N.I. & Shmarov, V.S. 1986. Rezul'taty issledovaniyaperekhodnykhprotsessov v zhidkometallicheskikhkonturakh s mgd-nasosami. *Trudy LPI* (374): 20–23.

Bolshakov, N.S., Krivoy, S.A. & Rakova, X.M. 2014. The "comfort in all respects" principle implementation by the example of an elementary school. *Advanced Materials Research*. Vols. 941–944: 895–900.

Borodinecs, A., Zemitis, J. & Prozuments, A. 2012. Passive use of solar energy in double skin facades for reduction of cooling loads. *World Renewable Energy Forum, WREF* Including World Renewable Energy Congress XII and Colorado Renewable Energy Society (CRES) Annual Conference6: 4181–4186.

Borodinecs, A., Zemitis, J., Kreslins, A. & Gaujena, B. 2012. Determination of optimal air exchange rate to provide optimal IAQ. *10th International Conference on Healthy Buildings* 2: 1114–1119.

Bukhartsev, V.N. & Petrichenko, M.R. 2012. Problem of filtration in a uniform rectangular soil mass is solved by variational principles. *Power Technology and Engineering*, 46 (3): 185–189.

Fidrikova A.S. & Grishina O.S. & Marichev A.P. & Rakova X.M. 2014. Energy-efficient technologies in the construction of school in hot climates. *Applied Mechanics and Materials*. Vols. 587–589: 287–293.

Gaevskaya Z.A. & Mityagin S.D. 2014. Capital construction and noosphere genesis. *Applied Mechanics and Materials*. T. 587–589: 123–127.

Gaevskaya Z.A. & Rakova X.M. 2014 Modern building materials and the concept of "sustainability project". *Advanced Materials Research*. Vols. 941–944: 825–830.

Kaklauskas, A., Rute, J., Zavadskas, E., Daniunas, A., Pruskus, V., Bivainis, J., Gudauskas, R. & Plakys, V. 2012. Passive House model for quantitative and qualitative analyses and its intelligent system. *Energy and Buildings*, 50: 7–18.

Murgul, V. 2014. Features of energy efficient upgrade of historic buildings (illustrated with the example of Saint-Petersburg). Journal of Applied Engineering Science. 12: 1–10.

Murgul, V. 2014. Solar energy systems in the reconstruction of heritage historical buildings of the northern towns (for example Sankt-Petersburg). Journal of Applied Engineering Science. Vol. 284: 121–128.

Nemova, D., Murgul, V., Pukhkal, V., Golik, A., Chizhov, E. & Vatin, N. 2014. Reconstruction of administrative buildings of the 70's: The possibility of energy modernization. *Journal of applied engineering science*, Vol. 12, No. 1: 37–44.

Nemova, D.V. 2012. Power effective technologies of external envelopes. *Construction of Unique Buildings and Structures*. 3: 78–82. (rus)

Petrichenko, M.R. 2012. Nonsteady filtration in a uniform soil mass. *Power Technology and Engineering*. 46 (3): 198–200.

Petrichenko, M.R. 1991. Convective heat and mass transfer in combustion chambers of piston engines. Basic results, Heat transfer. *Soviet research*. 23 (5): 703–715.

Petrichenko, M.R., Kanishchev, A.B., Zakharov, L.A. & Kandakzhi, B. 1990. Some principles of combustion of homogeneous fuel-air mixtures in the cylinder of an internal combustion engine, *Journal of Engineering Physics*. 59(6): 1539–1544.

Petrichenko, M.R. & Mezheritskii, A.D. 1989. Options in the quasi-stationary method of designing impulse turbines for supercharging internal combustion engine systems, *Soviet energy technology*. 5: 13–17.

Petrichenko, M.R. & Shabanov, A.Yu. 1985. Hydrodynamics of oil film under internal combustion engine piston rings, *Trudy LPI*. 411: 38–42.

Petrosova D.V. & Petrosov D.V. 2014. The energy efficiency of residential buildings with light walling. *Advanced Materials Research* Vols. T. 941–944: 814–820.

Vatin, N. & Gamayunova, O. 2014. Choosing the right type of windows to improve energy efficiency of buildings. *Applied Mechanics and Materials*. Vols. 633–634: 972–976.

Vatin, N. & Gamayunova, O. 2014. Energy saving at home. *Applied Mechanics and Materials*. Vols. 672–674: 550–553.

Vatin, N.I. & Gorshkov, A.S. & Nemova, D.V. & Staritcyna, A.A. & Tarasova, D.S. 2014. The energy-efficient heat insulation thickness for systems of hinged ventilated facades, *Advanced Materials Research*. 941–944: 905–920.

Vatin, N.I. & Mikhailova, T.N. 1986. Computation of cross correlation function of induced potential for developed turbulent flow with axisymmetric mean velocity profile, *Magnetohydrodynamics New York, N.Y.* 22 (4): 385–390.

Vatin, N.I., Nemova, D.V., Kazimirova, A.S. & Gureev, K.N. 2014. Increase of energy efficiency of the building of kindergarten. *Advanced Materials Research*. T. 953–954: 1537–1544.

Vatin, N.I., Nemova, D.V., Murgul, V., Pukhkal, V., Golik, A. & Chizhov, E. 2014. Reconstruction of administrative buildings of the 70's: The possibility of energy modernization. *Journal of Applied Engineering Science / Istrazivanja i Projektovanja za Privredu*. 1: 37–44.

Vatin, N.I., Nemova, D.V., Tarasova, D.S. & Staritcyna A.A. 2014. Increase of energy efficiency for educational institution building. *Advanced Materials Research*. Ò. 953–954: 854–870.

Vostrikova E.V. & Gayevskaya Z.A. 2014. Modernization of residental buildings of the 1960s. *Advanced Materials Research*. T. 941–944: 858–863.

A risk assessment and monitoring system using a total risk index

Jung ki Lee, Kang-Wook Lee, Yeonho Lee & Seung Heon Han
Yonsei University, Seoul, South Korea

ABSTRACT: Due to the high uncertainties incorporated in performing international construction projects, a risk-based intelligent agent system with a risk assessment and monitoring module is developed to support the practitioners in managing the overall construction lifecycle. The assessment of the risks are demonstrated through the Total Risk Index (TRI), which consists of the Analytic Hierarchy Process (AHP) for relative importance within each risk category and risk quantification method, in selecting the level of probability, impact, and the significance of the individual risk factors. Through the derived result of the TRI, the different phases of a construction project can be monitored, as well as deriving the risk factors that are to be managed in priority. Furthermore, in-depth survey databases of previous construction projects are supported to present the management guideline and the hedge strategy trend of each risk. Efficient management and monitoring of risks in the overall lifecycle of a construction project is expected to result in a successful project outcome.

Keywords: international construction, risk assessment, risk monitoring system, AHP, lifecycle of construction project.

1 INTRODUCTION

Risk management in international construction projects has been one of the key activities done by practitioners that lead to a successful project (Deng et al., 2014). A simple checklist was used to check continuously the risks that were exposed in the process of the construction projects. However, the checklist merely assesses the current status of the risk and does not offer a solution to the reduction of risks that have been exposed already. Also, no further application is done through a checklist where the past, current, and the future risk exposure must be managed for a successful project in the international market.

Therefore, this research aims to create an intelligent agent system that consists of risk assessment and monitoring of the whole construction project life cycle. The assessment is approached by academic methodologies such as the analytic hierarchy process (AHP) (Saaty, 1988) and risk assessment in three perspectives; probability, impact, and significance (Han et al., 2008). Furthermore, the assessed risks are presented with a Total Risk Index (TRI) that sums up the quantified risk exposure level to derive the critical risk factors that are to be managed in priority. Among the derived risks, historical records are shown for the users to review the assessment result. Also, the system proposes practical guidelines to each risk factor for reducing the impact of the risk in the process of the construction project.

2 PREVIOUS RESEARCH

Numerous researches into risk assessment have been conducted where models that systematically assess the risks are created. Baccarini and Archer (1999) proposed a methodology in creating a risk ranking of projects that consist of three steps; risk rating, risk management planning, and risk monitoring. In addition, Chapman (2001) focused on risk identification and the risk assessment process in construction design management that improves the project performance using semi-structured interviews, and created a model that links the relationship between each risk.

Different approaches have been employed to create agent-based management systems: Vinit and Viswanadham (2007) used case-based reasons for the experience-intensive problem for risk management, Kwon and Lee (2001) used a multi-agent intelligent system to create an Efficient Enterprise Resource Planning (ERP) system, and Wang (1997) created an intelligent agent-assisted decision support system by utilizing event-driven and task-driven data mining agents. In other research, a web-based system for international project risk management was proposed to support the practitioners in the decision-making process for the advancement of international construction projects (Han et al., 2008). This system included three modules which are the bid-decision model, the profit prediction model, and a risk scenario analysis and contract management guideline.

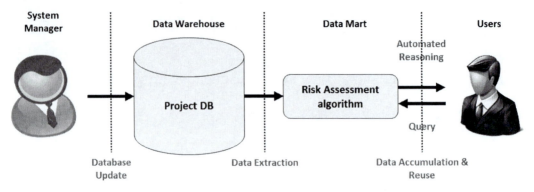

Figure 1.   Framework of the risk-based intelligent agent system.

This research differentiates from the previous studies where the risk assessment is achieved over the whole construction life cycle, as well as proposes a risk monitoring platform for users to manage and review risk response strategy. The development of the system uses a data-driven approach that is accumulated through previous research and surveys from professional engineers, where the database stores information of previous projects that is used to create the core of the system (Figure 1). Furthermore, a feedback system allows the accumulation of risk information and knowledge sharing between project participants. The system manager updates the database through additional data acquisition from the experts. The data mart accumulates new information made by the users and also reuses the data accumulated to be presented for users to use.

3   RISK-BASED INTELLIGENT SYSTEM

In developing the system, risk identification is done by utilizing the previous studies of Jung (2011) that derived the risk breakdown structure which is used in the risk assessment. The system is composed of risk assessment and monitoring procedures. The risk assessment requires the users to select the current stage of an entire project life cycle, assign the relative importance by risk type, and assess each risk for the final outcome that will be used in the risk monitoring procedures. Then the system proposes the risk monitoring platform by deriving the total risk index from the result of the risk assessment and proposes a risk response strategy and practical guidelines for project risk management. A web-based system is developed for users to access easily the system for risk management (where internet access is available) to use for international construction projects.

3.1   Risk hierarchy for assessment

The risk assessment consists of the selection of the project stage, the assignment of relative importance by risk type via the AHP technique, and the risk attribute

assessment in terms of probability of occurrence, impact, and significance of each risk.

At the beginning of the system, the users are to select one of the four construction stages for risk assessment. The four stages defined are the conceptual planning stage, the bidding and contractual stage, the construction stage, and the Operation and Maintenance (O&M) stage. The stages have been divided so that the user can assess each stage of the construction separately and select the risk that best suits that stage of the construction. By separately assessing the risk for each stage, the results can be compared and analysed according to the process of the construction, and the results can be monitored effectively. Also, the dynamic trends in the risk magnitude can be seen explicitly according to the assessment result of each process of the project stages. In addition, the users are to fill in information of the target project that is being assessed so that all projects can be monitored in the future. In the next step, the relative importance by risk type is assigned using the AHP technique (Saaty, 1988). The classifications of the risks are divided into three groups and are sub-categorized into eleven groups which cover the risks that occur throughout the entire life cycle of a construction project. The classification consists of the host country and owner's condition, the business environment of the project condition, and the business capability of the construction firm, where users are allowed to select the risks that are to be assessed and to be managed through the proposed system. The sub-categories consist of eleven detailed types where both the external conditions and internal conditions that affect the project performance are considered.

For the assessment of the relative importance, a five-point scale is used where the importance interval is divided into: 1) equally important, 2) 25% more important, 3) 50% more important, 4) 75% more important, and lastly 5) 100% more important.

A total of 71 risk attributes are embedded in the system. For the assessment of each risk, first the risk type is chosen, and whether or not, the risk affects the cost or the schedule in the whole construction project. Then, according to the type of risk, the selection of the probability of occurrence, the impact, and the significance is assessed. Each assessment is scored out of 5

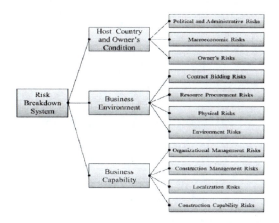

Figure 2.   Risk breakdown structure.

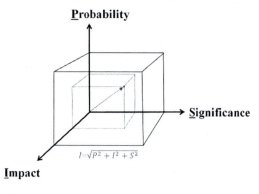

Figure 3.   3D risk quantification.

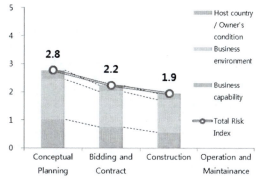

Figure 4.   Risk dynamics of the target project.

points where 1 represents the lowest level of influence and 5 represents the highest level of influence. Han et al. (2008) proposed the method in risk quantification through the Probability, Impact, Significance (PIS) method which is to incorporate the significance level of a risk that has been added from the exiting Probability Impact (PI) method. Due to the risk assessment having three dimensions (probability, impact, and significance), the final quantification of the risk is calculated by the Euclidean distance from the origin of each attribute in three-dimensional coordinates. In addition, the aforementioned relative importance of risk type is multiplied to formulate the final total risk index of the each stage, which will be explained in the following chapter.

3.2   Risk monitoring framework

After the assessment of the individual risks is completed, the risk monitoring procedures present the total risk index results, the appropriate risk response strategies, and practical guidelines for risk management. The following sections explain the monitoring steps of the system.

As described earlier, the summation of the risk scores is presented as a single integrated value which is called herein the "total risk index" (TRI). Han et al. (2008) proposed the risk quantification method for TRI as shown in Figure 3. Since the maximum value of the TRI is 5 points, and as the summation result is close to 0, this means that the current project is risk free. On the other hand, when the TRI value is close to 5, the project is exposed to an extremely high level of risk. The TRI approach allows the examination of individual risk attributes and the level of risk exposure of the sub-category, as well as the overall project's risk level. Figure 4 shows the graphical representation from the system where each bar represents each stage of the construction project. By representing each stage in a bar graph, the users are able to monitor the amount of changes according to the TRI result. A feedback system is employed to allow the users to continuously update and monitor the project

risk where the accumulation of the risk assessments is stored as a database for future references.

3.3   Strategy and guideline for risk management

In the process of deriving the TRI value, each risk attribute is quantified, whereas in this section, the top fifteen highest risk values for the cost or schedule performance are shown separately. Derived risks are the risks that have the highest impact on the overall construction where priority in managing these risks is crucial. By using the PIS method, each risk that has the highest probability, the highest impact, and the highest significance is derived, which numerically shows which risk requires priority treatment in order to reduce the impact.

Individual risks are given a countermeasure strategy to respond to the impact to the project performance. The system proposes seven strategies which are risk responses through a joint venture company (risk transfer or allocation); subcontractor (risk transfer); client (risk transfer through contract); financial hedge (insurance, futures, swap, etc.); additional consideration to contingency; additional consideration to the Bill Of Quantity (BOQ), and other methods. According to the risk characteristics, one of the seven strategies could be employed to respond to the risk in the future.

Being an intelligent agent system, the system composes an in-depth survey database of the previous

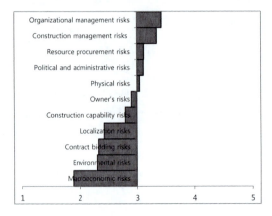

Figure 5. Risk priority assessment in conceptual planning stage.

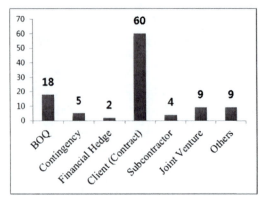

Figure 6. Risk priority assessment in conceptual planning stage.

140 projects' information. Database information is extracted, together with the current assessment results, which is shown to users to support their decision-making process. The frequency of the risk strategy chosen for the corresponding risk is presented in Figure 6, as well as a literal description of the risk management guideline chosen for the risk. The guideline is expected to support the practitioners in understanding the risk strategy the user chose and to apply the strategy the current project.

4 ILLUSTRATIVE CASE STUDY

This research presents an illustrative example of the system usage through assessing a plant project in Saudi Arabia. The target project is a petroleum plant project located in Jubail, Saudi Arabia, that took twenty-eight months to construct from 2006 to 2008. The owner of the project was a private company from Saudi Arabia and the main contractor of the project was a Korean EPC firm. The assessment is done by an expert who has worked in the field of engineering for over ten years. The user assessed the three stages of construction, which are the conceptual planning stage, the bidding stage, and the construction stage. Due to the fact that the project scope did not cover the operation and management stage, the assessment was carried out only to the construction stage.

The total risk index of the three stages is shown in Figure 4 where in the first stage, the conceptual planning stage had a result of 2.77 points out of 5 points. In the second stage, the bidding stage, the result was 2.24 points out of 5 points, and finally in the third stage, the construction stage, the result was 1.95 points. In all three stages, the highest risk was derived in the business environment risk, followed by the business capability, and lastly, the host country and owner's condition. However, as the project proceeded, the system shows that the reduction in the business capability risk reduced the most. The expert mentioned that as the project proceeded, more information on the project

was accumulated in the construction stage compared to the conceptual planning stage, which resulted in the reduction of uncertainties. The practitioners were able to hedge risks through the information which, in the end, helped the project to become a success. Among the sub-categories in the highest risk stage, the conceptual planning stage, organizational management risk was derived to have the highest risk, followed by construction management risk and resource procurement risk (Figure 5). The ranking of the sub-categories alter as the project proceeds, where in the bidding stage, the highest sub-category risk is derived as the localization risk, the contract bidding risk, macroeconomic risks, etc. In the construction stage, the highest sub-category risk was the localization risk, the environmental risk, the physical risk, etc.

Due to the space constraint, this paper exhibits only one of the top priority risks for the cost performance in the conceptual design stage where the historical data from the database is presented for risk response strategies. From a cost performance perspective, the highest risk value derived is the adequacy of claim and arbitration provision. The application to the contract to transfer the risk to the client has been the strategy used most in the past to respond to the risk of duration for bid preparation. Figure 6 is represented within the system where the user could use this result as a guideline for the current project that is being managed by the system.

As a result, the users were able to quantify the amount of risk occurring in the different stages of a construction project, and also derive individual risks that are highly critical which could be chosen as a priority in managing the risk. The guideline presented in Figure 7 would guide the practitioners to control the risk in the future.

5 CONCLUSION AND LIMITATIONS

Through the development of the risk assessment and monitoring system, practitioners are able to

▣ Adequacy of claim and arbitration provision

Possible damage from arbitrary interpretation of terminology	In the contracting document, the terms "sufficient" and "reasonable" must be excluded due to arbitrary interpretation of the owner and may not be clarified in advance. The spare parts list must be attached with the contracting document which not only include the quantity but also include the price of the supply. Additionally, a prior agreement between the parties must be met for the usage of the various experiments and testing materials in the commissioning stage. For the reference, the World Bank will include the following condition in the contract for the future project: World Bank Condition 25.1.2 "The Employer shall supply the operating and maintenance personnel and all raw materials, utilities, lubricants, chemicals, catalysts, facilities, services and other matters required for Commissioning."
Construction duration extension denied due to lack of awareness of claim notification date	According to the FIDIC standard contract form, the occurrence or recognition of a claim issue must be notified within 28 days due to the FIDIC Conditions of Contract for EPC Turnkey Projects 20.1[Contractor's Claim]. If the notification is not given within the deadline, the right to claim for construction duration extension is deprived and the owner is not responsible for the claim issue. If the contractor does not submit a claim document from the engineer's opinion, the claim can be rejected by the owner. It should be noted that oral agreement between parties during a process meeting may cause controversy. Therefore a writing claim document must be made according to the contract condition.
Unrealistic construction duration setting during bidding stage	During the bidding and contracting stage, by ignoring the massive risks of the constructor and not changing the contracting conditions, the constructor must be responsible for their own decision. The conditions that influences the profitability of the project must be taken in action for modification, where giving up the project due to high risk must be considered as well.

Figure 7. Guideline according to the risk.

quantify risk through the assessment from the proposed approach, and be able to manage risk in a systematic manner. Furthermore, the risk monitoring platform supports the users to review potential strategies for reducing the level of critical risk factors by providing the practical guidelines for risk countermeasures. The system not only obtains the input values, but also presents the output values through the algorithm of AHP and TRI. The accumulation of the data is stored within the system which allows for continuous monitoring of the target project, a feedback system through continuous assessment of individual risks as the project progresses, and finally earns lesson-learned information from the project assessment.

ACKNOWLEDGEMENT

This research was supported by a grant (14SCIP-C079124-01) from the Smart Civil Infrastructure Research Program funded by the Ministry of Land, Infrastructure, and Transport of the Korean government.

REFERENCES

Baccarini, D., & Archer, R. (2001). The risk ranking of projects: a methodology. International Journal of Project Management, 19(3), 139–145.

Chapman, R.J. (2001). The controlling influences on effective risk identification and assessment for construction design management. International Journal of Project Management, 19(3), 147–160.

Deng, X., Low, S.P., Li, Q., & Zhao, X. (2014). Developing Competitive Advantages in Political Risk Management for International Construction Enterprises. Journal of Construction Engineering and Management.

Han, S.H., Kim, D.Y., Kim, H., & Jang, W.S., (2008). A web-based integrated system for international project risk management. Automation in construction, Vol. 17, Issue 3., pp. 342–356.

Jung, W. (2011). Three-Staged Risk Evaluation Model for Bidding on International Construction Projects. Graduate School, Yonsei University: school of Civil and Environmental Engineering

KarimiAzari, A., Mousavi, N., Mousavi, S. F., & Hosseini, S. (2011). Risk assessment model selection in construction industry. Expert Systems with Applications, 38(8), 9105–9111.

Kwon, O.B., & Lee, J.J. (2001). A multi-agent intelligent system for efficient ERP maintenance. Expert Systems with Applications, 21(4), 191–202.

Mohamed, A. & Aminah, R.F., 2010. Risk Management in the Construction Industry Using Combined Fuzzy FMEA and Fuzzy AHP. Journal of Construction Engineering and Management, Vol. 136, Issue 9., pp. 1028–1036.

Navon, R., & Kolton, O. (2006). Model for automated monitoring of fall hazards in building construction. Journal of Construction Engineering and Management, 132(7), 733–740.

Pedro, M.S., (2005). Neural-Risk Assessment System for construction projects. Construction Research Congress 2005., pp. 1–11.

Saaty, T.L. (1988). What is the analytic hierarchy process? (pp. 109–121). Springer Berlin Heidelberg.

Vinit, K. & Viswanadham, N., 2007. A CBR-based Decision Support System Framework for Construction Supply Chain Risk Management. 3rd Annual IEEE conference on automation science and engineering Scottsdale., pp. 980–985.

Wang, H. (1997). Intelligent agent-assisted decision support systems: integration of knowledge discovery, knowledge analysis, and group decision support. Expert Systems with Applications, 12(3), 323–335.

Zavadskas, E.K., Turskis, Z., & Tamošaitiene, J. (2010). Risk assessment of construction projects. Journal of civil engineering and management, 16(1), 33–46.

Zhang, X. (2011). Web-based concession period analysis system. Expert Systems with Applications, 38(11), 13532–13542.

313

Advances in Civil Engineering and Building Materials IV – Chang et al (eds)
© 2015 Taylor & Francis Group, London, ISBN: 978-1-138-00088-9

The time-fractional diffusion equation for woven composite

Hang Yu & Chuwei Zhou
State Key Laboratory of Mechanics and Control of Mechanical Structures, Nanjing University of Aeronautics and Astronautics, Nanjing, China

ABSTRACT: The time-fractional diffusion equation for moisture uptake of woven composite exposed to hygrothermal environment was investigated in this research. There are two non-Fickian behaviors in woven composite based on moisture diffusion mechanics, one is caused by initial voids or cracks in fast diffusion stage. The other is induced by resin phase in slow diffusion stage. Therefore, in order to describe non-Fickian behaviors in fast diffusion, the percent moisture content of a homogeneous material under a moist environment was deduced by time-fractional diffusion equation. Then, a two-stage fractional diffusion model was developed to describe long-term slow diffusion. In this model, fractional order parameters were used to explain the nonlinear property of moisture diffusion in composite. The predicted moisture uptake were all in good agreement with results in references.

Keywords: fractional diffusion equation, woven composite, hydrothermal aging, non-Fickian behavior.

1 INTRODUCTION

Woven composites are widely used in aerospace, medical, chemical and civil industries due to their high strength-to-weight ratio. Despite the desirable properties of polymer matrix composites, a major drawback of them is their tendency to absorb water in hygrothermal environment [1–3]. Moisture absorption significantly degrades the composites performance. Since fiber absorbs a negligible amount of water, therefore, understanding the moisture diffusion mechanics of neat matrix is necessary to improve the long-term performance of advanced composites [4, 5].

For polymer matrix, both Fickian [6] and non-Fickian behaviors [7] caused by hygrothermal aging have been observed. It is widely accepted that transport in polymer involves two-stage diffusion, initial fast diffusion stage and long-term slow diffusion stage [8, 9]. The initial fast diffusion step is controlled by the concentration gradient and is almost Fickian diffusion. However, the long-term slow diffusion stage associated with the slow relaxation process is non-Fickian diffusion [10, 11].

However, the moisture uptake in fast diffusion of composite is not always linear while the weight gain of Fickian diffusion increases as a linear function of $t^{1/2}$. For example, the initial fast moisture uptake of unidirectional composite [13] shows Fickian behavior, but the fast diffusion moisture uptake of woven composite [12] initially increases rather rapidly then quickly slows down. The reason of this phenomenon may be associated with diffusion in cure-induced voids

and cracks in woven composite. Compared with short-term behavior, the weight gain of woven composite in long-term diffusion exhibits a linear function. The absorption behavior indicates that the structural relaxation of the matrix dominates the slow diffusion. Furthermore, after more than one year higher weight gain caused by cracks is observed due to hygrothermal aging environment.

Non-Fickian diffusion is actually an anomalous diffusion, and the fractional diffusion equations have been an useful tool to describe the anomalous diffusion phenomenon in recent years [14–18]. For example, chemistry physics [14], fluid flow in porous media [17] and fractional Brownian motions [18]. The normal diffusion is Markov process, and the current is history and space independent. However, the essential of anomalous diffusion is non-Markov process, and the current is related to the concentrations all over the spaces, also depends on the previous history and even the initial conditions [19]. The moisture absorption associated with concentration in polymer is time dependent, therefore, the memory or history effect of concentration should be taken into consideration. In contrasted to other model, the level of nonlinear property can be represented by the fractional order parameters in fractional model, and the memory or history effect can be described during different physical process [20–23].

In this paper, theoretical derivation of time-fractional diffusion equation was developed to describe moisture uptake of woven composite during hygrothermal aging. The total weight gain of woven composite was obtained by fast diffusion stage and slow diffusion stage. Different moisture behaviors of woven composite were explained by fractional diffusion model.

2 THEORY

2.1 Fractional diffusion model

The following problem was investigated. Since the moisture uptake exhibits nonlinear property as a function of $t^{1/2}$, a homogeneous material exposed to a moist environment was considered as shown in Fig. 1.

The objective is to determine the percent moisture content $M(t)$ (percent weight gain) of the material as a function of time. With the temperature and the diffusivity being constant inside the material, the problem is described by the time-fractional diffusion equation, which can describe the history effect and nonlinear property of moisture diffusion:

$$\frac{\partial^\alpha C(x,t)}{\partial t^\alpha} = D \frac{\partial^2 C(x,t)}{\partial x^2} \quad (0 \leq x \leq h, 0 < \alpha \leq 1) \quad (1)$$

where $C(x,t)$ is the moisture concentration of the material. D is the diffusivity of the material in the direction normal to the surface. α is fractional order parameter. The time-fractional derivative is intended in the Caputo sense:

$$\frac{\partial^\alpha C(x,t)}{\partial t^\alpha} = \frac{1}{\Gamma(1-\alpha)} \int_0^t \left[\frac{\partial}{\partial \tau} C(x,\tau) \right] \frac{d\tau}{(t-\tau)^\alpha} \quad (2)$$

For $\alpha = 1$, the diffusion equation degenerates to integral order diffusion equation:

$$\frac{\partial C(x,t)}{\partial t} = D \frac{\partial^2 C(x,t)}{\partial x^2} \quad (3)$$

The initial and boundary conditions [1] are:

$$C(x,0) = 0, \ 0 < x < h$$
$$C(0,t) = C_\infty, \ t > 0 \quad (4)$$
$$C(h,t) = C_\infty, \ t > 0$$

The solution of Eq.(1) is:

$$C(x,t) = C_\infty \left\{ 1 - \sum_{n=0}^\infty \frac{4}{(2n+1)\pi} * \right.$$
$$\left. E_{\alpha,1} \left[-D \left(\frac{2n+1}{h} \pi \right)^2 t^\alpha \right] \sin \left(\frac{2n+1}{h} \pi x \right) \right\} \quad (5)$$

where $n = 0, 1, 2 \ldots$, $E_{a,b}$ is generalized Mittag-Leffler function,

$$E_{a,b}(z) = \sum_{m=0}^\infty \frac{z^m}{\Gamma(am+b)} \quad (a > 0, b > 0) \quad (6)$$

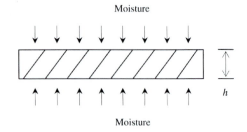

Figure 1. Fractional diffusion model for moisture uptake.

The total weight gain of the moisture in the material is obtained by integrating equation over the plate thickness:

$$\frac{M(t)}{M_\infty} = \left\{ 1 - \sum_{n=0}^\infty \frac{8}{\pi^2 (2n+1)^2} E_{\alpha,1} \left[-D \left(\frac{2n+1}{h} \pi \right)^2 t^\alpha \right] \right\} \quad (7)$$

M_∞ is maximum moisture content.

2.2 Application in hygrothermal aging

During hygrothermal aging process, some phenomena of neat resin and composite can be observed in initial fast diffusion and long-term slow diffusion [10–13]. The initial fast diffusion of resin is Fickian diffusion, but the fast diffusion of woven composite shows non-Fickian behavior. The long-term slow diffusion of neat resin and woven composite is non-Fickian diffusion. Since the initial fast diffusion is controlled by concentration gradient and long-term diffusion is related to slow relaxation process, thus a dual fractional model is developed to describe the weight gain during hygrothermal aging:

$$M(t) = M_f\left(\alpha, D_f, M_f, t\right) + M_s\left(\gamma, D_s, M_s, t\right) \quad (8)$$

where $M(t)$ represents the total weight gain in hygrothermal aging, M_f is the initial fast moisture uptake and D_f is the fast diffusivity. M_s is the long-term slow weight gain and D_s is the slow diffusivity. α and γ are fractional order parameters. M_f and M_s represent equilibrium moisture uptake of fast and slow diffusion, respectively.

Since the long-term diffusion in polymer is much slower than initial fast diffusion, D_f is much larger than D_s. If $\alpha = 1$, the diffusion in initial fast stage is Fickian diffusion, or the diffusion is anomalous which cannot be explained by Fick's law. The same situation is controlled by γ in long-term diffusion.

3 RESULT AND DISCUSSION

Figure 2 is the schematic diagram of woven composite and Fig. 3 shows the short-term diffusion behavior

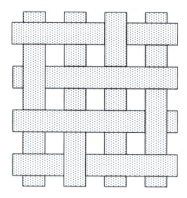

Figure 2.   Schematic diagram of the woven composite.

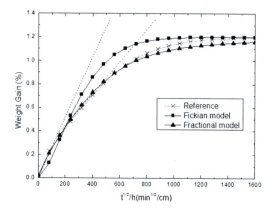

Figure 3.   Short-term diffusion behavior of woven composite at 70°C.

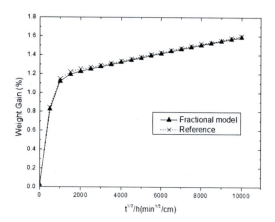

Figure 4.   Long-term diffusion behavior of woven composite at 70°C.

Table 1.   Parameters in time-fractional diffusion equation for woven composite.

| Material | Diffusivity | | M(%) | |
	D_f	D_s	M_f	M_s
Composite	2	0.0009	1.2	0.7

of woven composite [12] during hygrothermal aging at 70°C. Different from BMI resin, the fast diffusion of the woven composite cannot be explained by Fickian diffusion. Obviously, the initial moisture uptake is nonlinear while the weight gain of Fickian diffusion increases as a linear function of $t^{1/2}$. The moisture uptake increases rather rapidly in the early period but quickly slow down. Two dotted lines are depicted to show that the change of the slope in weight gain curve is nonlinear process.

The time-fractional diffusion equation is developed to describe the nonlinear behavior of the woven composite during the fast diffusion stage. $\alpha = 0.83$ is fitted to represents the nonlinear property during non-Fickian diffusion. Obviously, α is related to voids and cracks. As the decrease of α, the nonlinear level of the non-Fickian diffusion becomes larger and initial uptake becomes quicker.

Figure 4 shows the long-term diffusion behavior of woven composite during hygrothermal aging at 70°C. The outcome of fractional diffusion model is identical with the result in reference. At this point, fractional diffusion model is suitable for both neat resin and woven composite. However, after more than one year, higher weight gain of woven composite is observed at 70°C [12], the fractional diffusion model

cannot describe this phenomenon caused by cracks, because the weight gain becomes rather irregular. More complicated model is not developed to explain this phenomenon in this paper.

The parameters used in Eq. (8) to describe initial fast and long-term slow diffusion behavior of woven composite are listed in Table 1. Compared with neat resin, the average fast diffusivity D_f of woven composite is slower. However, the slow diffusivity is much slower than that in resin, because the micro reinforcement in woven composites may restrains the structural relaxation during hygrothermal aging. Obviously, since fiber cannot absorb moisture, the equilibrium moisture uptake in resin is much larger than that in woven composite. $\gamma = 1$ represents that the slow diffusion shows linear property controlled by resin phase

4   CONCLUSION

The time-fractional diffusion equation for moisture uptake of neat resin and woven composite during hygrothermal aging was studied in this paper. It is shown that there are three kinds of non-Fickian behaviors: long-term diffusion of neat resin, initial fast and long-term slow diffusion of woven composite. In order to explain non-Fickian behavior, the percent moisture content of a homogeneous material exposed to a moist environment was derived from time-fractional diffusion equation. Then, a two-stage fractional diffusion model was developed to describe total moisture diffusion of resin and woven composite, and the predicted weight gain curves were proved to match the results in references,. In this model, the fractional order parameter represents the nonlinear level of moisture diffusion. If $\alpha = 1$, the moisture uptake obeys Fick's law. If $0 < \alpha < 1$, the diffusion shows nonlinear property, and as the decrease of α, the nonlinear level of the non-Fickian diffusion becomes larger. The success of fractional diffusion model for non-Fickian behaviors of woven composite demonstrates that the time fractional diffusion equation can be used to explain some anomalous diffusion, which should consider the history effect.

ACKNOWLEGEMENT

The corresponding author of this paper is Chuwei Zhou. This paper is supported by Fund of Jiangsu Innovation Program for Graduate Education (CXZZ13_0149) and the Fundamental Research Funds for the Central Universities, National Natural Science Foundation of China (11272147,10772078), Aviation Science Foundation (2013ZF52074), Fund of State Key Laboratory of Mechanical Structural Mechanics and Control (0214G02), Project Funded by the Priority Academic Program Development of Jiangsu Higher Education Institutions (PAPD).

REFERENCES

[1] Shen. C.H & Springer. G.S. 1976. Moisture absorption and desorption of composite materials, Journal of Composite Materials, 10, 2–20.

[2] Cervenka. A. & Bannister. D.R. 1998. Moisture absorption and interfacial failure in aramid/epoxy composites, Composites Part A: Applied Science and Manufacturing, 29, 1137–1144.

[3] Patel. S. & Case. S. 2002. Durability of hygrothermally aged graphite/epoxy woven composite under combined hygrothermal conditions, International journal of fatigue, 24, 1295–1301.

[4] Zhou. J, & Lucas. J.P. 1999. Hygrothermal effects of epoxy resin. Part I: the nature of water in epoxy, Polymer, 40, 5505–5512.

[5] Apicella. A & Nicolais. L. 1979. Effect of thermal history on water sorption, elastic properties and the glass transition of epoxy resins, Polymer, 20 (1979) 1143–1148.

[6] Mijović. J. & Lin. K.F. 1985. The effect of hygrothermal fatigue on physical/mechanical properties and morphology of neat epoxy resin and graphite/epoxy composite, Journal of applied polymer science, 30, 2527–2549.

[7] Wong. T. & Broutman. L. 1985. Moisture diffusion in epoxy resins Part I. Non – Fickian sorption processes, Polymer Engineering & Science, 25, 521–528.

[8] Berens. A. & Hopfenberg. H. 1978. Diffusion and relaxation in glassy polymer powders: 2. Separation of diffusion and relaxation parameters, Polymer, 19, 489–496.

[9] Hodge. I.M. & Beren. A.R. 1982. Effects of annealing and prior history on enthalpy relaxation in glassy polymers. 2. Mathematical modeling, Macromolecules, 15, 762–770.

[10] Li. Y. & Miranda. J. 2001. Hygrothermal diffusion behavior in bismaleimide resin, Polymer, 42, 7791–7799.

[11] Bao. L.R. & Yee. A.F. 2001. Moisture absorption and hygrothermal aging in a bismaleimide resin, Polymer, 42, 7327–7333.

[12] Bao. L.R. & Yee. A.F. 2002. Moisture diffusion and hygrothermal aging in bismaleimide matrix carbon fiber composites: part II–woven and hybrid composites, Composites science and technology, 62, 2111–2119.

[13] Bao. L.R. & Yee. A.F. 2002. Moisture diffusion and hygrothermal aging in bismaleimide matrix carbon fiber composites—part I: uni-weave composites, Composites science and technology, 62, 2099–2110.

[14] Schlichter. J & Friedrich. J. 2000. Protein dynamics at low temperatures. The Journal of Chemical Physics, 112, 3045.

[15] Hilfer. R. 2000. Fractional time evolution, Applications of fractional calculus in physics, 87–130.

[16] Tadjeran. C. & Meerschaert. M.M. 2006. A second-order accurate numerical approximation for the fractional diffusion equation, Journal of Computational Physics, 213, 205–213.

[17] Meerschaert. M.M. 2004. Finite difference approximations for fractional advection–dispersion flow equations, Journal of Computational and Applied Mathematics, 172, 65–77.

[18] Jumarie. G. 2001. New results on Fokker–Planck equations of fractional order, Chaos, Solitons & Fractals, 12, 1873–1886.

[19] Chang. X.F. 2005. Anomalous diffusion and fractional advection-diffusion equation, 54, 1113–1117.

[20] Chen. W. 2006. Time–space fabric underlying anomalous diffusion, Chaos, Solitons & Fractals, 28, 923–929.

[21] Gorenflo. R. & Mainardi. F. 2007. Vivoli, Continuous-time random walk and parametric subordination in fractional diffusion, Chaos, Solitons & Fractals, 34, 87–103.

[22] Jumarie. G. 2004. Fractional Brownian motions via random walk in the complex plane and via fractional derivative. Comparison and further results on their Fokker–Planck equations, Chaos, Solitons & Fractals, 22, 907–925.

[23] Xu. Z. 2013. A fractional-order model on new experiments of linear viscoelastic creep of Hami Melon, Computers & Mathematics with Applications.

[24] Mainardi. F. 2003. The Wright functions as solutions of the time-fractional diffusion equation, Applied Mathematics and Computation, 141, 51–62.

Advances in Civil Engineering and Building Materials IV – Chang et al (eds)
© *2015 Taylor & Francis Group, London, ISBN: 978-1-138-00088-9*

Effect of external shading system based on coefficient outside residence of Hangzhou

Guanzhou Ji & Wei Zhu
Department of Architecture, Zhejiang University City College, Hangzhou, Zhejiang, China

ABSTRACT: In hot-summer and cold-winter regions, external shading systems can significantly improve indoor comfort. In this study, by analysing meteorological data, the time when an external shading system is needed will be determined. With a calculation about geometric principles, the size and shape of external shading systems are going to be discussed. It will provide a theoretical basis for design, and renovation of residential external shading systems in the future, and in what ways energy reduction can be achieved.

Keywords: shading coefficient, external shading system, effect of shading, residence.

1 INTRODUCTION

With the development of society, people start to pay much more attention to the energy consumption of buildings. In summer, people use mechanical refrigeration in order to cool down the indoor temperature. As an important carrier in the external structure of buildings, a window is the main cooling load factor of air conditions. In hot summer and cold winter zone, external shading system not only help residential users to block solar radiation during the summer, but also improve indoor thermal environment in winter effectively. A rational use of external shading system will improve thermal comfort the living environment greatly. Nonetheless, residential users currently fall into a misunderstanding of external shading system. Because of different regions and directions, the solar elevation angle and azimuth are different, it is significantly necessary to analyse the coefficient of the external shading system and consider effects of shading in depth.

In this study, the result can provide a theoretical basis for design and renovation of residential external shading systems in the future, which will bring a better implementation of passive energy-saving concepts to buildings.

2 THEORETICAL STUDY

During summer, the indoor thermal environment of the design specifications for indoor air temperature is 26.0 to 28.0°C. In winter, the indoor temperature is 18.0°C, and it is 26.0°C during summer. Therefore, in this study, 26.0°C is selected as standard of design and calculation. Referring to Typical Meteorological Database Handbook for Buildings, annual

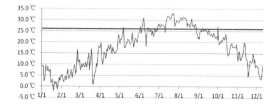

Figure 1. Annual average daily temperature data of Hangzhou.

average daily temperature data of Hangzhou are chosen in the standard meteorological year, as the line chart shows.

Hangzhou is a typical hot-summer and cold-winter region. With frequent temperature changes, the average diurnal temperature of Hangzhou exceeds 10.0°C. Thus, 26.0°C is determined as the standard for analyzing. The time when external shading system is need is from May to September approximately.

Select the data of five consecutive days whose average temperatures all exceed more than the 26.0°C steady as the critical value for calculation. Meanwhile select another data of five consecutive days when the temperatures are below 26.0°C continuously as another critical value.

As to the screening results, since June 28 the daily average temperature exceeds 26.0°C steady, while after 27 August, the daily average temperature is below 26.0°C continuously. In addition, located in the north of the Tropic of Cancer 23.5 N, Hangzhou is in the latitude of 30.3 N, the solar elevation angle on 27 August is smaller than on 28 June. Therefore, selecting the data on 27 August as the critical day is more applicable for calculations.

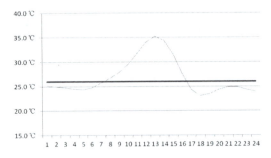

Figure 2. Daily average hourly temperature data in August.

Table 1. Daily solar elevation angle and azimuth data on 27 August.

Time	Elevation Angle	Azimuth
1	999	0
2	999	0
3	999	0
4	999	0
5	999	0
6	−98	4
7	−91	17
8	−83	30
9	−74	43
10	−60	55
11	−38	65
12	0	69
13	37	65
14	60	55
15	73	43
16	83	31
17	91	18
18	98	5
19	999	0
20	999	0
21	999	0
22	999	0
23	999	0
24	999	0

The average temperature exceeds 26.0°C after 8:00, and it is less than 26.0°C after 16:00. Therefore people need the external shading systems from 8:00 to 16:00. Because the symmetry points of the solar elevation angle and azimuth data are 12:00, the figures of them are the same at these time. Therefore, we select these figures of 8:00 as the critical value.

2.1 Calculation of southern shading systems

When the window is towards the south, i.e. the azimuth $\theta = 0°$, set the window width x, height y, (where x, y are greater than 0), assuming that the window edge is against the shading, the distance of overhangs $L = (\cos 30°/\tan 83°)y$, the Horizontal length of the shading system.

$$S = 2\left(\sin 30°/\tan 83°\right)y + x$$

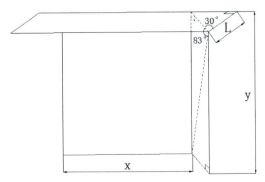

Figure 3. Schematic calculation of southern shading systems.

2.2 Calculation of southeastern and southwestern shading systems

When the window is towards the east or west, that azimuth $\theta = 90°$ or $-90°$, set the window width x, height y, (where x, y are greater than 0), using a three-dimensional shading strategy, assuming that the window edge is against the shading, the distance of overhangs $L = (\sin 30°/\tan 83°)y$. The vertical length of the three-dimensional shading.

$$H = y - \tan 83°\left(\left(\cos 30/\tan 83\right)^2 y^2 + x^2\right)^{-2}$$

2.3 Calculation of eastern and western shading systems

When the window is towards southeast or southwest, the azimuth $\theta = 45°$ or $-45°$, we are supposed to consider that the azimuth angle and the sun need to be superimposed, set windows wide x, height y, (where x, y are greater than 0), using three-dimensional shading strategy, assuming that the window edge is against the shading, the distance of overhangs $L = (\sin 75°/\tan 83°)y$. The vertical length of the three-dimensional shading.

$$H = y - \tan 83°\left(\left(\cos 30/\tan 83°\right)^2 y^2 + x^2\right)^{-2}$$

2.4 Calculation of other directions

Located in the north of the Tropic of Cancer 23.5 N, Hangzhou is in the latitude of 30.3 N, therefore for in the whole year, there is no need to consider external shading system for the direction of north, northwest and northeast.

Calculating methods of external shading coefficient:

$$SD_H = A_H PF^2 + B_H PF + 1$$

SD_H: external shading coefficient, A_H, B_H: calculation coefficient, PF: extended coefficient of the shading system, PF = A/B. When the calculated PF > 1, take PF = 1.

320

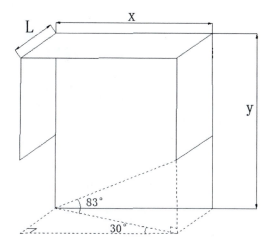

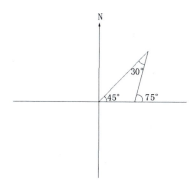

Figure 4. Schematic calculation of southeastern and south-western shading systems.

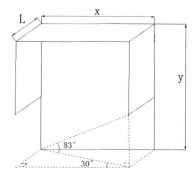

Figure 6. Schematic calculation of east and west.

Figure 7. Description on calculating methods.

Table 2. External shading coefficient of two types of shading system in different directions.

Direction	Horizontal		Vertical	
	A_H	B_H	A_H	B_H
E	0.35	−0.75	0.32	−0.65
SE	0.48	−0.83	0.42	−0.80
S	0.47	−0.79	0.42	−0.80
SW	0.36	−0.68	0.42	−0.82
W	0.36	−0.76	0.33	−0.66
NW	0.36	−0.68	0.41	−0.82
N	0.30	−0.58	0.44	−0.84
NE	0.48	−0.83	0.43	−0.83

South: $PF = A/B = \cos 30/\tan 83$, $A_H = 0.47$, $B_H = -0.79$, $PF = 0.1063$, $SD_H = 0.9203$.
Southwest: $PF = A/B = \sin 75/\tan 83$, $A_H = 0.36$, $B_H = -0.68$, $PF = 0.1186$, $SD_H = 0.9220$.
Southeast: $PF = A/B = \sin 75/\tan 83$, $A_H = 0.48$, $B_H = -0.83$, $PF = 0.1186$, $SD_H = 0.9119$.
East: $PF = A/B = \sin 30/\tan 83$, $A_H = 0.35$, $B_H = -0.75$, $PF = 0.0614$, $SD_H = 0.9522$.
West: $PF = A/B = \sin 30/\tan 83$, $A_H = 0.36$, $B_H = -0.76$, $PF = 0.0614$, $SD_H = 0.9547$.

Figure 5. Schematic calculation of angle variation.

3 SUMMARY

3.1 Analysis results

For existing south-facing windows, the lateral width of external shading systems should be extended. Design dimensions are shown in Figure 3.

For existing windows facing east, southeast, southwest, and west, if horizontal shading system has been equipped, it is better to be added with vertical shading. If vertical shading system has been already equipped, horizontal shading system is required. Design dimensions are shown in Figure 4.

In the future, for the windows facing east, southeast, and southwest, three-dimensional shading system is needed. Design dimensions are shown in Figure 6. Moreover, the windows facing south should be equipped with horizontal shading system. Design dimensions are shown in Figure 4.

3.2 Further study

In this study, the typical meteorological databases are processed, and then the type and size of shading systems in different directions are determined. It brings a more in-depth understanding of the effect of existing external shading system in Hangzhou. Because of different directions, each house has different directional windows and external shading systems. However, it can be found that almost all shading systems are the same. On the one hand, it is hard to know whether it meets the needs of shading to some extent. If the horizontal length of the shading system is too long, it may lead to a waste of materials. On the other hand, without a three-dimensional shading system, the excess length of the shading system does not reach a better effect. Only when we select different external shading systems reasonably, we will achieve the best shading objective.

It provides a theoretical basis for further design of shading systems. In the future, the calculations of each directions will be shown in designing and analyzing the external shading systems. When it comes to the actual use of external shading system and satisfaction of users, we still need to conduct field research. Therefore, further research will be carried out in the future.

REFERENCES

Jane YiWen, Wang Suying, Jiang Yi. Optimum overhang and vertical shading device dimensions for energy saving in Beijing. Xi'an University of Architecture & Technology. Vol. 33, No. 3. Sep. 2001 (212–217).

JGJ 75-2003, Design Standard for Energy Efficiency of Residential Buildings in Hot-Summer and Cold-Winter Zones.

Zhang Qingyuan & Yang Hongxing. Typical Meteorological Database Handbook for Buildings. China Building Industry Press. 2012.

Advances in Civil Engineering and Building Materials IV – Chang et al (eds)
© *2015 Taylor & Francis Group, London, ISBN: 978-1-138-00088-9*

Study of shaft mud film effect and final pile installation pressure estimation for a jacked pile

Yong-qiang Hu
School of Engineering, Sun Yat-Sen University, Guangzhou, China
School of Civil Engineering, Guangzhou University, Guangzhou, China

Lian-sheng Tang & Hai-tao Sang
School of Earth Science and Geological Engineering, Sun Yat-Sen University, Guangzhou, China

Zhi-zhong Li
Guangdong Yongji Building Foundation Co. Ltd., Shunde, China

ABSTRACT: The estimation of the final pile installation pressure for a jacked pile attracts the attention of researchers due to its value in the proper selection of a piling machine and the prediction of bearing capacity. The key to the estimation of the final pile installation pressure is the shaft resistance mobilization during pile installation. Some studies suggest that the degree of shaft resistance mobilization fluctuates and employs existing estimation formulas that use a fixed sensitivity or empirical coefficient to discount the shaft resistance, which fails to reflect the actual mobilization of shaft resistance. Based on previous studies and the results of a new type of Cone Penetration Test (CPT) and a model pile test, this paper analyses the formation of shaft mud film and its effect on shaft resistance mobilization during pile installation from the perspective of tribology. The total shaft resistance mobilization degree coefficient is introduced, and a novel estimation formula for final pile installation pressure is proposed. The formula is simple, practical, and highly precise.

Keywords: shaft mud film, pile installation pressure, jacked piles, shaft resistance, mobilization degree

With the continuous improvement of jacked pile construction technology and machines, and recent advancements in urban environmental protection requirements, the jacked pile has been extensively applied, especially in China. Compared with mature theoretical studies of the driven pile outside China, the theory of the jacked pile, which is a type of displacement pile, is underdeveloped. However, in contrast to the dynamic penetration of the driven pile, the jacked pile involves quasi-static penetration (Basu et al., 1989), which exhibits a better fit to the cone penetration test (CPT). Thus, this notion can be employed as a starting point for the study of piles in the same category (Zhang, 2004), which has attracted attention from a substantial number of researchers.

Final jacked pile pressure is a short-term axial force (Jackson, 2008), which comprises the pressure at the end of pile installation. Generally, it is the corresponding bearing capacity to time zero in a setup effect (also referred to as the time effect) study (Reddy & Stuedlein, 2014, Deng et al., 1989, Tarawneh, 2013, Zhang, 2002, and Wang, 2010), in which the capacity is typically less than the ultimate bearing capacity. Therefore, a precise final pressure prediction guarantees

smooth construction progress and facilitates not only the proper selection of construction machines, but also the estimation of a jacked pile's ultimate bearing capacity.

Few studies have addressed the jacked pile's final pile installation pressure outside China (Jackson, 2008, Sinitsyn & Loset, 2011, and Goncharov, 1966). However, numerous studies on this issue have been conducted in China[11−18] due to the extensive application of the jacked pile in China. Existing studies that different pile installation methods yield similar tip resistance values (Zhu, 2000, Zhen & Gu. 1998, Chu & Wang, 2000, Zhang & Deng, 2003, Liu et al., 2009, Fan, 2007, Li, 2010, Han, 1996, Samson & Authier, 1986, Skov & Denver 1988, and Gavin & Kelly, 2007); thus, the key to the estimation of a jacked pile's final pile installation pressure is the estimation of the shaft resistance. At the end of installation, the shaft resistance is typically less than the ultimate shaft resistance. Therefore, the shaft resistance should be deducted in the estimation. Mechanisms that offset shaft resistance primarily include pore water pressure and pore water seepage (along the pile), which is caused by shaft soil softening, shaft soil fatigue (Basu et al., 2014,

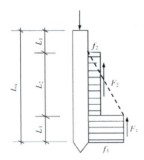

Figure 1. Shaft resistance distribution during installation (Yuan & Jin, 1988, Zhang & Deng, 2003).

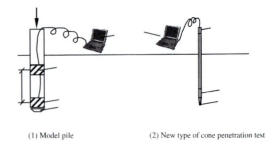

(1) Model pile   (2) New type of cone penetration test

Figure 2. Model pile structure and new type of cone penetration test.

Gavin & Kelly, 2007), sliding friction (Li, 2010, Han, 1996, Li et al., 2011, and Zhang & Deng, 2002), and partial drainage (Jackson, 2008). The majority of these formulas perform shaft resistance deduction at the pile trunk according to the shaft resistance distribution model for pile installation (Yuan & Jin, 1988) (Figure 1), which was proposed by Yuan & Jin (1988) (Zhu, 2000, Zhen & Gu, 1998, Chu & Wang, 2000, Zhang & Deng, 2003, Liu et al., 2009, Fan, 2007, and Li, 2010). For instance, Zhu (1992) used the pile penetration depth reduction coefficient for deduction. Zheng & Gu (1998) used the sensitivity and smear coefficient for deduction. Chu & Wang (2000) used the empirical reduction coefficient for calculation. Zhang & Deng (2003) used the linear increase in the dotted line in Figure 1 for their calculations. Fan (2007), and Li et al., (2010) used sensitivity for deduction. Other methods include the sharing coefficient by Han (1996, 2011), and the dynamic-static friction ratio by Li et al. (2011), for shaft resistance deduction. The majority of these formulas have regional limitations.

Many researchers have observed the effect of mud film or water film on shaft resistance. Zhang et al., (2009) observed that the "water film" that formed in the pile installation caused a "hard shell" to tightly attach around the pile after the resting phase and believe that this result proves the existence of mud film. Li (2010) conducted a pile installation test and discovered viscous soil attached to the side surface of the pile shoe when it was pulled out after the test. Li et al. (2011) discovered mud cake on the surface of the pile when it was pulled out after the pile installation test, which caused them to believe that the pile-soil friction during pile installation consists of soil-soil friction (Li et al., 2011). However, a mud film has not been investigated as a standalone phase, but is used as evidence of a qualitative explanation of a specific phenomenon. The investigation and testing of a tip-bearing jacked pile model in the Pearl River Delta region reveals that the shaft mud film significantly affects the shaft resistance. Therefore, this paper leverages the test results from the model pile test and the new type of CPT, incorporates tribology theory, and analyses the shaft mud film effect and its impact on pile installation pressure.

1 MODEL PILE TEST AND NEW TYPE OF CONE PENETRATION TEST

A pile tip pressure bearing board is installed at the tip of the model pile. Friction tubes are installed at several positions in the pile trunk to measure the variation in friction resistance at the pile tip and the shaft during pile installation and the load test (Figure 1). A hydraulic hoop pressure-piling machine is used to plant the model pile into the ground during the pile installation test. When pile installation is completed and after a short pause, the pile driver is used to continue the load test on the planted model pile. The dynamic variation of stress at the shaft and the pile tip is measured throughout the pile installation test and the load test. The new type of CPT (Patent number: 200420044127.5) introduces a total stress sensor on top of the conventional cone penetration test.

2 SHAFT MUD FILM EFFECT TEXT AND INDENTING

2.1 Formation of shaft mud film

Jacked pile penetration involves quasi-static penetration. During the entire penetration, the friction between the shaft surface and the soil surrounding the shaft is continuous. According to tribology theory (Weng & Huang, 2012, and Bhushan, 2013), friction and abrasion occur simultaneously. Because the pile material strength exceeds the strength of the surrounding soil, asperities on the pile surface constantly plow the soil surface, and a large amount of soil debris is plowed down. Because the shaft soil is under continuous pressure and shear, the pore water pressure in the soil increases, and the water in the soil flows to the pile-soil boundary and forms a mud film with the soil debris generated by the plowing effect. Under certain circumstances, such as sandy soil, a water film may be produced.

2.2 Effect of shaft mud film on shaft resistance

According to the lubrication theory, when lubrication exists between two objects, the friction force is calculated by the following formula (Weng & Huang, 2012, and Bhushan, 2013):

$$f_s = \tau_l = \frac{\eta_l v}{h} \quad (1)$$

In the formula, f_s and τ_l represent the friction force and the shear force, respectively; η_l is the viscosity of the lubricating layer; v is the relative velocity between two objects; and h is the thickness of the lubricating layer.

A mud film acts as a lubricating layer; therefore, the shaft resistance is calculated by Formula (1). According to Formula (1), the friction between the pile and the soil is affected by the viscosity and thickness of the mud film and the relative velocity between the pile and the soil.

(1) Effect of viscosity:

Viscosity is the most importance factor that influences the magnitude of the shaft resistance, and the value of viscosity is determined by the water content in the mud film. The shaft soil resembles a sponge saturated with water; water is squeezed out when constant shear is applied; once the shear force disappears, the stress release results in volume expansion and generates negative soil pore water pressure, which draws water from the pile-soil boundary into the soil. This situation explains why a brief pause in pile installation will cause a significant increase in pile driving pressure. The water content in the mud film is closely related to the pile installation speed: when the pile installation speed stabilizes, the shaft soil is constantly sheared to provide sufficient water supply to the mud film. The mud viscosity approaches the viscosity of water; thus, the friction force is reduced.

(2) Effect of thickness:

The shaft resistance is inversely proportional to the thickness of the mud film. The thickness of the shaft mud film differs at different depths. At the surface layer, because it endures the longest period of plowing and because it may detach from the pile due to construction-induced pile trunk horizontal sway[13], the maximum thickness of the surface layer is achieved, and the shaft stress is close to zero. At the layer near the pile tip, the plowing effect is short-lived; thus, the thickness of the mud film is minimal or nonexistent. Therefore, the shaft resistance is strongest at this location and approaches the ultimate soil friction resistance; in the middle, the situation is similar. This situation corresponds to the three-section model proposed by Yuan & Jin (1984).

(3) Effect of pile installation speed:

Formula (1) shows that the friction force is proportional to the pile installation speed, i.e., a higher speed produces a stronger friction force. However, the friction between the pile and the soil demonstrates a contradictory situation. Jackson (2007) discovered that the higher the pile installation speed is, the lower the pile installation pressure is. The reason for this deviation is that the increase in speed causes a decrease in the mud film viscosity, and more substantial decrease in the mud film viscosity. Under normal conditions,

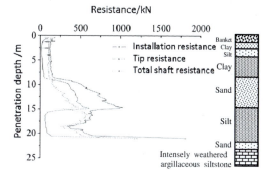

Figure 3.   Installation resistance of model pile.

the pile installation speed increases the disturbance on the shaft soil and the pore pressure; thus, more water enters the mud film, which reduces its viscosity. The reduction in pile installation speed produces opposite results. The variation in pile installation speed during pile installation is primarily caused by the variation in pile tip resistance: when the pile tip encounters a change in the soil layer, the pile installation speed changes.

2.3 Variation rule for model pile's total shaft resistance during pile installation

To determine the value of the shaft resistance, the shaft resistance distribution rule should be determined in advance. Figure 3 displays a diagram that explains the variation rule for a pile's actual total shaft resistance in the model pile test. According to the diagram, the pile tip resistance remains stable before the pile tip reaches the sand layer in the middle and the total shaft resistance remains almost unchanged. This finding indicates that this layer is soft, the pile installation speed is stable, the stable mud film exists, and its viscosity is low. Therefore, the increase is insignificant. When the pile tip proceeds into the sand layer, the pile tip resistance and the total shaft resistance significantly increase and certain abrupt changes occur when they break into the sand layer. This finding indicates a sudden increase in pile tip resistance, and an abrupt decrease in pile installation speed, which weakens the shear effect on the shaft soil and causes a reduction in pore pressure, shaft mud film water loss, an increase in viscosity, and a subsequent abrupt increase in total shaft resistance. When the pile tip passes the sand layer and enters the silt layer, the pile tip resistance and the total shaft resistance simultaneously decrease and an abrupt change occurs. At this moment, the pile installation speed increases, an enhanced shear effect on the shaft soil is enhanced, the pore pressure increases, and the water content in the shaft mud film increases, which decreases the viscosity, and total shaft resistance. When the pile tip extends into the bearing layer, the pile tip resistance and the total shaft resistance significantly increase; however, until the total shaft resistance increases in tandem with the pile tip resistance to a certain value and begins to decrease, the total shaft resistance increases with an increase in pile tip resistance. At this moment, the pile installation speed decreases, the viscosity of the mud film increases, the thickness decreases, the total shaft resistance increases, and the eventual total shaft resistance decreases, which indicate that the relative displacement between the pile and the soil decreases and begins to impede the mobilization of the shaft resistance.

3 CALCULATION OF FINAL PILE INSTALLATION PRESSURE

3.1 The relationship between the final pile installation pressure and the ultimate bearing capacity

The ideal ultimate bearing capacity typically refers to the long-term bearing capacity and the bearing capacity in completely drained shaft soil, whereas the bearing capacity obtained from the conventional load test after the resting phase should be referred to as the medium-term bearing capacity. The final pile installation pressure is a short-term axial force or the short-term bearing capacity. The bearing capacity obtained from the CPT is similar to the medium-term bearing capacity. Silva (2005) explained their relations. Cone penetration typically progresses at the speed of 15–25 mm/s[33]. According to the findings of studies on residue strength[34,35], when the shear rate exceeds 1.8 mm/s, the strength demonstrates a significant change. Therefore, the $q_t/q_{t,drained}$ is slightly less than 1. A jacked pile typically sinks at a rate of 30–50 mm/s; its dimensions are substantially larger than the cone penetration sensor bar, and the disturbance of the shaft soil is more significant; therefore, $q_b/q_{b,drained}$ is smaller.

3.2 Calculating the degree of shaft resistance mobilization and the final pile installation pressure

The mud film layer formed during pile installation is highly unstable: the viscosity, thickness, and pile installation speed change constantly, which causes a constant change in the unit shaft resistance of the jacked pile at a specific depth during the entire pile installation process. Chu Wang-Ying and Wang Neng-Min (2000) discovered that unit friction resistance is not constant during pile installation and shaft resistance is not a simple sum of data obtained from the cone penetration test[13]. Various reduction coefficients in existing final pile installation pressure calculation formulas are constant values, which cannot reflect this variation rule of shaft resistance. Silva (2005) showed that the shaft resistances obtained during the CPT and pile installation construction have not reached ultimate resistances. Therefore, the shaft resistance demonstrates a mobilization problem during pile installation. To reflect the shaft resistance

mobilization rule in pile installation, the mobilization degree coefficient η for the shaft resistance is introduced:

$$\eta = f_s / f_{s,drained} \tag{2}$$

In the formula, f_s and $f_{s,drained}$ represent the actual shaft resistance and ultimate shaft resistance, respectively, of the operational jacked pile.

According to the study of the mud film effect, $f_{s,drained}$ is equivalent to the pile-soil friction without mud film. When the cone penetration probe enters the soil layer at a specified speed, a lack of mud film is assumed. Thus, it is equal to the ultimate shaft resistance, whereas the actual shaft resistance is closely related to the viscosity and thickness of the mud film. Based on the previous discussion, the viscosity and thickness of the mud film at a specific depth are constantly changing during the entire pile installation process or the CPT, i.e., the corresponding layer's mobilization degree coefficient varies with time, which increases the difficulty of estimating the final pile installation pressure. For simplicity, this paper proposes the use of the total shaft resistance mobilization degree coefficient η_{total}:

$$\eta_{total} = Q_s / Q_{s,drained} \tag{3}$$

In the formula, Q_s is the actual total shaft resistance for a jacked pile in operation, and $Q_{s,drained}$ is the ultimate total shaft resistance, which can be obtained by the new type of CPT. This new type of cone $_{penetration}$ introduces a total resistance sensor on top of the conventional cone penetration. Thus, the ultimate total shaft resistance $Q_{s,drained}$ represents the difference between the cone penetration total resistance and the tip resistance. The standard curves for the cone penetration's actual total shaft resistance and the ultimate total shaft resistance are shown in Figure 4. The actual total shaft resistance curve is typically positioned under the ultimate total shaft resistance curve, especially in the case of a friction pile.

A similar layer combination has a similar impact on the cone penetration and pile penetration speed; therefore, the total shaft resistance mobilization degree coefficient obtained from the new type of CPT is used to estimate the final pile installation pressure of the jacked pile. The mobilization degree coefficient for the total shaft resistance at different penetration depths can be easily calculated from Figure 4. The impact of the dimension effect on the pile should be considered; therefore, the final pile installation pressure is calculated by the following formula:

$$Q_{EOJ} = \alpha \eta_{total} u \sum l_i f_{si} + \beta A_p q_c \tag{4}$$

In the formula, f_{si} is the unit ultimate shaft resistance at the ith layer, which can be obtained by the cone penetration test; l_i is the ultimate shaft resistance of the ith layer; A_p is the pile tip section area; q_c is the unit ultimate tip resistance, which is obtained from the CPT;

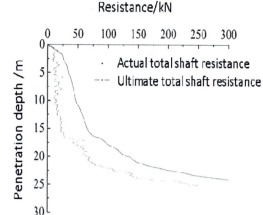

Resistance/kN

Figure 4. Actual total shaft resistance curve and ultimate total shaft resistance curve generated from the new type of cone penetration.

α and β are dimension effect correction coefficients, which are determined according to data published by other researchers.

The total shaft resistance mobilization degree coefficient is used to calculate the final pile installation pressure to prevent the difficulty of the situation in which the layer's mobilization degree coefficient at a specific depth varies throughout pile installation. This situation simplifies the calculation and provides high engineering practicability. In addition, the value is directly obtained from the new type of CPT, which differs from the empirical coefficient and does not have regional limitations.

The test in this paper is conducted in the Pearl River Delta region using the new type of cone penetration equipment. The total resistance function is manufactured by Guangdong Yongji Building Foundation Co., Ltd; α and β are close to 1. Its high precision has been proven by engineering practice. Additional studies in other regions are necessary, especially in the case of non-tip-bearing operation.

4 CONCLUSIONS

Based on theoretical analysis and field test results for the new type of CPT and the model pile test, the following conclusions are obtained:

1. The influence of mud film on the shaft resistance is analysed and discussed from the perspective of tribology, which provides a reasonable explanation of the total shaft resistance distribution rule.
2. The concept of the shaft resistance mobilization degree coefficient is proposed, and the total shaft resistance mobilization degree coefficient is introduced. This coefficient is calculated using data from the new type of CPT with the newly added total resistance sensor.

3. A new type of CPT that is based on the simple final pile installation pressure estimation formula with high precision is proposed. The formula has high engineering practicability and has been proven in actual field projects in the Pearl River Delta region.

REFERENCES

Basu, P., Loukidis, D., Prezzi, M., et al. 2011. Analysis of shaft resistance of jacked piles in sands. *International Journal for Numerical and Analytical Methods in Geomechanics* 35(15): 1605–1635.

Basu, P., Loukidis, D. & Prezzi, M. et al. 2014. The Mechanics of Friction Fatigue in Jacked Piles Installed in Sand. In Iskander, M., Garlanger, J. E. & Hussein, M. H. (ed.), *From Soil Behavior Fundamentals to Innovations in Ge-otechnical Engineering: Honoring Roy E. Olson:* 546–557; Geo-Congress 2014, Atlanta, Georgia, 23–26 February, 2014. ASCE.

Bhushan, B. 2013. *Introduction to tribology.* John Wiley & Sons.

Chu, W. & Wang, Y. 2000. Analysis and Estimation of Sinking Pile Resistance of Static Press Pile. Geotechnical Engineering Technique 1(1): 25–28.

Deng, T., Zhao, Z. & Gui, Y. 2013. Time Effect on Bearing Capacity of Jacked Piles Using the Back-Analysis Method. In Huang et al. (ed.), *Pavement and Geotechnical Engineering for Transportation:* 175–182; First International Symposium on Pavement and Geotechnical Engineering for Transportation Infrastructure, Nanchang, China, 5–7, June 2011. ASCE.

Fan, X. 2007. Study of sinking resistance and compaction effect of static-press pile. Shanghai: Tongji University.

Gavin, K. G. & Kelly, O. 2007. Effect of Friction Fatigue on Pile Capacity in Dense Sand. *Journal of Geotechnical and Geoenvironmental Engineering* 133(1): 63–71.

Goncharov, B. V. 1966. Soil resistance when jacking piles. *Soil Mechanics and Foundation Engineering* 3(4): 255–257.

Han, X. 1996. Experimental research on static pressure and bearing capacity of statically pressed piles. *Journal of building structures* 17(6): 71–77.

Jackson, A. 2008. Pile jacking in sand and silt. *Winner of the ICE's national Graduates and Students Papers Competition 2008.*

Li, Y., Li, J. & Zhao, Z. 2010. Estimation of Resistance of Static Pressure Pile Sinking Based on Cone Penetration Test. *Subgrade Engineering* 150(3): 67–69.

Li, Y., Li, J. & Zhao Z. et al. 2010. Model Test Research on Penetration Process of Jacked Pile in Layered Soil. *Journal of Jilin University: Earth science* 40(6): 1409–1414.

Li, Z., Tang, L. & Hu, Y. et al. 2011. A new estimation method for penetration depth of displacement pile based on a new type cone penetration test. *Applied Mechanics and Materials* (90):234–244.

Liu, J., Zhang, M. & Zhao H. et al. 2009. Computational simulation of jacking force based on spherical cavity expansion theory and friction fatigue effect. *Rock and Soil Mechanics* 30(4): 1181–1185.

Reddy S. & Studlein, A. 2014. Time-Dependent Capacity Increase of Piles Driven in the Puget Sound Lowlands. In Iskander, M., Garlanger, J. E. & Hussein, M. H. (ed.), *From Soil Behavior Fundamentals to Innovations in Geotechnical Engineering: Honoring Roy E. Olson:* 464–474; Geo-Congress 2014, Atlanta, Georgia, 23–26 February, 2014. ASCE.

Samson, L. & Authier, J. 1986. Change in pile capacity with time: case histories. *Canadian Geotechnical Journal* 23(2): 174–180.

Silva, M.F. 2005. Numerical and physical models of rate effect in soil penetration. London: Cambridge University.

Sinitsyn, A. O. & Loset, S. 2011. Obtaining the resistance of plastic frozen ground under pile driving by Jacking. *Soil Mechanics and Foundation Engineering* 48(2): 79–85.

Skov, R. & Denver, H. 1988. Time-dependence of bearing capacity of pile. *Proc. Third International Conference on the Application of Stress-Wave Theory to Piles, Ottawa*: 25–27

Tarawneh, B. 2013. Pipe pile setup: Database and prediction model using artificial neural network. *Soils and Foundations* 53 (4): 607–615.

Wang, J., Zhou, J. & Zhang, M. 2010. Computation of bearing capacity of jacked piles based on spherical cavity expansion theory and time effect. *Chinese Journal of Underground Space and Engineering* (05): 964–968.

Weng, S. & Huang, P. 2013. *Principle of tribology (fourth edition)*. Beijing, Tsinghua University Press.

Yuan, X. & Jin, C. 1984. Prediction of pile jacking installation resistance. *Construction Technology* (03): 7–12.

Zhang, M., Shi, W. & Wang, C. 2002. Time effect on the ultimate bearing capacity of static pressed pile. *Chinese Journal of Rock Mechanics and Engineering* 12(2): 2601–2604.

Zhang, M. & Deng A. 2002. Experimental study on sliding friction between pile and soil. *Rock and Soil Mechanics* 23(2): 246–249.

Zhang, M. & Deng A. 2003. A spherical cavity expansion-sliding friction calculation model on penetration of pressed-in piles. *Rock and Soil Mechanics* 24(5): 701–709.

Zhang, M. 2004. *Study and application of jacked pile*. Beijing: China building industry press.

Zhang, M., Liu, J. & Yu, X. 2009. Field test study of time effect on ultimate bearing capacity of jacked pipe pile in soft clay. *Rock and Soil Mechanics* 30(10): 3005–3008.

Zheng, G. & Gu, X. 1998. Analysis of some problems of statically pressed piles in soft soils. *Journal of building of structures* 19(4): 54–60.

Zhu, X. 2004. Prediction of pile installation resistance and pile installation possibility. In Gao, D. (ed.), *Theory and practice on soft soil Foundation*: 20–96. Beijing: China building industry press, Tongji university press.

Road engineering

Advances in Civil Engineering and Building Materials IV – Chang et al (eds)
© 2015 Taylor & Francis Group, London, ISBN: 978-1-138-00088-9

Prediction of subgrade settlement based on a fuzzy self-adaptable method of artificial intelligence

Xiangxing Kong
The First Highway Survey and Design Institute of China Communications Construction Company Ltd., Xi'an, China

ABSTRACT: In the forecasting of road foundation settlement, different methods can often provide valuable information. When different forecasting methods are combined properly, comprehensive forecast information is utilized in various ways. According to the principle of artificial intelligence and method of fuzzy mathematics, the fuzzy combination forecasting method of self-adaptive variable weight is proposed. Based on matching degrees of a past-recent period and actual value, the weights are automatically adjusted, so that forecasting results will be accurate. The fuzzy combination forecasting method of self-adaptive variable weight is applied to the subgrade settlement of the Changzhang expressway, which provides prediction accuracy and reliable applicability.

Keywords: prediction, subgrade settlement, fuzzy method, artificial intelligence

1 INTRODUCTION

The construction of expressways in mountainous areas, with a high fill subgrade construction of deep excavation, is often encountered [1–3]. Combined with the regular pattern of the development of high embankment settlement, the fuzzy combination forecasting method of self-adaptive variable weight is applied to the subgrade settlement of the Changsha Jishou expressway, which provides the prediction accuracy and reliable applicability [4, 5]. Faults in the roadbed settlement are prevented and safe operation is ensured.

2 PREDICTION METHOD ANALYSIS OF ROADBED SETTLEMENT

During the early period of construction, just after unloading, soil is in elastic status. In the later period of construction, with increasing loads, the loads imposed on the high embankment also increase. Pore water in the foundation soil is gradually discharged, whilst excess pore water pressure gradually decreases. As a result, compaction causes more deformation, and the soil in the embankment will be in elastic-plastic state. With the continuous development of the plastic zone, the settlement rate increases quickly, until the load no longer increases.

After construction, due to the pore water pressure almost disappearing completely, the consolidation process is not complete; settlement will continue with the passage of time, but with a significantly decreased sedimentation rate. If the time of settlement is infinity, settlement will be at the ultimate state.

At present, the prediction method of roadbed settlement is catalogued into two major types. Firstly, according to the constitutive model of consolidation theory, and combined with various soils, numerical methods are applied in settlement calculation [6]. However, there is a big gap between the actual model and the project, and a large number of soil tests to determine the soil parameters are needed [7]. Therefore, it is difficult to use a general engineering design. It can reflect the law of the development of high embankment settlement, which is an S-shape growth model. The Pearl and Gompertz curve model is the S-shaped prediction model for roadbed settlement:

(1) Pearl model:

$$y_t = \frac{k}{1 + ae^{-bt}} \qquad (1)$$

(2) Gompertz model:

$$y_t = e^{(k + ab^t)} \qquad (2)$$

3 FUZZY COMBINATION FORECASTING METHOD OF SELF ADAPTIVE VARIABLE WEIGHT

In the forecasting of road foundation settlement, different methods can often provide valuable information [8, 9]. When different forecasting methods are combined properly, comprehensive forecast information are utilized in various ways. According to the principle of artificial intelligence and the method of fuzzy mathematics, the fuzzy combination forecasting method of self-adaptive variable weight is proposed. Based on matching degrees of past-recent period and actual value, the weights are automatically adjusted, so that forecasting results will be accurate.

For the same prediction problem, there are n (n≥2) prediction models. Hypothesis: y(t) is the observed value; f(t) is the prediction value; k(t) is the weight coefficient; e(t) is the prediction error. For the i method, the matching degree and measured values (coefficients) are determined by two parameters: relative index E, which is the absolute average value of the last few sampling period prediction errors, and relative index EA, which is the absolute cumulative value of the rolling finite length prediction relative errors. The value can be calculated by the following formulae:

$$e_i(t) = y(t) - f_i(t) \qquad (3)$$

$$a_i(t) = \sum_{m=0}^{k-1} \frac{|e_i(t-m)|}{k} \qquad (4)$$

$$s_i(t) = \sum_{m=0}^{l-1} |e_i(t-m)| \qquad (5)$$

$$E_i(t) = \frac{a_i(t)}{\max_{1 \le i \le n}(a_i(t))} \qquad (6)$$

$$EA_i(t) = \frac{s_i(t)}{\max_{1 \le i \le n}(s_i(t))} \qquad (7)$$

where: t is the current sampling time, k is the average of the cycle number, l is the finite time length, ai is the absolute average value of prediction errors, and si is the absolute cumulative value of prediction errors.

$\tilde{E}_i(t), E\tilde{A}_i(t), \tilde{k}_i(t)$ are, respectively, the fuzzy values of $E_i(t), EA_i(t), k_i(t)$ on the same domain. The definition of fuzzy subsets can be the same as it, based on the fuzzy mathematics theory. In order to make a simple calculation, the fuzzy subset is respected by the nuclear element.

$$\tilde{k}_i(t) = -[a\,\tilde{E}_i(t) + (1-a)\,E\tilde{A}_i(t)] \qquad (8)$$

where a reflects the emphasis degree on recent situation prediction. Obviously, it reflects the fuzzy inference rules about weight coefficient, but also avoids the neutral reasoning rule definition or jump phenomenon. Then k(t) is normalized, so the weight coefficient can be obtained:

$$k_i(t) = \frac{k_i'(t)}{\sum_{i=1}^{n} k_i'(t)} \qquad (9)$$

$$k_i(T) = \sum_{t=1}^{N} k_i(t)/N \qquad (10)$$

Based on the prediction accuracy computing method of the existing research achievements [10], 0-N time weight should be used to N+1 time of single forecasting value.

$$f(t) = \sum_{i=1}^{n} k_i(t) f_i(t) \qquad (11)$$

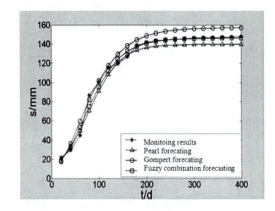

Figure 1. Measured and forecasting settlement-time curve.

4 ENGINEERING APPLICATION AND ANALYSIS

The data of K149 + 420 is selected as an example. The embankment fillers are low liquid limit clay and red sandstone, and the embankment heights are about 12 m and 17 m. Based on the above-mentioned principle, the corresponding calculation program is compiled by MATLAB 6.5.

In Figure1, the settlement of subgrade scatter is S-shaped. The first twelve points of the data are used as the sample; the relevant parameters can be obtained by the Pearl curve model and Gompertz curve model, and then a = 0.648, k = 2, and l = 3 are taken to fit and forecast the original sample. The weight coefficients of each model, as shown in Table 1, are computed. Finally, the results of the fuzzy combination forecasting method of self-adaptive variable weights are obtained.

The Pearl model and the Gompertz model can provide valuable information. When two forecasting methods are combined properly, comprehensive forecast information is used in various ways. The results of the fuzzy combination forecasting method of self-adaptive variable weight are much more accurate and reliable than any single method, such as Pearl and Gompertz, and the compared analysis and forecasting precision are shown in Tables1, 2, and 3.

5 CONCLUSION

In the forecasting of road foundation settlement, different methods can often provide valuable information. When different forecasting methods are combined properly, comprehensive forecast information is used in various ways.

(1) The developing pattern of high embankment settlement is similar to the S-shaped model. Therefore, a growth curve model can be applied to predict the settlement, such as the Pearl model and the Gompertz model.

Table 1. Settlement of Pearl and Gompertz forecasting.

Tine /d	self-adaptive variable weight		Pearl forecasting /mm	Gompertz forecasting /mm
	Pearl	Gompertz		
20	0.5215	0.4785	19.862	18.079
40	0.5897	0.4103	32.781	37.022
60	0.8536	0.1464	50.621	59.779
80	1.0000	0.0000	71.792	82.347
100	0.0000	1.0000	92.829	102.009
120	0.3129	0.6871	110.349	117.706
140	0.5414	0.4586	122.920	129.524
160	0.7961	0.2039	131.007	138.079
180	0.7481	0.2519	135.849	144.111
200	0.6496	0.3504	138.626	148.289
220	0.6525	0.3475	140.178	151.150
240	0.7073	0.2927	141.034	153.093
260	0.6144	0.3856	141.502	154.406
280	0.6144	0.3856	141.757	155.289
300	0.6144	0.3856	141.896	155.883
320	0.6144	0.3856	141.971	156.281
340	0.6144	0.3856	142.012	156.548
360	0.6144	0.3856	142.034	156.726
380	0.6144	0.3856	142.046	156.846
400	0.6144	0.3856	142.052	156.926

Table 2. Settlement of measuring in site and forecasting.

Tine /d	self-adaptive variable weight		Fuzzy combination forecasting /mm	Monitoring results /mm
	Pearl	Gompertz		
20	0.5215	0.4785	19.175	22.1
40	0.5897	0.4103	34.416	29.6
60	0.8536	0.1464	54.153	44.9
80	1.0000	0.0000	75.863	86.8
100	0.0000	1.0000	96.369	100.6
120	0.3129	0.6871	113.186	115.6
140	0.5414	0.4586	125.467	119.6
160	0.7961	0.2039	133.734	126.9
180	0.7481	0.2519	139.035	138.4
200	0.6496	0.3504	142.352	141.8
220	0.6525	0.3475	144.409	143.5
240	0.7073	0.2927	145.684	144.3
260	0.6144	0.3856	146.478	145.0
280	0.6144	0.3856	146.975	145.7
300	0.6144	0.3856	147.289	146.5
320	0.6144	0.3856	147.489	146.5
340	0.6144	0.3856	147.617	146.5
360	0.6144	0.3856	147.699	147.1
380	0.6144	0.3856	147.753	147.3
400	0.6144	0.3856	147.788	148.1

Table 3. Forecasting precision(S).

Pearl model	Gompertz model	Fuzzy combination forecasting model
607.9	1481.0	353.9

value, the weights are automatically adjusted, so that forecasting results will be accurate.

(3) The fuzzy combination forecasting method of self-adaptive variable weight is applied to the sub-grade settlement of the Changzhang expressway, which provides prediction accuracy and reliable applicability.

REFERENCES

[1] Xu Jiayu, Cheng Ka Kui. Road engineering [M]. Shanghai: Tongji University press, 1995.

[2] Highway traffic division. The highway engineering quality problems prevent guide [M]. Beijing: China Communications Press, 2002.

[3] Gu Xiaolu, Qian Hong Jin, et al. The foundation and foundation [M]. Beijing: China Architecture Industry Press, 1993.

[4] Zai Jin min, Mei Kunio. Settlement of the whole process of prediction method [J]. Rock and soil mechanics, 2000, 21 (4): 322–325.

[5] Zhao Minghua, Liu Yu, et al. The soft soil roadbed settlement development law and its forecasting [J]. Journal of Central South University, 2004, 35 (1): 157–161.

[6] PI Dao Ying, Sun Youxian. An algorithm of [J]. Control and decision of multi model adaptive control, 1996, 11 (1): 77–80.

[7] Bates J M, Granger C W J. The ecomination of forecasts [J]. Operations Research Quarterly, 1969 (12): 319–325.

[8] Tang Xiaowo, Wang Jing, et al. A new combination forecasting method of adaptive fuzzy variable weight [J]. Journal of University of Electronic Science and technology of the algorithm, 1997, 26 (3): 289–291.

[9] Yang Guanbiao. Fuzzy mathematics [M]. Guangzhou: South China University of Technology press, 2003.

[10] Tang Ji, Wang Jing. The combination forecasting method on [J]. Prediction, 1999 (2): 42–43.

(2) According to the principle of artificial intelligence and method of fuzzy mathematics, the fuzzy combination forecasting method of self-adaptive variable weight is proposed. Based on matching the degree of the past-recent period and actual

Advances in Civil Engineering and Building Materials IV – Chang et al (eds)
© 2015 Taylor & Francis Group, London, ISBN: 978-1-138-00088-9

Research and application on the prediction method of the Pearl model of high filling subgrade settlement

Xiangxing Kong
The First Highway Survey and Design Institute of China Communications Construction Company Ltd., Xi'an, China

ABSTRACT: According to the development law of high embankment settlement, the Pearl curve model is established. Combined with specific engineering examples, the Pearl curve reflects well the change process of high embankment settlement, and the prediction result of the model with the measured settlement value is very accurate and reliable.

Keywords: prediction method, Pearl model, high filling subgrade, settlement

1 INTRODUCTION

With the construction of expressways in the west of China, high filling and excavation of roadbed structures generally occur [1–2]. After the completed construction of high fill embankments, subgrade settlement often appears, and even causes the destruction of pavement structures and engineering accidents [3]. In order to ensure the safety of the normal operation of highways, according to the law of the development of high filling subgrade settlement observation data and the combined site, the Pearl model is applied to predict the settlement.

2 MECHANISM OF HIGH FILLING SUBGRADE SETTLEMENT

During the early period of construction, just after loading, the soil is in an elastic status [4–5]. In the later period of construction, with an increasing load, the load imposed on the high embankments becomes greater. The pore water in the foundation soil is gradually discharged; meanwhile, excess pore water pressure gradually decreases. Therefore, compaction causes more de-formation, and soil in the embankment will be in an elastic-plastic state. With the continuous development of the plastic zone, the settlement rate increases quickly, until the load no longer increases.

After construction, due to the pore water pressure almost disappearing completely, the consolidation process is not complete. Settlement will continue with the passage of time, but at a significantly decreased sedimentation rate [6–7]. If the time of settlement is infinity, the settlement will be at the ultimate state, and settlement rate is zero; the settlement at this time is the final settlement. In fact, for the highway time takes for 15 years plus filling time.

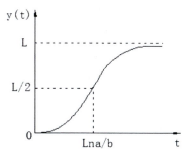

Figure 1. The curve of the pearl model.

3 THE FORECAST MODEL OF A PEARL CURVE

A Pearl curve is sometimes referred to as a 'logistic curve', which is based on the work of American biologist and demographer, Raymond Pearl [8]. Because the curve can reflect a biological growth process, the Pearl curve is widely applied in animal breeding, the development of population statistics and analysis of product life cycles, and so on. The Pearl curve prediction model is expressed as below:

$$y(t) = \frac{L}{1 + ae^{-bt}} \tag{1}$$

where L, a, and b are the three parameters of the model, in which a > 0, and b > 0. Figure 1 is the schematic diagram of a typical Pearl curve.

Seen from Figure 2, the change of the Pearl curve occurs at the point of the concave. The point is the inflection point of the Pearl curve, where the upper and lower halves are around the inflection point of symmetry. In addition, the Pearl curve graph is a long strip of 'S', so it is sometimes referred as the

'S' curve. Curve is relatively flat. The description of the Pearl curve actually reflects the process of things happening; the development, maturity. Because of the change of process and the Pearl curve, the settlement of a high embankment reflects accurately the development process Therefore, the Pearl curve model can reflect the law of development of high filling subgrade settlement, and is applied to predict the settlement.

4 CALCULATION OF PARAMETERS

Using the three parameters of the Pearl curve model method to solve problems [5], requires two things:

(1) the settlement data items in the N time series has to be a multiple of 3, then the calculation time sequence can be divided into 3 segments, each segment containing items;
(2) The time of T is numbered sequentially starting from 1, $t = 1, 2, 3, \ldots, n$. According to the requirement, the number of items in the time series is respectively $(i = 1, 2, \ldots, n)$. The time series is divided into three sections: in the first paragraph, $t = 1, 2, 3, \ldots$; in the second section, R, $t = r + 1$, $r = 2$, and $r + 3, \ldots$; in the third section, $t = 2r + 1$, $2r + 2, 2r + 3, 2R, \ldots$, and $3r$.

$$S_1 = \sum_{t=1}^{r} \frac{1}{y(t)} \tag{2}$$

$$S_2 = \sum_{t=r+1}^{2r} \frac{1}{y(t)} \tag{3}$$

$$S_3 = \sum_{t=2r+1}^{3r} \frac{1}{y(t)} \tag{4}$$

$$\frac{1}{y(t)} = \frac{1}{L} + \frac{ae^{-bt}}{L} \tag{5}$$

$$b = \frac{\ln \frac{(S_1 - S_2)}{(S_2 - S_3)}}{r} \tag{6}$$

$$L = \frac{r}{S_1 - \frac{(S_1 - S_2)^2}{(S_1 - S_2) - (S_2 - S_3)}} \tag{7}$$

$$a = \frac{(S_1 - S_2)^2 (1 - e^{-b}) L}{((S_1 - S_2) - (S_2 - S_3)) e^{-b} (1 - e^{-rb})} \tag{8}$$

So far, the three parameters are calculated by Formula (1), which can be obtained from the Pearl prediction model. The method of corresponding calculation program is compiled with MATLAB.

5 ENGINEERING EXAMPLES

Combined with the subgrade settlement observation data in Hunan province, the Changde Zhangjiajie expressway, the observation data of K103 + 380,

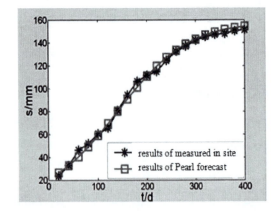

Figure 2. Results of Pearl and forecasting, measured in K103 + 380.

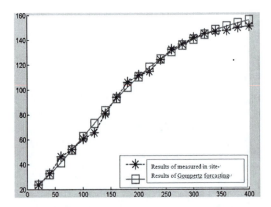

Figure 3. Results of Pearl and forecasting and measured in K123 + 560

K123 + 560 is chosen as the case study. The high fill subgrade filler is typical with low liquid limit clay and shale; the embankment height is about 13 m.

In Figure 3, the time-scattered settlement is 'S'-shaped. The points (20–240 days) data are selected as samples. MATLAB 6.5 is used to calculate the related parameters of the Pearl curve model (as shown in Table 1), then the results of the Pearl curve model for the settlement prediction of the roadbed (as shown in Table 2) are obtained, as well as the sum squared error (prediction precision) 168.13.

6 CONCLUSION

In the forecasting of road foundation settlement, different methods can often provide valuable information.

(1) The settlement of a high embankment has its own law of development, and a Pearl curve is a good reflection of the 'S'-shaped relationship between settlement and the time of the whole process of roadbed settlement; its settlement prediction is accurate and reliable.

Table 1. Results of Pearl and forecasting, measured in K103 + 380.

Time /d	Pearl forecasting/mm	Measured in site/mm	Absolute error/mm
20	26.3828	23.9	2.5
40	32.9468	32.9	0.0
60	40.6426	46.3	5.7
80	49.4299	52.1	2.6
100	59.1656	60.6	1.5
120	69.5986	65.7	3.9
140	80.3868	80.9	0.5
160	91.1385	94.4	3.3
180	101.4672	106.3	4.9
200	111.0450	111.6	0.6
220	119.6399	114.9	4.7
240	127.1288	124.6	2.5
260	133.4881	132.7	0.8
280	138.7712	137.0	1.8
300	143.0809	142.3	0.8
320	146.5447	145.2	1.3
340	149.2954	147.1	2.2
360	151.4591	148.0	3.5
380	153.1483	150.2	2.9
400	154.4594	151.4	3.1

Table 2. Results of Pearl and forecasting, measured in K123 + 560.

Time /d	Pearl forecasting/mm	Measured in site/mm	Absolute error/mm
20	21.48777	21.755	0.2813
40	29.61882	30.305	0.7223
60	38.71763	43.035	4.5446
80	48.45608	48.545	0.0936
100	58.49492	56.62	1.9736
120	68.52094	61.465	7.4273
140	78.27012	75.905	2.4896
160	87.53908	88.73	1.2536
180	96.18579	100.035	4.0518
200	104.1238	105.07	0.996
220	111.3138	108.205	3.2724
240	117.7523	117.42	0.3498
260	123.4631	125.115	1.7388
280	128.4877	129.2	0.7498
300	132.8785	134.235	1.4279
320	136.6934	136.99	0.3122
340	139.9921	138.795	1.2601
360	142.8328	139.65	3.3503
380	145.2708	141.74	3.7166
400	147.3569	142.88	4.7125

Table 3. Parameters of the Pearl method.

L	a	b
158.7895	6.5942	0.2730

(2) A Pearl curve model is used as a basis in forecasting in order to obtain good prediction-sufficient measured data.

(3) The aging situation also exists in the Pearl prediction model, and recent settlements should be added to the original sample, so that the prediction effect is better.

REFERENCES

[1] Xu Jiayu, Cheng Ka Kui. Road engineering [M]. Shanghai: Tongji University press, 1995.
[2] Highway traffic division. The highway engineering quality problems prevent guide [M]. Beijing: China Communications Press, 2002.
[3] Gu Xiaolu, Qian Hong Jin, et al. The foundation and foundation [M]. Beijing: China Architecture Industry Press, 1993.
[4] Zai Jin min, Mei Kunio. Settlement of the whole process of prediction method [J]. Rock and soil mechanics, 2000, 21 (4): 322–325.
[5] Zhao Minghua, Liu Yu, et al. The soft soil roadbed settlement development law and its forecasting [J]. Journal of Central South University, 2004, 35 (1): 157–161.
[6] PI Dao Ying, Sun Youxian. An algorithm of [J]. Control and decision of multi model adaptive control, 1996, 11 (1): 77–80.
[7] Bates J M, Granger C W J. The ecomination of forecasts [J]. Operations Research Quarterly, 1969 (12): 319–325.
[8] Tang Xiaowo, Wang Jing, et al. A new combination forecasting method of adaptive fuzzy variable weight [J]. Journal of University of Electronic Science and technology of the algorithm, 1997, 26 (3): 289–291.

Advances in Civil Engineering and Building Materials IV – Chang et al (eds)
© 2015 Taylor & Francis Group, London, ISBN: 978-1-138-00088-9

Research on thermoelectrics transducers for harvesting energy from asphalt pavement based on seebeck effects

Guoting Liang & Peilong Li
School of Highway, Chang'an University, Xi'an Shanxi, China

ABSTRACT: Asphalt pavement had been a popular practice for several decades. Part of the thermal energies inside the pavement was caused by hot pavement and the temperature gradient. In this paper, the laboratory research and outdoor test on a technique for harvesting energy from asphalt pavement based on seebeck effects was conducted. The distribution density and the depth of burying of the energy harvest system were determined. The durability of the harvester inside the pavement was analyzed by dynamic stability test and the effect of asphalt mixtures and generation chips on the generating efficiency of the harvester was investigated as well. Finally, an energy harvest system based on seebeck effects was developed, which could be installed safely in the pavement structure. And the outdoor test results verify the feasibility of harvesting energy from the temperature gradient inside the pavement, based on the seebeck effects.

Keywords: Asphalt pavement, thermoelectrics transducers, harvesting energy, seebeck effects

1 INTRODUCTION

A global renewable mix with resources and green technologies was proposed for modern civil engineering in recent years. And the energy harvesting technology, which was based on the seebeck (thermal) electromotive force inside the asphalt pavement, may be one of the most promising methods for collecting and transmitting thermal energies into electrical power using the temperature gradient along with the pavement depth. However, the validity and the energy conversion efficiency must be clear in order for this method to be viable for various practical applications. Meanwhile, hot temperature inside the asphalt pavements may to some extent contributes to the urban heat island effect and may also result in the potential of rutting failure in asphalt pavements. Therefore, researches on harvesting energy from asphalt pavement based on seebeck effects were of great theoretical and practical significance.

As Researchers (Wendel 1979) in America proposed a patent which named as 'paving and solar energy system and method', harvesting sunlight energy from the pavement and housetop, a number of studies begin to focus on harvesting the energies in different kinds of industries such as the transportation. Some researchers (Kyono T et al. 2003) in Japan had design a suitable arrangement of thermoelectric modules from the heat transfer theory in the cylindrical heat exchanger. A company in Israel proposed firstly an energy harvest system (Innowattech 2010) installed inside the pavement based on an Inno-wattech Piezo Electric Generator (IPEGTM) in 2008. Some

researchers in China acted as pioneers in proposing the method of harvesting energy from asphalt pavement. Zhao et al. (2010) have compared the performance of several piezoelectric transducers embedded in the pavement structure for harvesting energy from the asphalt pavement. Jiang et al. (2013) had design a novel linear permanent magnet vibration energy harvester.

In view of the researches on energy harvesting technology, this paper extended traditional practice to presents a novel seebeck-based pavement energy harvesting system to collect and transmit energy for the electrical equipment used in transportation infrastructures, such as the traffic light, etc. The burying depth and the distribution density of the seebeck-based pavement energy harvester were determined. And the effect of asphalt mixtures and generation chips on the generating efficiency of the harvester was investigated as well. Last but not least, durability of the harvester inside the pavement was analyzed by dynamic stability test in the laboratory and outdoor test on the generating efficiency of seebeck-based pavement energy harvester was conducted.

2 THE SEEBECK-BASED PAVEMENT ENERGY HARVESTER DESIGN

Fig. 1 schematically depicts the implementation of the proposed seebeck-based pavement energy harvester. In this paper, all the specimens were design as Marshall Specimens and preliminary experiment was conducted in the previous analysis, as depicted in Fig. 2.

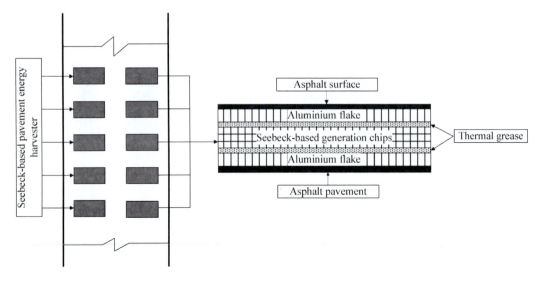

Figure 1. Seebeck-based pavement energy harvesting system.

Figure 2. Specimen design and preliminary experiment.

2.1 Analysis of the burying depth of the harvester in the pavement

Temperature and radiative extinction was existed inside the pavement structure. Temperature gradient was various in different depth of the pavement. And an optimal depth was determined in this paper.

Marshall specimens were designed with different burying depth of 10 mm, 20 mm, 30 mm, 40 mm and 50 mm, as depicted in Fig. 3. All the specimens were heated by infrared heat lamp and the voltage of harvester at various temperatures was measured by digital voltmeter, as depicted in Fig. 4.

The voltage of harvester in different depth at various temperatures was shown in Fig.4. It shown that the seebeck-based pavement energy harvester had well generating efficiency in the 20~30 mm depth. Therefore an optimal depth of the harvester in pavement structure was 20~30mm.

2.2 Analysis of the distribution density of the harvester in the pavement

Since the range of energy being absorbed by one harvester was limited, multiple harvesters should be laid in a given area. And an optimal distribution density was determined in this paper.

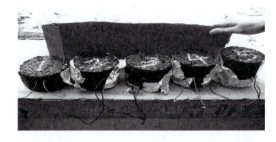

Figure 3. Marshall specimens with harvester in different burying depth.

Various specimens were designed in 30 cm × 30 cm asphalt mixture plate with different distribution density of 1, 4 and 9 harvesters in the same area, as depicted in Fig. 5. And the voltage of harvester at various distribution densities was measured by digital voltmeter, as depicted in Fig. 6.

The voltage of different numbers of harvesters at various temperatures was shown in Fig. 6. It shown that the seebeck-based pavement energy harvester had well generating efficiency at a condition that 4 pieces of harvesters being laid in a given area (30 cm × 30 cm). Therefore an optimal distribution density of the harvester in pavement structure to some extent was 44 pieces per square meter.

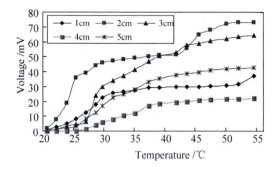

Figure 4. The voltage of harvester in different depth at various temperatures.

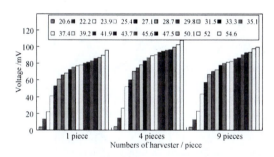

Figure 5. Specimens with harvester in different burying distribution density.

Figure 6. The voltage of harvester with different numbers of harvester at various temperatures.

2.3 The effect of asphalt mixtures on the generating efficiency of the harvester

Since temperature and radiative extinction existed inside the pavement structure were different when the type of asphalt mixtures was various. This paper presented the relationship between asphalt mixtures and the generating efficiency of the harvester, as depicted in Fig.7.

It was shown that the effect of different asphalt mixtures (AC-13, AC-20 and OGFC) on the generating efficiency of the harvester was no different with each other. Therefore, seebeck-based pavement energy harvesting system proposed in this paper could be used in different asphalt mixture pavements.

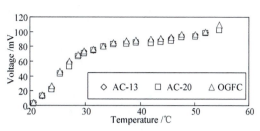

Figure 7. The effect of different asphalt mixtures on the generating efficiency of the harvester.

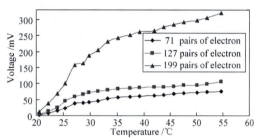

Figure 8. The effect of three kinds of generation chips on the generating efficiency of the harvester.

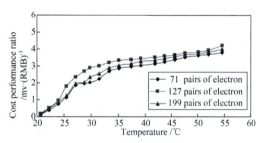

Figure 9. The cost performance of three kinds of generation chips on the generating efficiency of the harvester.

2.4 The effect of generation chips on the generating efficiency of the harvester

Materials required for the road construction was usually very huge, so the cost performance of the harvester was important as well. The effect of three kinds of generation chips (with 71, 127 and 199 pairs of electron) on the generating efficiency of the harvester was investigated, with relative prices of 20, 25 and 80 RMB per piece.

Three kinds of generation chips were buried in AC-13 asphalt mixture specimens, and the results were depicted in Fig.8 and Fig.9. Fig.8 was clearly depicted that generation chip with 199 pairs of electron had the best generating efficiency, but its price was high relatively. Fig.9 was indicated that generation chip with 127 pairs of electron had the highest cost performance ratio. So a harvester made by generation chips with 127 pairs of electron was appropriate for practical engineering purposes.

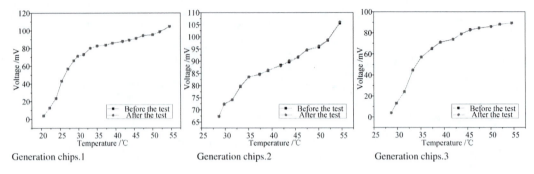

Generation chips.1　　　　　　　　Generation chips.2　　　　　　　　Generation chips.3

Figure 10. Comparison between the generating efficiency of the harvester before the test and those measured after the test.

3 DYNAMIC STABILITY TESTS ON THE HARVESTER INSIDE THE PAVEMENT STRUCTURE

Durability to the traffic was the most basic services for the pavement. The durability of the harvesters inside the pavement structure was analyzed by the dynamic stability tests in the laboratory. The specimens were designed as 30 cm × 30 cm asphalt mixture plate with 3 pieces of harvester at a depth of 20 mm inside the pavement. Dynamic stability tests was conducted an hour at 50°C. After that, the voltage of harvester at various temperatures was measured by digital voltmeter and the results were depicted in Fig. 10, compared with those measured before the test.

It was shown that the generating efficiency of the harvester before the test and those measured after the tests have no different with each other and it was indicated that harvesters at a depth of 20 mm inside the pavement have a good durability.

4 OUTDOOR TEST ON THE GENERATING EFFICIENCY OF SEEBECK-BASED PAVEMENT ENERGY HARVESTER

The generating efficiency of the seebeck-based pavement energy harvesting system under the sunlight was investigated in this paper. The specimen was design as 30 cm × 30 cm asphalt mixture plate with 4 pieces of harvester at a depth of 20 mm inside the pavement. And the specimen was placed under the nature circumstance. Then, the voltage of harvester at various temperatures was measured by digital voltmeter every 15 minutes from 10:25 am to 4:25 pm in a day. Finally, the outdoor test results were depicted in Fig. 11.

It was shown that the voltage approximately reaches a maximum at noon (12:30 pm) and specimen temperature becomes the highest at 14:30 pm. This phenomenon indicated that the generating efficiency of the seebeck-based pavement energy harvesting system had no relationship with the temperature but the temperature gradient. Since the resistance value of generation chips was 0.03 Ω, based on the infinitesimal dividing method, the output of generated electricity

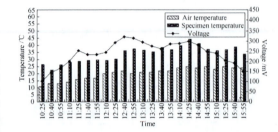

Figure 11. Comparison between the generating efficiency of the harvester, the specimen temperature and air temperature.

during the day for a harvester was 0.00072 kW·h (2592 J).

5 CONCLUSIONS

Based on the seebeck effect, a novel seebeck-based pavement energy harvesting system to collect and transmit energy for the electrical equipment used in transportation infrastructures was proposed in this paper. Relevant research finding of the study were:

(1) Seebeck-based pavement energy harvesting system proposed in this paper could be used in different asphalt mixture pavements.
(2) An optimal depth of the harvester in pavement structure was 20–30 mm.
(3) According to the study, an optimal distribution density of the harvester in pavement structure was 44 pieces per square meter.
(4) Harvesters at a depth of 20 mm inside the pavement have a good durability.
(5) Based on the analysis, a harvester made by generation chips with 127 pairs of electron was appropriate for practical engineering purposes.

REFERENCES

Innowattech, 2010. Innowattech's solution- The innowattech piezo electric generator (IPEGTM). Israel, [2010-09-10]. http://www. innowattech. co. il/index. aspx.

Kyono T, Suzuki R Ono K, 2003. Conversion of unused heat energy to electricity by means of thermoelectric generation in condenser. Transactions on Energy Conversion, 18(1): 330–334.

Wendel I.L., 1979. Paving and solar energy system and method: United States, 4132074.

Hong-duo Zhao, Jian Yu, Jian-ming Ling, 2010. Finite element analysis of cymbal piezoelectric transducers for harvesting energy from asphalt pavement. Journal of the Ceramic Society of Japan, 118(10): 909–915.

Hasebel M, Kamikawa Y, Meiarashi S, 2006. Thermoelectric generators using solar thermal energy in heated toad pavement . Proceedings of 25th International Conference on Thermoelectrics, New York: 697–700.

H. C. Xiong, L. B. Wang, D. Wang, and C. Druta, 2012. Piezoelectric energy harvesting from traffic induced deformation of pavements. Int. J. Pavement Res. Technol. 5(5): 333–337.

X. Z. Jiang, J. Wang, Y. C. Li, and J. C. Li, 2013. Design of a novel linear permanent magnet vibration energy harvester. International Conference on Advanced Intelligent Mechatronics, Wollongong: 1090–1095.

Industry standard of PRC, 2006. Specifications for design of highway asphalt pavement. Ministry of Communications of PRC.

The effect of granular BRA modifier binder on the stiffness modulus of modified asphalt

Muhammad Karami & Hamid Nikraz

Department of Civil Engineering, Curtin University, Perth, Western Australia, Australia

ABSTRACT: This study is aimed to evaluate the effect of using a granular Buton Rock Asphalt (BRA) modifier binder for improving the stiffness modulus performance of asphalt mixtures. The stiffness moduli of BRA modified asphalt mixtures and unmodified asphalt mixtures (as control mixtures) for a dense grading of 10 mm were determined using an Indirect Tensile Stiffness Modulus (ITSM) test at five different temperatures (5, 15, 25, 40, and 60°C), three different rise times (40, 60, and 80 ms), and three pulse repetition periods (1000, 2000, and 3000 ms). Results from this study indicated that stiffness moduli for BRA modified asphalt mixtures were higher than that for control mixtures at any test temperature, rise time, and pulse repetition period. The change in temperature is the biggest influence on changing the stiffness modulus compared to the change in rise time and the pulse repetition period for both asphalt mixtures.

Keywords: Buton Rock Asphalt (BRA), asphalt mixtures, Indirect Tensile Stiffness Modulus (ITSM)

1 INTRODUCTION

The need to increase the capability of asphalt mixtures to withstand increasingly heavy loads as well as to withstand the effect of climate is very important in order to prevent the rapid deterioration of asphalt mixtures. Therefore, the use of modified asphalt mixtures in road structures with the aim to improve the performance and service life of road surfaces is greatly needed.

Asphalt mixtures are formed by bitumen and aggregate. In a certain conditions, such as high temperature and low loading times, bitumen behaves as a viscous material, whereas at low temperature and high loading times, bitumen behaves as an elastic solid (Yilmaz & Celoglu, 2013). Asphalt mixtures behave like bitumen because the main function of the bitumen in the mixtures is as a binder, where the properties of bitumen have a direct and significant effect on the properties of asphalt mixtures. As a result, at a typical traffic speed and temperature, asphalt mixtures almost behave as an elastic material, where the deformation under loading conditions is considered recoverable (Hartman et al., 2001, Ahmedzade & Sengoz, 2009, Mokhtari & Nejad, 2012). Since the binder characteristic is among the factors that affect the elastic stiffness of an asphalt mixture (Tayfur et al., 2007), a stiffness modulus is considered as an important performance characteristic of asphalt mixtures. A stiffness modulus is used as a measure on resistance to bending and the capability to spread the applied load for asphalt mixtures (Hartman et al. 2001).

In order to improve the stiffness modulus of asphalt mixtures, this research considered the use of a granular BRA modifier binder. The form of this modifier binder is granular (pellets), and is composed of two parts, natural binder and fine mineral, which are known together as mastic asphalt. It is well known that mastic asphalt has a favourable influence on the performance of asphalt mixtures (Wang et al, 2011, Silva et al., 2003). Limited study reported about using BRA modifier binder products for improving stiffness modulus and other performances of asphalt mixtures. Affandi (2012) reported that the stiffness modulus is increased and the temperature sensitivity of asphalt mixtures is improved by using BRA modifier binder semi-extraction. Other BRA modifier binder products, for example, filler, were studied and resulted in an increase in the resilient modulus and rutting potential (Subagio et al., 2003), and also in the fatigue performance of asphalt mixtures (Subagio et al., 2005).

The objective of this study is to perform a laboratory investigation on unmodified and BRA modified asphalt mixtures to identify their effect on the stiffness modulus performance.

2 MATERIALS AND METHODS

2.1 *Asphalt binder*

A type of asphalt cement, C-170, from a regional supplier in Western Australia was used as the base

Table 1. Properties of base binder.

Properties	Test Method	Value
Density, kg/L	ASTM D-1298	1.03
Flash point, °C	ASTM D-93	>250
Viscosity at 60°C, Pa.S	ASTM D-445	170
Viscosity at 135°C, Pa.S	ASTM D-445	0.35

Figure 1. The form of granular BRA modifier binder.

Table 2. Particle size distribution of BRA mineral.

Sieve size (mm)	2.36	1.18	0.60	0.30	0.15	0.075
Passing (%)	100	97	92	81	61	36

Table 3. Aggregate gradation used in this study.

| | Dense grading 10-mm, Passing (%) | | | | Lower-upper limit***, Passing (%) |
| Sieve size (mm) | Control* Crushed aggregate | BRA modified** | | | |
		Crushed aggregate	BRA mineral	Final	
13.20	100	100		100	100
9.50	97.5	97.5		97.5	95–100
6.70	83.0	83.0		83.0	78–88
4.75	68.0	68.0		68.0	63–73
2.36	44.0	44.0	100	44.0	40–48
1.18	28.5	28.6	99.9	28.5	25–32
0.60	21.0	21.2	99.8	21.0	18–24
0.30	14.5	15.0	99.5	14.5	12–17
0.15	10.0	11.0	99.0	10.0	8–12
0.075	4.0	5.7	98.3	4.0	3–5
(%)	100	97.4	2.6	100	

*Control asphalt mixtures;
**BRA modified asphalt mixtures;
***Limit values in accordance with Specification 504

asphalt binder for control asphalt mixtures; it is classified according to the Australian Standard AS-2008. The properties of the base asphalt binder are listed in Table 1. In this study, BRA modified asphalt mixtures were made by replacing 20% of the base asphalt binder (by weight of total asphalt binder) with 20% of the natural binder, containing the granular BRA modifier binder, with the purpose of improving the stiffness modulus of asphalt mixtures. The form of BRA modifier binder used in this study was granular (pellets) with a size of 7 to 10 mm in diameter, as shown in Figure 1. Three samples of granular BRA modifier binder were extracted in accordance with Standard WA730.1-2011. These test results showed that the granular BRA modifier binder was composed of 70% mineral and 30% binder by total weight. Table 2 shows the particle size distribution for BRA mineral. Based on a size of less than 4.75 mm, the mineral is considered as fine aggregate and the granular BRA modifier binder can be concluded as mastic binder based on a mineral size of less than 2 mm (Silva et al., 2003).

2.2 Aggregate

The aggregate gradation of dense graded, as shown in Table 3, was used for the control and the BRA modified asphalt mixtures. The gradation had a maximum aggregate size of 9.5 mm in accordance with

Specification 504 used in Australia. One sourced of crushed granite was used for all of the asphalt mixtures. In this study, the aggregate gradation for BRA modified asphalt mixtures below 2.36 mm was adjusted to accommodate the mineral content in the granular BRA modifier binder. The weight of the crushed aggregate was reduced by the BRA mineral in order to minimize the variance of aggregate gradation.

2.3 Mix design and specimen preparation

The Marshall Mix design method was used for determining the optimum binder content for the control asphalt mixtures in accordance with specification 504. Three specimens of 101 mm diameter and 63.5 mm height were produced with compacting energy of seventy-five blows applied to each side. The binder range region was 5.0% to 6.0% with 0.5% increment. The optimum binder content was determined as 5.4% by weight of the mixtures. In this study, the BRA modified asphalt mixtures used the same binder content as the control asphalt mixtures in order to maintain consistency for comparison purposes.

The BRA modified asphalt mixtures were fabricated as follows: 1.5% of hydrated lime (by weight of the total aggregates) was included in blended aggregates and mixed manually (Specification 504). The blended aggregates were then heated in a controlled temperature oven at 160°C for at least 12 hours. The base asphalt binder and the granular BRA modifier binder altogether were put in a bowl and then heated in another oven at 150 ± 5 for 30 min to 1 hour, with frequent manual stirring in order to blend and incorporate the two binders as the BRA modified asphalt binder. The blended aggregates were then put in the same bowl and mixed with the BRA modified asphalt binder for

1.5 minutes at a mixing temperature of 150°C. The mixing temperature for the control and the BRA modified asphalt mixtures were 140°C and 150°C, based on viscosity testing results.

3 INDIRECT TENSILE STIFFNESS MODULUS TEST

The resilient modulus tests were conducted in accordance with the test procedures Australian Standard AS2891.13.1-1995, to evaluate elastic properties in the form of stress-strain measurement. In this study, the Universal Testing Machine (UTM25) was used for the test to determine resilient modulus asphalt mixture in accordance with AS2891.13.1-1995. A gyratory compactor was used to fabricate specimens of 100 ± 2 mm in diameter and between 35 and 70 mm in height, following AS2891.2.2-1995. A number of gyrations were applied to achieve target air voids of $5 \pm 0.5\%$. The compactions were performed at a gyratory angle of 2° and vertical loading stress of 240 kPa. Fifteen specimens were prepared for the control asphalt mixtures, and fifteen specimens were also prepared for the BRA modified asphalt mixtures (three specimens for each level of temperature).

The specimens were placed in the temperature–controlled cabinet at the required test temperature, and the temperature in the specimen was allowed to reach equilibrium before the testing was performed. The load pulse was applied vertically in the vertical diameter of a cylindrical specimen through a curved loading strip. Two Linier Variable Differential Transformers (LVDTs) were attached at the mid thickness at each end of the horizontal diameter to measure horizontal deformation. Initially, the test specimens were conditioned through the application of five load pulses with the specified rise time to the peak load at the specified pulse repetition period, and then the calculation of the modulus was done based on the average of a further five load pulses.

Specimens were conducted with control and BRA modified asphalt mixtures at five different temperature (5, 15, 25, 40, and 60°C), three different rise time (40, 60, and 80 ms), and three different pulse repetition period (1000, 2000, and 3000 ms). During testing, recovered horizontal strain were set to 50 $\mu\varepsilon$. The rise time is the time used for the applied load to increase from 10% to 90% while the pulse repetition period is a distance between 10% applied load in one pulse and 10% applied load in the next pulse. For simulating the volume and speed of traffic, a pulse repetition period of 1000 and 3000 ms was chosen for high and low trafficked volume roads respectively; for vehicle speeds, a rise time of 40 and 80 ms was chosen for high and low speeds respectively (Tayfur et al., 2007).

The average values of the stiffness modulus that were conducted to correlate the three identical specimens using power regression are shown in Table 4. Equations are obtained, where y is the average of the modulus in MPa and x is the rise time in mili-seconds (ms). The coefficient of correlation (R^2) was used to verify the accuracy of the equation.

Table 4. Variation of experimental parameters.

Temp. (°C)	Asphalt mixtures	Pulse rep. period (ms)	Equation	R^2
5	Control	1000	$y = 27692.309x^{-0.155}$	0.898
	Control	2000	$y = 25220.375x^{-0.128}$	0.840
	Control	3000	$y = 22982.688x^{-0.092}$	0.697
	BRA Modified	1000	$y = 34567.427x^{-0.128}$	0.832
	BRA Modified	2000	$y = 31427.543x^{-0.099}$	0.865
	BRA Modified	3000	$y = 30358.117x^{-0.080}$	0.912
15	Control	1000	$y = 21281.490x^{-0.219}$	0.713
	Control	2000	$y = 21235.835x^{-0.230}$	0.729
	Control	3000	$y = 21746.324x^{-0.239}$	0.763
	BRA Modified	1000	$y = 25320.082x^{-0.178}$	0.828
	BRA Modified	2000	$y = 24982.692x^{-0.180}$	0.809
	BRA Modified	3000	$y = 25721.289x^{-0.186}$	0.812
25	Control	1000	$y = 17773.527x^{-0.405}$	0.817
	Control	2000	$y = 13897.884x^{-0.375}$	0.780
	Control	3000	$y = 14391.167x^{-0.393}$	0.788
	BRA Modified	1000	$y = 19080.082x^{-0.288}$	0.929
	BRA Modified	2000	$y = 19812.853x^{-0.316}$	0.914
	BRA Modified	3000	$y = 20146.454x^{-0.328}$	0.853
40	Control	1000	$y = 4152.254x^{-0.508}$	0.939
	Control	2000	$y = 2577.166x^{-0.412}$	0.787
	Control	3000	$y = 2286.208x^{-0.389}$	0.749
	BRA Modified	1000	$y = 6898.153x^{-0.438}$	0.924
	BRA Modified	2000	$y = 5559.128x^{-0.411}$	0.963
	BRA Modified	3000	$y = 5271.487x^{-0.405}$	0.954
60	Control	1000	$y = 348.243x^{-0.221}$	0.585
	Control	2000	$y = 370.539x^{-0.249}$	0.550
	Control	3000	$y = 415.445x^{-0.290}$	0.648
	BRA Modified	1000	$y = 824.259x^{-0.258}$	0.817
	BRA Modified	2000	$y = 726.250x^{-0.237}$	0.762
	BRA Modified	3000	$y = 677.955x^{-0.233}$	0.878

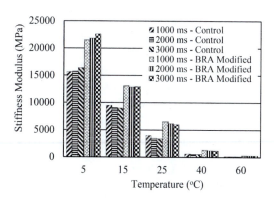

Figure 2. ITSM values for asphalt mixtures at a rise time of 40 ms.

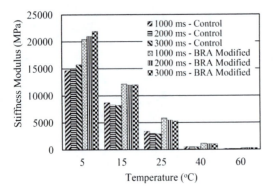

Figure 3. ITSM values for asphalt mixtures at a rise time of 60 ms.

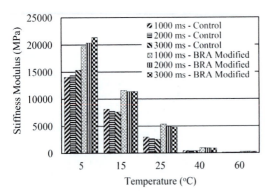

Figure 4. ITSM values for asphalt mixtures at a rise time of 80 ms.

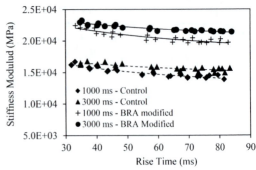

Figure 5. ITSM values for asphalt mixtures at test temperature of 5°C.

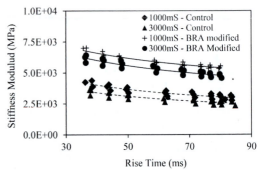

Figure 6. ITSM values for asphalt mixtures at a test temperature of 25°C.

Figures 2 to 4 present ITSM values for mixtures at any level of temperature, rise time and pulse repetition period. The ITSM results for control and BRA modified asphalt mixtures decreased with increasing temperature at all levels of rise time and pulse repetition periods. The ITSM values in BRA modified asphalt mixtures increased and was significantly higher at all levels of temperature, rise time and pulse repetition periods compared to control mixtures.

The ratio of the ITSM values from the BRA modified and the control asphalt mixtures was increased when the temperature increased. At the same rise time and pulse repetition period, the ITSM values of BRA modified were between 38% and 126% higher than that of the control asphalt mixtures.

It can be seen that the ITSM values for both mixtures were affected most by temperature. The ITSM values were decreased as much as 85% and 82% for control and BRA modified asphalt mixtures respectively in between two adjacent test temperatures. At the same rise time and repetition period, the decrease in the value of ITSM for control asphalt mixtures was between 2% and 3% higher than that of BRA modified asphalt mixtures, except for the change in temperature from 40°C to 60°C.

Pulse repetition period also affected the ITSM values. Figure 5 shows that at 5°C, control and BRA modified asphalt mixtures with pulse repetition period of 3000-ms had ITSM values higher than that of asphalt mixtures with pulse repetition period of 1000 ms. ITSM values for both asphalt mixtures at the same level of rise time increased up to 9.4% with increasing pulse repetition period from 1000 to 3000 ms. In contrast, at other test temperatures, the ITSM values for both asphalt mixtures with pulse repetition period of 3000 ms were lower than that of asphalt mixtures with pulse repetition period of 1000 ms. The ITSM values for both asphalt mixtures decreased up to 15.4% with increasing pulse repetition period from 1000 to 3000 ms, as presented in Figure 6.

The ITSM results for control and BRA modified asphalt mixtures decreased with increasing rise time at any level of test temperature and pulse repetition period. The change in rise time from 40 to 80-ms decreased ITSM values up to 29.7% and 26.2% for control and BRA modified asphalt mixtures respectively. The decrease in ITSM values due to the increase in rise time for BRA modified was lower than that for control asphalt mixtures at any test temperature and the same pulse repetition period.

4 CONCLUSIONS

Based on the testing result, the substitution of 20% of a base asphalt binder with a natural binder, containing a granular BRA modifier binder, resulted in an increase in ITSM values for the BRA modified asphalt mixtures, indicating that the cracking resistance for BRA modified mixtures increased.

The ITSM values for the control and BRA modified asphalt mixtures were affected by any level of temperature, rise time and pulse repetition period. However, the ITSM values for both asphalt mixtures were most affected by the temperature. The increase in temperature resulted in a decrease of the ITSM values for both mixtures. The ITSM results for BRA modified were higher than for control asphalt mixtures at any testing level temperature. A higher of test temperature resulted in an increase in the ratio of ITSM values of BRA modified and control asphalt mixtures increased.

The pulse repetition period also affected the ITSM results for control and BRA modified asphalt mixtures. However, the change in the ITSM values due to a change in the pulse repetition period is different for any given temperature. The increase in the pulse repetition period from 1000 ms (high trafficked volume roads) to 3000 ms (low trafficked volume roads), resulted in increase of the ITSM values for both the control and the BRA modified asphalt mixtures at a test temperature of 5°C, while at any other test temperature, the value decreased. Moreover, the increase in the rise time from 40 ms (high speed) to 80 ms (low speed) resulted in decreased ITSM values for both the control and the BRA modified asphalt mixtures at any level of test temperature and pulse repetition period.

ACKNOWLEDEGMENTS

The first author is grateful for financial support from the Directorate of Higher Education (DIKTI) – Indonesian Ministry of Education within the scholarship programme, and for the support received in undertaking this research from the University of Lampung, Indonesia. The authors acknowledge Mr. Arief Santoso of Olah Bumi Mandiri Pty Ltd. for supplying the granular BRA modifier binder for this project. Acknowledgement must also be made of the Curtin Geomechanics and Pavement Laboratory for providing laboratory equipment, aggregate and bitumen materials during the research.

REFERENCES

Affandi F. 2012. The performance of bituminous mixes using Indonesia natural asphalt. *25th ARRB Conference*, Perth, Australia: 1–12.

Ahmedzade P. & Sengoz B. 2009. Evaluation of steel slag coarse aggregate in hot mix asphalt concrete. *Journal of Hazardous Materials* 165: 300–305.

Hartman A.M, Gilchrist M.D. & Walsh G. 2001. Effect of mixture compaction on indirect tensile stiffness and fatigue. *Journal of Transportation Engineering* (127) 5: 370–378.

Main Roads Western Australia (MRWA) standards. 2010. Specification 504 Asphalt Wearing Course: 1–45.

Main Roads Western Australia (MRWA) standards. 2011. WA 730.1-2011: Bitumen Content and Particle Size Distribution of Asphalt and Stabilised Soil-Centrifuge Method: 1–4.

Mokhtari A & Nejad F.M. 2012. Mechanistic approach for fiber and polymer modified SMA mixtures. *Journal of Construction and Building Materials* 36: 381–390.

Silva H, Pais J, Pereira P. & Picardo-Santos L. 2003. Evaluation of the bond between mastic and coarse aggregates. *3rd International Symposium on Maintenance and Rehabilitation of Pavement and Technological Control, Guimaras, Portugal*: 465–474.

Standards Australia. 2008. AS 2008-Residual Bitumen for Pavements.

Standard Australia. 1995. AS 2891.2.2-1995: Method of Sampling and Testing Asphalt-Sample Preparation-Compaction of Asphalt Test Specimens Using a Gyratory Compactor: 1–8.

Standard Australia. 1995. AS 2891.13.1-1995: Methods of sampling and testing asphalt-Method 13.1: Determination of resilient modulus of asphalt-Indirect tensile method: 1–8.

Subagio B.S, Siswosoebrotho B.I. & Karsaman R.H. 2003. Development of laboratory performance of Indonesian rock asphalt (Asbuton) in hot rolled asphalt mix. *Proceedings of the Eastern Asia Society for Transportation Studies* 4: 436–449.

Subagio B.S, Karsaman R.H, Adwang J. & Fahmi I. 2005. Fatigue performance of HRA (hot rolled asphalt) and superpave mixes using Indonesian rock asphalt (asbuton) as fine aggregates and filler. *Journal of the Eastern Asia Society for Transportation Studies* 6: 1207–1216.

Tayfur S, Ozen H. & Aksoy A. 2007. Investigation of rutting performance of asphalt mixtures containing polymer modifiers. *Journal of Construction and Building Materials* 21: 328–337.

Wang H, Al-Qadi I.L, Faheem A.F, Bahia H.U, Yang S.H. & Reinke G.H. 2011. Effect of mineral filler characteristic-son asphalt mastic and mixture rutting potential. *In Transportation Research Record: Journal of the Transportation Research Board* 2208: 33–39.

Yilmaz, M. & Celoglu, M.E. 2013. Effects of SBS and different natural asphalts on the properties of bituminous binders and mixtures. *Journal of construction and Building Materials* 44: 533–540.

Structural engineering

Advances in Civil Engineering and Building Materials IV – Chang et al (eds)
© 2015 Taylor & Francis Group, London, ISBN: 978-1-138-00088-9

Experimental study of the ductility of prestressed concrete frame beams

Feng Gao
School of Civil Engineering and Architecture, University of Jinan, Jinan, China

Shaohong Zhang
Engineering Construction Standards Quota Station of Shandong Province, Jinan, China

ABSTRACT: A Prestressed Steel-Reinforced Concrete (PSRC) structure is a kind of composite structure incorporating merits of both a prestressed concrete structure and a Steel-Reinforced Concrete (SRC) structure. A vertical low-cyclic reversed load experiment was carried out on one PSRC frame with steel sections in columns and another PSRC frame with steel sections in beams and columns; ductility properties of the PSRC frame beam were studied. It is found from the experiment that the hysteresis curve of the PSRC beam is plumper than that of an ordinary prestressed concrete beam, and that the downward displacement ductility coefficient of the PSRC frame beam is 4.77, whereas that of the ordinary prestressed concrete frame beam is 3.92. Therefore, the ductility of a PSRC frame beam is better than that of an ordinary prestressed concrete frame beam. The experimental study demonstrates that a PSRC frame has excellent vertical seismic behaviour and it is completely safe and feasible to apply it to long-span, heavy-load structures, etc.

Keywords: prestressing, steel-reinforced concrete structure, ductility, seismic behaviour, frame.

1 INTRODUCTION

The prestressed steel-reinforced concrete (PSRC) structures are composite prestressed concrete structures for arranging rolled or welded steel sections, mostly applied in large-bay long-span architectural structures such as multi-storey public buildings, bases and tops of tall buildings, garages, etc. The static behaviour of a PSRC structure has been studied worldwide (Xueyu *et al.* 2011, Jun *et al.* 2009, Chuanguo *et al.* 2007), but there are few studies on ductility. The results of a low-cyclic reversed load experiment on four PSRC and SRC simply supported beams (Weichen, 2007) demonstrate that the hysteresis curves of the PSRC beam exhibit a distinct shuttle shape, with good displacement ductility, energy dissipation capacity, and deformation capacity. Liu Bingkang et al. (Bingkang, 2007) conducted an experimental study of two double-span prestressed precast concrete frames which were subjected to a vertical low-cyclic reversed load. This demonstrated that the beam-end sections that resist bending depending only on prestressed reinforcement have plumper hysteresis curves and excellent energy dissipation capacity, without obvious reduction in section bearing capacity when curvature ductility coefficient arrives at 4. Thus, they can meet the requirements of bending moment adjustment. However, there are still no studies of the basic issues of a PSRC frame structure, including ductility under vertical seismic effect, and failure mode in the context of rare earthquake.

2 EXPERIMENTAL DESIGN

Specimens in this experiment are numbered XGKJ1 and XGKJ2 respectively, among which specimen XGKJ1 has steel in frame columns only, and no steel in frame beams; specimen XGKJ2 has steel sections in both beams and columns. The frame columns have a height h = 2100 mm (starting from the top of foundation beam) and a span of 8200 mm. The steel in columns of specimen XGKJ1 are welded I-bars I280 × 120 × 8 × 14 (H × B × t_1 × t_2). The steel sections in columns of specimen XGKJ2 are the same as in specimen XGKJ1, while the steel sections in its beams are I290 × 100 × 8 × 10 (H × B × t1 × t2).; The steel type of the steel sections is Q235 and the focal points at the steel sections of the beams and columns are stud welded with class-8.8 M20 high-strength friction bolts. Two rows of Φ19@200 pegs are mounted on both upper and lower flanges of the steel sections in the columns, and on the upper flange of the steel section at the beam end. The reinforcement distribution and steel section focal point connection of the test structures are shown in Figure 1.

All prestressed reinforcements are 2Φ^s 15.2 (fptk = 1860 N/mm2) high-strength stranded steel wires laid in three parabolic sections of beam length, with inflection points located at 0.1 times beam axis span from respective beam end (see detailed locations in Figure 2). The tensile control stress of a prestressed reinforcement is 0.75 fptk, and tensioned ends adopt one-hole OVM two-piece anchorage, beneath which

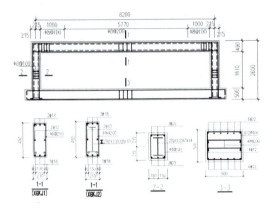

Figure 1. Basic dimensions and detailed reinforcement diagram of specimens XGKJ1 and XGKJ2 (Unit: mm).

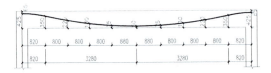

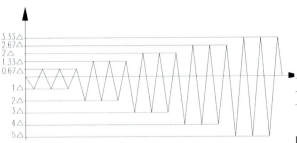

Figure 2. Locations of prestressed stranded steel wire curve for XGKJ1 and XGKJ2.

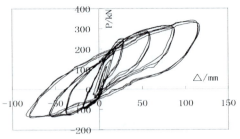

(a)Specimen XGKJ1

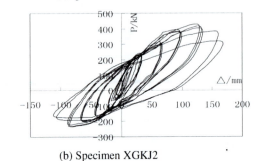

(b) Specimen XGKJ2

Figure 4. Hysteresis loops at mid-span.

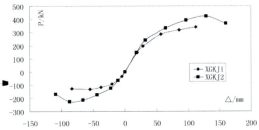

Figure 5. Skeleton curves of load-displacement at midspan.

Figure 3. Vertical loading regime.

pressure sensors are mounted to measure effective tensile force.

The loading was conducted with a hybrid control of load and displacement, in other words, a structural member is controlled by load before yielding, and by displacement at trisection points after yielding. Yield points were determined with a geometric diagramming method (Guo, 2006). Figure 3 is the schematic of the loading regime.

Since the normal load of either of the two specimens is 1.5 times the reverse load, a concrete crushing phenomenon did not occur at both the upper ends and the lower midspan of the specimen beam upon reverse loading. Three or four major cracks formed in the upper ends and the lower midspan of the beam respectively after the specimen yielded; the number and scope of the opening cracks and crack propagation rate of specimen XGKJ2 are smaller than those of XGKJ1, indicating that the presence of the steel section enables effective suppression of crack opening.

3 HYSTERESIS CURVES

The P-hysteresis curves of frame beam mid-span of two specimens are shown in Figure 4.

It can be observed in Figure 4 that:

(1) Before the frame beams of the two specimens crack, hysteresis curves envelop a small area; the P-relationship is basically linear. During cyclic loading, stiffness degradation is insignificant and the specimens remain in an elastic operating state. As cracks in the frame beams of specimens XGKJ1 and XGKJ2 emerge and propagate and the plasticity develops, the hysteresis curves approach the transverse axis.

(2) With regard to the hysteresis curves of specimen XGKJ1, before the steel bars at the ends of the frame beam yield, unloaded concrete in the compressed zone has a small residual deformation and good deformation recovery; the hysteresis curves approach the origin. Affected by prestressed reinforcement, the frame beam has good capability of

Table 1. Results of the main stages of the experiment.

Specimen ID	Loading direction	Initial cracking point		Yield point		Maximum loading point		Failure point		Ductility coefficient $\dfrac{\Delta_u}{\Delta_y}$	Residual deformation Δ_e	Rate of residual deformation $\dfrac{\Delta_e}{\Delta_u}$
		Load/ kN	Δ/ mm	Load/ kN	Δ/ mm	Load/ kN	Δ/ mm	Load/ kN	Δ/ mm			
XGKJ1	Normal	32.8	3.5	152.3	30.0	318.8	114	318.8	114	3.92	32.46	0.29
	Reverse	22.0	2.4	71.6	18.2	125.0	80.8	125.0	80.8	4.44	55.20	0.68
XGKJ2	Normal	40.0	3.6	214.6	33.3	421.1	128.0	366.5	159	4.77	69.67	0.44
	Reverse	28.6	3.2	125.0	22.7	225.1	87.2	168.6	110.0	4.83	68.30	0.62

deformation recovery; the phenomenon of the hysteresis curve being pinched appears when a normal load is applied to the frame beam.

(3) Specimen XGKJ2 does not exhibit the phenomenon of being pinched together, and its hysteresis curves exhibit shuttle shape and are relatively plump, demonstrating an excellent energy dissipation capacity.

4 SKELETON CURVES AND DUCTILITY

Figure 5 shows skeleton curves of the specimens, and Table1 lists results of the main stages of the experiment.

It can be observed in Figure 5 and Table 1 that:

(1) The frame beams of specimens subjected to low-cyclic reversed load undergo elastic stage, yielding stage, and limit stage.

(2) In the case where normal crack displacement and yield displacement of frame beams of both specimens are basically equal, the loads of specimen XGKJ2 rise by 22% and 41% respectively, indicating that the presence of the steel section enhances frame beam stiffness and suppresses propagation and deformation of cracks in the frame beam;

(3) The skeleton curve of specimen XGKJ1 has no load falling stage after arriving at the maximum load, whereas XGKJ2 has a falling load stage after the maximum load is reached, indicating that specimen XGKJ2 has good ductility and its failure is a ductile failure. When the load amounts to about 60% of the maximum load, the longitudinal bars or steel sections in the frame beams of both specimens yield.

(4) The mean of normal and reverse displacement ductility coefficients for frame beams of specimen XGKJ1 is 4.18; the mean of displacement ductility coefficients for frame beams of specimen XGKJ2 is 4.8, all of which demonstrate good displacement ductility. Owing to the prestress effect at the bottom of the midspan of the frame beam, normal ductility coefficients of frame beams of both specimens are smaller than the reverse ductility coefficients.

(5) For frame beams of both specimens, normal residual deformation rates are obviously lower than reverse ones, indicating that the prestress effect enables effective reduction of residual deformation. Residual deformation rates of specimen XGKJ2 are 0.44–0.62, with a deformation recovery capability which is lower than that of specimen XGKJ1.

5 CONCLUSIONS

(1) Under vertical low-cyclic reversed load, the failure mechanism of a PSRC frame is "three-hinge" beam-hinge failure Owing to the "arching effect" of the frame, the limit bearing capacities of the frame beams of both specimens are greatly raised.

(2) A PSRC frame exhibits shuttle-shape and relatively plump hysteresis curves, and has lower stiffness degradation than a prestressed concrete frame; the means of both normal and reverse displacement ductility coefficients are 4.8, demonstrating excellent seismic behaviour and deformation capacity.

ACKNOWLEDGMENT

This study is supported by Research Project of Department of Housing and Urban-Rural Development (2014-K2-029); Technology Project of Shandong Province Construction Department of Urban and Rural Housing (KY027).

REFERENCES

Chuanguo Fu and Shuding Liang, 2007. Experimental study on bending behavior of simply supported prestressed steel reinforced concrete beams. Journal of Building Structures, 28(3), 62–73.

Guo, Z. H. and Shi, X. D, 2006. Reinforced concrete theory and analyses, Tsinghua University Press, Beijing, 336–337.

Jun Wang, Dan Wu, Wenzhong Zheng, 2009. Mechanical behavior of normal section of simply supported prestressed H-steel reinforced concrete beam. Journal of Harbin Institute of Technology, 41(6), 22–27.

Liu, B. K. and Tian, J. F., 2007. Ductility behavior and energy dissipation capacity of prestressed precast concrete frame under low-cyclic reversed load. J. Build. Struct. 28(3), 71–84.

Xue, W. C., Yang, F. Su, X. L. et al. 2007. Experimental study on prestressed steel reinforced concrete beams under low-cyclic reversed load. J. Harbin Inst. Tech., 39(8), 1185–1190.

Xueyu Xiongand Feng Gao, 2011. Experimental investigation and analysis on large scale prestressed steel reinforced concrete frame. Journal of Sichuan University, 43(6), 1–8.

Xueyu Xiong and Feng Gao, 2011. Experimental investigation and crack resistance analysis on large scale prestressed steel reinforced concrete frame. Industrial Construction, 41(12), 16–20.

Xueyu Xiong and Feng Gao, 2011. Experiment and calculation of crack control on prestressed steel reinforced concrete frame beams. Industrial Construction, 41(12), 20–24.

Xueyu Xiong and Feng Gao, 2011. Experiment and calculation of bearing capacity of normal section of prestressed steel reinforced concrete frame beams. Industrial Construction, 41(12), 24–30.

Advances in Civil Engineering and Building Materials IV – Chang et al (eds)
© 2015 Taylor & Francis Group, London, ISBN: 978-1-138-00088-9

The transfer matrix method for the analysis of the steady-state forced vibrations of pier-pile-foundation systems

Yaping Li
School of Civil & Engineering, Central South University, Changsha, China

Meng Yang
Hunan Provincial Architectural Design Institute, Changsha, China

ABSTRACT: The transfer matrix method has been used for analysing the steady-state forced vibrations of pier-pile-foundation system in the paper. Based on the beam's transfer matrix method, the dot matrixes of the spring-damper and the lumped masses have been deduced to analyse the steady-state forced vibrations of simplified model of pier-pile-foundation system. The main contents of this paper include free vibration characteristics of the structural systems, the law of sectional elements change with the exciting forces' frequency, and the vibration attenuation of pier-pile-foundation system under the steady-state forced vibration. Finite element analysis software ANSYS has been used to analyse the harmonic response of the same model to test the validity of the transfer matrix.

Keywords: transfer matrix, pier-pile-foundation system, steady-state forced vibrations, ANSYS.

1 INTRODUCTION

The analysis of the steady-state forced vibrations can predict the dynamic characteristics of structures and avoid resonance, fatigue, and other adverse effects caused by forced vibrations [1].

In the transfer matrix method, the structure is divided into several elements and the state vector of each element transfers from one to the next, and the sectional displacements and inner forces can be calculated by solving the structure's differential equations. The transfer matrix method can be used in computer programming and the data input is simple, while at the same time, the calculation result is accurate [2].

In this paper, the harmonic force was supplied on the top pier of a railway bridge, and the steady-state forced vibration of the pier-pile-foundation system was analysed by the transfer matrix method to obtain the steady-state forced vibration law of the system.

2 THE TRANSFER MATRIX METHOD OF THE BEAM UNDER STEADY-STATE FORCED VIBRATION

2.1 *The field transfer matrix of the beam under steady-state forced vibration*

Take a micro part i from the beam, represent the state vector of the left side of section *i*, and represent the state vector of the right side of section $(i-1)$. Suppose

$$U_i = e^{A_i}$$

$$
A_i =
\begin{bmatrix}
0 & 1 & 0 & \dfrac{EJ}{GSl^2} \\
0 & 0 & -1 & 0 \\
0 & \dfrac{J_m l^2}{EJ}\omega^2 & 0 & 1 \\
-\dfrac{ml^4}{EJ}\omega^2 & 0 & 0 & 0
\end{bmatrix}
+
\begin{bmatrix}
0 & 0 & 0 & 0 \\
0 & 0 & 0 & 0 \\
0 & 0 & 0 & 0 \\
\dfrac{l^4\omega c}{EJ} & 0 & 0 & 0
\end{bmatrix}
\tag{1}
$$

where ω is the frequency of harmonic vibration force, m is the mass per unit length, c is the viscous damping coefficient per unit length, J_m is the moment of inertia, l is the length of beam, and EJ is the flexural stiffness of the beam's section.

Using power series expansion of exponent matrix, then

$$
\begin{aligned}
U_i &\cong I + A_i + A_i^2/2 + A_i^3/3! + A_i^4/4! \\
&= I + A + A_i^2 \times \left[I + A_i/3 + A_i^2/12 \right]/2
\end{aligned}
\tag{2}
$$

U_i is a matrix with complex number, and it can be represented by real part and imaginary part, that is

$$U_i = UR_i + UI_ij$$

Suppose $Z = [Z^r Z^i 1]^T$, where $Z^r = [Y^r, \theta^r, M^r, Q^r]^T$ represent the real part of the state section of the

357

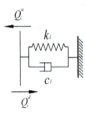

Figure 1. The spring-damper support.

Figure 2. The lumped masses.

beam, and $Z^i = \begin{bmatrix} Y^i & \theta^i & M^i & Q^i \end{bmatrix}^T$ represent the imaginary part. So the transfer relation of the two sides of the micro part i is

$$\begin{bmatrix} Z^r \\ Z^i \\ 1 \end{bmatrix}^L = \begin{bmatrix} UR & -UI & 0 \\ UI & UR & 0 \\ 0 & 0 & 1 \end{bmatrix} \begin{bmatrix} Z^r \\ Z^i \\ 1 \end{bmatrix}^R_{i-1} \quad (3)$$

Analysis of the coefficient matrix in the Eq. (3) is the dimensionless state matrix of the beam element [2].

2.2 The dot transfer matrix of the beam under steady-state forced vibration

Suppose there are concentrated force $Pe^{j\omega t}$ and concentrated moment $Be^{j\omega t}$ supply on the section i, and consider the equilibrium conditions of the section i, then

$$Z^R_i = \begin{bmatrix} I & 0 & RR \\ 0 & I & RI \\ 0 & 0 & 1 \end{bmatrix} Z^L_i$$

$$RR = \begin{bmatrix} 0 & 0 & -\dfrac{B^r l}{EJ} & -\dfrac{P^r l^2}{EJ} \end{bmatrix}^T \quad (4)$$

$$RI = \begin{bmatrix} 0 & 0 & -\dfrac{B^i l}{EJ} & -\dfrac{P^i l^2}{EJ} \end{bmatrix}^T$$

In Eq. (4), I is a identity matrix (4×4), and the square matrix in the Eq. (4) is the dimensionless dot matrix of the beam element [2].

2.3 The dot matrix of spring-damper under steady-state forced vibration

The spring-damper support is shown as Figure 1, and the transfer relation of two sides of the support is

$$Q^d_i = Q^u_i - P_i - m_i\omega^2 v$$

Using a dimensionless matrix to represent the transfer relation of two sides of the support, then

$$\begin{bmatrix} v \\ \theta \\ M \\ Q \\ 1 \end{bmatrix}^d_i = \begin{bmatrix} 1 & 0 & 0 & 0 & 0 \\ 0 & 1 & 0 & 0 & 0 \\ 0 & 0 & 1 & 0 & 0 \\ -\dfrac{(k_i + i\omega c_i)l^3}{EJ} & 0 & 0 & 1 & 0 \\ 0 & 0 & 0 & 0 & 1 \end{bmatrix} \begin{bmatrix} v \\ \theta \\ M \\ Q \\ 1 \end{bmatrix}^u_i \quad (5)$$

The square matrix in the Eq. (5) is the dimensionless dot matrix of the spring-damper support.

2.4 The dot matrix of the lumped masses under steady-state forced vibration

The lumped masses i is shown as Figure 2, and the mass of it is m_i. Suppose the excited force $P_i \cos \omega t$ supplied on it and the transfer relation of two sides of the lumped masses is

$$Q^d_i = Q^u_i - P_i - m_i\omega^2 v$$

Using dimensionless matrix to represent the transfer relation of two sides of the lumped masses, then

$$\begin{bmatrix} v \\ \theta \\ M \\ Q \\ 1 \end{bmatrix}^d_i = \begin{bmatrix} 1 & 0 & 0 & 0 & 0 \\ 0 & 1 & 0 & 0 & 0 \\ 0 & 0 & 1 & 0 & 0 \\ -\dfrac{m_i\omega^2 l^3}{EJ} & 0 & 0 & 1 & -P_i \\ 0 & 0 & 0 & 0 & 1 \end{bmatrix} \begin{bmatrix} v \\ \theta \\ M \\ Q \\ 1 \end{bmatrix}^u_i \quad (6)$$

The square matrix in the Eq. (6) is the dimensionless dot matrix of the lumped masses.

3 THE ANALYSIS OF THE STEADY-STATE FORCED VIBRATIONS OF A RAILWAY BRIDGE'S PIER-PILE-FOUNDATION SYSTEM

3.1 The simplified model for the analysis

Take a railway bridge as the research subject: the height of the pier is 11.6 m, the thickness of the pile cap is 2 m, and the length of pile is 32 m. Soil parameters are shown in Tabale 1. When analysing the steady-state forced vibrations of the pier-pile-foundation system, a virtual pile is used to represent the pile group foundation and the simplified model is shown as in Figure 3. Elastic coefficient of the top pier's elastic constraint $k_d = 3.0 \times 10^6$ kN/m [3]. The lumped masses on the top pier are 350 ton (i.e. the dead load of the upper beam). The spring-dampers are used to simulate the interaction between pile and soil. The stiffness coefficient k_w and damping coefficient c_w of spring-damper [4] are shown in Table 1.

Table 1. Soil parameters, the stiffness coefficient and damping coefficient of spring-dampers.

	Thickness of soil/m	Density ρ_s/kg.m^{-3}	The shear wave velocity v_s/m.s^{-1}	Poisson's ratio λ	The stiffness coefficient k_i/N.m^{-1}	The damping coefficient c_i/N.s.m^{-1}
Artificial fill	7	1900	120	0.4	8.417e8	9.883e6
Coarse sand	12	1600	150	0.3	2.024e9	8.658e6
Strong-weathered rock	12	2400	300	0.2	1.167e10	2.382e7
Moderately weathered rock	4	2400	500	0.2	1.080e10	3.970e7

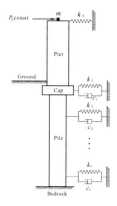

Figure 3. The simplified model.

Table 2. The natural frequencies of the system and the relative errors.

Natural frequency	1	2	3	4
The transfer matrix method	7.8	9.3	28.9	41.9
ANSYS	7.2	9.1	28.7	42.2
Relative errors (%)	8.9	2.7	0.5	0.7

3.2 The analysis results of the steady-state forced vibrations of the pier-pile-foundation system

Suppose the harmonic force in transverse direction supplied on the top pier, the amplitude of which is 62.95 KN [5], and the range of frequency is $\rho_s v_s \lambda$ 1–50 Hz. The transfer matrix method has been used to calculate the sections' vibration displacement, rotation, bending moments, and shear force when the pier-pile-foundation system was under the harmonic force. It also obtained the law of the sections' vibration displacement, rotation, bending moments, and shear force changes with the frequency of load. In order to test the validity of transfer matrix method, the harmonic response of the same model has also been carried out by using ANSYS. Figure 4 shows the section displacement and inner force of top pier change with the frequency of harmonic force. Figure 4 also indicates that the free vibration characteristics of the system, calculated by the transfer matrix

method, are close to the results calculated by ANSYS. The first four natural frequencies of system and the relative errors calculated by two methods are shown in the Table 2.

3.3 The vibration attenuation of pier-pile-foundation system under the steady-state forced vibration

A harmonic force is supplied on the top pier, the amplitude of it is 62.95 KN and the frequency is 3 Hz. Using the transfer matrix method and ANSYS to carry on the harmonic analysis to the pier-pile-foundation system and obtain the vibration displacement, rotation, bending moments and shear force on several sections, including the top pier, the top pile cap, and piles, etc. The vibration attenuation of the pier-pile-foundation system is shown as Figure 5. In the Figure 5, the origin of y-axis is the section of the cap top. Figure 5 indicates the vibration displacement and rotation decrease from the top pier to the bottom of pile, while bending moments increase from the top to bottom. The shear force changes on the sections where spring-damper exist. The vibration displacement, rotation, bending moments and shear force calculated by transfer matrix method are less than ANSYS, because ANSYS cannot consider the nonlinear factors of pile-soil interaction (i.e. it only considers the elastic of spring and neglect the damping of the spring-damper). Damping will not affect the vibration period and mode of vibration of structure [6], however, it affect the structure's displacement and inner force. Damping can dissipate vibration and help to reduce structure vibration.

4 CONCLUSION

The transfer matrix method and ANSYS have been used for analysing the steady-state forced vibrations of pier-pile-foundation system in the paper. The natural frequencies of system calculated by two methods are similar, indicating the correct of the dot matrixes of spring-dampers and the lumped masses deduced in the paper.

The sections' vibration displacement, rotation, bending moments, and shear force change with the frequency of load when the pier-pile-foundation system under the steady-state forced vibrations. It is clear that vibration attenuates from the top pier to the bottom of pile.

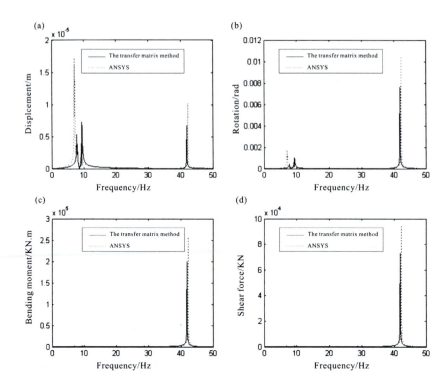

Figure 4. The inner force and displacement on top pier section change with frequency of exciting force (a) Transverse displacement: (b) Section rotation: (c) bending moment: (d) shear force.

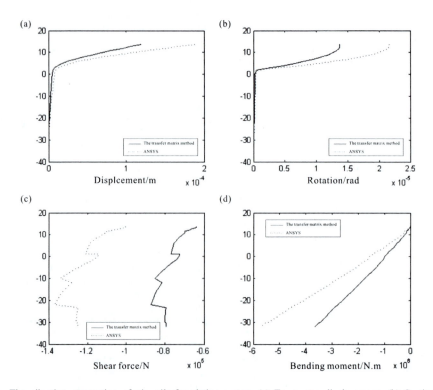

Figure 5. The vibration attenuation of pier-pile-foundation system (a) Transverse displacement: (b) Section rotation: (c) bending moment: (d) shear force.

Damping will not affect the vibration period and mode of vibration of the structure system, however, it will affect the structure's displacement and inner force.

Through the transfer matrix method and taking damping into consideration, more correct results can be obtained for analysing the steady-state forced vibrations of a pier-pile-foundation system.

REFERENCES

[1] Gu Kouxiu. Encyclopedia of Vibration Engineering. *Machinery Industry Press,* 1983.

[2] Xue Hujue. Exact transfer matrix method for computing stable forced vibration of changeable cross-section beams. *Journal of Suzhou University (Natural Science),* 2002, 18:49–54.

[3] Luo Rudeng, Ye Meixin, Mo Chaoqing. The comparison study on valuing method of the stiffness on the direction of horizontal static constraint of support in seismic finite element analysis on bridges. *Journal of railway science and engineering,* 2008, 2(5):23–28.

[4] Liu Dong-jia, Wang Jian-guo. Winkler parameters k and c for transient lateral vibrating piles. *Rock and Soil Mechanics,* 2003, 24(6):922–926.

[5] Li Yaping, Yang Meng. Identification of the dynamic load supplied on top pier of high speed railway bridge. *Advances in Civil Engineering and Building Materials III,* 2014:348–353

[6] Skinner R I, Robinson W H. Introduction to vibration isolation of engineering. *Beijin: Earthquake Press,* 1996, 438.

Advances in Civil Engineering and Building Materials IV – Chang et al (eds)
© 2015 Taylor & Francis Group, London, ISBN: 978-1-138-00088-9

Cable truss topology optimization for prestressed long-span structure

V. Goremikins, K. Rocens, D. Serdjuks & L. Pakrastins
Riga Technical University, Riga, Latvia

N. Vatin
St. Petersburg State Polytechnical University, St. Petersburg, Russia

ABSTRACT: A suspension bridge is one of the most rational structural solutions for long-span bridges because its main load carrying elements are under tension. However, on the other hand, increased deformability, which is conditioned by the appearance of elastic and kinematic displacements, is the main disadvantage of suspension bridges.

The problem of increased kinematic displacements under the action of a non-symmetrical load can be solved by prestressing. A prestressed suspension bridge with a span of 200 m was considered as an object of investigation. A cable truss with a cross web was considered as the main load carrying structure of the prestressed suspension bridge. All elements of the cable truss were tensioned.

The considered cable truss was optimized by 47 variable factors using genetic algorithms and the FEM program ANSYS.

It was stated that the maximum total vertical displacements are reduced up to 29.9 per cent by using the cable truss with the rational characteristics instead of a single cable in case of the worst situated load.

Keywords: cable truss, kinematic vertical displacements, non-symmetrical load, genetic algorithm.

1 INTRODUCTION

Limited raw materials and energy resources are actual problems. A decrease in weight and an increase in span and durability of load carrying structures are possible solutions to these problems. The increase of structural efficiency could be achieved by applying the tensioned structures, where stress distribution by cross-section is close to uniform. Therefore, a decrease of dead weight for long-span structures, such as roofs and bridges, is actual (Goremikins 2013a).

The largest structural spans were achieved by application of structures, where the main load carrying elements are tensioned. The hierarchic cable roof, which allows covering of extremely big spans without intermediate supports, is an example of such structures (Figure 1) (Pakrastinsh et al. 2006). The hierarchic structures are formed by the saddle-shaped cable roofs by suspending the separate corners of the roof to higher level cable structures. Determination of rational geometric parameters of higher level cable structures and parameters of separate saddle-shaped cable roof allows increasing structural efficiency of the whole hierarchic structures (Pakrastinsh et al. 2006, Serdjuks & Rocens 2008a). Rational structure in this case means the structure with the minimal material consumption. The long-span bridges is another group of structures, which are of special interest due to their ability to

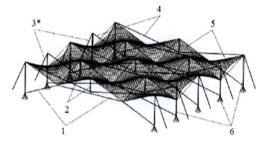

Figure 1. Simplified variant of hierarchic cable structure module. 1 – Bottom level tie net, 2 – Vertical posts, 3 – Guyropes, 4 – Higher level cable structures, 5 – Primary saddle-shaped elements, 6 – Supports. *Guyropes in longitudinal direction of module have not been conventionally displayed (Pakrastinsh et al. 2006).

cover extremely long spans, which can be measured in hundreds and thousands of meters. The maximum structural span achieved for a bridge is equal to 1991m (Chen & Duan 2000).

A suspension bridge is the most suitable type of prestressed long-span structure. Long spans can be achieved because the main load carrying cables are subjected to tension, and distributions of normal stresses in the cable cross-section are close to uniform (Juozapaitis et al. 2010).

However, on the other hand, increased deformability, which is conditioned by the appearance of elastic and kinematic displacements, is the main disadvantage of structures, where main load carrying elements are tensioned, namely suspension bridges (Walther et al. 1999).

Therefore, methods for a decrease in deformability of suspension bridges must be analysed. Rational parameters of the main load carrying elements of the suspension bridge should be evaluated. Rational structure in this case and in the further text means structure with minimal vertical displacements.

2 DECREASE OF SUSPENSION BRIDGE DEFORMABILITY

Suspension bridges are structures where the deck is continuously supported by the stretched catenary cable (Chen & Lui 2005). Suspension bridges are the most important and attractive structures possessing a number of technical, economical, and aesthetic advantages (Grigorjeva et al. 2010).

The kinematic displacements of suspension bridges are caused by the initial parabolic shape change, resulting from non-symmetrical or local loads (Juozapaitis & Norkus 2005, Kadisov G.M. & Chernyshov V.V.2013). These displacements are not connected with the cable elastic characteristics. A serviceability limit state dominates for suspension cable structures.

The elastic displacements can be reduced by applying low strength steel structural profiles, increasing elastic modulus, reinforcing concrete application, and increasing the cable camber (Kirsanov 1973).

The problem of increased kinematic displacements can be solved by increasing dead weight and imposed load relation, which is achieved by the addition of cantledge. However, this method causes the increase of material consumption. The stiffness of the suspended structure can be increased also by increasing the girder stiffness, increasing the main cable camber, connecting the main cable and girder at the centre of span, applying diagonal suspenders or inclined additional cables, and applying two chain systems, stiff chains and stress ribbons (Kirsanov 1973, Strasky 2005, Bahtin et al. 1999, Kacurin et al. 1971). Nevertheless, these systems are characterized also with an increased material consumption, and the system stiffness is not sufficient in many cases.

The use of prestressed cable truss is another method for solving the problem of increased kinematic displacements under the action of an unsymmetrical load (Serdjuks & Roens 2004, Goremikins et al. 2011). Several types of cable trusses are used for bridges, but the concave cable truss is the most efficient and convenient (Goremikins et al. 2011, Schierle 2012). Cable truss usage allows for bridges with reduced requirements for girder stiffness, where overall bridge rigidity will be ensured by prestressing of the stabilization cable (Kirsanov 1973). The deck can be made of light composite materials (Goremikins et al. 2012b,c). The kinematic

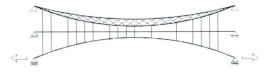

Figure 2. Suspension bridge stabilization by using a prestressed stabilization cable and cable truss.

displacements of a prestressed suspension bridge can be decreased by replacing the main single cable with a cable truss with a cross web (Fig. 2) (Goremikins et al. 2012a,b,c).

Cables can be cambered in a horizontal plane to increase structure stiffness in the same plane (Kirsanov 1973).

Decrease of displacements can be achieved by rational positioning of cable truss elements and rational material distribution between them. Topology optimization of the cable truss web is presented in this paper.

3 DESCRIPTION OF THE INVESTIGATIONS OBJECT

A cable truss with a cross web, which is a main load carrying element of prestressed suspension bridge, was considered as an object of investigation (Figure 4) (Goremikins et al. 2012a).

The main span l of the considered bridge is equal to 200 m. The distances from the top of the pylon and from the connection of the stabilization cable up to the deck are equal to 21 m and 11 m, respectively. The bridge has two lines in each direction, two pedestrian lines and their total width is equal to 18.2 m (Figure 3). The chamber of the cable truss bottom chord f_b is equal to 20 m. The bridge is prestressed in horizontal and vertical planes by the stabilization cables. The stabilization cable camber is equal to 10 m. The deck is connected with the main load carrying cables by the suspensions with step a equal to 5 m (Figure 4). The cable string is placed between suspensions to minimize the horizontal prestressing force effects acting in the deck. Prestressed horizontal cables are placed along the deck to minimize the effects of the horizontal braking force. The deck of the bridge is made of pultrusion composite trussed beams, pultrusion composite beams with a step of 1 m, and a pultrusion composite plank with a height of 40 mm that is covered with an asphalt layer (Fiberline Composites A/S 2002, Goremikins et al. 2010). It is assumed that cables are covered with high density polyethylene and are heated with electricity to reduce the influence of temperature effects (Xiang et al. 2009). Possible loss of prestressing is reduced by active tendons (EN3-1-11). It is possible to reduce requirements for girder stiffness by bridge prestressing. This aspect allows the use of composite pultrusion materials in the deck structure and makes it possible to develop construction of bridges with large

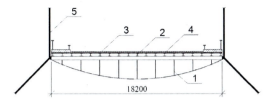

Figure 3. The bridge deck structure. 1 – Composite trussed beam, 2 – Composite I type beams, 3 – Composite plank, 4 – Cover of the bridge, 5 – Suspensions.

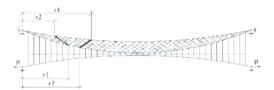

Figure 4. Design scheme of suspension bridge. q – imposed load, g – dead load, P – prestressing, fb – bottom chord camber, ft – top chord camber, l – main span, b – width, a – suspension step.

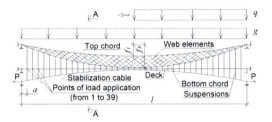

Figure 5. Position of web elements.

spans and reduced dead weights in comparison with steel or concrete bridges (Bahtin et al. 1999).

A design scheme of the investigation object is shown in Figure 4. The structural material is pre-stressed steel rope (Fiberline composites A/S 2002, Feyrer 2007). The dead load g that is applied to the structure is equal to 51.1 kN/m. The bridge is loaded by the imposed load q, which is equal to 82.2 kN/m (EN1-2). An imposed load can be applied to any place of the span. The distributed load is reduced to the point load and is applied to the connections of the deck and suspensions. There are 39 possible points of load application (Figure 5).

The position of each web element of the cable truss is defined by the distance from the pylon to the connection of the web element with the top chord, depending on the distance from the pylon to the connection of the same element with the bottom chord (Figure 5). The web elements are divided into two groups – the elements inclined to the centre of the cable truss and the elements inclined to the edges of the cable truss. Each element of the web may have its own angle on

inclination. The second order polynomial equation is assumed to express the position of each web element and to minimize the amount of variable factors.

The position of the web elements, which are inclined to the edges of the cable truss is expressed by:

$$x_2 = x_1 - (root1 \cdot x_1^2 + root2 \cdot x_1 + root3) \quad (1)$$

The position of the web elements, inclined to the edges of cable truss, is expressed by:

$$x_4 = x_3 + (root4 \cdot x_3^2 + root5 \cdot x_3 + root6) \quad (2)$$

where x_2 and x_4 – distances from the pylon to the connection of web element and top cord; x_1 and x_3 – distances from the pylon to the connection of web element and bottom cord; root1…root6 – roots of the system of equations:

$$\begin{cases} s_1 = root1 \cdot a_1^2 + root2 \cdot a_1 + root3 \\ s_2 = root1 \cdot a_2^2 + root2 \cdot a_2 + root3 \\ s_3 = root1 \cdot a_3^2 + root2 \cdot a_3 + root3 \end{cases} \quad (3)$$

$$\begin{cases} s_4 = root4 \cdot a_1^2 + root5 \cdot a_1 + root6 \\ s_5 = root4 \cdot a_2^2 + root5 \cdot a_2 + root6 \\ s_6 = root4 \cdot a_3^2 + root5 \cdot a_3 + root6 \end{cases} \quad (4)$$

where s_1 – distance between x_1 and x_2 if $x_1 = a_1$;
s_2 – distance between x_1 and x_2 if $x_1 = a_2$;
s_3 – distance between x_1 and x_2 if $x_1 = a_3$;
s_4 – distance between x_3 and x_4 if $x_3 = a_1$;
s_5 – distance between x_3 and x_4 if $x_3 = a_2$;
s_6 – distance between x_3 and x_4 if $x_3 = a_3$;
a_1 – distance from the pylon to the connection of the first web element with the bottom chord, counting for the middle of the span;
a_2 – distance from the pylon to the connection of the middle web element with the bottom chord, counting for the middle of the span;
a_3 – distance from the pylon to the connection of the last web element with the bottom chord, counting for the middle of the span.

Distribution of the material among the cable truss elements can be expressed by:

$$g = g_b + g_t + g_w$$

$$g_w = \sum_{i=01}^{39} g_{w,i} \quad (5)$$

$$g_t = g - g_b - g_w$$

where g – material consumption of the cable truss; g_b – material consumption of the bottom chord; g_t – material consumption of the top chord; g_w – material consumption of all web elements; $g_{w,i}$ – material consumption of i-th web element.

The web elements of the cable truss, which are inclined to the supports of the cable truss, are numbered from 1 to 20, starting from the support. The web elements are numbered from 21 to 39, starting from the support.

365

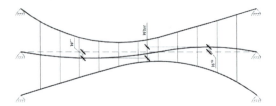

Figure 6. Deformed shape of prestressed suspension bridge in non-symmetrical loading case.

4 TOPOLOGY OPTIMIZATION OF CABLE TRUSS WEB

4.1 Definition of optimization problem and optimization method

The aim of optimization is to evaluate rational characteristics of the cable truss for the prestressed suspension bridge.

The bottom chord camber f_b, the material consumption of cable truss g, the material consumption of stabilization cable, level of prestressing, and the bridge geometrical parameters: pylon height, main span, and suspension step, are considered as constants of the optimization.

The relation of the top and bottom chord cambers f_t/f_b, the distances s_1, s_2, s_3, s_4, s_5, s_6, the rations g_b/g and $g_{w,1}/g - g_{w,39}/g$ are variable factors for the optimization; 47 factors in all.

The optimization problem is to minimize the objective function:

$$w_{tot}\left(s_1, s_2, s_3, s_4, s_5, s_6, \frac{g_b}{g}, \frac{g_{w.01}}{g}, ..., \frac{g_{w.39}}{g}, \frac{f_t}{f_b}\right) \quad (6)$$

subject to:

$$[K(U)]\cdot\{U\} = [F(U)] \quad (7)$$

and Equations (1)-(6), where $[K(U)]$ is the stiffness matrix, $\{U\}$ is the displacement vector, and $[F(U)]$ is the force vector.

The total displacements w_{tot} are found by summing displacements upwards w^+ and displacements downwards w^- (Figure 6). The maximum vertical displacements for suspended cable structures appears under the action of load applied to different parts of the span, therefore 39 different loading cases were analysed. The problem has to be solved in static and in non-linear stages.

The optimization of the cable truss by 47 variable factors is done by genetic algorithm (Goremikins et al. 2012a, Lute et al. 2009, Šešok & Belevičius 2008)

The genetic algorithm is a method for solving both constrained and unconstrained optimization problems that are based on natural selection, the process that drives biological evolution. The genetic algorithm repeatedly modifies a population of individual solutions. At each step, the genetic algorithm selects individuals at random from the current population to be parents and uses them to produce the children

for the next generation. Over successive generations, the population "evolves" towards an optimal solution. Genetic algorithms are used to solve a variety of optimization problems that are not well suited for standard optimization algorithms, including problems in which the objective function is discontinuous, non-differentiable, stochastic, or highly nonlinear (Math Works 2001).

The genetic algorithm uses three main types of rules at each step to create the next generation from the current population:

– Selection rules select the individuals, called parents, which contribute to the population at the next generation.
– Crossover rules combine two parents to form children for the next generation.
– Mutation rules apply random changes to individual parents to form children (Math Works 2001).

The GA Toolbox of mathematical software MATLAB was used in the optimization. A special program was written in a MATLAB programming environment to calculate fitness using FEM. FEM program ANSYS was used to calculate the displacements of the suspension bridge. Specially written MATLAB function called ANSYS and ANSYS returned to MATLAB vertical displacements. The cable truss is modelled by two-node link type compressions less finite elements (LINK10 in ANSYS). The analysis type is geometrically nonlinear static including large-deflection effects, because suspension cable structures are characterized with large deflections before stabilization (EN3-1-11, Semenov et al. 2005).

4.2 Rational characteristics of cable truss

Topology optimization with the genetic algorithm was realized, using 40 generations, the population size was equal to 100, the elite child number was equal to 10. The rational characteristics of the cable truss were evaluated. It was stated, rational relation of the top chord camber/the bottom chord camber and material consumption of the bottom chord/material consumption of the whole truss are equal to 0.48 and 0.40, respectively. The rational values of the distances, which define position of web elements s_1, s_2, s_3, s_4, s_5 and s_6, are equal to 4.95, 19.98, 17.80, 0.77, 15.14, and 18.25 m, respectively. The rational relation of material consumption of the bottom chord and the whole truss is equal to 0.40. The rational relations of material consumption of the separate web elements and the whole truss changes within the limits of 0.00039 to 0.00489. Optimized structure by 47 variable factors was compared with optimized structure by 9 variable factors. It was determined, that displacements of the cable truss optimized by 47 variable factors are smaller by 4.5 per cent, compared with displacements of the cable truss, optimized by 9 variable factors.

The displacements of the prestressed suspension bridge with the rational cable truss and a single cable were compared. The material consumption of the cable truss was the same as the material consumption of the single cable. The analysis were carried out by the FEM software ANSYS.

The maximum total displacements are reduced by up to 29.9 per cent by using the cable truss instead of the single cable in the case of the worst situated load.

5 CONCLUSIONS

A possibility to decrease deformability of a prestressed suspension bridge is provided. Rational structure of the cable truss was developed. Rational relation of the top chord camber/the bottom chord camber and the material consumption of the bottom chord/material consumption of the whole truss are equal to 0.48 and 0.40, respectively. The topology optimization of the cable truss web was realized. It was stated that displacements of the cable truss optimized by 47 variable factors are smaller by 4.5 per cent, compared with displacements of the cable truss, optimized by 9 variable factors.

The maximum total displacements are reduced by up to 29.9 per cent by using the cable truss instead of the single cable in the case of the worst situated load.

REFERENCES

Achkire, Y.A. & Preumont, A. 1996. Active tendon control of cable-stayed bridges, *Earthquake Engineering and Structural Dynamics*, 25(6): 585–597.

Bahtin, S.; Ovchinnikov, I. & Inamov, R. 1999. *Visjačije i vatovje mosti*. Saratov: Saratov state technical university.

Chen, W. F. & Duan, L. 2000. *Bridge engineering handbook*. New York: CRC Press.

Chen, W. F. Lui, E. M. 2005. *Handbook of structural engineering*. New York: CRC Press.

European Committee for Standardization. 2003. *Eurocode 3: Design of steel structures – Part 1–11: Design of structures with tensile components*, Brussels.

European Committee for Standardization. 2004. *Eurocode 1: Actions on structures – Part 2: Traffic loads on bridges*, Brussels.

Feyrer, K. 2007. *Wire ropes*. Berlin: Springer-Verlag Berlin Heidelberg.

Fiberline composites A/S. 2002. *Design Manual*, Middelfart: Fiberline Composites A/S.

Goremikins, V. 2013. *Rational large span prestressed cable structure. Doctoral Thesis*. Riga: RTU.

Goremikins, V.; Rocens, K. & Serdjuks, D. 2010. Rational Large Span Structure of Composite Pultrusion Trussed Beam, *Scientific Journal of RTU. Construction Science* 11: 26–31.

Goremikins, V.; Rocens, K. & Serdjuks, D. 2011. Rational structure of cable truss, *World Academy of Science, Engineering and Technology*, 76: 571–578.

Goremikins, V.; Rocens, K. & Serdjuks, D. 2012a. Decreasing displacements of prestressed suspension bridge, *Journal of Civil Engineering and Management* 18(6): 858–866.

Goremikins, V.; Rocens, K. & Serdjuks, D. 2012b. Decreasing of displacements of prestressed cable truss, *World Academy of Science, Engineering and Technology* 63: 554–562.

Goremikins, V.; Rocens, K. & Serdjuks D. 2012c. Cable truss analyses for suspension bridge, *in Proc. of 10th International Scientific Conference Engineering for Rural Development, 24–25 May, 2012, Jelgava, Latvia*: 228–233.

Goremikins, V.; Rocens, K. & Serdjuks, D. 2012d. Analysis of hybrid composite cable for prestressed suspension bridge, *in Proc. of the 17th International Conference Mechanics of Composite Materials, 28 May–1 June, 2012, Riga, Latvia*: 93.

Grigorjeva, T.; Juozapaitis, A. & Kamaitis, Z. 2010. Static analyses and simplified design of suspension bridges having various rigidity of cables, *Journal of Civil Engineering and Management*, 16(3): 363–371.

Juozapaitis, A. & Norkus, A. 2004. Displacement analysis of asymmetrically loaded cable, *Journal of Civil Engineering and Management*, 10(4): 277–284.

Juozapaitis, A.; Idnurm, S.; Kaklauskas, G.; Idnurm, J. & Gribniak, V. 2010. Non-linear analysis of suspension bridges with flexible and rigid cables, *Journal of Civil Engineering and Management*, 16(1): 149–154.

Kadisov G.M. & Chernyshov V.V. 2013. *Konechno-elementnoye modelirovaniye dinamiki mostov pri vozdeystvii podvizhnoy nagruzki. Magazine of Civil Engineering*. No. 9(44): 56–63.

Kačurin, V.; Bragin, A. & Erunov, B. 1971. *Projektirovanije visjačih i vantovih mostov*. Moskva: Transport.

Kirsanov, N. 1973. *Visjačije sistemi povišennoi žestkosti*. Moskva: Strojizdat.

Kirsanov, N. 1981. *Visjačie i vantovie konstrukciji*. Moskva: Strojizdat.

Lute, V.; Upadhyay, K. & Singh, K. 2009. Computationally efficient analysis of cable-stayed bridge for GA-based optimization, *Engineering Applications of Artificial Intelligence* 22: 750–758.

MathWorks 2011. *MATLAB User's manual. What Is the Genetic algorithm?* MathWorks.

Pakrastinsh, L.; Rocens, K. & Serdjuks, D. 2006. Deformability of hierarchic cable roof, *Journal of Constructional Steel Research* 62: 1295–1301.

Schierle, G. G. 2012. *Structure and Design*. San Diego: Cognella.

Semenov, A.S.; Melnikov, B.E. & Gorokhov, M.Y. 2005. About the causes of cyclical instability at computations of large elasto-plastic strains. *In Proc. of SPIE – The International Society for Optical Engineering*: 167–173.

Serdjuks, D. & Rocens, K. 2004. Decrease the displacements of a composite saddle-shaped cable roof, *Mechanics of Composite Materials* 40(5): 675–684.

Serdjuks, D. & Rocens, K. 2008a. Hybrid composite cable with steel component as a structural element, *An International Journal Advanced Steel Construction* 4(3): 184–197.

Serdjuks, D.; Rocens, K. & Pakrastins L. 2008b. Hybrid Composite cable with an increased specific strength for tensioned structures, *The Baltic Journal of Road and Bridge Engineering* 3(3): 129–136.

Strasky, J. 2005. *Stress ribbon and cable supported pedestrian bridge*. London: Thomas Telford Publishing.

Straupe, V. & Paeglitis, A. 2012. Analysis of interaction between the elements in cable stayed bridge. *The Baltic Journal of Road and bridge Engineering* 7(2): 84–91.

Šešok, D. & Belevičius, R. 2008. Global optimization of trusses with a modified Ggenetic Aalgorithm. *Journal of Civil Engineering and Management*, 14(3): 147–154.

Walther, R.; Houriet, B.; Isler, W.; Moia, P. & Klein, J. F. 1999. *Cable Stayed Bridges. Second Edition*. London: Thomas Telford.

Xiang, R.; Ping-ming, H.; Kui-hua, M. & Zhi-hua, P. 2009. Influence of temperature on main cable sagging of suspension bridge. *Journal of Zhengzhou University Engineering Science*, 30(2): 22–25.

Advances in Civil Engineering and Building Materials IV – Chang et al (eds)
© 2015 Taylor & Francis Group, London, ISBN: 978-1-138-00088-9

Investigation of the behaviour of beam-to-column pinned connections of H-beams with corrugated webs

X. Fan
China State Construction Technical Centre, Beijing, China
Harbin Institute of Technology, Harbin, China

J.Y. Sun & M. Li
China State Construction Technical Centre, Beijing, China

ABSTRACT: In recent years, H-beams with corrugated webs have been widely used in steel structures. Compared with H-beams with flat webs, there are many advantages such as higher buckling resistance and thinner webs for beams with corrugated webs. There has been much research on behaviours of H-beams with corrugated webs, but little research on the connections of these beams has been done. In steel structures; the performance of connections has an important effect on the safety and economy of the entire structure. Therefore, it is necessary to investigate the behaviours of beam-to-column connections of H-beams with corrugated webs. In this paper, two tests of beam-to-column pinned connections of H-beams with corrugated webs were carried out. The test results were compared with the results calculated from the theoretical formulas for pinned connections of H-beams with flat webs, and the results proved the effectiveness of the practical design formulas also for the connections of H-beams with corrugated webs.

Keywords: pinned connections, static behaviour, experimental research, H-beams with corrugated webs, and finite element analysis.

1 EXPERIMENTAL RESEARCH

1.1 *Design of specimens*

The connections were designed based on the shear resistance of the H-beams, and two specimens were prepared. H-beams with corrugated webs CWA500–200 × 10 were used. CWA500-200 × 10 indicates that the flanges are 200 mm wide and 10 mm thick, and the webs are 500 mm high and 2 mm thick. The geometric parameters of the web corrugation are shown in Figure 1.

Since the webs are corrugated, common pinned connections could not be used for H-beams with corrugated webs. A type of pinned connections is proposed for H-beams with corrugated webs. An endplate is welded at the end of the H-beam, and then the beam connecting plate is welded on the endplate. The column connecting plate is welded on the column's flange.

High-strength bolts are used to connect the beam and column connecting plates. The dimensions of the connection are shown in Figure 2. The thickness of the endplate, beam connecting plate and column connecting plate t_e, t_b and t_c for each specimen are shown in

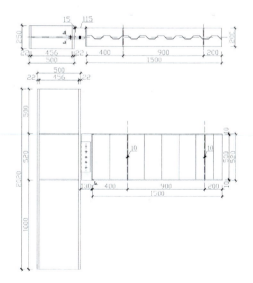

Figure 2.   Dimensions of the specimens.

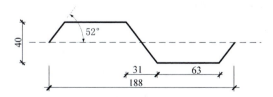

Figure 1.   Geometric parameters of the web corrugation.

Table 1. Geometric parameters of the test connections.

Specimens	Thickness of the endplate t_e (mm)	Thickness of the endplate t_e (mm)	Thickness of the endplate t_e (mm)
Joint1	12	12	18
Joint2	10	6	10

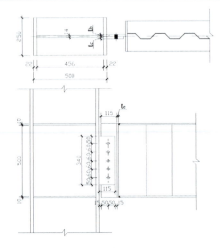

Figure 3. Schematic diagram of the specimens.

Table 2. Material test results.

Batch	Thickness (mm)	Yielding Strength σ_y (MPa)	Tensile Strength σ_u (MPa)	Elongation (%)
1	2	244	394	34.7
1	10	297	413	35.7
1	12	318	443	28.3
1	18	303	435	34.7
2	2	320	457	38.5
2	6	292	426	40.5
2	10	278	428	40.5

Table 1. The details of the specimens are shown in Figure 3. High-strength bolts M16 of grade 10.9 are used, and steel grade Q235 is used in the specimens, except for bolts.

1.2 Material test results

Rectangular specimens for material property tests were prepared following Chinese standard GB/T 2975-1998 and GB 6397-86. The specimens were loaded following GB/T 228-2002. The specimens Joint 1 and Joint 2 were not made from steel of the same batch, so the rectangular specimens were prepared individually. The test results are shown in Table 2.

1.3 Loading scheme

The experiments were carried out in the structure laboratory of Tongji University. The test equipment

Figure 4. Photo of the test site.

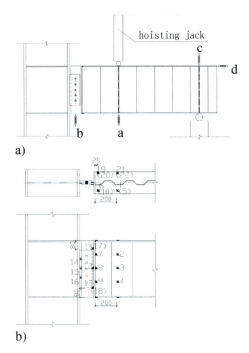

Figure 5. a) Displacement meters; b) Right angle strain flowers.

included a vertical reaction frame, hoisting jack, sensors, and data collection systems.

The test setup is shown in Figure 4. The column is fixed at the testing platform by anchors, which restrict the displacements and rotations of the end of the column. The beam was loaded vertically by the hoisting jack, so that the beam bore shear force.

1.4 Arrangements of measuring points

The displacements of the loading and supporting points were monitored by displacement meters, and the strain contribution at each section was measured using a right angle strain flower. The arrangement of the measuring points is shown in Figure 5.

370

Figure 6. a) After loading of Joint 1; b) After loading of Joint 2.

Table 3. Material test results.

Specimens	Yielding Load (kN)	Ultimate Load (kN)	Displacement at ultimate load (mm)
Joint1	230	290	4.5
Joint2	250	311	6.8

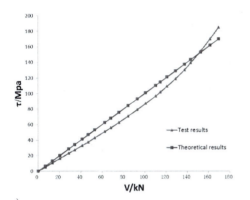

a)

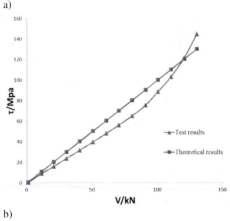

b)

Figure 7. a) Shear stress near fillet welds of Joint 1; b) Shear stress near fillet welds of Joint 2.

2 TEST RESULTS AND COMPARISONS WITH PRACTICAL FORMULAS

2.1 Test results

The failure modes of the two specimens are both shear buckling of the beams' corrugated webs, which are

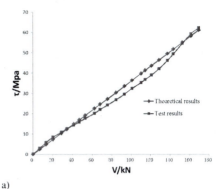

a)

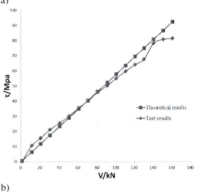

b)

Figure 8. a) Shear stress of Point 12 of Joint 1; b) Shear stress of Point 12 of Joint 2.

shown in Figure 6, and the test results are shown in Table 3.

2.2 Shear stress of each section

2.2.1 Shear stress near fillet welds between endplate and beam

Figure 7 shows the average shear stress of the section compared with the stress calculated from the practical formula Equation 1. The predicted results agree well with the test results.

$$\tau = \frac{V}{h_w t_w} \qquad (1)$$

2.2.2 Shear stress near fillet welds between beam connecting plate and endplate

The measuring point 12 at the section near fillet welds between beam connecting plate and endplate. Figure 8 shows the shear stress of point 12 compared with the stress calculated from the practical formula Equation 2. Good agreements are achieved.

$$\tau = \frac{VS}{It_w} \qquad (2)$$

2.2.3 Shear stress near fillet welds between column connecting plate and column flange

The measuring point 15 at the section near fillet welds between column connecting plate and column flange.

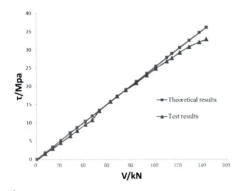

a)

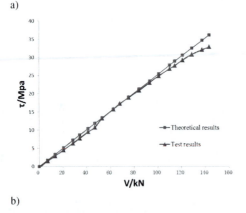

b)

Figure 9. a) Shear stress of Point 15 of Joint 1; b) Shear stress of Point 15 of Joint 2.

Figure 9 shows the shear stress of point 15 compared with the stress calculated from the practical formula Equation 3. There are good agreements between the predicted and measured results.

$$\tau = \frac{V}{S} \tag{3}$$

3 FINITE ELEMENT ANALYSES

3.1 *Finite element model*

The FE software ABAQUS was used to carry out FE analyses. The material property was assigned test results. Solid element C3D8 was used to build the FE model. Boundary conditions were set the same as for the tests. The FE model is shown in Figure 10.

3.2 *Load-displacement curves compared with the test results*

A comparison of the load-displacement curves from the tests and FE analyses of Joint 1 is shown in Figure 11. It can be seen that before peak load FEA results are close to the test results, therefore the FE model can predict the behaviour of the connection at the elastic stage with precision.

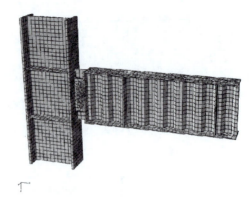

Figure 10. Finite element model of the connection.

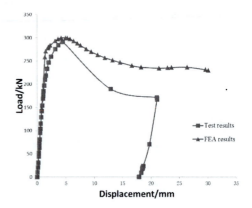

Figure 11. Comparison of load-displacement curves of Joint 1.

4 CONCLUSIONS

In this paper, two static tests were carried out on pinned connections of H-beams with corrugated webs. Then the test results were compared with the calculation from design formulas. Good agreements have been obtained. FE analyses were carried out using ABAQUS and the load-displacement curves were compared with the test results. The comparison proved the reliability of the FE model during the elastic stage.

REFERENCES

Elgaaly M, Seshadri A, 1997. "Girders with corrugated webs under partial compressive edge loading", Journal of Structural Engineering, ASCE, Vol. 123: Issue 6. pp. 783–91.
Gerstle K H, 1988. "Effect of Connections on Frames", Journal of Constructional Steel Research, Vol. 10: pp. 241–267.
Zhang Z, Li GQ, Sun FF, 2008. "Summary of investigation of the H-beam with corrugated web", Progress in Steel Building Structures, Vol. 6, pp. 41–45.
Zhang Z, 2009. "Theoretical and Experimental Research on the H-beams and the composite beams with corrugated webs". Tongji University, pp. 84–112.

Advances in Civil Engineering and Building Materials IV – Chang et al (eds)
© 2015 Taylor & Francis Group, London, ISBN: 978-1-138-00088-9

A probability assessment of the carrying capacity of round timber joints

A. Lokaj & K. Klajmonová
Department of Building Structures, Faculty of Civil Engineering, VŠB-Technical University of Ostrava,
Czech Republic

ABSTRACT: Present European standards for timber structures (Eurocode 5 2004) are based on the semi-probabilistic approach to the structural reliability assessment represented by Partial Factors Design (PFD). Developments in computer technology allow for considering a transition to applications of higher levels of assessment (fully probabilistic methods). One of these methods is Simulation-Based Reliability Assessment (SBRA) documented in publications (Marek et al. 1995, Marek et al. 2003). The subject of this paper is a demonstration of the potential of SBRA assessment of the reliability of timber components and joints on examples of round timber bolted joints with slotted-in steel plates and their carrying capacity in tension.

Keywords: Timber structures, Structural reliability assessment SBRA assessment

1 INTRODUCTION

The deterministic approach to structural reliability assessment has been replaced by a semi-probabilistic method using Partial Factors Design (PFD), which represents individual variables affecting their reliability. Each random variable is represented by a characteristic value, which is modified by partial factors. This approach does not allow any direct evaluation of the probability of failure. The PFD method is used in current national and international specifications in Europe (e.g. in Eurocodes). The development in computer technology allows for considering a transition to the application of fully probabilistic concepts. Such a transition would require reengineering of the entire design procedure, re-arrangement of standards, and development of databases and more. One of these methods is Simulation-Based Reliability Assessment (SBRA) documented in textbooks (Marek et al. 1995, Marek et al. 2003) containing numerous examples of applications related to the assessment of carrying capacity and serviceability. The application of SBRA method in the assessment of round timber bolted joint carrying capacity is demonstrated. Contemporary timber constructions made of round timber elements have become increasingly popular. It concerns footbridges, bridges, watchtowers, or playground equipment. If these constructions are designed with truss supporting systems, element connections are often made of bolts with slotted-in steel plates. Round timber element connections do not have sufficient support in existing European standards (Eurocode 5 2004).

2 RELIABILITY ASSESSMENT USING SBRA METHOD

In the SBRA method, all input values are expressed by variables represented by bounded histograms. The loading is expressed by load duration curves. The reliability function ($RF = R - S$) is analysed using the direct Monte Carlo method and the reliability is expressed by comparing the calculated probability of failure P_f and the target probability P_d given in standards (see Tab. 1 and ČSN 73 1401 1998).

Each action is expressed by a load duration curve and a bounded histogram with the design load values

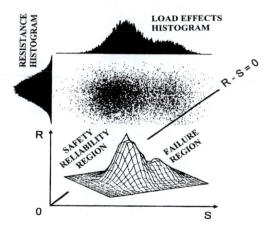

Figure 1. SBRA method assessment scheme.

Table 1. Target probability according [6].

Importance of the structure	Ultimate limit states	Serviceability limit states
Low	0.0005	0.16
Common	0.00007	0.07
High	0.000008	0.023

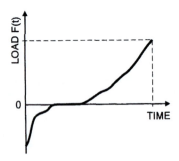

Figure 2. Load time history.

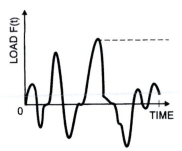

Figure 3. Sorted load time history – "Duration curve".

in SBRA method. Corresponding bounded histograms are developed on the basis of the correlation between action F(t) and time of the service life span (see Fig. 2). This "Load time history curve" is rearranged to the "Load duration curve" (see Fig. 3). Than the "Load duration curve" is transformed into relevant bounded histogram (see Fig. 4).

Material properties are expressed by bounded histograms obtained mainly from short-time laboratory tests of specimens. Figure 5 contains example of round timber density histogram made of 47 laboratory specimens.

3 LABORATORY TESTS

3.1 Tested specimens

The equipment in the laboratory of the Faculty of Civil Engineering at VŠB – Technical University of Ostrava allows the testing of specimens of real construction sizes. The spruce round timber of 120 mm in diameter and specimens of 450 mm long were used. The bolts

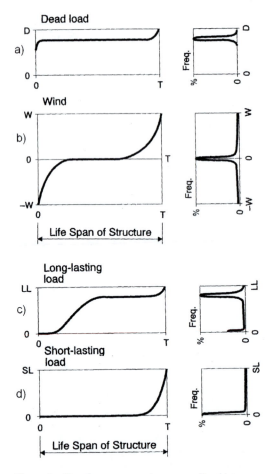

Figure 4. Duration curves and corresponding histograms for selected types of loads.

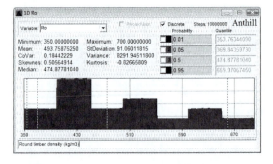

Figure 5. Round timber density histogram based on 47 specimens test.

made of HS (High Strength Steel) steel category 8.8 ($f_y = 640$ MPa, $f_u = 800$ MPa) of 20 mm in diameter were used. Connection plates made of steel S235 with 8 mm thickness and 80 mm width were used. Holes for bolts in steel plates had diameter 22 mm. Holes for bolts 20 mm in diameter were made in the round timbers. Forty seven test specimens were produced (some

Figure 6. Round timber specimens before testing.

Figure 7. Specimen in Press EU100 (left), damage of round timber joint specimen in tension (right).

of them – see in Fig. 6). Tensile tests were conducted on the press EU100 with a recording system (e.g. Fig. 7).

3.2 Static tests

Testing was proceeded on the press EU100, while tension force was increasing gradually. The selected displacement rate of the press jaws seems to be optimal, because destruction of all tested specimens appeared in time-boundary 300 ± 120 sec, which corresponds to interval of laboratory tests for short-time strength according to current European standards for timber structures (Eurocode 5 2004).

The relation between the displacement (e.g. elongation) of the joint and the tension force working on the joints of several specimens can be seen in Figure 8.

The weakest part of the joint should be a steel bolt, according to relations for double shear joints steel to timber type with steel plate inside, even if this bolt is designed from high strength steel. Destruction of the joint should have been caused by the achievement of plastic carrying capacity of bolt in bending and by setting of plastic hinge. Deceleration of increase of force in relation to increase of displacement can be observed on Figure 7. It indicates plastic reshaping of

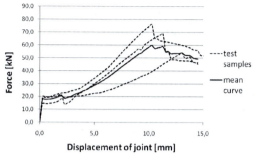

Figure 8. Relation between tension force and displacement of the joint for specimens having a density of 400 ± 20 kg/m³.

the bolt. All testing samples collapsed by disruption of the sample (see Fig. 6 and 8). Disruption of the sample was caused by exceeding of timber strength in tension perpendicular to the grains, but a block shear collapse was not observed. Fracture of the bolt was not observed in any test.

3.3 Results

The results of the carrying capacity of the joints obtained from laboratory tests and the values calculated by (Eurocode 5 2004) with the actual values of timber density are shown in Tab. 1. It is not possible to draw explicit conclusions due to the limited number of samples, but the response from all the tested samples of joints to loading shows some similar signs. After an initial displacement of the joint (displacement was about 5 mm), which was caused by different diameter between a bolt and a hole in a steel plate, there followed an almost linear phase of "working diagram" of the joint up to 80% of maximal carrying capacity. Audible cracking was observed over this border and a "plastic" phase of the joint displacement occurred, e.g. displacement of the joint was increasing more than an adequate increasing of force (see Fig. 8). Rapid disruption of timber specimen in the area between bolt and the end of the round timber occurred in the final phase (see Fig. 7 and 9). The carrying capacity of all forty-seven tested specimens in tension shows a relatively large variability (from 42 kN to 111 kN), as shown in Fig. 9. For more information – see (Lokaj & Klajmonová 2013, Lokaj & Klajmonová 2014).

Characteristic value of carrying capacity of bolted joint with slotted-in steel plate ($F_{V,Rk}$) according to (Eurocode 5 2004) can be calculate:

$$F_{V,Rk} = 2 \cdot f_{h,1,k} \cdot t_1 \cdot d \cdot \left[\sqrt{2 + \frac{4 \cdot M_{y,Rk}}{f_{h,1,k} \cdot d \cdot t_1^2}} \right] \qquad (1)$$

where $f_{h,1,k}$ = characteristic value of embedment strength of timber [N/mm²].

$$f_{h,1,k} = 0.082 \cdot (1 - 0.01 \cdot d) \cdot \rho_k \qquad (2)$$

Figure 9.   Disruption of round timber specimen in tension.

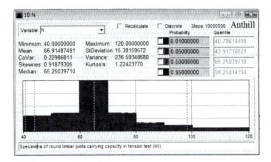

Figure 10.   Histogram with results of 47 round timber bolted joints tested in tension.

where d = diameter of the bolt [mm], ρ_k = characteristic value of timber density [kg/m^3], t_1 = thickness of the timber part [mm], $M_{y,Rk}$ = plastic bending moment of bolt [Nmm].

$$M_{y,Rk} = 0.3 \cdot f_{u,k} \cdot d^{2.6} \qquad (3)$$

where $f_{u,k}$ = ultimate strength of bolt [N/mm^2].

Design value of carrying capacity of bolted joint with slotted-in steel plate ($F_{V,Rd}$) according to (Eurocode 5 2004) can be calculate:

$$F_{V,Rd} = \frac{k_{mod} \cdot F_{V,Rk}}{\gamma_M} \qquad (4)$$

where k_{mod} = modification factor (k_{mod} = 1.0 for short term laboratory test), γ_M = material partial factor (γ_M = 1.3 for solid timber).

The characteristic value of the carrying capacity of tested bolted joint with slotted-in steel plate ($F_{V,Rk}$) according to (Eurocode 5 2004) is $F_{V,Rk}$ = 47.6 kN – see Eq. 1 (for round timber density 370 kg/m^3 – see 5% value in histogram in Fig. 5). The design value of the carrying capacity of a tested bolted joint with slotted-in steel plate ($F_{V,Rd}$) according to (Eurocode 5 2004) is $F_{V,Rd}$ = 36.62 kN – see Eq. 4.

The carrying capacity of a tested bolted joint with a slotted-in steel plate ($F_{V,Rd}$) calculated according to fully probabilistic SBRA method using AntHill^{TM} program for 10 mil. steps (Marek & Guštar 1988–2001) is shown in Fig. 10. Carrying capacity varies between 46 and 70 kN. 5%-value is 47.6 kN.

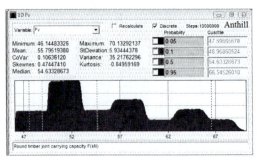

Figure 11.   Histogram with calculated round timber bolted joints carrying capacity in tension.

No correlation between the carrying capacity of the joint and timber density was observed during laboratory tests (correlation coefficient was 0.04). Correlation between timber density and thickness of annual rings was high (correlation coefficient was −0.75).

4   CONCLUSION

The results from the static testing of round timber bolted joints with slotted-in steel plates indicate that they correspond well with calculated values, in spite of a relatively large variability in measured values. Due to this relative high dispersion it is suitable to consider using some fully probabilistic method. SBRA method introduces a qualitatively new approach in the structural design corresponding to the computer era. It allows the designer to have a better understanding of the actual safety and serviceability of timber components and joints. The application of Monte Carlo simulations technique corresponds to the potential of computers available to all designers.

ACKNOWLEDGMENT

This outcome has been achieved with the financial support of the Ministry of Education, Youth and Sports of the Czech Republic – funds of Conceptual development of science, research and innovation assigned to VŠB-TU Ostrava in 2014 under identification numbers SPP IP2214441.

REFERENCES

ČSN 73 1401, 1998. *Design of steel structures*, (in Czech), CNI, Prague.
Eurocode 5, 2004. *Design of timber structures – Part 1.1: General – Common rules and rules for buildings*, European Committee for Standardization.
Lokaj, A. 2001. Timber Structures Assessment using SBRA Method, Ph.D. thesis, VSB-Technical University of Ostrava, Czech Republic.
Lokaj, A. & Marek, P. 2009. Simulation-based reliability assessment of timber structures, in: Proceedings of

the 12th International Conference on Civil Structural and Environmental Engineering Computing. Funchal, Madeira, ISBN 978-190508830-0.

Lokaj, A. & Klajmonová, K. 2013. Carrying Capacity of Round Timber Bolted Joints with Steel Plates under Static Loading. Transactions of the VŠB – Technical University of Ostrava, Civil Engineering Series. Volume XII, Issue 2, Pages 100–105, ISSN (Online) 1804-4824, ISSN (Print) 1213-1962, DOI: 10.2478/v10160-012-0023-5.

Lokaj, A. & Klajmonová, K. 2014. Round timber bolted joints exposed to static and dynamic loading, *Wood Research*, Vol. 59, Issue 3/2014, pp. 439–448.

Marek, P., Guštar, M. & Anagnos, T. 1995. *Simulation-Based Reliability Assessment for Structural Engineers*, CCR Press, Boca Raton, Florida, ISBN 0-8493-8286-6.

Marek, P. et al., 2003. *Probabilistic Assessment of Structures using Monte Carlo Simulation. Background, Exercises and Software, (2nd edition)*, ITAM CAS, Prague, Czech Republic, ISBN 80-86246-19-1.

Marek, P. & Guštar, M. 1988 – 2001 Computer programs DamAc™, M-Star™, AntHill™ (Copyright), Distr. ARTech, Nad Vinicí 7, 143 00 Praha 4.

Advances in Civil Engineering and Building Materials IV – Chang et al (eds)
© *2015 Taylor & Francis Group, London, ISBN: 978-1-138-00088-9*

A two-step unconditionally stable explicit method with controllable numerical dissipation for structural dynamics

Shuenn-Yih Chang & Ngoc-Cuong Tran
National Taipei University of Technology, Taiwan, Republic of China

ABSTRACT: Although the family methods with unconditional stability and numerical dissipation have been developed for structural dynamics, they are all implicit methods and thus an iterative procedure is generally involved for each time step. In this paper, a new family method is proposed. It involves no nonlinear iterations in addition to unconditional stability and favourable numerical dissipation, which can be continuously controlled. In particular, it can have a zero damping ratio. The most important improvement of this family method is that it involves no nonlinear iterations for each time step and thus it can save many computational efforts when compared with the currently available dissipative implicit integration methods.

Keywords: unconditional stability, numerical dissipation, explicit formulation, dynamic analysis.

1 PROPOSED FAMILY METHOD

Consider the equation of motion for a single degree of freedom system

$$m\ddot{u} + c\dot{u} + ku = f \qquad (1)$$

where m, c, k and f are the mass, viscous damping coefficient, stiffness and external force, respectively; and u, $\dot{u}$ and $\ddot{u}$ are the displacement, velocity and acceleration, respectively. A family method is proposed for the step-by-step solution of equation (1) and it can be written as

$$ma_{i+1} + c_0 v_{i+1} + \frac{2p}{p+1} k_{i+1} d_{i+1} - \frac{p-1}{p+1} k_i d_i$$

$$= \frac{2p}{p+1} f_{i+1} - \frac{p-1}{p+1} f_i \qquad (2)$$

$$d_{i+1} = \beta_0 d_{i-1} + \beta_1 d_i + \beta_2 (\Delta t) v_i + \beta_3 (\Delta t)^2 a_i$$

$$v_{i+1} = v_i + \frac{3p-1}{2(p+1)} (\Delta t) a_i - \frac{p-3}{2(p+1)} (\Delta t) a_{i+1}$$

where d_i, v_i, a_i and f_i are the nodal displacement, velocity, acceleration and external force at the i-th time step, respectively, Δt is a step size and k_i is the stiffness at the end of the i-th time step. The coefficients β_0 to β_3 are found to be

$$\beta_0 = -\frac{1}{D} \left[\frac{p-1}{8} \left(\frac{2}{p+1} \right)^3 \Omega_0^2 \right]$$

$$\beta_1 = 1 + \frac{1}{D} \left[\frac{p-1}{8} \left(\frac{2}{p+1} \right)^3 \Omega_0^2 \right]$$

$$\beta_2 = \frac{1}{D} \left(1 - \frac{p-3}{p+1} \xi \Omega_0 \right) \qquad (3)$$

$$\beta_3 = \frac{1}{D} \left\{ \frac{1}{2} - \frac{1}{2} \left[\left(\frac{2}{p+1} \right)^2 + \frac{p-3}{p+1} \right] \xi \Omega_0 \right\}$$

where ξ is a viscous damping ratio; $\Omega_0 = \omega_0(\Delta t)$ and $\omega_0 = \sqrt{k_0/m}$ is the natural frequency of the system determined from the initial stiffness of k_0. In addition, p is the parameter to govern the numerical properties and D is defined as

$$D = 1 - \frac{p-3}{p+1} \xi \Omega_0 + \frac{p}{4} \left(\frac{2}{p+1} \right)^3 \Omega_0^2 \qquad (4)$$

The development details of this method are similar to those of the previously published algorithms (Chang 2009, 2010) and thus will not be elaborated herein.

For computational efficiency, express $\xi \Omega_0$ and Ω_0^2 in terms of the initial structural properties and step size for a structure-dependent integration method (Chang 2009, 2010). After substituting the relations of $c_0 = 2\xi \omega_0 m$ and $\Omega_0^2 = (\Delta t)^2 (k_0/m)$ into equations (3) and (4), they become

$$\beta_0 = -\frac{1}{D} \left[\frac{p-1}{8} \left(\frac{2}{p+1} \right)^3 (\Delta t)^2 k_0 \right]$$

$$\beta_1 = 1 + \frac{1}{D}\left[\frac{p-1}{8}\left(\frac{2}{p+1} \right)^3 (\Delta t)^2 k_0 \right]$$

$$\beta_2 = \frac{1}{D}\left[m - \frac{p-3}{2(p+1)}(\Delta t)c_0 \right]$$

$$\beta_3 = \frac{1}{D}\left\{ \frac{1}{2}m - \frac{1}{4}\left[\left(\frac{2}{p+1} \right)^2 + \frac{p-3}{p+1} \right](\Delta t)c_0 \right\} \tag{5}$$

$$D = m - \frac{p-3}{2(p+1)}(\Delta t)c_0 + \frac{p}{4}\left(\frac{2}{p+1} \right)^3 (\Delta t)^2 k_0$$

These coefficients remain invariant for a whole integration procedure since they are determined from the initial structural properties of c_0 and k_0 if a fixed time step is also employed.

It is important to note that the two difference equations of the proposed family method (PFM) are structure dependent since their coefficients are functions of the initial structural properties m and k_0. This method also has the advantage similar to the method developed by Zhou & Tamma (2004, 2006) in that all the coefficients of the difference equations are constant values. However, the method of Zhou & Tamma is implicit, while PFM is explicit. PFM is exactly the same as that for the Newmark explicit method (NEM), where the equation of motion for the $i + 1$-th time step is not involved in determining the displacement d_{i+1}. As a result, they involve no nonlinear iterations for a nonlinear systems.

2 RECURSIVE MATRIX FORM

For free vibration, the proposed family algorithm can be succinctly expressed in a recursive matrix form (Bathe & Wilson 1973; Hilber et al. 1977; Hughes 1987) of

$$\mathbf{X}_{i+1} = \mathbf{A}\mathbf{X}_i \tag{6}$$

where $\mathbf{X}_{i+1} = [d_{i+1}\ (\Delta t)v_{i+1}\ (\Delta t)^2 a_{i+1}]^T$ is defined, and $\mathbf{A}$ is an amplification matrix. The characteristic equation of $\mathbf{A}$ can be obtained from $|\mathbf{A} - \lambda\mathbf{I}| = 0$ and is found to be

$$\lambda^3 - A_1\lambda^2 + A_2\lambda - A_3 = 0 \tag{7}$$

where λ is an eigenvalue of the matrix $\mathbf{A}$ and the coefficients, A_1, A_2 and A_3 are found to be

$$A_1 = 2 - \frac{2\xi\Omega}{D} + \frac{1}{D}\left[\frac{p-1}{8}\left(\frac{2}{p+1} \right)^3 - \frac{4p}{(p+1)^2} \right]\Omega^2$$

$$A_2 = 1 - \frac{2\xi\Omega}{D} + \frac{1}{D}\left[\frac{p-1}{4}\left(\frac{2}{p+1} \right)^3 + \frac{2(p-1)^2}{(p+1)^2} \right]\Omega^2 \tag{8}$$

$$A_3 = \frac{1}{D}\left[\frac{p-1}{8}\left(\frac{2}{p+1} \right)^3 + \left(\frac{p-1}{p+1} \right)^2 \right]\Omega^2$$

where B is further defined as

$$B = 1 - \frac{p-3}{p+1}\xi\Omega \tag{9}$$

for brevity. It is worth noting that for a linear elastic system $\Omega_0 = \Omega$ is taken in the corresponding equations.

3 CONVERGENCE

The convergence of a computational method is implied by the consistency and the stability based on the Lax equivalence theorem (Lax and Richmyer 1956). The consistency is in terms of the qualitative measure such as the order of accuracy determined from the local truncation error. In general, an algorithm is said to be convergent if it is both consistent and stable.

3.1 Consistency and local truncation error

A local truncation error is defined as the error committed in each time step by replacing the differential equation with its corresponding difference equation (Belytschko & Hughes 1983; Bathe 1986; Hughes 1987). The approximating difference equation for PFM can be obtained from equation (6) after eliminating velocities and accelerations and is found to be

$$d_{i+1} - A_1 d_i + A_2 d_{i-1} - A_3 d_{i-2} = 0 \tag{10}$$

Consequently, after replacing equation (1) by equation (10), the local truncation error for PFM is:

$$E = \frac{1}{(\Delta t)^2}\left[u(t + \Delta t) - A_1 u(t) + A_2 u(t - \Delta t) \right. \tag{11}$$

$$\left. - A_3 u(t - 2\Delta t) \right]$$

In addition, if $u(t)$ is assumed to be continuously differentiable up to any required order, the terms of and $u(t + \Delta t)$, $u(t - \Delta t)$ and $u(t-2\Delta t)$ can be expanded into finite Taylor series at t. After substituting A_1, A_2 and A_3 into the resultant of equation (11), the local truncation error for PFM is found to be

$$E = \frac{2\xi}{BD}\left(\frac{p-1}{p+1} \right)\Omega\omega^2 u + \frac{4\xi^2}{BD}\left(\frac{p-1}{p+1} \right)\Omega\omega\dot{u}$$

$$+ \frac{1}{BD}\left[\frac{1}{12} - \frac{1}{4}\left(\frac{2}{p+1} \right)^2 + \left(\frac{p-1}{p+1} \right)^2 + \frac{1}{3}\xi^2 \right]\Omega^2\omega^2 u \tag{12}$$

$$+ \frac{\xi}{BD}\left[2\left(\frac{p-1}{p+1} \right)^2 - \frac{1}{2}\left(\frac{2}{p+1} \right)^2 + \frac{2}{3}\xi^2 \right]\Omega^2\omega\dot{u}$$

$$+ O\left[(\Delta t)^3 \right]$$

for a linear elastic system. This equation reveals that PFM has a minimum order of accuracy 1 and thus its consistency is verified for any values of p and ξ. In addition, an order of accuracy 2 can be generally achieved for either $p = 1$ or $\xi = 0$.

3.2 Stability

Stability analysis of PFM is very complicated since it has three non-zero eigenvalues due to $A_3 \neq 0$, as shown in equation (8). Alternatively, stability conditions for the cases of $\Omega \to 0$ and $\Omega \to \infty$ are cautiously examined and are applied to find out the restrictions of the parameters p and ξ to have unconditional stability. As a result, in the limiting case of $\Omega \to 0$, equation (7) reduces to

$$\lambda \left(\lambda - 1\right)^2 = 0 \tag{13}$$

It is apparent that $\lambda_{1,2} \to 1$ and $\lambda_3 \to 0$; and these eigenvalues are independent of the parameters p and ξ. On the other hand, in the limit $\Omega \to \infty$, it is found to be

$$\left(\lambda - \frac{p-1}{2p}\right)\left\{\lambda^2 - \left[2 - 4\left(\frac{p+1}{2}\right)\right]\lambda\right.$$
$$\left. + \left[1 + 2(p-1)\left(\frac{p+1}{2}\right)\right]\right\} = 0 \tag{14}$$

for any viscous damping ratio. The roots of this equation are plotted in Figure 1 as functions of p. It is apparent that PFM is stable in the limit $\Omega \to \infty$ if $1/3 \leq p \leq 1$ is satisfied since in this range the spectral radius is always less than or equal to 1. This figure shows that the decrease of p below $1/2$ will increase the spectral radius. Thus, it is implied that the range of $1/2 \leq p \leq 1$ is of interest.

After considering the limiting cases of $\Omega \to 0$ and $\Omega \to \infty$, the stability properties of PFM with $1/2 \leq p \leq 1$ for a general value of Ω are further evaluated by the Routh-Hurwitz criterion which gives necessary and sufficient conditions for the roots of a polynomial to have negative real parts and is following the procedure given by Lambert (1973). Hence, a necessary and sufficient condition for the roots of equation (7) to lie within or on the circle $|\lambda| = 1$ is the satisfaction of the following inequalities:

$$1 - A_1 + A_2 - A_3 \geq 0 \qquad 3 - A_1 - A_2 + 3A_3 \geq 0$$
$$3 + A_1 - A_2 - 3A_3 \geq 0 \qquad 1 + A_1 + A_2 + A_3 \geq 0 \tag{15}$$
$$1 - A_2 + A_3 \left(A_1 - A_3\right) \geq 0$$

After substituting equation (8) into this equation, it is found that all the five inequalities will be met if $\xi \geq 0$ holds. This proves stability for PFM with the condition of $\xi \geq 0$. This fact of stability property in conjunction with the previous proof of consistency implies convergence for PFM.

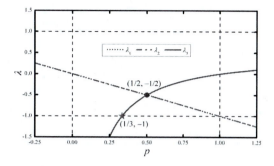

Figure 1. Eigenvalues of amplification matrix as $\Omega \to \infty$.

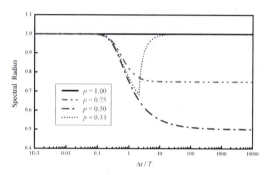

Figure 2. Variation of spectral radius with $\Delta t/T$ for SPFM.

4 NUMERICAL PROPERTIES FOR PFM

After the proof of the convergence of PFM, it is of great interest to further investigate its numerical properties for a linear elastic system. Since the evaluation techniques for an integration method can be easily found in the related references (Zienkiewicz 1977; Belytschko & Hughes 1983; Hughes 1987), they will not be explained here again.

4.1 Spectral radius

The variation of spectral radius with $\Delta t/T$ is shown in Figure 2 for $p = 1$, 0.75, 0.5 and 1/3. The spectral radius is always equal to 1 for $p = 1$ and thus it is indicated that zero damping can be achieved for PFM. For each curve, the spectral radius is almost equal to 1 for a small value of $\Delta t/T$. Subsequently, it decreases gradually and finally tends to a certain value, which is smaller than 1 for large $\Delta t/T$ for $p = 0.75$ and 0.50. It is also found that this value generally decreases with the decrease of p until $p = 0.5$. This implies that PFM with $p = 0.5$ can provide the largest numerical dissipation for high frequency modes. It is clear that $p = 1/3$ cannot have the favourable dissipative property since the spectral radius approaches 1 for a large value of $\Delta t/T$.

4.2 Relative period error

Similarly, the variation of the relative period error with $\Delta t/T$ for PFM for various cases is shown in Figure 3 while that for the numerical damping ratio is plotted

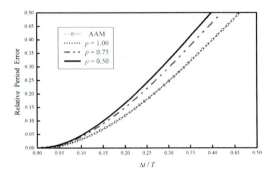

Figure 3. Variation of relative period error.

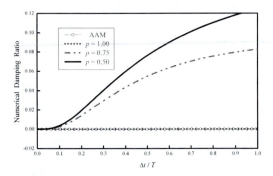

Figure 4. Variation of numerical damping ratio.

in Figure 4. In addition, those for the constant average acceleration method (AAM) are also plotted in the figures correspondingly for comparisons. Apparently, the relative period error increases with the decrease of p for a given value of $\Delta t / T$. In general, the relative period error is very small for a small value of $\Delta t / T$, say $\Delta t / T \leq 0.5$, as $1/2 \leq p \leq 1$. Hence, its corresponding time step will lead to insignificant period distortion during the step-by-step integration. It is interesting to find that the curve for the case of $p = 1$ is almost coincided with that of AAM in Figure 3. This implies that PFM with $p = 1$ will have exactly the same period distortion property as that of AAM.

4.3 Numerical damping

It is manifested from Figure 4 that the continuous control of numerical damping is evident. In addition, the desired numerical damping properties are also achieved since there is a zero tangent at the origin and subsequently a controlled turn upward for the curves with $p = 0.75$ and 0.50. Hence, the higher modes can be suppressed or eliminated by numerical dissipation and at the same time the lower modes are almost unaffected. It is apparent that the case of $p = 1$ leads to no numerical dissipation and thus it has the least period distortion as shown in Figure 3. Figures 3 and 4 reveal that the increase of numerical dissipation for PFM will sacrifice its period distortion as it is found for a general dissipative integration method (Belytschko & Hughes, 1983).

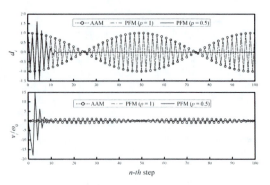

Figure 5. Comparisons of overshoot responses.

4.4 Overshooting

To evaluate the tendency of an integration method to overshoot the exact solutions (Goudreau & Taylor 1972; Hilber & Hughes 1978) one can compute the free vibration response of a single degree of freedom system for the current time step based on the previous step data. The behaviour as $\Omega \to \infty$ gives an indication of the behaviour of the high frequency modes. Using equation (6), the following equations can be obtained for the limiting condition of $\Omega \to \infty$.

$$d_{i+1} \approx \left[1 - \frac{(p+1)^3}{4p} \right] d_i$$

$$(16)$$

$$v_{i+1} \approx -(p-1)^2 \Omega \omega d_i + \left[\frac{(p-1)^2}{2} - 1 \right] v_i$$

It is manifested from the first line of this equation that there is no overshoot in displacement for PFM while it has a tendency to overshoot linearly in Ω in the velocity equation due to the initial displacement term except for $p = 1$.

To confirm the overshoot behaviour of PFM, the discrete displacement and velocity responses are obtained from PFM with $p = 1$ and 0.5. The free vibration to the initial conditions $d_0 = 1$ and $v_0 = 0$ is considered. A time step of $\Delta t = 10T$ is used. Numerical solutions are shown in Figure 5. In addition, the results obtained from AAM are also plotted in the figure for comparison. The velocity term is normalized by the initial natural frequency of the system in order to have the same unit as displacement. Figure 5 shows that the curves are overlapped together for AAM and PFM with $p = 1$ and exhibit no overshoot both in displacement and velocity. On the other hand, there is almost no overshooting in displacement for PFM with $p = 0.5$ while a significant overshoot in velocity is found. These numerical results are consistent with analytical results.

5 IMPLEMENTATION DETAILS

The implementation details of PFM for a multiple degree of freedom system are sketched next and thus some dynamic analyses can be performed to confirm its favourable numerical properties. At first, the displacement vector $\mathbf{d}_{i+1}$ can be calculated by the following equation:

$$\left[\mathbf{M} - \frac{p-3}{2(p+1)}(\Delta t)\mathbf{C}_0 + \frac{p}{4}\left(\frac{2}{p+1}\right)^3 (\Delta t)^2 \mathbf{K}_0 \right] \mathbf{d}_{i+1}$$

$$= \frac{p-1}{8}\left(\frac{2}{p+1}\right)^3 (\Delta t)^2 \mathbf{K}_0 \left(\mathbf{d}_i - \mathbf{d}_{i-1}\right)$$

$$+ \left[\mathbf{M} + \frac{p}{4}\left(\frac{2}{p+1}\right)^3 (\Delta t)^2 \mathbf{K}_0 \right] \mathbf{d}_i \qquad (17)$$

$$+ \left[\mathbf{M} - \frac{p-3}{2(p+1)}(\Delta t)\mathbf{C}_0 \right](\Delta t)\mathbf{v}_i$$

$$+ \left\{ \tfrac{1}{2}\mathbf{M} - \tfrac{1}{4}\left[\left(\frac{2}{p+1}\right)^2 + \frac{p-3}{p+1}\right](\Delta t)\mathbf{C}_0 \right\}(\Delta t)^2 \mathbf{a}_i$$

Next, the velocity vector can be found as:

$$\mathbf{v}_{i+1} = \left[\mathbf{M} - \frac{p-3}{2(p+1)}(\Delta t)\mathbf{C}_0 \right]^{-1}$$

$$* \left\{ \mathbf{M}\left[\mathbf{v}_i + \frac{3p-1}{2(p+1)}(\Delta t)\mathbf{a}_i \right] - \frac{p-3}{2(p+1)}(\Delta t) \qquad (18) \right.$$

$$* \left. \left[\frac{2p}{p+1}\mathbf{f}_{i+1} - \frac{p-1}{p+1}\mathbf{f}_i - \frac{2p}{p+1}\mathbf{R}_{i+1} - \frac{p-1}{p+1}\mathbf{R}_i \right] \right\}$$

Finally, the acceleration vector can be calculated by using the equations of motion as:

$$\mathbf{a}_{i+1} = \mathbf{M}^{-1} * \left[\frac{2p}{p+1}\mathbf{f}_{i+1} - \frac{p-1}{p+1}\mathbf{f}_i \right.$$

$$\left. -\mathbf{C}_0 \mathbf{v}_{i+1} - \frac{2p}{p+1}\mathbf{Kd}_{i+1} + \frac{p-1}{p+1}\mathbf{Kd}_i \right] \qquad (19)$$

6 CONCLUSIONS

In this paper, a parameter p is applied to develop a new family of integration methods for structural dynamics, in that numerical properties of the proposed family method are controlled by p. An appropriate selection of $1/2 \leq p \leq 1$ will lead to unconditional stability and favourable numerical dissipation, which can be continuously controlled by p and it is possible to achieve zero numerical damping if $p = 1$ is adopted. This numerical damping can be used to suppress or even eliminate the spurious participation of high frequency modes while the low frequency modes can be accurately integrated. Compared to the currently available dissipative

integration methods, the most important improvement of this family method is that it involves no nonlinear iterations for each time step in addition to unconditional stability and desired numerical dissipation. As a result, it is computationally very efficient for solving an inertial-type problem.

ACKNOWLEDGEMENT

The authors is grateful to acknowledge that this study is financially supported by the National Science Council, Taiwan, R.O.C., under Grant No. NSC-99-2221-E-027-029.

REFERENCES

Bathe, K.J. 1986. *Finite element procedure in engineering analysis*. Prentice-Hall, Inc., Englewood Cliffs, NJ, USA.
Bathe, K.J. and Wilson, E.L. 1973. Stability and accuracy analysis of direct integration methods. *Earthquake Engineering and Structural Dynamics*. Vol. 1: 283–291.
Belytschko, T. and Schoeberle, D.F. 1975. On the unconditional stability of an implicit algorithm for nonlinear structural dynamics. *Journal of Applied Mechanics*, Vol. 17: 865–869.
Belytschko, T. and Hughes, T.J.R. 1983. *Computational methods for transient analysis*. Elsevier Science Publishers B.V., North-Holland.
Chang, S.Y. 2009. An explicit method with improved stability property. *International Journal for Numerical Method in Engineering*, Vol. 77, No. 8: 1100–1120.
Chang, S.Y. 2010. A new family of explicit method for linear structural dynamics. *Computers & Structures*, Vol. 88, No. 11–12: 755–772.
Goudreau, G.L. and Taylor, R.L. 1972. Evaluation of numerical integration methods in elasto-dynamics. *Computer Methods in Applied Mechanics and Engineering*, Vol. 2: 69–97.
Hilber, H.M., Hughes, T.J.R. and Taylor, R.L. 1977. Improved numerical dissipation for time integration algorithms in structural dynamics. *Earthquake Engineering and Structural Dynamics*, Vol. 5: 283–292.
Hilber, H.M. and Hughes, T.J.R. 1978. Collocation, dissipation, and 'overshoot' for time integration schemes in structural dynamics. *Earthquake Engineering and Structural Dynamics*, Vol. 6: 99–118.
Hughes, T.J.R. 1987. *The Finite element method*. Prentice-Hall, Inc., Englewood Cliffs, N.J., U.S.A.
Lambert, J.D. 1973. *Computational Methods in Ordinary Differential Equations*. John Wiley, London.
Lax, P.D. and Richmyer R.D. 1956. Survey of the stability of linear difference equations. *Communications on Pure and Applied Mathematics*, Vol. 9: 267–293.
Newmark, N.M. 1959. A method of computation for structural dynamics. *Journal of Engineering Mechanics Division*, ASCE, Vol. 85: 67–94.
Zhou, X. and Tamma K.K. 2004. Design, analysis and synthesis of generalized single step single solve and optimal algorithms for structural dynamics. *International Journal for Numerical Methods in Engineering*; Vol. 59: 597–668.
Zhou, X. and Tamma K.K. 2006. Algorithms by design with illustrations to solid and structural mechanics/ dynamics. *International Journal for Numerical Methods in Engineering*; Vol. 66: 1841–1870.
Zienkiewicz, O.C. 1977. *The Finite Element Method*. McGraw-Hill Book Co (UK) Ltd. Third edition.

Advances in Civil Engineering and Building Materials IV – Chang et al (eds)
© 2015 Taylor & Francis Group, London, ISBN: 978-1-138-00088-9

Progressive collapse analysis of space steel frames considering influence of joint rotational stiffness

Hui Yang & Lai Wang
Department of Civil Engineering, Shandong University of Science and Technology, Qingdao, China

ABSTRACT: To study the influence of different types of joint rotational stiffness on the steel frames during their progressive collapse, the lumped plastic hinge model in SAP2000 and the alternate path method are adopted to perform the nonlinear static and dynamic analysis of a multi-story space steel frame. The analytical results show that the decrease of joint rotational stiffness would evidently weaken the structural progressive collapse resistance performance; and the influence of different joint rotational stiffness has the bigger impact on the structure where the corner column was removed.

Keywords: multi-story steel frame; progressive collapse; joint rotational stiffness; nonlinear analysis.

1 INTRODUCTION

In recent years, the method of preventing progressive collapse of structure has been the hotspot of research. A progressive collapse of a building is initiated by an event that causes local damage that the structural system cannot absorb or contain, and that subsequently propagates throughout the structural system, or a major portion of it, leading to a final damage state that is disproportionate to the local damage that initiated it[1].

A serious of researches and several design code, standard have been carried out, such as Eurocode1[2], GSA (2003)[3] and DoD (2010)[4], there are four typical analysis methods in present: concept design, tie-force design, alternate path method and enhance local resistance method. Alternate path method is the most used method in the analysis of progressive collapse. Alternate path method does not consider the source of the effect and it is relatively simple and universal. After the column is removed, the axial force above the removed column will be redistributed in a very short time, the remaining structure will find a new equilibrium path to transmit this force. According to consider the nonlinear and dynamic effect, Alternate path method can be divided into linear static analysis, nonlinear static analysis, linear dynamic analysis and nonlinear dynamic analysis[5]. In nonlinear dynamic analysis, apply increased gravity loads on floor areas above removed column, the dynamic increase factor is 2.

Studies on dynamic effect mainly concentrate on aspects of the failure time, damping ratio, building size, dimension and so on. There have been few studies to consider the influence of the joint rotation stiffness in progressive collapse analysis. In 2010, Jingsi-Huo[6] used equivalent plastic angle and equivalent vertical displacement, increasing the coefficient to consider the joint rotation stiffness, and draw a conclusion "the influence of joint rotational stiffness has the greater influence on the structure where the side column was removed."

This paper will use the alternate path method to perform the nonlinear static and dynamic analysis of a 6-story space steel frame considering joint rotation stiffness. Research the influence of different joint rotational stiffness on the steel frames during progressive collapse by analyzing the internal force and vertical displacement at failure point.

2 FINITE ELEMENT MODEL

Set up a steel frame with 4×4 bays 6 floors, the span is 6 m in both directions, the floor height is 3.6 m for each floor, the floor plan as shown in Figure 1. Beams and columns are all welded I-section, material specifications as shown in Table 1. Does not consider the effect of the slab, and the bottom columns are fixed, an elastic modulus of steel is 2.06×10^5 MPa, the density is 7.85×10^3 kg/m^3, and yield strength is 345 MPa. Poisson ratio is 0.3.

Use the lumped plastic hinge model in SAP2000[7]. Install the PMM hinge at both ends of column and M3 hinge at the beam ends and middle parts, plastic hinge parameters recommended to FEMA356[8]. Set up the finite element model as shown in Fig. 2.

For nodes, to simulate node rotational stiffness by applying spring on beam-to-column connections area, the rotational stiffness considering rigid, 20 K_b, 10 K_b, 5.0 K_b (K_b is the beam linear stiffness) and analyze two situations that removing the ground corner column and the side column.

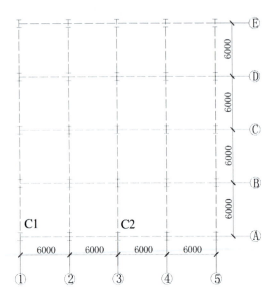

Figure 1. Construction plan.

Table 1. The size of columns and beams section.

Element	Height	Width	The web thickness	Flange thickness
Beam	300	300	10	16
The peripheral column	400	300	10	16
internal column	500	300	10	16

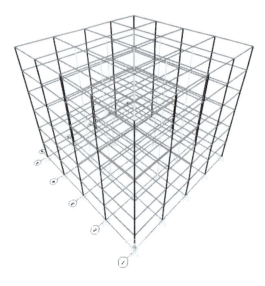

Figure 2. The finite element modal.

3 ANALYSIS METHOD

3.1 Nonlinear static analysis

Use the nonlinear static analysis method for progressive collapse analysis by removing the collapsed

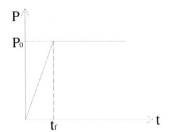

Figure 3. Load-Time curve.

column directly from the frame model. According to DoD (2010) design criteria, failure column effect region need to consider the different load combinations, For example, column C1 effect is limited to the intersection floor between A-B axis and (1)–(2) axis, influence in the regional range is two times the value of $1.2(D_L + D_W) + 0.5L_L$ load combination, influence outside the range is the value of $1.2(D_L + D_W) + 0.5L_L$.

3.2 Nonlinear dynamic analysis

Nonlinear dynamic analysis for load combination:

$$1.2(D_L + D_W) + 0.5L_L \qquad (1)$$

Of which: D_L is the floor dead load, $5 \, kN/m^2$; D_W is the wall line load, $8 \, kN/m$; L_L is the floor live load, $2 \, kN/m^2$.

In nonlinear dynamic analysis, time history function technique is used to simulate the scenarios of column loss and calculate the residual response of structure. This can be divided into three steps:

1. apply loads on the frame according to formula (1) and calculate the axial force at the top of each removal column;
2. delete these columns respectively in finite element models, apply nodal loads equal to the axial force at the failure connections and the remaining structure reaches a same balance stage.
3. define these nodal loads to conform to the time history showing in Fig. 3. In Fig. 3, t_f is the time period of column removal. Here define t_f equals 0.1 times vertical vibration period of the remaining structure [4]. Rayleigh damping is used in this paper and the lumped plastic hinge model in SAP2000 is adopted to perform the nonlinear static and dynamic analysis of the space steel frame.

4 ANALYSIS RESULTS

4.1 Modal analysis

Using SAP2000 to do modal analysis of the remaining structure after the loss of column C1 and C2, vertical vibration period T_2 can be used to determine the failure time, and w_1, w_2 are used to determine the

Table 2. Analysis parameters.

Rotational stiffness	Case	T_1/s	T_2/s	T_f/s	t/s	Δt/s	w_1/(rad/s)	w_2/(rad/s)	α	β
rigid	failure of C1	0.632	0.104	0.0104	1	0.005	9.935	60.406	0.341	0.000569
	failure of C2	0.63	0.098	0.0098	1	0.005	9.961	64.377	0.345	0.000538
20K_b	failure of C1	0.676	0.12	0.012	1	0.005	9.295	52.359	0.316	0.000649
	failure of C2	0.674	0.111	0.111	1	0.005	9.322	56.549	0.32	0.000607
10K_b	failure of C1	0.716	0.13	0.013	1	0.005	8.774	48.371	0.297	0.0007
	failure of C2	0.714	0.125	0.0125	1	0.005	8.801	50.119	0.299	0.000679
5.0K_b	failure of C1	0.788	0.157	0.0157	1	0.005	7.969	39.978	0.266	0.000834
	failure of C2	0.786	0.145	0.0145	1	0.005	7.996	43.248	0.27	0.000781

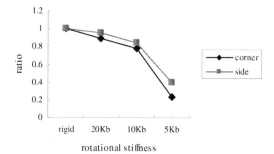

Figure 4. Moment of beam end.

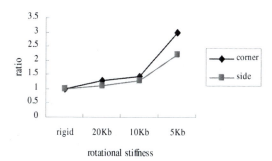

Figure 5. Axial force of column.

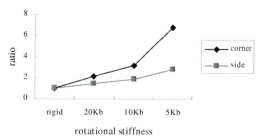

Figure 6. Displacement of joint.

proportion of rayleigh damping coefficient, damping ratio is 0.02. Analysis of parameters as shown in Table 2, among them: t is the time history analysis time; Δt is the time step; α, β is respectively the coefficient of mass damping and the stiffness damping.

As seen in 2, following the increment of joint rotational stiffness, the fundamental vibration period and vertical vibration period are decreasing.

4.2 Nonlinear static analysis

Analyze and compare the moment at the end of beams, the axial force at the top of columns, and the vertical displacement of failure connections.

(1) Internal forces analysis

As can be seen from Fig. 4, following the decrement of the joint rotational stiffness, the ratios of moments at the end of beams are reducing, it means that lower

rigidity of joint will distribute lower bending moment. After the corner column removed, the joint stiffness with 5.0 K_b only supports 22% moment of rigid joint. In addition, with the decrement of the joint rotational stiffness, the moment at the end of beams after corner column failure are smaller than that of the side columns.

Fig. 5 is the influence of the joint stiffness on axial forces at the top of adjacent columns under two column failure cases, with the decrement of the joint rotational stiffness, the adjacent column axial force becomes larger, that is lower joint rigidity lead to higher column axial force. In addition, the column axial force after corner column failure is higher than that of the side columns .

(2) Vertical displacement analysis

As seen in Fig. 6, with joint stiffness being rigid, 20K_b, 10K_b, 5.0 K_b, the peak vertical displacement of failure point after the corner column loss is respectively 0.031 m, 0.066 m, 0.097 m, 0.215 m while failure point after the side column loss is respectively 0.021 m, 0.032 m, 0.039 m, 0.060 m with the decrement of the joint rotational stiffness, the vertical displacement of the failure point are increasing, it means that the connection with lower rotational stiffness has the better rotation performance, and the vertical displacement of connections at the top of corner columns increases more obviously.

4.3 Nonlinear dynamic analysis

As can be seen from Fig. 7, Fig. 8, with joint stiffness being rigid, 20 K_b 10 K_b 5.0 K_b, the timing of peak

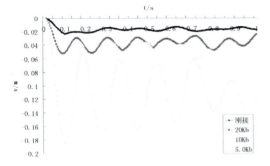

Figure 7.   Corner column removal.

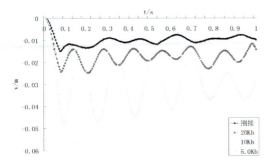

Figure 8.   Side column removal.

displacement after corner columns removed is respectively 0.085 s, 0.090 s, 0.095 s, 0.10 s and the peak displacements is respectively 0.024 m, 0.052 m, 0.077 m, 0.181 m, while the timing of peak displacement after side columns removed is respectively 0.075 s, 0.080 s, 0.085 s, 0.095 s and the peak displacements is respectively 0.015 m, 0.024 m, 0.030 m, 0.048 m. With the decrement of the joint rotational stiffness, the timing of peak displacement turn up a little later, which means a longer failure time for structure to resist progressive collapse, studies have shown that the longer failure time the better for structure, and despite the peak displacement will be greater, the greater performance to prevent progressive collapse should not be ignored with the energy dissipation capacity considered.

Compared with the side column loss, the corner columns loss leads to longer period of peak displacement and greater value of displacement.

5   CONCLUSIONS

Based on the analysis above, the following conclusions were made:

(1) With the decrement of the joint rotational stiffness, the capacity of preventing progressive collapse significantly decline. Lower joint stiffness will increase the axial column force, enlarge the joint vertical displacement and support less bending moment.

(2) With the decrement of the joint rotational stiffness, the failure time would be longer and the peak displacement would be greater, which means that the longer failure time is good for weakening the damage of remaining structure, and greater displacement is contribute to dissipate energy so that component is not destroyed. Therefore, it is very necessary to consider the influence of joint rotational stiffness in progressive collapse on the dynamic effect of structure.

(3) The influence of joint rotational stiffness after corner columns loss is more obvious than after side columns are lost with the internal forces and vertical displacement considered. Therefore, the corner column design needs more attention.

(4) Comparing the static nonlinear analysis with dynamic nonlinear analysis for the vertical displacement at failure point, it is relatively conservative to take 2 as dynamic increase factor.

REFERENCES

Elliingwood B. R. 2006. Mitigating risk from abnormal loads and progressive collapse [J]. Journal of Performance of Constructed Facilities, 20(4):315–323.

European Committee for Standardization. EN 1991-1-7: Eurocode 1: actions on structures. Part 1–7: general actions-accidental actions, (2006).

US General Services Administration. Progressive collapse analysis and design guidelines for new federal office buildings and major modernization projects [S]. Washington DC: GSA, 2003.

DoD 2010, Design of structures to resist progressive collapse [S]. Department of Defense, January 2010.

Marjanishvili S M. Progressive analysis procedure for progressive collapse [J], ASCE Journal of Performance of Constructed Facilities, 2004, 18(2):79–85.

Jing-si Huo, Cong-ling Hu. Study on dynamic effects of steel frame during progressive collapse [J], Disaster science, 2010, 25(supplement): 89–93.

Computers and Structures, Inc. SAP2000 analysis reference manual [M]. Berkeley, California, 2004.

FEMA 356, Prestandard and commentary for the seismic rehabilitation of buildings [S]. Federal Emergency Management Agency, November, 2000.

Advances in Civil Engineering and Building Materials IV – Chang et al (eds)
© *2015 Taylor & Francis Group, London, ISBN: 978-1-138-00088-9*

Sectional curvature computational analysis on a prestressed steel reinforced concrete frame beam

Feng Gao
School of Civil Engineering and Architecture, University of Jinan, Jinan, China

Shaohong Zhang
Engineering Construction Standards Quota Station of Shandong Province, Jinan, China

ABSTRACT: Based on the plane section assumption and simplified stress distribution of composite sections, the resultant normal forces of cross-sections of Prestressed Steel Reinforced Concrete (PSRC) frame beams under different load levels were presented in accordance with the stress-strain relationships of composite sections. Then, the curvature and moment of the prestressed steel reinforced concrete frame beams of every loading phase were deduced respectively. The comparison of the results between the calculations and tests shows that the calculated curvature is in good agreement with the test data. It is said that the method and the expressions of the curvature proposed can represent well the actual character of prestressed steel reinforced concrete frame beams.

Keywords: prestressing, steel reinforced concrete structure, frame, beam, curvature.

1 INTRODUCTION

The prestressed steel reinforced concrete (PSRC) structure is a configured, rolled, or welded steel composite structure in prestressed concrete structures, and is used for multistorey public buildings, high-rise buildings at the bottom and the top, and large bay garage span buildings. Static characteristics of prestressed steel reinforced concrete structures including failure mode, and development and distribution of cracks and stiffness were recently researched by domestic scholars (Xueyu et al., 2011, Jun et al., 2009, Chuan Guo et al., 2007). Based on a three two-span prestressed concrete continuous composite steel H beam test, moment coefficients of not more than 0.44 were obtained, and a moment modulation factor formula of a prestressed steel reinforced concrete continuous beam regarding relatively plastic rotation as independent variables, was obtained (Wenzhong 2010).

This paper is based on the plane section assumption and simplified stress distribution of the composite section, and the resultant normal forces of the cross-section of prestressed steel reinforced concrete (PSRC) frame beams under different load levels were in accordance with the stress-strain relationships of the composite section. Then, the curvature and moment of the prestressed steel reinforced concrete frame beams of every loading phase were deduced respectively. The comparison of the results between the calculations and tests shows that the calculated curvature is in good agreement with the test data. It is said that the method and agreement with the test data. It is said that the method and the expressions of the curvature proposed can represent well the actual character of prestressed steel reinforced concrete frame beams.

2 BASIC ASSUMPTION

To facilitate the analysis, basic assumptions based on mechanical characteristics of prestressed steel reinforced concrete frame beams are as follows:

(1) Throughout the whole process, the influence of the tensile stress of the concrete tension zone is ignored, and the average strain distribution complies with the flat cross-section (or amendment plane section) assumption.

(2) When the compression zone limit states of prestressed steel reinforced concrete frame beams were reached, the concrete compressive stress area is rectangular distribution and the compression zone height is taken as 0.8C, where C is the height of the actual concrete compression zone.

(3) The stress-strain relationship of the concrete is shown in Figure 1.

$$\sigma_c = \begin{cases} f_c[2(\frac{\varepsilon_c}{\varepsilon_0}) - (\frac{\varepsilon_c}{\varepsilon_0})^2], & \varepsilon_c \leq \varepsilon_0 \\ f_c, & \varepsilon_0 \leq \varepsilon_c \leq \varepsilon_{cu} \end{cases} \tag{1}$$

(4) The stress-strain relationship of a steel bar and steel is shown in Figure 2.

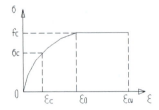

Figure 1. Stress-strain relationship of concrete.

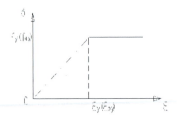

Figure 2. Stress-strain relationship of steel bar and steel.

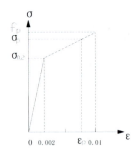

Figure 3. Stress-strain relationship of prestressing tendon.

$$\sigma_s(\sigma_a) = \begin{cases} E_s(E_a)\varepsilon, & \varepsilon \leq \varepsilon_y(\varepsilon_{ay}) \\ f_y(f_{ay}), & \varepsilon > \varepsilon_y(\varepsilon_{ay}) \end{cases} \tag{2}$$

(5) The stress-strain relationship of a prestressing tendon is shown in Figure 3.

The stress-strain relationship remains linear when the stress is less than $\sigma0.2$.

$$\sigma_p = E_p \varepsilon_p \tag{3}$$

The stress of a prestressing tendon can be obtained by interpolation when it is between $\sigma0.2$ and fp.

3 COMPUTATIONAL ANALYSIS ON YIELD CURVATURE AND YIELD MOMENT

In general, the stress-strain distribution of a prestressed steel reinforced concrete frame beam section when tensile reinforcement yielding is shown in Figure 4. As can be seen from the figure, member is in the elastic stage and the calculations of yield curvature (ϕ_y) and yield moment (M_y) is in line with the theory of elasticity.

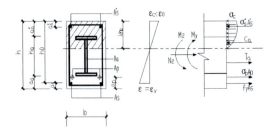

Figure 4. The stress-strain distribution of a section in yielding state.

From Figure 4, force C_c of the concrete compression zone has the following formula under the condition of tension reinforcement yielding:

$$C_c = \int_0^{kh_0} b\sigma_c dx = b\int_0^{kh_0} f_c[2\frac{x\varepsilon_y}{h_0(1-k)\varepsilon_0} \tag{4}$$

$$-(\frac{x\varepsilon_y}{h_0(1-k)\varepsilon_0})^2]dx$$

To obtain:

$$C_c = \frac{h_0 bf_c \varepsilon_y k^2}{(1-k)\varepsilon_0} - \frac{h_0 bf_c \varepsilon^2_y k^3}{3(1-k)^2 \varepsilon^2_0} \tag{5}$$

force C_s of the compression steel bar is approximated as:

$$C_s = \sigma'_s A'_s = E_s \frac{k}{1-k}\varepsilon_y A'_s \tag{6}$$

C_a is the steel force of the compression part approximated as:

$$C_a = \sigma'_{af} A'_{af} + \frac{1}{2}\sigma'_{af} t_w (kh_0 - a'_a) \tag{7}$$

$$= E_a \frac{kh_0 - a'_a}{h_0(1-k)}\varepsilon_y A'_{af} + E_a \frac{(kh_0 - a'_a)^2}{2h_0(1-k)}\varepsilon_y t_w$$

The tension part of the steel force T_a is:

$$T_a = \sigma_{af} A_{af} + \frac{1}{2}\sigma_{af} t_w (h_a - kh_0 + a'_a) \tag{8}$$

$$= E_a \frac{h_a - kh_0 + a'_a}{h_0(1-k)}\varepsilon_y A_{af} + E_a \frac{(h_a - kh_0 + a'_a)^2}{2h_0(1-k)}\varepsilon_y t_w$$

Force T_p of the prestressing strand is taken as the following:

$$T_p = \sigma_p A_p = E_p \frac{h - a_p - kh_0}{h_0(1-k)}\varepsilon_y A_p \tag{9}$$

By $\sum X = 0$ the following formula can be obtained:

$$C_c + C_s + C_a - T_a - T_p - f_y A_s - N_2 = 0 \tag{10}$$

390

So, we obtain:

$$
\begin{aligned}
& -2h_0^2bf_c(\varepsilon_0 + \tfrac{1}{3}\varepsilon_y)\varepsilon_y k^3 + 2h_0\varepsilon_0(h_0bf_c\varepsilon_y \\
& -\varepsilon_0 f_y A_s' - \varepsilon_0 f_y A_s - \varepsilon_0 N_2 - E_a\varepsilon_0\varepsilon_y A_a - \\
& E_p\varepsilon_0\varepsilon_y A_p)k^2 + [2h_0\varepsilon_0 f_y A_s' + 4h_0\varepsilon_0^2 f_y A_s \\
& + E_a\varepsilon_0^2\varepsilon_y A_a(2h_0 + 2a_a' + h_a) + 4h_0\varepsilon_0^2 N_2 \\
& + 2E_p\varepsilon_0^2\varepsilon_y A_p(h + h_0 - a_p)]k - [2h_0\varepsilon_0^2 f_y A_s \\
& + E_a\varepsilon_0^2\varepsilon_y A_a(2a_a' + h_a) + 2h_0\varepsilon_0^2 N_2 \\
& + 2E_p\varepsilon_0^2\varepsilon_y A_p(h - a_p)] = 0
\end{aligned}
\tag{11}
$$

The coefficient k can be solved by the above formula, so the yield curvature of the prestressed steel reinforced concrete frame beams is as follows:

$$
\varphi_y = \frac{\varepsilon_v}{h_0(1-k)} = \frac{f_y}{E_s h_0(1-k)}
\tag{12}
$$

The yield moment can be obtained by taking the moment on the compression zone longitudinal reinforcement force points as follows:

$$
\begin{aligned}
M_y = {}& M_{yTs} + M_{yTp} + M_{yTa} \\
& + M_{yN_2} + M_2 - M_{yc} - M_{yCa}
\end{aligned}
\tag{13}
$$

where M_{yTs} is the moment of tension reinforcement in the case of tension longitudinal reinforcement yielding, expressed as follows:

$$
M_{yTs} = f_y A_s (h_0 - a_s')
\tag{14}
$$

M_{yTp} is the moment of the prestressing strand in the case of tension longitudinal reinforcement yielding, expressed as follows:

$$
M_{yTp} = E_p\varepsilon_y A_p \frac{(h - a_p - kh_0)(h - a_p - a_s')}{h_0(1-k)}
\tag{15}
$$

M_{yTa} is the moment of the tension section of steel in the case of tension longitudinal reinforcement yielding, expressed as follows:

$$
\begin{aligned}
M_{yTa} = {}& \sigma_{af} A_{af} (h - a_a - a_s') + \tfrac{1}{6}\sigma_{af} t_w (h_a + a_a' \\
& - kh_0)(2h_a + 2a_a' + kh_0 - 3a_s')
\end{aligned}
\tag{16}
$$

M_{yN_2} is the moment of the secondary axial force N_2 in the case of tension longitudinal reinforcement yielding, expressed as follows:

$$
M_{yN_2} = N_2(0.5h - a_s')
\tag{17}
$$

M_{yc} is the moment of the concrete force of the compression zone in the case of tension longitudinal reinforcement yielding, expressed as follows:

$$
M_{yc} = C_c(\tfrac{1}{3}kh_0 - a_s')
\tag{18}
$$

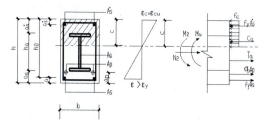

Figure 5. The stress-strain distribution of the section in ultimate state.

M_{yCa} is the moment of the compression section of steel in the case of tension longitudinal reinforcement yielding, expressed as follows:

$$
\begin{aligned}
M_{yCa} = {}& \sigma_{af} A_{af}' (a_a - a_s') + \tfrac{1}{6}\sigma_{af} t_w (kh_0 \\
& - a_a')(2a_a' + kh_0 - 3a_s')
\end{aligned}
\tag{19}
$$

4 COMPUTATIONAL ANALYSIS ON ULTIMATE CURVATURE AND ULTIMATE MOMENT

The stress-strain distribution of a prestressed steel reinforced concrete frame beam section is shown in Figure 5, while the concrete strain of compression zone reach ultimate compressive strain .

From Figure 5, force Cc of the concrete compression zone has the following formula in the limit states:

$$
C_c = 0.8Cbf_c
\tag{20}
$$

The force Cs of the compression steel bar is approximated as:

$$
C_s = f_y A_s'
\tag{21}
$$

Ca is the steel force of the compression part approximated as:

$$
C_a = f_a' A_{af}' + f_a' t_w (C - a_a' - \frac{\varepsilon_{ay}}{\varepsilon_{cu}} C)
\tag{22}
$$

The tension part of the steel force Ta is approximated as:

$$
T_a = f_a A_{af} + f_a t_w (h - a_a - C - \frac{\varepsilon_{ay}}{\varepsilon_{cu}} C)
\tag{23}
$$

The force T_p of the prestressing strand is taken as follows:

$$
T_p = f_p A_p
\tag{24}
$$

By $\sum X = 0$ the following formula can be obtained:

$$
C_c + C_s + C_a - T_a - T_p - f_y A_s - N_2 = 0
\tag{25}
$$

The actual compression zone height C can be obtained as follows:

$$C = \frac{N_2 + f_a A_{af} + f_p A_p + f_y A_s - f_y' A_s'}{0.8(bf_c + 2.5t_w f_a)}$$

$$+ \frac{-f_a' A_{af}' + t_w(f_a h + f_a' a_a - f_a a_a)}{0.8(bf_c + 2.5t_w f_a)} \quad (26)$$

The limit curvature can be obtained as follows:

$$\varphi_u = \frac{\varepsilon_{cu}}{C} \quad (27)$$

The ultimate moment can be obtained by taking the moment on the compression zone longitudinal reinforcement force points, as follows:

$$M_u = M_{uTs} + M_{uTp} + M_{uTa} + M_{uN_2} + M_2 - M_{uc} - M_{uCa} \quad (28)$$

where M_{uTs} is the moment of tension reinforcement in the limit states,

$$M_{uTs} = f_y A_s (h_0 - a_s') \quad (29)$$

M_{uTp} is the moment of the prestressing strand in the limit states:

$$M_{uTp} = f_p A_p (h - a_p - a_s') \quad (30)$$

M_{yTa} is the moment of the tension section of steel in the limit states:

$$M_{uTa} = f_a A_{sf} (h - a_a - a_s') + 0.5 f_a t_w (h$$
$$- a_a - C - \frac{\varepsilon_{ay}}{\varepsilon_{cu}} C)(h + C + \frac{\varepsilon_{ay}}{\varepsilon_{cu}} C - a_a - a_s') \quad (31)$$

M_{uN_2} is the moment of the secondary axial force N2 in the limit states:

$$M_{uN_2} = N_2(0.5h - a_s') \quad (32)$$

M_{uc} is the moment of the concrete force of the compression zone in the limit states:

$$M_{uc} = 0.4Cbf_c(C - 2a_s') \quad (33)$$

M_{uCa} is the moment of the compression section of steel in the limit states:

$$M_{uCa} = f_a' A_{af}'(a_a - a_s') + 0.5 f_a' t_w (C -$$
$$a_a' - \frac{\varepsilon_{ay}}{\varepsilon_{cu}} C)(C + a_a' - \frac{\varepsilon_{ay}}{\varepsilon_{cu}} C - 2a_s') \quad (34)$$

Table 1. Calculation on yield curvature, limit curvature and curvature ductility factor.

Frame Number	Limit curvature/ϕ_u	Yield curvature/ϕ_y	ϕ_u/ϕ_y
XGKJ1	2.62×10^{-5}	0.94×10^{-5}	2.78
XGKJ2	2.62×10^{-5}	1.00×10^{-5}	2.62

5 CALCULATION EXAMPLE

The yield curvature, the limit curvature, and the curvature ductility factor of two prestressed steel reinforced concrete frame beam specimens which come from literature (Xueyu et al., 2011), were calculated using the calculation method of this paper. The results are shown in Table 1.

As can be seen from Table 1, the curvature ductility factor of the prestressed steel reinforced concrete frame beam meets the needs of earthquake engineering. With increasing degrees of prestressing, the ductility factor decreases.

6 CONCLUSIONS

The calculation expression of constituent materials' force, the sectional curvature, and the moment of prestressed steel reinforced concrete frame beam sections in the yield and ultimate states, have been obtained by using analytical methods this paper. The calculation examples show that the calculation method of this paper can reflect the actual performance of PSRC structuresr. This method can be used to estimate and review the ductility of prestressed steel reinforced concrete frame beams.

ACKNOWLEDGEMENT

This study is supported by the Research Project of the Department of Housing and Urban-Rural Development (2014-K2-029) and the Technology Project of Shandong Province Construction Department of Urban and Rural Housing (KY027).

REFERENCES

Xueyu Xiong, Feng Gao, 2011. Experimental investigation and analysis on large scale prestressed steel reinforced concrete frame. Journal of Sichuan University, 43(6), 1–8.
Xueyu Xiong, Feng Gao, 2011. Experimental investigation and crack resistance analysis on large scale prestressed steel reinforced concrete frame. Industrial Construction, 41(12), 16–20.
Xueyu Xiong, Feng Gao, 2011. Experiment and calculation of crack control on prestressed steel reinforced concrete frame beams. Industrial Construction, 41(12), 20–24.

Xueyu Xiong, Feng Gao, 2011. Experiment and calculation of bearing capacity of normal section of prestressed steel reinforced concrete frame beams. Industrial Construction, 41(12), 24–30.

Jun Wang, Dan Wu, Wenzhong Zheng, 2009. Mechanical behavior of normal section of simply supported pre-stressed H-steel reinforced concrete beam. Journal of Harbin Institute of Technology, 41(6), 22–27.

Fu C.G., Li Y.Y. 2007. Experimental study on simply supported prestressed steel reinforced concrete beams. Journal of Building Structures, 28 (3): 62–73.

Zheng W.Z., Wang J.J. 2010. Experimental research on mechanical behavior of continuous prestressed composite concrete beams with encased H-steel. Journal of Building Structures, (7): 23–31.

Advances in Civil Engineering and Building Materials IV – Chang et al (eds)
© 2015 Taylor & Francis Group, London, ISBN: 978-1-138-00088-9

The foundation bearing capacity test of different forms of pipe pile based on PFC^{2D} simulation

Rui Huang & Xiangxing Kong

The First Highway Survey and Design Institute of China Communications Construction Company Ltd.,
Xi'an, China

ABSTRACT: In this paper, in order to compare the differences between an open-ended boots pipe pile and a closed boots pipe pile in sandy soil, a discrete element method is applied to investigate the foundation bearing capacity test based on the numerical simulation software PFC^{2D}, obtaining both the foundation bearing capacity, the scope of reinforcement in sand, and the failure pattern. The analysis has a certain guiding importance on optimization design, construction of pile foundation, and reduction in the engineering costs. Additionally, it provides certain reference significance for an understanding of the soil failure mechanism.

Keywords: foundation, bearing capacity, pipe pile, PFC, reinforcement scope.

1 INTRODUCTION

PFC^{2D}, the granular flow procedure in two-dimension, is applied to simulate the movement and interaction of circular particle medium (Cundall, 1999 & Cundall, 1979). With the outstanding economic performance, open-ended pipe piles have been widely used in engineering construction field. Based on particle flow theory, the laws of soil plug formation, mesa-structural changes of sandy soil particles, and stress field distributions of soil around pile are studied by using PFC^{2D} to simulate and reconstruct the process for different types of open-ended pipe piles driven into sandy soil (Zhan, 2013). In this paper, the foundation bearing capacity of two types of pipe piles were simulated by using PFC^{2D} for analysing the reinforcement zone and the failure pattern.

2 PFC^{2D} MODEL

2.1 Basic information

This paper aims to get the soil compaction effect regularity of open-ended boots pipe pile, and closed boots pipe pile. Therefore, the soil is supposed to be isotropic and homogeneous. The three-dimensional issue is simplified into the two-dimensional problem based on the axial symmetry. At the same time, considering the influence of the calculation efficiency, the foundation width and depth is set as 2.0 m and 7.0 m, respectively. The material parameters of this sandy soil stratum are shown in Table 1.

The numerical test simulates the body material of both open-ended boots pipe pile, and closed boots pipe

Table 1. Material parameters of sandy gravel.

Soil type	Rmin (m)	Rmax (m)	kn (Pa)	ks (Pa)	fric	e
sand	0.002	0.03	5.0e8	5.0e8	0.6	0.2

pile by soil clump composed of multiple particles, and the soil clump stiffness kn and ks is used to simulate the elastic modulus of pile. The friction coefficient fric of soil clump is used for the simulation of the internal friction angle of the pipe pile. The parameters of the pipe pile model are selected as shown in Table 2. The model generates about 14000 particles unit. For sand, the cohesive force is 0.

2.2 The ground model and the pipe pile model

The ground model and the pipe pile model were generated through the PFC^{2D} (Figures 1 and 3), and the gravity stress distribution is shown in Figure 2.

3 SIMULATION AND ANALYSIS OF THE PIPE BEARING CAPACITY TEST

3.1 The simulation of the pile foundation bearing capacity test

After the pile sinking, the bearing capacity test of the pile pipe is supposed to be accomplished for pile reinforcement effect evaluation. This paper used 1500 kN of static pressure to sink the pile into the sand ground,

Table 2. The parameters of the pipe pile model.

Pipe pile types	Length outside (m)	Diameter inside (m)	Diameter (m)	Pipe lining (m)	Pipe boots (mm)			Kn (Pa)	Ks (Pa)	Fric
					H	t1	t2			
Open-ended boots	4.0	0.5	0.3	0.1	450	–	12	5.0e8	5.0e8	0.6
Closed boots	6.0	0.5	0.3	0.1	120	10	15	5.0e8	5.0e8	0.6

Note: Where H is the length of boots, and t1 and t2 is bottom plate thickness of closed pile and the thickness of boots plate respectively.

Figure 1. Model of ground.

Figure 2. Gravity stress field.

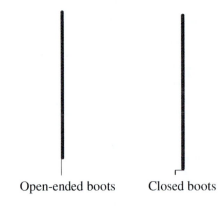

Open-ended boots   Closed boots

Figure 3. Model of the pipe pile.

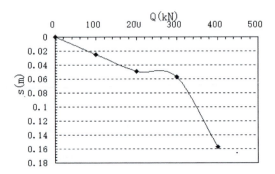

Figure 4. The pile foundation bearing capacity Q-s curve of the open-ended boots pipe pile.

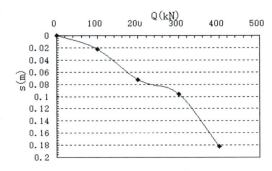

Figure 5. The pile foundation bearing capacity Q-s curve of the closed boots pipe pile.

and then the pressure was removed to make the pile rebound stable. Lastly, the pipe pile bearing capacity test was carried out.

In this test, the increment of load is 100 kN for each stage loading and, in total, ten stages were conducted. The direction of the load is vertical down. It is time to stop loading when the settlement reaches 0.1 m, in order to simulate the pipe foundation bearing capacity test after the pile construction is completed. The settlement of pile was recorded as the subsidence remained stable under each state loading, and the loading-displacement curve was shown in Figure 4 and Figure 5.

With the purpose of comparing the influence region and the soil failure pattern of these two forms of pipe pile in sandy soil, the distribution of interparticle contact force in the ground was observed, as shown in Figure 6.

The black line is used to represent the contact force between particles, and the width of line indicates the magnitude of contact force.

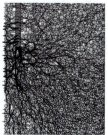

Before loading   After loading

Open-ended Boots Pipe Pile

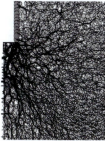
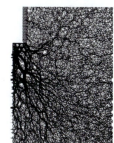

Before loading   After loading

Closed Boots Pipe Pile

Figure 6.   Inter-particle contact force.

Table 3.   The comparison of ultimate bearing capacity between different forms of pipe pile.

Pipe pile types	Open-ended boots pipe pile	Closed boots pipe pile
Ultimate bearing capacity	260 kN	230 kN

3.2   Analysis of the pipe pile foundation bearing capacity

The ultimate bearing capacity of pipe pile foundation was set equal to the value corresponding to 80 mm subsidence in Figure 4 and Figure 5, and was listed in Table 4.

It can be seen from Table 4 that the ultimate bearing capacity of different forms of pipe pile is noticeable, that is to say the ultimate bearing capacity of open-ended boots pipe was greater than that of closed boots pipe pile.

3.3   The main reinforcement region of pipe pile and the failure pattern of soil

As shown in Figure 6, the soil reinforcement effect of different forms of pipe pile was evident chiefly within two to three times of the pipe pile diameter. The soil compaction effect of the open-ended boots pipe pile within the pile diameter was on the trend of symmetric form, and the soil compaction degree of closed

Open-ended Boot Pipe Pile   Closed Boot Pipe Pile

Figure 7.   The soil particle velocity vector diagram as the pile was destroyed.

boots pipe pile decreased rapidly from the pipe wall to ambient soil.

Therefore, from the perspective of soil reinforcement effect of pipe pile, open-ended boots pipe pile was better than closed boots pipe pile.

It was shown in Figure 7 that the maximum speed was all 1.5 m/s when the foundation bearing capacity test of different open-endings forms of pipe pile was carried out until the pile was destructed. The length of the line denoted the magnitude of particles movement speed. The area of soil can be considered to be damaged already when the movement speed of soil particles were sufficiently large. Compared with the closed boots pipe pile, the soil damage scope of the open-ended boots pipe pile was the biggest and the damage zone appeared as triangles along the pile body. In addition, the damage area was greatest around the pile end for both of them.

The damage area reflected the sphere of influence of different forms of open-endings pipe pile to a certain extent.

4   CONCLUSIONS AND THE PROBLEMS

The following conclusions can be obtained by simulation and analysis above:

(1) From the point of pile foundation bearing capacity, open-ended boots pipe pile was superior to closed boots pipe pile.

(2) From the point of soil reinforcement area, the area of both forms of pipe pile was 2–3 times pile diameter. The soil compaction effect of open-ended boots pipe pile within the pile diameter was on the trend of symmetric form, and then it gradually degraded around. Besides, the soil compaction

degree of closed boots pipe pile decreased rapidly from the pipe wall to ambient soil.

(3) From the point of failure pattern, compared with the closed boots pipe pile, the soil damage scope of the open-ended boots pipe pile was the biggest, and the damage zone appeared as triangles along the pile body. In addition, the damage area was greatest around the pile end both of them.

In this paper, the foundation bearing capacity of the two forms of pipe pile, the characteristics of reinforcing zone, and the variety law of soil stress before and after destruction were simulated, and analysed using the theory of particle flow. The analysis results have a certain guiding significance for pile foundation design and construction, optimal design, reducing costs and so on. Furthermore, it has definite reference significance, and value for a deep understanding of soil failure mechanism.

REFERENCES

Cundall P A, Strack O D L. Particle flow code in 2 Dimensions [A]. Itasca Consulting Group, Inc, 1999.

Cundall P A, Strack O D L. A discrete numerical model for granular assemblies [J]. Geotechnique, 1979, 29(1): 47–65.

Zhan Yongxiang, Yao Hailin, Dong Qipeng et al. Study of process of open-ended pipe pile driven into sand soil by particle flow simulation [J]. Rock and Soil Mechanism, 2013(01): 283–289.

Transportation engineering

Advances in Civil Engineering and Building Materials IV – Chang et al (eds)
© 2015 Taylor & Francis Group, London, ISBN: 978-1-138-00088-9

A mode choice behaviour study of bicycle travelling

Liu Yang, Yuanqing Wang & Chao Li
School of Highway, Chang'an University, Xi'an, China

ABSTRACT: We established a binary logistic regression model to analyse bicycle mode choice behaviours, and the model estimation results are also discussed. Age, annual income, and trip distance are found to be significant impact factors of the bicycle mode. People less than 18 years old, with an annual income of less than RMB 40,000, and trip distance of between 3 km and 5 km, prefer to choose the bicycle mode. In addition, the reasons for the bicycle mode choice behaviour are also analysed in this paper. Almost all people think of danger as being the first reason for not choosing the bicycle mode, secondly, the inconvenience, and thirdly, the discomfort. Time-saving and convenience are the two main reasons for the bicycle mode choice except for those less than 18 years old, who believe bicycles are advantageous in time-saving and safety.

Keywords: mode choice, binary logistic regression, cycling, travel behaviour, low-carbon transport.

1 INTRODUCTION

In contrast to driving, cycling is one of the most sustainable forms of transport, and cycling offers benefits to both individual and environmental health (Jason G. Su et al., 2010, Elliot Fishman et al., 2012). However, cycling is under decline with continuous economic growth, fast motorizations, and urbanizations in most Chinese cities in recent decades. To try to reverse this tendency, many cities began to promote cycling, and thus bicycle mode choice behaviour studies have become the foundations for policymakers. Yang Chen divided the impact factors of the bicycle mode choice into individual socio-economic characteristics and trip characteristics (Yang Chen et al., 2007). Shan Xiaofeng also believed that age, job, income, and trip distance would affect people choosing the mode of the bicycle (Shan Xiaofeng et al., 2006), and Ma Pei considered bicycle facilities as a significant factor influencing the mode choice behaviour (Ma Pei et al., 2011). Thus, based on the binary logistic regression models established by using the data collected in the Economic and Technological Development Zone in Xi'an, this paper works on analysing the socio-economic and travel characteristics that will affect the cycling. In addition, the reasons for bicycle mode choice behaviour are also analysed in this paper. Policy suggestions are proposed based on the model results and data analysis, and the findings in this paper can be of reference value for the policymakers in order to promote cycling in Xi'an and other cities in China as well.

2 METHODOLOGY AND DATA COLLECTION

The binary logistic regression models are established with individual socio-economic characteristics (age and annual income) and travel distance as independent variables, and the dependent variable is whether to choose the bicycle for commuting. The logistic regression module in SPSS 19.0 was applied in this paper.

In a binary logistic regression, the cumulative logistic probability function is used, which assumes if the possibility of $Y = 1$ is P, then the one of $Y = 0$ is $1 - P$. The dependent variable Y could be regarded as a result of a Bernoulli trial because Y is either 0 or 1. To logit transfer P means to $\ln(P/1 - P)$, marked as Logit$P$. The value of Logit P is between infinity and infinitesimal (Mark Wardman et al., 2007, Yang Liu, 2010). Then the equation of linear regression with Logit P as the dependent variable can be built as below:

$$\text{Logit} P = \alpha + \sum_{k=1}^{K} \beta_k x_k \, (k = 1, 2 \cdots K) \qquad (1)$$

$$P = \frac{Exp(\alpha + \beta_1 x_1 + \beta_2 x_2 + \beta_3 x_3 + \cdots + \beta_k x_k)}{1 + Exp(\alpha + \beta_1 x_1 + \beta_2 x_2 + \beta_3 x_3 + \cdots + \beta_k x_k)} \qquad (2)$$

The maximum likelihood estimation method is used for model estimation. In the model, α is the constant, which means the value of Logit$P$ when each independent variable is equal to 0, and β_k is called logistic regression coefficient, which means the change of Logit$P$, when other independent variables remain still and its x_k increases by one unit.

Table 1. Sampling rate in the neighbourhoods.

Household number in the neighbourhoods	Sampling rate
100–300	>50%
300–800	50%–30%
>800	30%–10%

Table 2. Results of the bicycle mode choice models.

	Model 1	Model 2	Model 3	Model 4
constant	−1.558***	−3.932***	−2.732***	−3.922***
Age 0~18 years old	0.711			1.266*
Age 18~35 years old	−0.989*			−0.589
Age 35~55 years old	−0.120			0.273
Annual income less than RMB40,000		1.949**		1.697*
Annual income of RMB40,000–80,000		1.563*		1.488*
Trip distance less than 3 km			0.338	0.138
Trip distance 3 km–5 km			0.991***	0.961***

*$p < 0.1$, **$p < 0.05$, ***$p < 0.01$.

The survey was conducted in the Economic and Technological Development Zone in Xi'an, and the sampling rate is determined according to the number of households in the neighbourhoods as shown in Table 1. The questionnaire included socio-economic and travel characteristics, people's assessment, and their subjective perceptions of the transport facility and its service, and the reasons for the mode choice behaviours.

3 RESULT ANALYSIS AND DISCUSSIONS

The logistic regression module in SPSS 19.0 was applied and the model estimation and test results are shown in Table 2.

The negative constants for all the four binary logistic models imply that commuters would not prefer to cycle even with all other factors including age, annual income, and travel distance held equal. Model 1 shows that people less than 18 years old prefer to cycle rather than people between 18 and 55 years of age. Model 2 shows that people with an income less than RMB 40,000 prefer to cycle more than ones with an annual income of RMB 40,000–80,000. Model 3 shows that people who travel between 3 km and 5 km are more likely to use bicycles. Model 4 combined with all variables, shows that all variables maintain a

Table 3. Reasons for not choosing the bicycle mode for different ages, annual incomes, and trip distances.

	Inconvenience	dangerous	discomfort
Age 0~18 years old	25.00%	55.00%	20.00%
Age 18~35 years old	32.52%	52.52%	14.96%
Age 35~55 years old	31.08%	57.43%	11.49%
Annual income less than RMB40,000	34.29%	48.69%	17.02%
Annual income of RMB40,000–80,000	28.62%	60.86%	10.53%
Trip distance less than 3 km	33.75%	53.75%	12.50%
Trip distance 3 km–5 km	32.31%	53.71%	13.97%

good stability of impact on the bicycle mode choice except between the ages of 35 and 55 years. Model 4 shows again that people less than 18 years old, with an annual income of less than RMB40,000, and who travel between 3 km and 5 km, tend to prefer the bicycle. These factors are strong impact factors and significant predictors of the bicycle mode choice.

Reasons for bicycle mode choice behaviours from the aspects of convenience, safety, and comfort, are analysed in this paper. From the statistical results in Table 3, 48.69%–60.86% of the surveyed samples think that bicycles are dangerous, 25%–33.75% do not choose the bicycle because of the inconvenience, and 10.53%–20% do not use bicycles because of the discomfort factor. The reasons are almost the same for why people do not choose the bicycle as a mode of transport; the majority of people think that first, they are dangerous, second, they are inconvenient, and third, they are uncomfortable.

From the statistical results in Table 4, there are consistent reasons for the bicycle mode choice except for those less than 18 years old. 22.09%–30.86% choose the bicycle mode for time-saving, 12.06%–18.31% for economic reasons, 25.56%–34.75% for convenience, 14.29%–20.93% for safety, and 8.14%–13.41% for comfort. The statistical results show that the proportions of the five reasons (time-saving, economy, convenience, safety, comfort) for the bicycle mode choice have little difference among the people of different ages, annual incomes, and trip distances, except for those less than 18 years old. The proportions of the five reasons are 29.89% for time-saving, 13.63% for economic reasons, 26.97% for convenience, 19.61% for safety and 9.90% for comfort; time-saving and convenience are the two main reasons. However, those less than 18 years old are exceptions. 50% of them choose the bicycle for timesaving, 37.5% for safety, while only 12.5% for convenience.

Table 4. Reasons for choosing the bicycle mode for different ages, annual incomes, and trip distances.

	Time-saving	Economy	Convenience	Safety	Comfort
Age 0~18 years old	50.00%	0.00%	12.50%	37.50%	0.00%
Age 18~35 years old	29.27%	13.66%	28.54%	15.12%	13.41%
Age 35~55 years old	22.09%	17.44%	31.40%	20.93%	8.14%
Annual income less than RMB40,000	29.63%	18.52%	25.56%	14.81%	11.48%
Annual income of RMB40,000-80,000	23.40%	12.06%	34.75%	17.73%	12.06%
Trip distance less than 3 km	30.86%	15.43%	28.57%	14.29%	10.86%
Trip distance 3 km–5 km	23.94%	18.31%	27.46%	16.90%	13.38%

4 CONCLUSION

The binary logistic regression model of bicycle mode choice is established and the model estimation results are analysed in this paper. We find that people's age, annual income, and trip distance have significant impacts on bicycle mode choice behaviour. People less than 18 years old, with an annual income of less than RMB 40,000, and trip distances between 3 km and 5 km, are much more likely to choose the bicycle mode. In addition, the reasons for the mode choice behaviour are discussed in this paper as well. The majority of people not choosing the bicycle mode share the same reasons: safety being the first, second is the inconvenience, and third is the discomfort. Except for those less than 18 years old, other people's preferences for choosing the bicycle are almost the same, and time-saving and convenience being the two prior reasons. Thus, attracting more middle-aged people with a high income to use bicycles is the crucial point in promoting cycling, which can be realized by having more cycle lanes and parking sites to improve the safety and convenience for cyclists. These findings can be references for the policy makers to encourage more bicycle mode choices in Xi'an and other Chinese cities.

ACKNOWLEDGEMENT

This study was funded by the Natural Science Foundation of China (No.51178055-E0807).

REFERENCES

Su, J. G., Winters, M., Nunes, M., & Brauer, M. (2010). Designing a route planner to facilitate and promote cycling in Metro Vancouver, Canada. Transportation research part A: policy and practice, 44(7), 495–505.

Yang C., Lu J., Wang W., & Wan Q. (2007). A Study on the influencing factors of bicycle transportation based on individual mode choice. Journal of Transportation Systems Engineering and Information Technology, 7(4), 131–136.

Shan X. F., Wang W., Wan Q., & Yang C. (2006). Research on the mode choice of bicycle base on the behavior analysis. Modern Urban Research, 11, 81–88.

Ma P., & Wu H. (2011). Research on the Bicycle Transfer rail traffic behavior mechanism and model. Journal of Beijing University of Civil Engineering and Architecture, 27(2), 36–40.

Wardman M., Tight M., & Page M. (2007). Factors influencing the propensity to cycle to work. Transportation research part A: policy and practice, 41, 339–350.

Yang L. (2010). Study on the travel behaviors and characteristics. Thesis for the Master Degree of Chang'an University.

Fishman E., Washington S., & Haworth N. (2012). Barriers and facilitators to public bicycle scheme use: A qualitative approach, Transportation Research Part F: Traffic Psychology and Behaviour, 15, 686–698.

Advances in Civil Engineering and Building Materials IV – Chang et al (eds)
© 2015 Taylor & Francis Group, London, ISBN: 978-1-138-00088-9

Urban expressway density analysis based on the cusp catastrophe theory

Jian Gu & Shuyan Chen
School of Transportation, Southeast University, Nanjing, China

ABSTRACT: Catastrophe characteristics are represented in the evolution of the traffic flow actual data records, which were insufficiently supported by the traffic fundamental continuous models. Catastrophe theory was used to explore the mathematical relations among the traffic data collected from highway conditions, however, the results could not be used for the urban expressway conditions. Traffic flow data collected from the 3rd Ring Road Expressway in Beijing was used to build flow-density models and speed-density models. Then the density was discussed based on the traffic wave speed function with cusp catastrophe theory, which was used to analyse the traffic freeway characteristics. Density conditions between median lanes and shoulder lanes were discussed in depth, and the results indicated that the catastrophe critical density value in the shoulder lanes was a little smaller than that in the median lanes.

Keywords: traffic flow, catastrophe theory, density, actual data analysis.

1 INTRODUCTION

Operation characteristics under the condition of the urban expressway are becoming important objects for discussion concerning urban traffic flow. There is substantial research mainly on traffic parameter relations under the urban expressway condition and many methods are implemented, such as models based on traffic flow fundamental diagrams with mathematical statistics, and models that are more complex, derived from dynamical system theory. Dong et al., built the regression models including the traffic flow speed model, the traffic flow-occupancy model, and the speed-occupancy model based on actual collected data by the RTMS (Remote Traffic Microwave Sensor) fixed on the 3rd Ring Road Expressway in Beijing (Dong et al., 2013). Zhao et al., analysed traffic flow characteristics for each lane of the 3rd Ring Road Expressway in Beijing using a Van Aerde single-regime model (Zhao et al., 2010). Zhu et al., discussed the macro traffic operations and congestion affecting factors of the 3rd Ring Road Expressway in Beijing using macro fundamental diagram methods (Zhu et al., 2012).

Models based on the typical traffic fundamental diagram could describe traffic parameters better; however, complex and stochastic factors, such as driver behaviour, weather, and the environment, cause difficulties in explaining and building precise models, especially the discontinuous or 'kick' phenomena, which is hard to explain by the three-parameter traffic fundamental continuous models in two-dimensional planes. If the traffic parameters are treated as a dynamic system, it could be a catastrophe system, such as Guo et al., discussed traffic models with cusp catastrophe theory (Guo et al., 2007, 2008). Cusp and swallowtail catastrophe theory models had also been selected to build highway traffic parameters models in substantial research respectively (Zhang et al., 2000, Tang et al., 2005, Forbes et al., 1990, Hall et al., 1987, Acha-Daza et al., 1994). Conclusions from the research above were mainly inferred from highway conditions, but these are different from urban expressway conditions, which have smaller distance between ramps, and vehicles could be easily influenced when driving near the ramps; traffic operation in the inner lanes could be disturbed as well. Therefore, the conclusion would not be accurate for expressway conditions. However, the methods of using the cusp catastrophe theory may be used for discussion.

This paper focuses on analysing expressway traffic conditions by using the cusp catastrophe model; parameters of density will particularly be discussed in depth.

2 CATASTROPHE THEORY

René Thom set up the Catastrophe Theory based on discontinuous research, an evolution from stability theory and bifurcation theory, which can be used to explain the discontinuous transition of a system between equilibrium states. Using differential equations and functions, seven types of catastrophe models were induced from the singularities of a gradient system. Five characteristics (Ling F.H., 1987) including bimodality, hysteresis, inaccessible behaviour, catastrophe, and divergence are also observed in a dynamical system of traffic parameters, so the catastrophe theory can be applied to solve the traffic model problem, and one of the models, the cusp catastrophe model, is selected for analysing due to its wide use. Its basic equations are shown in Equations 1 to 3:

Potential function: $\quad P(x) = x^4 + Ax^2 + Bx \quad$ (1)

Balance curved surface: $4x^3 + 2Ax + B = 0 \quad$ (2)

Catastrophe set: $\begin{cases} 4x^3 + 2Ax + B = 0 \\ 12x^2 + 2A = 0 \end{cases} \quad$ (3)

where, x is the state variable, and A, B are control factors (bifurcation/splitting factor and normal/asymmetry factor respectively).

3 TRAFFIC FLOW MODELS

Flow Q, speed V, density K, are used to describe traffic operation, and the basic relationship among them is $Q = K \cdot V$. Meanwhile, generalized car-following theory function in Equation 4 when $r = 0, s = 3$, it can be the widely used model as in Equation 5 (May et al., 1967):

$$V^{1-r} = V_f^{1-r}\left[1 - \left(\frac{K}{K_j}\right)^{s-1}\right] \quad (4)$$

$$V = V_f\left[1 - \left(\frac{K}{K_j}\right)^2\right] \quad (5)$$

The first differential of Equation 5 on K is shown as Equation 6:

$$\frac{dV}{dK} = -2\frac{V_f}{K_j}\left(\frac{K}{K_j}\right) \quad (6)$$

In Equation 7 V_ω is deduced as the traffic shock wave function:

$$V_\omega = \frac{\Delta Q}{\Delta K} = \frac{dQ}{dK} = K\frac{dV}{dK} + V = K\frac{dV}{dK} + \frac{Q}{K} \quad (7)$$

The traffic shock wave function can be adapted as Equation 8 with density K and flow Q based on Equation 6 and Equation 7:

$$K^3 + \frac{K_j^2}{2V_f} \cdot V_\omega \cdot K - \frac{K_j^2}{2V_f} \cdot Q = 0 \quad (8)$$

Equation 8 is similar to the cusp catastrophe balanced curved surface function form in Equation 2, and the potential function for traffic parameters may be designed as Equation 9 by integral, where density K is the state variable:

$$P = \frac{1}{4}K^4 + (\frac{1}{4}\frac{K_j^2}{V_f} \cdot V_\omega) \cdot K^2 - (\frac{K_j^2}{2V_f} \cdot Q) \cdot K \quad (9)$$

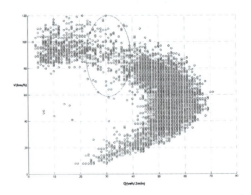

Figure 1. Scatter Illustration of Speed (v)-Volume (Q).

Then the critical wave speed V_ω and density K_{cusp} are deduced as Equation 10 and Equation 11 (Guo et al., 2007, 2008):

$$V_\omega = -\left(\frac{27V_f}{2K_j^2} \cdot Q^2\right)^{\frac{1}{3}} \quad (10)$$

$$K_{cusp} = \left(\frac{2K_j^2}{V_f} \cdot Q\right)^{\frac{1}{3}} \quad (11)$$

Based on Equation 11, the catastrophe critical density K_{cusp} is deduced from the traffic flow Q, and when vehicles pass by, there will be a maximum volume, Q_{max}, which would be selected to estimate the K_{mcusp} as Equation 12:

$$K_{mcusp} = \left(\frac{2K_j^2}{V_f} \cdot Q_m\right)^{\frac{1}{3}} \quad (12)$$

4 IMPLEMENTATION AND RESULTS

In this part, actual data collected in two-minute intervals from the RTMS (Remote Traffic Microwave Sensor) in Beijing is used, mainly from the 3rd Ring Road Expressway including flow volume, speed, and occupancy on workdays in a month. The discussion focuses on the traffic operation in the median lanes and the shoulder lanes. Traffic interruption in the shoulder lanes is usually observed due to vehicles driving in and out the ramps. And the primary discussion point focused on one direction traffic operation.

4.1 Data characteristics

Data cleaning was firstly applied to the raw data set by correcting abnormal data, and adding lost data according to time stamps was also used. Traffic propagation is shown as Figure 1 and Figure 2 based on the modified data.

Figure 1 descripts the speed (v) – traffic flow volume (Q) characteristics, illustrating that the point zone

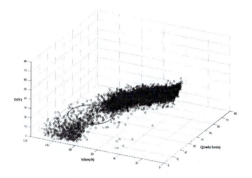

Figure 2. Scatter Illustration of Volume (Q)-Speed (v)-Occupancy (O).

with high velocity and low volume, namely free-flow phase, was discontinuously changed into the congested phase and a 'kick' phenomenon could exist in the evolution. In addition data points scatter in the free and congested zones. However, data is sparse in the joint circled by the red line. Figure 2, Volume (Q) – Speed (v) – Occupancy (O), shows the same feature including the occupancy with the same data set.

4.2 Parameter statistics

4.2.1 O-K function

Catastrophe critical density is deduced in Equation 12 based on the maximum value of traffic flow Q_{max}, free-flow speed V_f, and congested density K_j, however, only occupancy was recorded in the raw data, therefore it has to be changed to the density. Based on several former works, the formulation between occupancy and density is shown as Equation 13 (Roess et al., 2004, Zhao et al., 2009), where K is the density, C is the constant, O is the occupancy, $length_{vehicle}$ is the average length of vehicles, $length_{detector}$ is the sensor property:

$$K = \frac{CgO}{length_{vehicle} + length_{detector}} \quad (13)$$

In the condition of the Ring Road Expressway in Beijing, the average length of vehicles, $length_{vehicle}$, in Equation 13 is 5.78 meters in median lanes, 6.03 metres in middle lanes, and 6.33 metres in shoulder lanes (Zhao et al., 2009), and the constant C in Equation 13 is chosen as 10 for the expressway condition in Beijing. Meanwhile, the sensor property, $length_{detector}$, in Equation 14 is 2 metres based on the RTMS property. Then the median lanes O-K function and shoulder lanes O-K function are shown as Equation 14 and Equation 15 respectively:

Median lanes condition:

$$K_{median} = \frac{10}{7.78} \cdot O_{median} \quad (14)$$

Shoulder lanes condition:

$$K_{shoulder} = \frac{10}{8.33} \cdot O_{shoulder} \quad (15)$$

Table 1. Coefficients and Significance Test of Q-K Quadratic Regression Model in Median Lanes.

Coefficients results					Significance Test	
					t test	
m_1	n_1	p_1	R	F test	Quadratic	Linear
−0.02	1.89	18.43	0.78	139201.46 0.00	−475.33 0.00	526.51 0.00

Table 2. Coefficients and Significance Test of Traffic Q-K Quadratic Regression Model in Shoulder Lanes.

Coefficients results					Significance Test	
					t test	
m_2	n_2	p_2	R	F test	Quadratic	Linear
−0.01	1.39	14.63	0.73	98260.66 0.00	−400.48 0.00	441.53 0.00

4.2.2 Q-K quadratic regression

The quadratic regression model was used to build Q-K regression function in both median lanes condition and shoulder lanes condition. The results of parameters and model significance testing were shown as following based on the actual data. Quadratic Q-K regressions in median lanes and shoulder lanes are shown as Equation 16 and Equation 17 respectively.

$$Q_{median} = m_1 \cdot K_{median}^2 + n_1 \cdot K_{median} + p_1 \quad (16)$$

$$Q_{shoulder} = m_2 \cdot K_{shoulder}^2 + n_2 \cdot K_{shoulder} + p_2 \quad (17)$$

From Table 1, R value equals to 0.78, and F testing results are smaller than 0.05; the relation between Q_{median} and K_{median} is statistically authentic. The t testing results for regression coefficients are smaller than 0.05, meaning statistically significant.

From Table 2, R value equals to 0.73, and F testing results are smaller than 0.05; the relation between $Q_{shoulder}$ and $K_{shoulder}$ is statistically reliable. The t testing results for regression coefficients are smaller than 0.05, meaning statistically significant.

4.3 K_{cusp} Calculation

Equation 12 is applied to calculate critical catastrophe density, K_{mcusp}. According to Equation 16 and Equation 17, maximum Q can be calculated. Free-flow velocity value, V_f, could be inferred by building a speed-flow Greenshields model with congested density from actual data. Finally, the critical catastrophe density K_{mcusp} can be calculated and the results in median lanes and shoulder lanes are recorded in Table 3.

Table 3. K_{cusp} Calculation in Median Lanes and Shoulder Lanes.

	Median lanes	Shoulder lanes
K (veh/km)	48	52
Q_{max} (veh)	64	51
V_f (km/h)	77.6	69.2
K_j (veh/km)	87	89
K_{mcusp} (veh/km)	23	22

4.4 Results analysis

From the values in Table 3, several results can be inferred. Firstly, the critical density is smaller than congested density in both median and shoulder conditions, and the data description shows that when the density value reaches the critical value, the traffic flow phase would be transformed from unconstrained status to constrained status with the 'kick' phenomenon.

Therefore, when traffic operation is being monitored, once the density parameter rises to the critical level, several effective control measures should be actualized to reduce the volume in shoulder lanes near ramps in order to keep a safe driving and avoid traffic accidents.

Secondly, the critical density, K_{mcusp}, in different lane conditions shows different values, and the results show that K_{mcusp} in shoulder lanes is smaller than that in median lanes. This says that the conditions of shoulder lanes could reach the catastrophe phase earlier than the conditions of median lanes in terms of density.

In fact vehicles driving in or out of ramps will change lanes across the shoulder lanes, for example, when a vehicle drives across the shoulder lanes in or out of ramps, the drivers following could be influenced by lane-changing, and local density may be increased causing a traffic jam. Therefore traffic flow may be disturbed and the density will be changed as a wave spreads backwards with certain speed, like a 'shock wave' causing the flow-density catastrophe phenomenon, in particular traffic flow in shoulder lanes near ramps.

5 CONCLUSIONS

In this paper, actual data collected from the 3rd Ring Road Expressway in Beijing was applied to develop flow-density models and speed-density models. The density parameter was analysed in particular, and the traffic wave speed function with cusp catastrophe theory was applied to analyse the freeway traffic characteristics. In particular, conditions both in median lanes and shoulder lanes were discussed in depth and the results indicated that the critical catastrophe density value, K_{mcusp}, in the shoulder lanes was a little smaller than that in the median lanes. Driving in-out ramps needs to be controlled when the density reaches the critical value by effective measures.

This paper only focused on the condition of the 3rd Ring Road Expressway in Beijing, the results may be restricted by the raw data from RTMS, and several measures will be used to improve the model accuracy. Meanwhile, temporal and spatial factors will be considered in future research.

ACKNOWLEDGEMENT

This work is part of a research project supported by Jiangsu Province University Graduate Student Research and Innovation Program NO. CXLX13_109.

REFERENCES

Acha-Daza J. A. & Hall F L., 1994. Application of catastrophe theory to traffic flow variables. Transportation Research Part B: Methodological 28(3), 235–250.

Dong C.J., Shao C.F., Zhuge C.X. & Li H.X., 2013. Estimating traffic flow models for urban expressway based on measurement data (in Chinese). Journal of Transportation Systems Engineering and Information Technology (Chinese Edition) 3, 46–52.

Forbes G. J. & Hall F. L., 1990. The applicability of catastrophe theory in modelling freeway traffic operations. Transportation Research Part A: General 24(5), 335–344.

Guo J., Chen X.L. & Jin H.Z., 2007. Based on the cusp catastrophe to research the relationship among traffic flow three parameters, 2007 IEEE International Conference on Automation and Logistics, 858–861.

Guo J., Chen X.L. & Jin H.Z., 2008. Research on model of traffic flow based on cusp catastrophe (in Chinese), Control and Decision 2, 237–240.

Hall F L., 1987. An interpretation of speed-flow-concentration relationships using catastrophe theory. Transportation Research Part A: General 21(3), 191–201.

Ling F.H. Catastrophe Theory and Application (in Chinese). Shanghai Jiaotong University Press, 1987.

May A. D., & Harmut, E. M., 1967. Non-integer car-following models. Highway Research Record, issue no.199, 19–32.

Roess, Roger P., Elena S. Prassas, & William R. McShane., 2004. Traffic Engineering. Prentice Hall, 2004.

RTMS, http://www.chinahighway.com/cp/cp_info.php?id=2025

Tang T.Q. & Huang H.J., 2005. The discussion of traffic flow forecast by using swallowtail catastrophe theory (in Chinese). Journal of Mathematical Study 1, 112–116.

Zhang Y.P.& Zhang Q.S., 2000. The application of cusp catastrophe theory in the traffic flow forecast (in Chinese). Journal of Systems Engineering 3, 272–276.

Zhao N.L., Yu L., Zhao H., Guo J.F. & Wen H.M., 2009. Analysis of traffic flow characteristics on ring road expressways in Beijing. Transportation Research Record: Journal of the Transportation Research Board 2124(1), 178–185.

Zhao N.L., Yu L., Chen X.M, Guo J.F. & Wen H.M., 2010. A multidimensional study on traffic flow characteristics of expressways in Beijing (in Chinese). Journal of Beijing Jiaotong University 6, 35–39.

Zhu L., Yu L. & Song G.H., 2012. MFD-based investigation into macroscopic traffic status of urban networks and its influencing factors (in Chinese). Journal of South China University of Technology (Natural Science Edition) 11, 138–146.

Advances in Civil Engineering and Building Materials IV – Chang et al (eds)
© 2015 Taylor & Francis Group, London, ISBN: 978-1-138-00088-9

An evolution model of a navigation system

Zhao Liu, Jing-Xian Liu, Yuan-Qiao Wen, Jiao Zhu & Ya-Kun Fang
Hubei Inland Shipping Technology Key Laboratory, School of Navigation,
Wuhan University of Technology, Wuhan, Hubei, China

Ryan Wen Liu
Department of Mathematics, Wuhan University of Technology, Wuhan, Hubei, China

ABSTRACT: The navigation environment is an important basic condition for ship sailing, and it is related to vessel traffic flow. The navigation environment and vessel traffic flow are two factors required to investigate navigation safety, navigation efficiency, and port design. Firstly, navigation environment elements are analysed according to the characteristics of ship sailing. Secondly, a navigation system model is developed, which is composed of several sub-systems, such as natural conditions, navigation waters, service management, and vessel traffic flow. Thirdly, the driving mechanism of vessel traffic flow, the revulsive mechanism of navigation waters, the restriction mechanism of natural conditions, and the optimal adaptation mechanism of service management are analysed. Finally, a mathematical model of the evolution of a navigation system is developed by combining the four sub-systems. The study of navigation systems has an important theory value for exploring the interaction evolution mechanism between berth, channel, anchorage, service management, vessel traffic flow, and the natural environment. In practice, it is important to navigation efficiency improvement, navigation safety enhancement, and port design optimization.

Keywords: Navigation system; vessel traffic flow; evolution mechanism; evolution model.

1 INTRODUCTION

With the development of water transportation, the limited navigation resources will exacerbate the problem of navigation safety and efficiency. How to improve navigation efficiency and navigation safety is one of the key factors in ensuring the healthy development of navigation traffic engineering.

In current literature, many efforts have been made to use the navigation environment to evaluate and analyse navigation safety. For example, Li et al. [1] divided the navigation environment into five aspects of hydrological and meteorological factors, that is, port conditions factors, channel factors, traffic factors, and surface or underwater operation factors, and analysed their influence on navigation safety. Li et al. [2] divided the navigation environment into natural environmental factors, navigational factors, and human environmental factors, and the affecting factors of the navigation environment were analysed by using the interpretative structural model. Kiyoshi [3] analysed the potential risk to water transportation from two aspects, the probability of ship encounter, and the consequences of collision, and evaluated maritime traffic environment quantitatively. Jiang et al. [4] analysed the navigation environment factors of channels, and evaluated the navigation environment based on the source data. There is no literature which studies the navigation

system based on complex navigation systems, composed of natural conditions, navigation waters, service management, and vessel traffic flow.

In this study, a complex navigation system model is constructed by using system engineering based on influence factors of navigation. In order to construct a navigation system model, ship sailing elements are identified, and then the influence factors are divided into four sub-systems, that is, natural conditions, navigation waters, service management, and vessel traffic environment. In this system model, the sub-system of vessel traffic flow is the core of the complex navigation system. Study on the evolution model of the navigation system has an important guiding significance to improve the traffic efficiency, ensure navigation safety, and optimize the port design.

2 NAVIGATION SYSTEM

2.1 *Analysis of the navigation environment*

The navigation environment refers to the sum of natural and social conditions of a ship in a navigation process, including hydrology and meteorology, channel, anchorage, berth, navigation facilities, management rules and regulations, and so on.

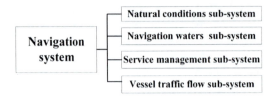

Figure 1. Navigation system model.

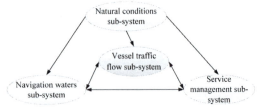

Figure 2. Navigation system relationship model.

In the navigation environment, hydrology and meteorology conditions are the sum of the natural conditions that have an effect on ship navigation safety, mainly including wind, flow, fog, rain (snow), wave, tide, and so on. The channel conditions mainly include the depth, width, and length of the channel. The anchorage conditions mainly include the capacity, depth, and quality of the bottom. The berth conditions include the grade, type, and operating efficiency. Navigational facilities and attachment equipment include beacons, navigation aids, and dam headlights. Traffic management includes the management regulations, navigation requirements, and so on.

2.2 Navigation system model

The vessel traffic flow is the sum of a ship sailing in the navigation environment and its characteristics, including the number of ships, traffic flow structure, spatial and temporal distribution, behaviour characteristics, and so on.

In order to analyse the relationship between the navigation environment and the vessel traffic flow, a navigation system is constructed. The navigation system is a system of integrity, inner–correlation, structure, and dynamic balance.

A navigation system is divided into a natural conditions sub-system, a navigation waters sub-system, a service management sub-system, and a vessel traffic flow sub-system according to the characteristics of navigation environment conditions and vessel traffic flow. Because the vessel traffic flow has a characteristic of self-organizing and a navigation system can be controlled by humans, the vessel traffic flow sub-system is regarded as the core of the navigation system. The navigation system model is shown in Figure 1.

In Figure 1, the natural conditions sub-system is composed of wind, flow, fog, rain (snow), waves, tides, ice, and other entities. The navigation waters sub-system is composed of channel, anchorage, and berth. The service management sub-system includes provisions for shore-based facilities and navigation rules and requirements. The vessel traffic flow sub-system includes the number of ships, type, scale, tonnage, structural characteristics, and a large number of the temporal and spatial distribution characteristics and behaviour characteristics of the ships.

2.3 Navigation system relationship model

In the navigation system, the natural conditions sub-system is the basal sub-system of the navigation

system, and has a key influence on the navigation waters sub-system, the service management sub-system, and the vessel traffic flow sub-system. The service management sub-system and the navigation waters sub-system are the foundations of the ship's navigation, and the service management sub-system, the navigation waters sub-system, and the natural conditions sub-system together influence the vessel traffic flow sub-system. The vessel traffic flow sub-system information is fed back to improve the navigation waters sub-system and the service management sub-system, and drive the evolution of the navigation system. The navigation system relationship model is shown in Figure 2.

In Figure 2, the core function of the vessel traffic flow sub-system and the interrelationship between each sub-system are reflected. The natural conditions sub-system acts on the navigation waters sub-system and the service management sub-system. The three sub-systems, which are the natural conditions sub-system and navigation waters sub-system and service management sub-system, affect vessel traffic flow sub-system together. The navigation waters sub-system and the natural conditions sub-system act on the service management sub-system together, and accept the feedback from the service management sub-system. The vessel traffic flow sub-system and the navigation waters sub-system and service management sub-system act upon each other.

3 EVOLUTION MECHANISM OF THE NAVIGATION SYSTEM

The essence of the evolution of the navigation system is the interactive and changing process of the natural conditions sub-system, the service management sub-system, and the navigation waters sub-system in order to meet the needs of vessel traffic flow change. The change of vessel traffic flow drives the evolution of the navigation system, and the evolution of the navigation system reflects the restriction of the natural conditions sub-system, the revulsive mechanism of navigation waters, and the adaptability of service management.

3.1 Vessel traffic flow sub-system driving mechanism

The self-organization of the navigation system is such that the service management sub-system and the navigation waters sub-system must evolve according to the development of the vessel traffic flow sub-system. In this process, when the vessel traffic flow is far greater than the capacity of the navigation waters and service management sub-systems, the vessel traffic flow drives the evolution of the navigation waters sub-system and the service management sub-system, and provides the power for the navigation system evolution.

3.2 Natural conditions sub-system restriction mechanism

The natural conditions sub-system acts on the navigation waters sub-system, the management sub-system, and the vessel traffic flow sub-system directly. When the natural conditions are well, ship can sail in navigation waters, otherwise the ship cannot sail. And the time and the place of navigation are provided according to limiting conditions for ship sailing in port.

3.3 Navigation waters sub-system revulsive mechanism

When the capacity of the navigation waters and service management sub-systems is far greater than the vessel traffic flow, the navigation waters will induce ship to reach the port. And the bigger capacity the navigation waters sub-system, the stronger the magnetism to ships. So the navigation waters sub-system leads the vessel traffic flow sub-system to develop. The stronger the magnetism to

3.4 Service management sub-system optimal adaptation mechanism

The vessel traffic flow sub-system and the navigation waters sub-system are interrelated and interact on each other, and develop constantly under natural conditions. Therefore, a service management sub-system is needed to adapt the vessel traffic flow sub-system. In order to ensure that vessel traffic flow evolves normally and navigation waters sub-system is economized, the service management sub-system must optimally match with the natural conditions' sub-system, the vessel traffic flow sub-system, and the navigation waters sub-system.

4 EVOLUTION MODEL OF NAVIGATION SYSTEM

4.1 Mathematical model of the four sub-systems

A complex navigation system evolution model should not only reflect the driving mechanism of the vessel traffic flow sub-system, but also reflect the restriction mechanism of the natural conditions sub-system, the induced mechanism of the navigation waters sub-system, and the optimal adaptation mechanism of the service management sub-system, and reflect the nonlinear characteristics of the system's development.

Navigation system evolution parameters' settings:

(1) Vessel traffic flow sub-system: set ship number as the evolution parameters of the vessel traffic flow sub-system.
(2) Navigation waters sub-system: set the through capacity of channel, berth, and anchorage as the evolution parameters of the navigation waters sub-system.
(3) Natural conditions sub-system and service management sub-system: the restriction mechanism of the natural conditions sub-system and the optimal adaptation mechanism of the service management sub-system are reflected.

The calculation model on through capacity of the navigation waters sub-system can be shown as Equation 1:

$$V' = f_{\text{service}}(f_{\text{natural}}(C_c, C_b, C_a)) \tag{1}$$

where,
V': the through capacity of the navigation waters sub-system in unit time, ships/day.
C_c: the through capacity of the channel in unit time, ships/day.
C_b: the through capacity of the berth in unit time, ships/day.
C_a: the through capacity of the anchorage in unit time, ships/day.
f_{natural}: the comprehensive effect of natural conditions acting on the channel, berth, or anchorage.
f_{service}: the optimal adaptation of service management.

The calculation model on ship numbers of the vessel traffic flow sub-system is shown as Equation 2:

$$V = f_s(V_{11}, V_{12}, \cdots, V_{ij}, \cdots, V_{mn}) \tag{2}$$

where,
V: standard ship number of vessel traffic flow in unit time, ships/day.
V_{ij}: ship number of the j scale of the i types in unit time, ships/day.
f_s: the conversion function of the different types and different scales.

4.2 Construction of evolution model

In the navigation system, the through capacity is limited. When vessel traffic flow reaches a maximum, the vessel traffic flow no longer increases, and the blocking effect of environmental conditions on vessel traffic flow growth increases in proportion to vessel traffic flow.

The development of the navigation system is the growth and change of the vessel traffic flow, the change of the navigation waters sub-system, and the improvement of the service management sub-system. In the ideal state, the growth of vessel traffic flow is "J" growth, and the ship number of vessel traffic flow has doubled and redoubled. However, the growth of

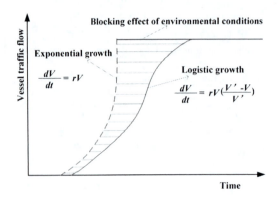

Figure 3. The "*J*" growth and "*S*" growth of vessel traffic flow (*r* is a constant).

vessel traffic flow is limited by the navigation waters sub-system, the natural conditions sub-system, and the service management sub-system, so the growth of vessel traffic flow will be "*S*" growth.

Based on a mathematical model of the four sub-systems and the "*S*" growth of vessel traffic flow, the evolution model of vessel traffic flow can be expressed as Formula 3, and in the ideal state the "*J*" growth of vessel traffic flow can be expressed as Formula 4:

$$\frac{dV}{dt} = rV\left(\frac{V'-V}{V'}\right) \tag{3}$$

$$\frac{dV}{dt} = rV \tag{4}$$

where,

V': the through capacity of the navigation waters sub-system in unit time, ships/day.

V: standard ship number of vessel traffic flow in unit time, ships/day.

r: a constant.

The "*J*" growth and "*S*" growth of vessel traffic flow are shown in Figure 3.

In Figure 3, the "*S*" growth is vessel traffic flow growth. With the growth of vessel traffic flow, the blocking effect of environmental conditions will increase, and when $V = V'$, the blocking effect of environmental conditions is largest, so vessel traffic flow cannot increase. Then measures will be taken to decrease the blocking effect of environmental conditions, and the navigation system will evolve.

The essence of navigation system evolution is the matching process of the navigation waters sub-system through capacity with vessel traffic flow under the restriction mechanism of the natural conditions sub-system and optimal adaptation mechanism of the service management sub-system. Therefore, the relationship between the navigation waters sub-system and vessel traffic flow can be shown as Formula 5:

$$\alpha = \frac{V}{V'} \tag{5}$$

where,

V': the through capacity of the navigation waters sub-system in unit time, ships/day.

V: standard ship number of vessel traffic flow in unit time, ships/day.

α: the evolution power coefficient of the navigation system. $0 \leq \alpha \leq 1$. The bigger α is, the greater the navigation waters and the service management should be developed.

Actually, there is a value related to the vessel traffic flow, in which navigation safety can be protected and navigation efficiency is higher. The value is set as from α_1, and $0 \leq \alpha_1 \leq 1$. When $\alpha = \alpha_1$, the navigation system is optimal and the evolution power is small. When $\alpha \leq \alpha_1$, the through capacity of the navigation waters sub-system can meet the needs of ship better, so the evolution model of the navigation system is Formula 3. When $\alpha > \alpha_1$, the through capacity of the navigation waters sub-system cannot meet the needs of ship, so the evolution model of the navigation system is Formula 6.

$$f_{service}\left(f_{natural}\left(C_c, C_b, C_a\right)\right) = V_1 \tag{6}$$

where, V_1 is the future vessel traffic flow, which can be got from the prediction.

Therefore, the evolution model of the navigation system can be expressed as Formula 7:

$$\begin{cases} \dfrac{dV}{dt} = rV\left(\dfrac{V'-V}{V'}\right) & \alpha \leq \alpha_1 \\ \\ f_{service}\left(f_{natural}\left(C_c, C_b, C_a\right)\right) = V_1 & \alpha > \alpha_1 \end{cases} \tag{7}$$

where,

V': the through capacity of the navigation waters sub-system in unit time, ships/day.

V: standard ship number of vessel traffic flow in unit time, ships/day.

r: a constant.

V': the through capacity of the navigation waters sub-system in unit time, ships/day.

C_c: the through capacity of channel in unit time, ships/day.

C_b: the through capacity of berth in unit time, ships/day.

C_a: the through capacity of anchorage in unit time, ships/day.

$f_{natural}$: the comprehensive effect of natural conditions acting on the channel, berth, or anchorage.

$f_{service}$: the optimal adaptation of service management.

Otherwise, the variables in Formula 7 should be correct, when the navigation waters or natural conditions are changed.

5 CONCLUSIONS

A navigation system is designed according to the characteristics of the navigation environment and vessel traffic flow. The navigation system is composed of a natural conditions sub-system, a navigation waters sub-system, a service management sub-system, and a vessel traffic flow sub-system. Then the relationships of the four sub-systems are analysed, and the mathematical model of evolution of the navigation system is constructed based on the "S" growth.

The evolution mechanism and model for the navigation system have an important theory value for exploring the interaction evolution mechanism between berth, channel, anchorage, service management, vessel traffic flow, and natural environment,. This study has an important guiding significance for improving navigation efficiency, ensuring navigation safety, and optimizing port design in waters.

ACKNOWLEDGEMENT

The work described in this paper was supported by a grant from the National Natural Science Foundation of China (Project No. 51179147), and a grant from the Fundamental Research Funds for the Central Universities ("Financially supported by self-determined and innovative research funds of Wuhan University of Technology", Project No. 2013-YB-004).

REFERENCES

[1] China Institute (2007): Li Y., Zheng B., Chen H. Analysis and Evaluation of Navigation Safety Influenced by Navigation Environment in Port. Journal of Waterway and Harbor, 5: 342–347.
[2] China Institute (2013): Li Z., Li Y., Yu Q. Analysis of the Arctic Route Navigation Environment Factors. World Regional Studies, 2:11–17.
[3] Japan Institute (1995): Kiyoshi H. A Comprehensive Assessment System for the Maritime Traffic Environment. Safety Science, 19:203–215.
[4] China Institute (2011): Jiang, F., Zhang, H., and Guo, W. Navigation Environment Safety Research on Channel Expansion Project Using PAWSA. ICTIS 2011: 2813–2821.

Advances in Civil Engineering and Building Materials IV – Chang et al (eds)
© 2015 Taylor & Francis Group, London, ISBN: 978-1-138-00088-9

Research on early-warning of navigation adaption under complex weather in the Three Gorges Reservoir area

Dan Jiang, Liwen Huang & Guozhu Hao
Department of Navigation, Wuhan University of Technology, Wuhan, China
Hubei Inland Shipping Technology Key Laboratory

ABSTRACT: In view of great influence on navigational safety which is caused by the complicated navigation environment in the Three Gorges Reservoir Area, the complex weather phenomenon is taken as the research object. By combining the method of Preliminary Hazard Analysis (PHA) and operation risk assessment (LECD), the risk source of the reservoir complex weather is identified and evaluated. Then the early warning index system for navigation under complex weather conditions is built and uses the analytic hierarchy process method (AHP) and fuzzy comprehensive evaluation method to assess the navigation safety risk of ro-ro and container ships. According to standard ship type classification for Chuanjiang river and the Three Gorges Reservoir area, combined with the complex weather conditions and loading condition of the ship, warning hierarchies for dry bulk ship, ro-ro and container ship are determined.

Keywords: Three Gorges Reservoir area, complex weather, navigation adaption, risk identification.

1 INTRODUCTION

Since the impoundment of Three Gorges Reservoir reached 135 meters, the reservoir river became wider and the water vapour evaporation capacity increased within the special geographic environment which aggravated the formation of weather such as rain and fog. The complicated reservoir navigable environment and significant fog and wind led to abundant accidents. Therefore, to carry out early warning and prevention research under complex weather conditions plays an important role for the security of the shipping safety, the development of the Three Gorges shipping economy and the protection of the water environment. Currently, research on navigation safety of the Three Gorges Reservoir mainly focuses on the early warning management, most of which conduct research from the angle of external management (Xu 2005, Li 2006, and Xiong 2009). Thus analysing the effect of complex weather on the ship navigation safety from the perspective of the ship and study on ship-based navigation warning, which helps the navigator take a rational action according to the ship's condition which enhances the pertinence and effectiveness of the early warning.

2 THE COMPLEX WEATHER AND DISTRIBUTION OF THE THREE GORGES RESERVOIR

2.1 Wind characteristics of the Three Gorges Reservoir area

According to the statistical analysis of channel meteorological data between January 2003 and December 2008, the monthly average wind speed in the Three Gorges Reservoir Area is about 1 m/s, (maximum is 1.7 m/s) which is measured in the Three Gorges Station. Maximum wind speed in 10 minutes of Fengjie station reaches 13.7 m/s which appears in April. Except for the three gorges station, peaks of average wind speed of other sites basically locate between February and May, July and September. Maximum wind speed peak values are mostly concentrated between April and August. The wind speed of the Three Gorges Reservoir increases gradually from the upstream to the downstream. The wind speed distribution is as shown in Figure 1.

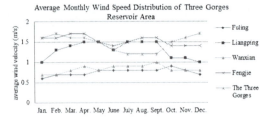

Figure 1. Wind speed distribution on the Three Gorges Reservoir.

Table 1. Statistical table of visibility frequency for each level in the Three Gorges Reservoir area.

Visibility	≤50 m	50–100 m	100–500 m	500–999 m	≥1000 m
Jan.	2	0	8	27	350
Feb.	2	0	11	30	408
Mar.	0	2	5	20	323
Apr.	1	0	4	19	237
May	1	0	3	21	254
June	0	0	9	32	211
July	0	0	3	17	331
Aug.	0	0	0	9	530
Sept.	0	1	6	49	474
Oct.	0	1	25	70	443
Nov.	0	2	10	46	472
Dec.	0	1	16	67	456

Table 2. Distribution of frequency of visibility level in each observation point.

Visibility	≤50 m	50–100 m	100–500 m	500–999 m	≥1000 m
Fuling	0	1	19	50	486
Fengdu	0	1	6	23	526
Wanzhou	0	0	0	19	541
Zhongxian	5	3	14	54	479
Yunyang	0	1	9	7	539
Fengjie	0	0	0	0	556
Wushan	0	1	0	0	556

2.2 Analysis of visibility in the Three Gorges Reservoir area

(1) Temporal distribution

Poor visibility (the visibility is less than 1000 m) emerges more in the winter half of the year, mostly in October and least frequently in summer. Visibility within 500 and 999 m occurs nine times in August, which is consistent with the changing pattern rule of the fog season (Wang & Chen, 2008). Distribution of visibility frequency for each level in the Three Gorges Reservoir area is shown in Table 1.

(2) Spatial distribution

From space, there is more poor visibility of Yangtze River in Chongqing western part. Poor visibility mainly appears around the Zhongxian station which accounts for about 35% of the fog, and secondly near Fuling and Fengdu. There is rarely visibility below 1000 m in Wushan and Fengjie station. Poor visibility mainly appears in the upper reach over Wanzhou (Wang & Chen 2008). Occurrences of every level of visibility in each station are shown in Table 2.

2.3 Characteristics of precipitation in the Three Gorges Reservoir area

There is abundant rainfall around the Three Gorges Reservoir Area, annual precipitation of most station is more than 1000 mm. From the regional distribution, annual precipitation from west to east shows the pattern of much, then less and much. Influenced by the monsoon, the climate characteristics of the reservoir area is a dry winter and summer rain. Precipitation mainly concentrated between April and October. Summer precipitation accounts for 40% to 50% of annual rainfall of which the peak occurs in June which is 17518 mm.

In terms of the rainy days at all levels, there are 35.1 days in average of which the rainfall is more than 10 mm. They mainly gather during May and August. There are 12.5 days in average of which the rainfall exceed 25 mm and the peaks emerge in July and August. The number of rainstorm days of which the precipitation surpasses 50 mm is 3.8 in average and mainly appear among June and August (Chen et al. 2009, Sun et al. 2006).

3 RISK IDENTIFICATION OF COMPLEX WEATHER IN THE THREE GORGES RESERVOIR

Through the analysis of the historical data combined with the questionnaire, complex weather risk source of the Three Gorges Reservoir is determined. The risk source recognition of complex weather in the Three Gorges Reservoir area based on PHA is shown in Table 3.

Use LECD to evaluate the risk level of complex weather in the Three Gorges Reservoir, and use the expert investigation method to determine the risk values of indicators. Analyse the data by recycling the 17 copies of effective questionnaires and then the result of risk assessment and recognition can be achieved (Wang, 2000), as shown in Table 4.

4 AN EARLY WARNING INDEX SYSTEM OF COMPLEX WEATHER IN THE THREE GORGES RESERVOIR

4.1 Construction of index system

Using the analysis of influence factors for ship navigation and traits of complex weather in the Three

Table 3. A preliminary hazard analysis of complex weather in the Three Gorges Reservoir.

Risk source	Accident conditions	Indication/Accident cause	Consequence
Heavy fog	A sudden fog during navigation and visibility reduction	① Lack of visibility ② Anchor difficulty ③ Inharmonious communication while encounter ships ④ Delayed awareness of dangerous omen ⑤ Going the wrong route	Collision/ Touch damages/ Grounding/ Stranding
High wind	A sudden high wind during navigation	① Instability, yaw, drifting of a ship ② Ship manoeuvring difficulties	Sink/ Capsize/ Collision/ Stranding/ Grounding
Heavy rainfall	Sudden heavy rainfall during navigation	① Poor visibility ② Surge of water level, complex flow regime, manoeuvring difficulties ③ The pilot's psychologically affected and decision-making errors	Sink/ Collision

Table 4. The result of risk assessment and recognition of complex weather in the Three Gorges Reservoir area.

Risk source	Risk assessment				Dangerous level
	L	E	C	D	
High wind	5.21	1.86	12.57	121.81	Second level
Heavy fog	5.71	2.26	12.71	164.02	Second level
Heavy rainfall	4.89	1.91	12.31	114.37	Second level

Table 5. Early warning indicators for navigation safety in the Three Gorges Reservoir.

1st class index	2nd class index	3rd class index
Ship factors	Ship type Rate of equipping Loading condition	Draft Wind area
Weather factors	Visibility Wind and rain	Wind scale Precipitation

Gorges Reservoir, early warning indicators for navigation safety can be divided into two first class indexes, five second class indexes and four third class indexes. The composition of early warning indicators is shown in Table 5.

4.2 A risk assessment of early warning indicators operation

i. Establishment of the factor set

According to the early warning indicators in Table 5, early warning factor set of navigation safety is built as below:

First level factor set $U^{(1)}:\{U_1^{(1)}\ U_2^{(1)}\} = \{$ship, and weather$\}$

Second level factor set $U_1^{(2)}:\{U_{11}^{(2)}\ U_{12}^{(2)}U_{13}^{(2)}\} = \{$ship type, rate of equipping, and loading condition$\}$, $U_2^{(2)}:\{U_{21}^{(2)}\ U_{22}^{(2)}\} = \{$visibility, wind, and rain$\}$

Third level $U_{13}^{(3)}:\{U_{131}^{(3)}\ U_{132}^{(3)}\} = \{$draft, and wind area$\}$, $U_{22}^{(3)}:\{U_{221}^{(3)}\ U_{222}^{(3)}\} = \{$wind scale, and precipitation$\}$

ii. Establishment of the evaluation set

According to the safety level of navigation under complex weather in the Three Gorges Reservoir, evaluation set is determined as follows:

$V = \{$relatively safe, critical, relative risk, risk, and high risk$\}$

iii. Establishment of fuzzy subsets of membership of each indicator

(1) Membership degree of ship factors:

Ship factors mostly include ship type, rate of equipping, and loading condition.

1) Membership degree of ship type

Combined with standard ship type developing trend of the Three Gorges Reservoir area, the bulk carrier, ro-ro, and container ship are selected.

2) Membership degree of equipping rate

"Equipment" refers to the ship's auxiliary navigation equipment. Auxiliary navigation equipment on inland river ships mainly covers radar, AIS, GPS, depthometer, magnetic compass etc. Whilst equipped with radar and AIS, the equipping rate is defined as high. Whilst equipped with AIS but without radar, the equipping rate is defined as ordinary. Whilst equipped with neither radar nor AIS is defined as low.

3) Membership degree of loading condition

Under complex weather, impact on navigation safety by ship loading mainly reflects in ship's draft and wind area.

a) Membership degree of draft

Ship draft varies from different tonnage. Then the ship's draft is divided into full load, half load and in ballast.

b) Membership degree of wind area

On the basis of ship type and the calculation of wind area, wind area is divided into three levels as large, medium, and small.

(2) Membership degree of weather factors

Weather factors mostly cover visibility, wind, and rain.

1) Membership degree of visibility

Factors which have impact on visibility mostly include weather conditions such as fog, rainfall, and snowfall. Visibility is determined by four levels: inferior visibility (100 m to 500 m), restricted visibility (500 m to 1000 m), moderate visibility (1000 m to 1500 m), and good visibility (more than 1500 m).

2) Membership degree of wind, and rain

a) Membership degree of wind

Generally speaking, wind above Beaufort force 4 produces certain effect on navigation in an inland river. Thus, the wind is divided into three levels which is Beaufort force 3 to 4, 5 to 6, and 7 to 8.

b) Membership degree of rain

Impact caused by rainfall on shipping is largely related to precipitation and intensity of rainfall. According to the distribution of rainfall in the Three Gorges Reservoir, influence on navigation made by rainfall is measured in slight, medium and obvious. The corresponding daily rainfall of the level is 10 mm to 25 mm, 25 mm to 50 mm, and more than 50 mm.

On the basis of a membership degree, by processing the seventeen pay back effective questionnaires, the fuzzy subsets of membership degree for each indicator are decided (MacDonald & Cain, 2000).

iv. Determination of the weight

According to the data of expert questionnaire, the weights for all levels are shown below:

The third level weights,
$$A_{13}^{(3)} = [0.4\ 0.6]\ A_{22}^{(3)} = [0.75\ 0.25]$$
The second level weights,
$$A_{1}^{(2)} = [0.3\ 0.33\ 0.37]\ A_{2}^{(2)} = [0.5\ 0.5]$$
The first level weights,
$$A^{(1)} = [0.43\ 0.57\ ]$$

4.3 Early warning analysis of designed weather

Take ro-ro and container ships as examples, conduct risk assessment of navigation safety under designed weather conditions.

1) Ship type: ro-ro
   Loading condition: full load
   Wind scale: Beaufort 7
   Daily rainfall: 10–25 mm
   Visibility: moderate

By calculating, the result is $B = A^{(1)} \cdot R^{(1)} = [0.034\ 0.125\ 0.475\ 0.234\ 0.132]$. According to the results, the maximum value of B is 0.475, thus the result of risk assessments for navigation under designed weather condition is relatively dangerous. Therefore it is known that, under the condition of medium visibility and slightly affected by rainfall, wind of Beaufort 7 have certain influence on navigation safety of ro-ro.

2) Ship type: container ship
   Loading condition: full load
   Wind scale: Beaufort 5
   Daily rainfall: more than 50 mm
   Visibility: restricted

By calculating, the result is $B = A^{(1)} \cdot R^{(1)} = [0.064\ 0.143\ 0.205\ 0.495\ 0.103]$. According to the results, the maximum value of B is 0.495, thus the result of risk assessments for navigation under designed weather condition is dangerous. Therefore, it is known that, under the coefficient condition of poor visibility and heavy rainfall, wind of Beaufort 5 may cause certain threat to navigation safety of a container ship.

5 SUMMARY

In this paper, according to the characteristics of complex weather in the Three Gorges Reservoir area, the three main risk sources that are fog, strong wind, and rainfall are identified by PHA combined with LECD. Taking into account comprehensive consideration of various influencing factors to navigation safety in the Three Gorges Reservoir area, the early warning index system is established. Taking ro-ro and container ship as examples, the analytical hierarchy process (AHP) and fuzzy comprehensive evaluation are used to conduct a risk assessment for navigation safety under designed weather condition, the result of which indicates the relation between navigational safety and weather conditions. The results can provide reference for navigation management under complex weather.

Furthermore, detection, recognition, and transmission of complex weather combined with the exploitation of ship-based navigation equipment can provide a timely and effective early warning for navigation in the Three Gorges Reservoir area, so as to improve the efficiency of shipping.

ACKNOWLEDGMENT

This work is supported by the Fundamental Research Funds for the Central Universities (Grant Nos. 2014-VII-026 and 2014-JL-010). The authors would like to thank the funds for the financial support for this project.

REFERENCES

Chen, X.Y. et al. 2009. The Local Climate Changes of the Three Gorges Reservoir Area. Chinese Journal of Mechanical Engineering 18(1):47–51.

Li, J.H. 2006. Early Warning and Emergency Management Mechanism in the Three Gorges Reservoir Area. Journal of Wuhan University of Technology (01):114–117.

MacDonald, A. & Cain, M. 2000. Marine Environmental High Risk areas (MEHRAS) for the UK Trans. ImarE 112(2):61–71.

Sun, S.X. et al. 2006. Analysis of Climate Characteristics of the Three Gorges Reservoir Area. China Three Gorges Construction 24(3):22–24.

Wang, J.A. 2000. Subjective modeling tool applied to formal ship safety assessment. Ocean Engineering (27): 1019–1035.

Wang, Z. & Chen, Y.Y. 2008. Analysis of Shipping Meteorological Condition in the Three Gorges Reservoir Area. Chinese Journal of Mechanical Engineering 17(1):79–82.

Xu, K.J. 2005. Research on Water Safety Early Warning Management Based on Crisis Management. Transportation Science & Technology (04):124–125.

Xiong, B. & Zhang, S.Y. 2009. Early Warning Management Mechanism for Fog Navigation in the Three Gorges Reservoir Area. Shipping Management 31(9):36–38.

Advances in Civil Engineering and Building Materials IV – Chang et al (eds)
© 2015 Taylor & Francis Group, London, ISBN: 978-1-138-00088-9

The response characteristics of a buried pipeline under moving constant load of vehicles

Yang Chen, Dawei Liu & Xiangyin Liu
Qingdao University, Qingdao, China

ABSTRACT: In order to study the influence of the moving constant load of vehicles on the performance of a buried pipeline under the semi-rigid asphalt pavement, the vehicle load applied to the road surface is simplified to moving constant load. Based on this, using finite element analysis software ABAQUS to establish a semi-rigid asphalt pavement-buried pipeline dimensional Finite Element (FE) analysis model under vehicle loads, dynamic response of buried pipelines under moving constant load of vehicles was analysed. The results showed that the stress of a buried pipe showed two peaks when the three-axis vehicle passed the determining location. The vertical stress of a pipeline is less than the longitudinal stress and is greater than the transverse stress. The transverse strain of a pipeline is greater than the longitudinal strain and is less than the vertical strain.

Keywords: ABAQUS, buried pipelines, finite element analysis, moving constant load.

1 INTRODUCTION

With the expansion of urban functions, buried pipelines have become a cheap and efficient transport system. Buried pipelines are directly or indirectly subjected to the impact of the traffic load, variable load and accidental load. Among them, the traffic load is one of the most frequent variable loads on buried pipelines. With the rapid development of the economy and the transport industry, the failure and damage of buried pipelines that are caused by vehicle loads are increasing. A long period effect of repeated traffic loads will create a permanent unrecoverable settlement deformation of soil subgrade, which leads to the pipeline in soil subgrade to produce uneven settlement or buried local stress concentration, which causes pipes cracking or leaking and other accidents, seriously affecting the normal operation of pipelines. For the dynamic response of buried pipelines under traffic loads, many scholars have done a lot of research, but it lacked dynamic response of buried pipelines under multiple rounds loads of the vehicle. Based on this, the vehicle load applied to the road surface is simplified to moving constant load. Dynamic response of buried pipelines under multiple rounds loads of the vehicle was analysed by using finite element analysis software ABAQUS, to establish a semi-rigid asphalt pavement-buried pipeline dimensional finite element analysis model under multiple rounds loads of the vehicle.

Table 1. The parameters of semi rigid asphalt pavement model.

	Asphalt surface	Semi-rigid base	Semi-rigid sub base	Soil base
Thickness/m	0.18	0.2	0.2	4.42
Density/kg/m^3	2613	2083	1932	1926
Modulus of elasticity/MPa	1200	1100	400	50
Poisson's ratio/v	0.35	0.35	0.4	0.4
Damping rate/%	0.05	0.05	0.05	0.05
Friction angle/°	–	–	–	30
Cohesion/kPa	–	–	–	17

2 ESTABLISHMENT OF PAVEMENT-BURIED PIPELINE FINITE ELEMENT MODEL (FEM)

2.1 Pavement-pipeline FEM

When the pavement-pipeline finite element model was built, the structure of semi-rigid asphalt pavement was simplified to four layers; they are: asphalt surface, semi-rigid base, semi-rigid sub base, and soil base. Each layer of the materials has different material properties and every layer adopts linear elastic model. Their parameters are shown in Table 1.

Table 2. The model parameters of the buried pipeline.

Parameters	Value
Diameter/mm	700
Wall thickness/mm	10
Density/kg/m^3	7850
Modulus of elasticity/GPa	210.7
Poisson's ratio/ν	0.3
Yield Strength/MPa	490

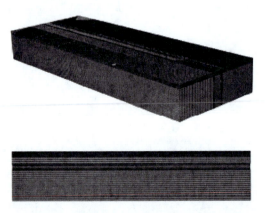

Figure 1. The finite element model of pavement – pipeline.

There is a steel pipeline of the water supply in this paper. Its performance parameters are shown in Table 2.

The continuous system of elastic layers both in horizontal or vertical direction is infinite. It is impossible to divide units in the infinite domain by the finite element. Therefore, we assume that a buried pipeline lays on the left side of the road about 1.2 m at the wheel tracks and take length of the road model 50 m, 18 m wide, and 5 m high, both using C3D8R hexahedral element of finite element software ABAQUS to mesh mode. In order to improve accuracy and reduce the computer calculation, grids surrounding pipelines and action lines of the road are divided smaller. The number of units is 552,956 and the number of nodes is 589750. Pavement – pipeline finite element model is shown in Figure 1.

2.2 The simulation of pipe-soil interaction

For contact problems of pipes and soil, some varied problems need to be considered, for example: elasticity, plasticity, the relative sliding friction between the contact surfaces, various physical factors of two different structures of objects, and other problems. ABAQUS/Explicit has a significant capacity to deal with the contact problem that can automatically identify the contact between the components of the model and define the surface-surface contact, self-contact and general contact, which can efficiently simulate complex contact between them. This article will define the buried pipelines in contact with the soil for surface – surface contact. Surface-surface contact is

a contact pair including a principal face and a subordinate face, where the principal face can penetrate into the subordinate face, but the subordinate face cannot penetrate into the principal face. Selecting the principal and subordinate face should follow four principles:

- When a contact pair includes a deformable surface and a rigid surface, the rigid surface is the principal face;
- When there is a rigid part in contact, the surface of rigid part should be used as the principal face;
- Select the surface with more granular grids as he principal face; when the grid density closes, select the greater rigidity material as the main surface;
- For the finite slip, the nodes of subordinate face should try not to fall outside the principal face in the analysis.

Based on the above principles, the surface of pipelines is selected as the principal face and the surface of soil as the subordinate face to build a contact pair. Contact properties include two parts: the normal behaviour and tangential behaviour between the contact surfaces. For the normal behaviour, the default relationship of contact pressure and space is hard contact in ABAQUS. For the tangential behaviour, penalty friction model, coulomb friction model, Lagrange friction model, kinetic friction model, or other models are often used in ABAQUS. In this paper, adopt coulomb friction model (using the coefficient of friction expresses the friction characteristics between the contact surface) to build finite element analysis model of pipe-soil structure. The friction coefficient of soil-pipe is 0.3 in the article.

In the course of loading, the constraint between buried pipeline and the soil may cause the elastic and plastic deformation. Therefore, it needs to properly set the contact constraints between a pipe and soil. The constraint types in ABAQUS include Tie, Rigid body, display body, coupling, shell-to-solid coupling, and embedded region, Equation. For the constraint between buried pipe and soil, tie constraint is used to bind the two regions of model together, so that there is no relative movement between them. Tie constraint can contact the pipe and soil as a whole, and it can provide different methods of three-dimensional mesh for them. The method of selecting the principal and subordinate face is the same as above.

2.3 Loading method of the vehicle load

The 6 × 4 heavy-duty dump truck of a company is used, the main parameters are shown in Table 3. The pavement of buried pipelines is grade B, and the vehicle speed is 60 km/h. The dynamic loads of the left, middle, and rear wheels respectively, are 1.37 MPa, 0.99 MPa, 1.22 MPa; right side respectively are 1.86 MPa, 1.06 MPa, and 1.14 MPa.

In the tire contact patch, the pressure of wheels is asymmetrically distributed in both the longitudinal and transverse. The tire pressure is assumed to

Table 3. Vehicle basic performance parameter.

Parameters	Value
Full load quality (kg)	27000
Front axle load (kN)	70
Middle and rear axle load (kN)	20
Wheelbase (mm)	3825 + 1360
Front wheel distance (mm)	2220
Rear wheel distance (mm)	1830
Stiffness coefficient of front spring (N/mm)	200
Stiffness coefficient of rear spring (N/mm)	2200
Tire type	11.00R20

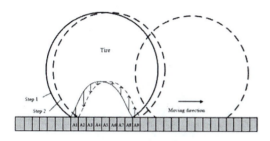

Figure 2. Schematic of wheels moving on the road.

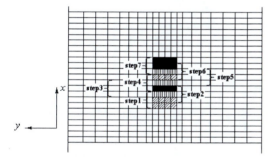

Figure 3. Loading process.

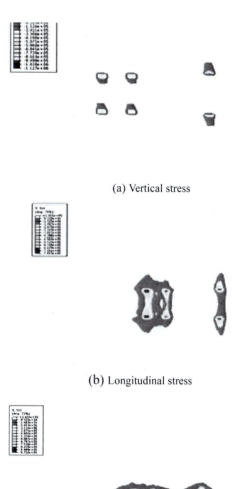

(a) Vertical stress

(b) Longitudinal stress

(c) Transverse stress

Figure 4. Stress nephograms.

be uniformly distributed within the tire contact patch while calculating. When the tire pressure moves on the road, the size of contact area does not change at each momentary state. It is shown in Figure 2.

Except when the position of the vehicle load changes, the tire pressure of vertical over the entire analysis step constantly changes when the tires are moving. Vehicle load should be determined by step load or slope load within the range of every time increments in the analysis step. According to the actual situation of the vehicle dynamic load, load would be loaded in accordance with the step load (step load refers to the first incremental step initially applied directly in accordance with the load amplitude, and remains unchanged throughout the incremental step). Loading process is shown in Figure 3.

To simulate the effect of dead load during vehicle movement, use vdload subprogram to apply the moving loads on semi-rigid asphalt pavement finite element model. Set the load time and the time increment in the analysis step and control the initial position of loading in the subroutine. The loading time of the vehicle whose speed is 60 km/h is set to 1.8 s. In the subroutine it controls load on the road at 10 m to 40 m by the starting position of load.

3 FEM RESULTS AND ANALYSIS

3.1 Stress nephograms

When a vehicle passed through a determining position of the road, the stress nephograms of vertical stress, longitudinal stress and transverse stress of asphalt layer are shown in Figure 4a, b, and c.

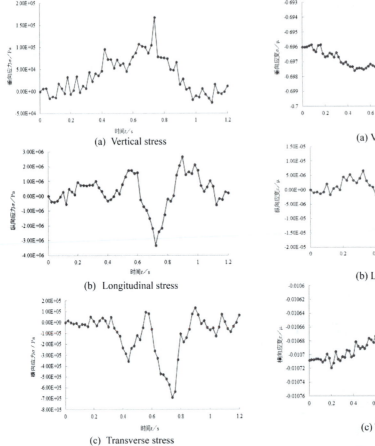

(a) Vertical stress

(b) Longitudinal stress

(c) Transverse stress

Figure 5.   The time history curves of buried pipeline stress.

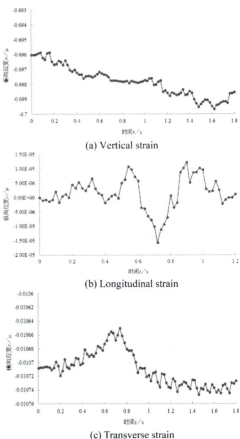

(a) Vertical strain

(b) Longitudinal strain

(c) Transverse strain

Figure 6.   The time history curves of buried pipeline strain.

It is shown in Figure 4 that the stress around the wheels of each axes asymmetrically distributes under the two-wheeled incentive. Because there are two tires on both sides of the mid-axle and rear axle and there is a single tire on both sides of the front axle, the range of the pressure that the front axle wheels produce is smaller, thus resulting in greater stress. Comparing transverse stress with the vertical stress, because the wheels of each axis and the wheels on both sides produce a coupling action, the distribution of transverse compressive stress is greater than the vertical stress and the longitudinal stress.

3.2   The time history curves of buried pipeline stress

When a vehicle passed through a determining position of the road, time history curves of vertical stress, longitudinal stress and transverse stress of the top of buried pipeline are shown in Figure 5a, b, and c.

It is shown in Figure 5 that there are two peaks in the time history curves of vertical stress and longitudinal stress of buried pipeline. They are the superimposed stress when the mid-axle and rear axle ran over and the stress when the front axle wheels ran over. The

superimposed stress is greater than the stress when the front axle wheels ran over.

It can also be seen in Figure 5 that vertical tensile stress of the specific location of buried pipeline is greater than the vertical compressive stress under moving dead load. Because the vehicle load that applied to the side of a pipeline is greater than that applied to the top of the pipeline, the top of the pipeline becomes stretched. The longitudinal tensile stress becomes small, as the wheels pass a specific position. When the mid-axle and rear axle run over this point, the longitudinal tensile stress of the pipeline becomes vertically compressively stressed. There has been a state of tension and compression alternating stress, so that the top of the pipe can easily to produce fatigue failure. The trend of time history curve of transverse stress is consistent with longitudinal stress. Its value is less than the longitudinal stress and greater than the vertical stress.

3.3   The time history curves of buried pipeline strain

When a vehicle passed through a determining position of the road, time history curves of vertical strain, longitudinal strain and transverse strain of the top of a buried pipeline are shown in Figure 6a, b, and c.

It is shown in Figure 7 that the vertical strain and transverse strain are negative strains when the specific location of buried pipeline is under the moving dead load. Longitudinal strain shows positive and negative alternating. Vertical negative strain of a buried pipeline presents an increasing trend with the advance of the vehicle. There are two peaks in the time history curve of longitudinal strain. When the vehicle is approaching, it is a positive strain. The strain gets smaller while the wheels passed by the position, and it gets larger while the wheels left. Transverse strain of buried pipeline is negative strain and it becomes smaller with the advance of the vehicle. It also can be seen in Figure 7 that the value of transverse strain is less than the vertical strain and greater than the longitudinal strain.

4 CONCLUSION

A semi-rigid asphalt pavement – pipe coupling finite element analysis model was established by the finite element analysis software ABAQUS. The relative stiffness of the soil and pipe, pipe-soil contact, constraints and other issues are fully considered. The simulation calculation of the dynamic response of buried pipeline under the moving dead load analysed the variation of road stress and strain. It provides a reference for the damage study and safe operation of a buried pipeline.

ACKNOWLEDGEMENTS

The research is supported by Shandong provincial Natural Science Foundation and Qingdao University Excellent Postgraduate Thesis Nurture Project (Grant No. 51475248)

REFERENCES

GB50423-2007 Code for Design of Oil and Gas Transportation Pipeline Crossing Engineering(s).Chinese Planning Press, 2008.

Rajani B, Abdel-Akher A. Re-assessment of Resistance of Cast Iron Pipes Subjected to Vertical Loads and Internal pressure [J]. Engineering Structures, 2012, 45: 192–212.

T Bajcar, F Cimerman, B Širok, M Ameršek. Impact Assessment of Traffic-Induced Vibration on Natural Gas Transmission Pipeline [J]. Journal of Loss Prevention in the Process Industries, 2012, 25(6): 1055–1068.

Khavassefat P, Jelagin D, Birgisson B. A Computational Framework for Viscoelastic Analysis of Flexible Pavements under Moving Loads [J]. Materials and structures, 2012, 45(11): 1655–1671.

Ambassa Z, Allou F, Petit C, et al. Fatigue Life Prediction of an Asphalt Pavement Subjected to Multiple Axle Loadings with Viscoelastic FEM [J]. Construction and Building Materials, 2013, 43: 443–452.

Hyuk Lee, Finite Element Analysis of a Buried Pipeline [D], 2010: 18–20.

Sun L, Kenis W, & Wang W. Stochastic Spatial Excitation Induced by a Distributed Contact on Homogenous-Gaussian Random Fields [J]. Journal of Engineering Mechanics, 2006, 132(7): 714–722.

Gong Longying. On the Use of ABAQUS for Analyzing the Problem of Contacts [J]. Chinese Coal, 2009, 7: 66–68.

Zhao Shiping, Zeng Xiangguo, Yao Anlin, Wang Qingyuan, Shi Xiaoshuang, and Xing Yifeng. Dynamic Response Numerical Simulation of Buried Gas Pipeline During Third-Party Interference. Sichuan Building Science, 2009, 35(1):135.

Dawei Liu, Huanming Chen, Rongchao Jiang, and Song Wang, Simulation of Dynamic Load for Heavy Vehicle under Bilateral Tracks' Road Excitation [J]. Civil Engineering and Building Materials, 2012.

Advances in Civil Engineering and Building Materials IV – Chang et al (eds)
© 2015 Taylor & Francis Group, London, ISBN: 978-1-138-00088-9

Database design of expressway electronic toll collection system

Fengquan Yu & Jianhua Guo
Intelligent Transportation System Research Center, Southeast University, Nanjing, China

Bo Zhao
Jiangsu Lianxu Expressway Co. Ltd., Xuzhou, China

ABSTRACT: Regarding the provincial toll management center, toll management sub-center, and toll station as three layers of electronic tolling system, the organization and business processes are analyzed for the highway electronic tolling system of Lianxu expressway. In addition, through applying the basic principles of database design, detailed analysis of the relations between entities and E-R model for the tolling system are conducted, and the logical framework and physical framework of the database are proposed for the system. Finally, complete database design of the electronic system is presented for supporting automatic vehicle tolling.

Keywords: database, ETC, requirement analysis, concept design, logical framework design, physical framework design.

1 INTRODUCTION

With the increase of motor vehicles as well as the gradual improvement of the highway network, defects and shortcomings of traditional highway toll collection technique have become increasingly prominent. Currently, the ratio of standard passenger cars for Lianxu expressway is rising gradually. Meanwhile, although only two electronic tolling collection (ETC) lanes are included in a ordinary toll station, the electronic collected toll has occupied a significant part of the entire toll. Therefore, the promotion of ETC lane is of great practical significance according to changes in traffic component [1].

The United States was the first country to use ETC, and its famous E-Zpass system contributes 43% of the average monthly trading volume in just one year and a half. Italy's most massive system (Telepass) can handle 30 cars per minute and the number of toll transactions reached 500,000 times per day. Hundreds of ETC lanes have been conducted in a dozen of cities in China [2]. It is proved that most of the ETC system is secure, stable, and reliable [3]. Database of toll collection system has great significance for the achievement of the functions. According to business processes of Lianxu expressway as well as the basic principles of database design, a database design of ETC system is presented in this paper to support automatic charging.

2 REQUIREMENT ANALYSIS

The mission of database design requirements analysis phase is accomplished by a detailed investigation and analysis of information and operating requirements in the field of each application requirements. The relationship between different data items of applications and data operations are given to determine their requirements.

2.1 Organization requirement

Organization refers to the division of departments within the organization and the relationship between them [4]. Database design in this article is based on ETC system of Lianxu expressway. The overall organization of Lianxu highway toll management system has three layers of provincial management center, toll management sub-center, and toll station.

Provincial management center is specifically responsible for the management and development of regulations and charging standards of toll collection. Then it is able to fully grasp the entire situation of charging in the network and coordinate various business of the toll center. Charging data and traffic data can be collected and aggregated by the center. Simultaneously, statistics of traffic and charging fees can be transmitted to a computer monitoring system.

Toll management sub-center is mainly responsible for the management of toll collection in the specific district. Through the computer network monitoring, charging and demolition account for each toll station can be obtained reliably. Fees and traffic data can be collected and aggregated by the sub-center. Also it is able to monitor and control certain toll videos. Regional blacklist is generated and instructions from the management center are received.

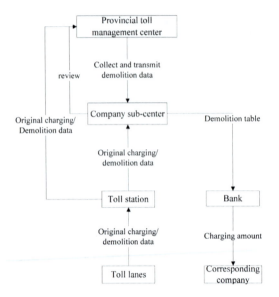

Figure 1. Electronic toll settlement and split process.

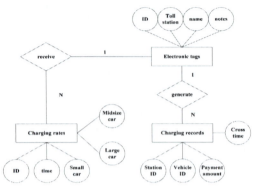

Figure 2. Partial E-R diagram of ETC basic process.

Toll station is the most basic toll charging management system administration. Tolls of vehicles are collected by toll lane in the station. According to the mileage that vehicles run in the road network, fees are calculated by control unit of toll lane. Finally, the data will be uploaded to toll station computers. Toll station computers summarize data information of each toll lane, gather the deserved share of each section, and upload statistical information to the central computer of the company.

2.2 Business process analysis

Business process of Lianxu expressway toll collection system involves charging and management etc. The data flow is analyzed in detail to provide the basis for the conceptual design of the database. The specific business data flow is shown in Fig. 1.

According to each highway fees approved by the provincial government, first, split the proceeds of each section. Then split the proceeds based on route length ratio within the different sections of the owners. Toll split process can be divided as the following steps:

2.2.1 Upload dynamic transactions

Original charging and demolition data generated by each lane is unified and uploaded to Lianxu highway toll management center and the road toll management sub-centers.

2.2.2 Data verification

According to the uploaded data from toll station, tolls are collected and the accuracy of the data is checked based on toll station unit. After checking the demolition formulation, statements can be received by toll management sub-centers. While there are some issues of demolition data, they can be handled effectively with consultation.

2.2.3 Fee payment settlement and allocation

Based on profit-sharing reports of provincial toll management center, companies notify their corresponding settlement banks that the amount of revenue allocated to other settlement bank accounts. Meanwhile, other companies are also allocated to accept payments. In This way the entire data flow is completed.

3 CONCEPT DESIGN

View integration method is used in this paper, which is to define the core processes of highway toll collection and the partial E-R diagrams [5]. Then following the charging data flow process, the entire E-R diagram can be generated for toll station, company sub-center and toll management center at all levels.

3.1 Highway toll partial E-R diagram

Entities of expressway electronic toll collection include electronic tags, toll stations, and the corresponding charging records. The basic process of ETC is achieved by receiving charging rates information through electronic tags and generating the charging records. Related entities involve charging rates, electronic tags, and charging records. The resulted E-R diagrams are shown in Fig. 2.

3.2 Partial E-R diagram of ETC management agencies

The basic process of ETC system is managed by the corresponding departments to make billing records transfer from one to the other. Entities of management departments at all levels including three grades in total: toll station, company sub-center, and provincial management center. In this paper, the data flow between different administrations is analyzed. Provincial toll management center is divided into several sub-centers and the company sub-center consists of several toll stations. Fig. 3 shows the partial E-R diagram of ETC system management agencies.

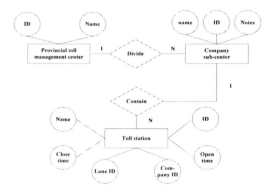

Figure 3. Partial E-R diagram of ETC management agencies at all levels.

Figure 4. The entire E-R model of ETC.

3.3 E-R diagram integration

ETC data and reports between the electronic tags, toll station, company sub-centers, and provincial toll management center and their interaction process are designed. Thus the complete result of E-R integration of ETC system database design is shown in Fig. 4.

4 LOGICAL FRAMEWORK DESIGN

According to electronic toll collection process, the basic database tables of ETC system are divided into two categories: static and dynamic tables. Logical database design gives the table number of ETC system database and related table illustrations which are shown in Tab. 1.

5 PHYSICAL FRAMEWORK DESIGN

Storage efficiency and system configuration are mainly considered while designing physical framework of the ETC system database. Relying on the implementation of the actual database software such as Oracle, MS SQL Server, etc., relationships, indexing, clustering, logging, and other storage arrangements and storage structures are built. The configuration of the system is then determined. The number of open objects, buffer size, time slice size, filling factor, and other parameters should be adjusted reasonably to achieve optimal system performance.

Table 1. Descriptions of ETC database and corresponding tables.

Category	Table name	Descriptions
Static tables	Electronic tags table	Store the basic information of electronic tags
	Toll station table	Store basic properties and information of toll station
	Company sub-center table	Describe the name and ID of company sub-center
	Toll management center table	Store the name and ID of toll management center
Dynamic tables	Charging rates table	Store charging rates of different types of vehicles at different times or toll collection area
	Payment record sheet	Store ETC payment records such as payment amount, basic vehicle information and access time

6 CONCLUSIONS

The use of modern computer and communication technology in ETC system not only achieves the information transmission between the management departments at all levels, but greatly reduces the intensity of the work of toll staff and managers. In this paper, using the ETC system of Lianxu expressway as an example, database which is the core component of ETC system is analyzed and designed in detail. Illustrations of the basic tables are given based on the management model.

ACKNOWLEDGMENT

This paper was supported in part by the National Science Foundation of China under the grant No. 71101025.

REFERENCES

[1] H. Jia, Z. Juan and H. Yao, 2004. Socio-Economic Analyses for Electronic Toll Collection Projects Systems Engineering-Theory & Practice, 2004(7):121–134. (In Chinese). China.

[2] D. Levinson, E. Chang, 2003. A model for optimizing electronic toll collection systems. Transportation Research Part A, 37:293–314.

[3] J. Chen, B. Shan, 2010. Survey on 5.8 GHz Electronic Toll Collection Technologies. Journal of Tongji University (Natural Science), 38(11):1675–1681. (In Chinese). China.

[4] R. Elmasri, S.B, 2007. Navathe, Fundamentals of Database Systems, fifth ed., Addison-Wesley.

[5] J. Guo, W. Huang, 2001. Database Design of Expressway Progress Management Systems. Journal of Southeast University (Natural Science), 31(2):94–99. (In Chinese). China.

Advances in Civil Engineering and Building Materials IV – Chang et al (eds)
© *2015 Taylor & Francis Group, London, ISBN: 978-1-138-00088-9*

Road design problems solved by affine arithmetic

G. Bosurgi, O. Pellegrino & G. Sollazzo
DICIEAMA Department, University of Messina, Italy

ABSTRACT: In this paper we propose the use of techniques based on range numbers for solving problems of road designing. We applied the Affine Arithmetic to verify the driver's visibility in a horizontal curve, providing very interesting results and confirming its usefulness even in the most complex analytical expressions. The preparation of charts such as those proposed in this paper also makes this method directly accessible to those not so familiar with this technique.

Keywords: interval analysis, affine arithmetic, road standard, road design

1 INTRODUCTION

Generally, a final satisfactory solution of a road design is quite difficult, since it involves technical regulations, increasing respect for environment and new attention to economic aspect in the construction phase.

Traditional and well-known analytical procedures allow the above problems to be solved, but only after having carried out numerous calculations representative of all potentially possible combinations. Instead, it is appropriate to identify a methodology that allows the computational cost to be drastically reduced within the limits imposed.

These problems can only be overcome with an analytical tool suited to the nature of the available data. In particular, to solve this situation, we applied some techniques derived from Interval Analysis, though this was born and developed to manage uncertainties when the available information does not allow the use of more complex procedures for example, probabilistic analysis (Moore R.E. et al. 2009; Muscolino G. & Sofi A. 2012).

The results confirmed the usefulness of the proposed methodology in order to reduce the computational burden of the designer and to attain all the potential solutions for the investigated problem.

2 METHOD

In the following we briefly illustrate the main technique (Stolfi J. & de Figueredo L.H. 2003) that can handle range numbers in a better way, that is, Affine Arithmetic (AA). In order to evaluate their effectiveness in road design, we applied them to a check regarding the visibility in a horizontal curve.

As is known, Interval Analysis considers the quantities represented as intervals as being perfectly independent, i.e. without any correlations among them (Bosurgi et al. 2011). This peculiarity could generate excessive output intervals and this kind of error is proportional to the recursivity of the variables and to the complexity of the analytical expressions.

Affine Arithmetic deals with and solves the aforementioned problems, although with and at the expense of greater computational cost, as will be appreciated from the brief summary reported below. More specifically, each intermediate result is expressed by a first-degree polynomial that describe the quantities analyzed:

$$\hat{x} = x_0 + x_1 \cdot \varepsilon_1 + x_2 \cdot \varepsilon_2 + ... + x_n \cdot \varepsilon_n \quad (1)$$

where:

- x_0 is the central value of the examined quantities;
- x_i are coefficients that represent the partial deviations and, therefore, the deviation in respect to the central value of the i-th component;
- ε_i coefficients are the noise symbols, have the values -1 or 1 and represent an independent component of the total uncertainty of the x quantity.

The most important feature of AA concerns the possibility that the same ε can be assigned to two or more quantities belonging to different affine forms (x and y, for example), representing, thus, a partial dependency between them whose magnitude and sign depend on the coefficients x_i and y_i.

Any operation involving affine forms (x and y, for example) will return, in turn, an affine form $\hat{z}$ able to preserve the information contained in the original quantities $\hat{x}$ and $\hat{y}$, apart from overflows and roundoff errors.

The error committed in the approximation of the affine form depends quadratically on the extent of the ranges of the input variables, but it has been demonstrated (Stolfi J. & de Figueiredo L.H. 2003) that it is essentially zero if the function f is univariate.

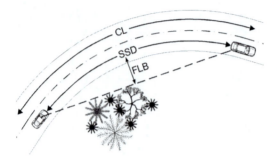

Figure 1. Importance of the free lateral board (FLB) to satisfy the sight distance demanded.

2.1 Formulation of the problem

In order to evaluate the effectiveness of this kind of analysis so applied to road problems and to ascertain the convenience of AA with respect to Interval Analysis, we have examined a well-known and sufficiently simple problem, regarding the driver's vision in a horizontal curve.

In a bend, eventual obstacles located internally can represent a great problem for the visibility of drivers. It is therefore necessary to provide a proper free lateral board (in this case indicated as FLB in Figure 1), whose amplitude can be analytically evaluated considering two possible scenarios:

- The Stopping Sight Distance SSD is less than the Curve Length CL (SSD<CL).
- The SSD is greater than CL (SSD>CL).

The first step must involve, therefore, the evaluation of the required free lateral board (FLB), according to the following expressions:

$$FLB1 = R \cdot \left(1 - \cos\frac{SSD}{2 \cdot R}\right) \quad \text{for SSD} < CL \qquad (2)$$

$$FLB2 = R \cdot \left(1 - \cos\frac{CL}{2 \cdot R}\right) + \left(\frac{SSD - CL}{2}\right) \cdot \sin\frac{CL}{2 \cdot R} \quad (3)$$

for SSD > CL

where $CL = R \times \gamma$ with γ centre angle of the circular arc and also, in the absence of a transition curve, deflection angle between the two tangents.

The offered distance can be evaluated via the tracking of a series of lines joining the driver's eyes with the target to be observed. Since this method does not consider the third dimension of the space, it is reliable only in the case of flat areas. This assumption simplifies the treated problem and, therefore, brings out the potential of the proposed technique.

Therefore, the variables affecting the phenomenon are:

The radius of the curve R.
The centre angle of the bend γ.
The stopping sight distance SSD.

Table 1. Input data useful to define the affine forms R, SSD and γ.

	R [m]	SSD [m]	γ [rad]
Mid	45	55	0.58905
Rad 1	3	0	0.11781
Rad 2	2	3	0
Rad 3	0	3	0

The mentioned formulas must be reported in interval form in order to be able to derive the unknowns of the problem.

Here, for brevity, we have illustrated only the example of the equation (2) with which we get FLB2 (SSD > CL), since the other case (SSD < CL) is resolved with greater ease.

3 RESULTS

The procedure illustrated in detail in the previous paragraphs provided a very interesting outcome. Predictably, the technique based on Arithmetic Affine produced better results and the difference with respect to the classical Interval Analysis is evident especially when the analytical formulation of the problem is particularly complex, as in the case of the equation (2).

The assumed data and the obtained results, reported in the tables and graphs of this section, although they did not refer to a real situation, represent very likely design conditions. In Table 1 we have included the values of the affine quantities R, SSD and γ represented by the central value (Mid) and deviation (Rad 1, Rad 2, Rad 3).

In order to assess quantitatively the potential advantages offered by AA with respect to IA and to highlight the real usability of the procedure as an aid to the ordinary activities of a road designer, we summarized the results through some charts.

With three input variables in the interval form, there would be several ways of representing the results. We chose, therefore, to report in the following figures the deviation of the output FLB in function of R and of SSD. For example, Figure 2 illustrates the variation of FLB2 (SSD < CL) with respect to R in terms of inferior and superior limit, calculated both with the traditional IA and with AA. It should be noted that this Figure 4 refers to the formula (3) which should be used when SSD > CL.

We also added that, in the legend, the subscript M refers to the function FLB calculated with the traditional IA (M for Moore), while the subscript S for Stolfi) indicates that the FLB was calculated with the AA.

Similarly, Figure 3 illustrates the variation of FLB2 (SSD < CL) with respect to the SSD, always in an inf-sup form, calculated with IA (subscript M) and AA (subscript S). It should be noted that this figure refers to the formula (3), where the variables R

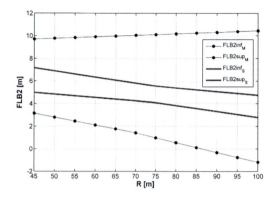

Figure 2. Trend of the inferior and superior limits of FLB2 (SSD > CL) in function of R. The subscript M means that analysis was performed with Interval Analysis, while the subscript S implies the use of Affine Arithmetic.

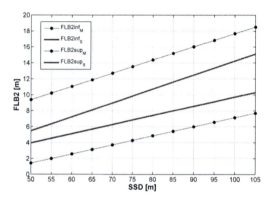

Figure 3. Trend of the inferior and superior limits of FLB2 (SSD > CL) in function of SSD. The subscript M means that analysis was performed with Interval Analysis, while the subscript S implies the use of Affine Arithmetic.

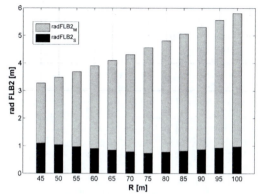

Figure 4. Deviation of FLB2 (SSD > CL) in function of R, inferred with Interval Analysis (subscript M) and with Affine Arithmetic (subscript S).

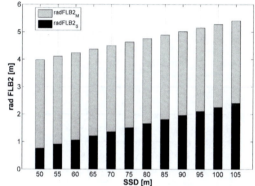

Figure 5. Deviation of FLB2 (SSD > CL) in function of SSD, inferred with Interval Analysis (subscript M) and with Affine Arithmetic (subscript S).

and γ remained constant in order to evaluate only the influence of SSD on FLB, although they are always expressed in terms of interval respectively equal to [54.00, 66.00] m and [0.53014, 0.64795] rad.

In Figures 2–3 we reported the results in this way in order to highlight the differences between the two techniques IA and AA. In contrast, in Figures 4–5 we represented only the deviation of the variable investigated (FLB) with respect to R and SSD without considering the central value. Even in this case, we assumed the same position about the constant values assigned to R-γ and SSD-γ.

4 DISCUSSION

The numerical application based on the determination of a free lateral board in a horizontal curve allowed us to highlight the two main advantages of the proposed procedure:

– Firstly, the principles of range analysis can be applied not only to assign some uncertainty to a

quantity, but also to explore quickly the range of possible solutions in a multivariable problem.
– It is possible to control the explosion of the errors. It is known that excessive overestimation of the final result can be caused by the well-known problem of dependence between the variables involved when applying IA in the traditional form. The phenomenon was greatly mitigated by applying Affine Arithmetic, which, while resulting in greater analytical complexity, led to results with intervals close to the real ones.

The control of FLB in the two conditions (SSD < CL and SSD > CL) allowed us to test these techniques with two different levels of complexity. In fact, the equation (2), being more simple, is less affected by the problem of the dependence among the variables and, in particular, the error does not amplify with R and increases only slightly when SSD becomes greater. In contrast, when the quantity in range form is present more than once in the same formula, as in the case of equation (3), the solution proposed by the traditional IA is not reliable. In contrast, Affine Arithmetic produces results with very realistic intervals. As a matter of fact, the errors are very small and, in any case, are located in the last component of the deviation of the approximated affine form.

Moreover, for this type of application, the problem of dependencies can be eliminated or, at least, mitigated as much as possible. In fact, if we applied range analysis to uncertainty, a result with a wider range than the real one could be seen as excessively precautionary, but not harmful. Instead, the search for all possible solutions, as in this case, could lead to catastrophic events, since not all these results are potential solutions to the problem.

Therefore, the right conclusion is that, even for very complex situations, the use of these techniques to design problems can be employed only to minimize the output interval and this can only be achieved by means of Affine Arithmetic.

However, also if this technique clearly shows its difficulty, compared to that of IA, it is easily automatable with ordinary scientific software. In addition, the preparation of opportune diagrams and graphs such as those shown in Figures 2–3 can definitely help the designer during his activity, without requiring knowledge and application of Affine Arithmetic.

Furthermore, the use of range analysis, in general, considerably reduces the computation time for a designer. Indeed, the intervals assigned to the input variables contain a very large number of different combinations that represent all the possibilities offered to the designer in that particular condition. In this regard, the input variables and their potential range may represent the "degrees of freedom" of a designer in the resolution of a specific problem. The solution can be obtained by balancing, again not so homogeneous, the input intervals. This suggestion can come from any safety or accident analysis that may clarify the role of the more dangerous variables.

5 CONCLUSIONS

During the design of a new road or the adaptation of an existing one, there are some steps that lead to an increased focus by engineers. In general, all the problems of alignment composition are easily solved since they are controlled by an equation with a certain number of unknowns fixed by designers in order to obtain a satisfactory solution. When there are some constraints, the number of potential solutions is greatly reduced and this occurs when:

- There are critical environmental issues;
- It is necessary to limit the cost of construction;

- The limits imposed by the standard are very strict;
- In the case of existing roads, the old design was prepared in compliance with obsolete regulations;
- An accident analysis has highlighted some critical variables that must, therefore, be carefully controlled.

Under these conditions, designers cannot "randomly" choose one of the possible solutions, but must investigate the role played by all the variables that contribute to the solution of the problem, confining them within acceptable ranges. This problem, in our opinion, cannot be addressed through optimization procedures, since there is not an "optimal solution", but we need to evaluate all the potential solutions so constrained.

The better way, therefore, is to represent the structure of the equations of interest in terms of range, so that the final result is also reported as a range and is representative of all the possible solutions to the problem. Unfortunately, the classical approach of Interval Analysis, in cases of complex analytical expressions where the variables are not perfectly independent of each other, leads to significant errors. This was demonstrated in this paper by a simple application that the simplified expression of the FLB (when $SSD < CL$) could still be treated with the classic IA, while the more detailed equation (with $SSD > CL$) necessitated the approach introduced by Affine Arithmetic. This procedure, though slightly more complex, allows the control of the dependencies between variables and also identifies with precision the eventual approximate terms of the affine forms, allowing the quality of the final results to be verified.

REFERENCES

Bosurgi, G., D'Andrea, A. & Pellegrino, O. 2011. Context Sensitive Solution Using Interval Analysis. *Transport*, 26(2), 171–177.

Moore, R.E., Kearfott, R.B. & Cloud, M.J. 2009. *Introduction to Interval Analysis*. SIAM: Philadelphia, 2009.

Muscolino, G. & Sofi, A. 2012. Stochastic analysis of structures with uncertain-but-bounded parameters via improved interval analysis. *Probabilistic Engineering Mechanics* 28, 152–163.

Stolfi, J. & de Figueiredo, L.H. 2003. An introduction to affine arithmetic, *TEMA Tend. Matematica Aplicada e Computacional* 4, 297–312.

Urban planning

Advances in Civil Engineering and Building Materials IV – Chang et al (eds)
© 2015 Taylor & Francis Group, London, ISBN: 978-1-138-00088-9

Research on age-friendly community planning based on aging in place

Xiaoyun Li
Jiangxi Normal University, Nanchang, Jiangxi Province, China

ABSTRACT: With the ongoing trends of an increased ageing population and accelerated urbanization, this paper attempts to construct a supporting system involving social and physical conditions, public policy, and technology, etc., to standardize the contents. This consists of social and physical environmental planning of the age-friendly community, which accommodates the unique characteristics of physiology, psychology, and society of the elderly. With the purpose to integrate the social planning and physical environmental planning of age-friendly communities, this paper promotes planning implementation from social goals to space resource allocation, by planning coordination and the regulation of interests throughout the whole process.

Keywords: ageing in place; age-friendly community; community supporting system; integration.

1 INTRODUCTION

The continuing increase of the ageing population has been an irreversible global trend. The ageing population of China, by the end of 2012, had reached 194 million, accounting for 14.3% of the overall population. According to the Research on the Developing Trend of China's Population Aging, released by the Chinese National Committee on Aging in December 2007, the ageing population will have amounted to 248 million by 2020, and 400 million by 2050, with a growth rate of 30%, and by then, the more ageing population would live in cities due to the gathering pace of urbanization. Bearing in mind all these considerations, it is going to be essential to solve elderly inhabitants' problems and to create better living conditions for them during future city constructions.

When facing the problems of the elderly, letting them grow old in the same place has been proved to be a shared target for most countries (Pastalan, 1990; Wagnild, 2001; Bookman, 2008; Wiles et al., 2012; Lehning, 2012; Costa-Font et al., 2009; Sixsmith et al., 2008), while difficulties exist still in realizing ageing in place in terms of the health condition of older people, housing, environment support, supporting facilities, etc. (Pynoos et al., 2008; Kloseck et al., 2010). To achieve the goal of ageing in place, problems of housing shortages, ineffective health and supporting services, inconvenient transportation, etc., are due to be surmounted (Merrill et al., 1990).

In 2005, the concept of "Age-friendly Cities" was first presented in 'Global Age-friendly Cities: A Guide', published by the World Health Organization (WHO). In 2009, aiming to better satisfy the elderly's needs in community construction,

the Chinese National Committee on Aging initiated numbers of pilot projects of age-friendly communities based on the framework in Huangpu District, Shanghai, Xuanwu District, Nanjing, Jiangsu Province, and Jianhua District, Qiqihar, Heirongjiang Province. Despite integrating elements to meet with the distinctive situations of China, the projects, with macro-political, economic, and social objectives, served little use to domestic age-friendly community construction because they were short of specific measures to solve the problems of housing, space environment, etc. Other than that, communities, as the essential units of cities, are the principle places of activities for the elderly. Therefore, only through creating high calibre age-friendly communities can we achieve the goal of age-friendly cities.

Overall, against the backdrop of an ageing society whose population is increasing, this paper attempts to enhance the theoretical system of housing for the elderly and enrich the contents of age-friendly cities by both analysing the needs of the elderly in cities, and standardizing the contents of age-friendly communities. For further purposes, the paper may help to instruct the construction of age-friendly communities, adjust the social mechanism facing the ageing population, and solve the housing problems for the elderly.

2 INTERNATIONAL RESEARCH ON PLANNING & CONSTRUCTION OF AGE-FRIENDLY COMMUNITIES

The definition of an "age-friendly community" varies in different groups with distinct purposes, but it can be roughly interpreted as "a friendly community, with a convenient living environment, accommodating the special needs of the elderly for housing, health, social engagement, etc., and providing infrastructures and services for the living, medical care, education, entertainment, activity, etc. of the elderly." It is foremost age-friendly whether it is mix of ages or sexes or not. Moreover, it is a harmonious state of inhabitation between the living environment and the elderly, who are both the subjects and the objects of age-friendly communities.

The elements and contents of age-friendly communities cannot be unanimously agreed upon by scholars and institutes. Alley and Liebig articulated that an age-friendly community was comprised of housing, healthcare, security, affordable transportation, social engagement, etc. (Alley et al., 2007). It had been pointed out by Lehning and Scharlach that an age-friendly community should consist of land-use policy and community design with an increase in social engagement, and a decrease in reliance upon cars, and having access to the choice of affordable housing, diversified transport, accessible medical care and activities promoting social engagements (Lehning et al., 2010; Scharlach, 2012). However, according to 'The Advantage Initiative*, an age-friendly community requires four essential elements: (1) Addressing basic needs; (2) Optimizing physical and mental health and well-being; (3) Maximizing independence for the frail and disabled; and (4) Promoting social and civic engagement. In addition, these four elements can be further deducted into thirty-three specific norms to evaluate a community's age-friendliness. This interpretation and evaluation system by The Advantage Initiative focuses much more on medical care and nursing than on the physical environment.

The multi-dimensions of international research imply that both physical and social environments shall be integrated into the planning and construction of age-friendly communities (Lui et al., 2009). Aside from the physical and technical optimization, policy-making should draw attention on the social integration of the age-friendly community and emphasize the improvement of infrastructure and services, which can reinforce social engagement of the elderly. In addition, the planning and construction of an age-friendly community consists of a process of management. As both the beneficiary and the builders of age-friendly communities, the elderly shall be encouraged to participate from the bottom, up, in the creation of an integrated and interactive social environment, which leads to an inter-department cooperation between the elderly and various interest groups, such as service providers, volunteer groups, private businesses, organizations, etc. Such kind of cooperation is the goal of ageing in place and it requires more on the planner's role orientation.

---

*The AdvantAge Initiative is a community-building effort focused on creating vibrant and elder-friendly, or "AdvantAged," communities that are prepared to meet the needs and nurture the aspirations of older adults

Together with theoretical research, international communities have been making efforts in age-friendly community construction. Results from a survey on the framework of an "age-friendly city", conducted by the city of Thunder Bay, Canada, indicated that a more age-friendly community is achieved by housing options, more access to health and community services, improved transportation and increased safety (Kelley et al., 2010). In addition, the Waihi Beach Community of New Zealand initiated an age-friendly community programme through the update of outdoor spaces and buildings, transport, housing, respect and inclusion, social participation, communication and participation, civic participation and employment, and health and community services (Gordon et al., 2009). The conclusion can be extracted from those experimental programmes that age-friendly communities are essentially composed of accessible housing, safe public spaces, affordable transport, health support, public security, education and employment, social participation, etc., despite differences in regions, cities, and standards by various organizations.

3 DOMESTIC AGE-FRIENDLY COMMUNITY CONSTRUCTION

3.1 Ageing in place—analysis of the demands of the elderly

Sharing the characteristics of the general inhabitants in cities, the elderly possess identical features. They will experience the degeneration of physical wellness, senses and motor skills when they grow old, accompanied by the modified features in physiology, psychology, social economy, etc. These modified features will lead to dynamic and diversified demands by the elderly, and only when those demands are satisfied can we guarantee quality of life for the elderly and ageing in place.

Meanwhile, "ageing in place" accommodates the majority rather than a specific group of elderly people. According to Maslow's Hierarchy of Needs, urban elders' needs can be categorized into physiological, health, safety, psychological and social needs, reflecting a hierarchical gradation from physical to spiritual, object to obscurity, individuality to the public. In the mode of "ageing in place," the urban elderly's general residential behaviour will result in needs for living space. As illustrated in Figure 1, different needs will be alluded to by different behaviour, depending on whether the circumstances are private or semi-private, public or semi-public. Moreover, needs for living space will, as well, be raised by elders' solitude, repose and entertainment, communication, walking, education, medical treatment, etc. It is fundamental for helping the elderly to create graceful, comfortable, and pleasant living environments, guaranteeing their quality of life, and realizing through ageing in place, that we satisfy those dynamic and diversified needs for residential spaces.

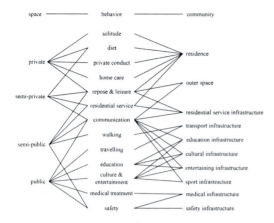

Figure 1. Elderly requirements and social association.

Figure 2. Supporting system of an ageing-friendly community.

Figure 3. Content system of ageing-friendly community planning.

3.2 Contents of age-friendly communities fulfilling the diversified needs of the elderly

Combining international research and practice on age-friendly cities or communities, and the analysis of the needs of the elderly in the mode of ageing in place, it can be articulated that four major contents: social, physical, policy, and information technology environments, formulate the supporting system of an age-friendly community, as shown in Figure 2.

The theory of community planning defines comprehensive community planning as social development planning, economic development planning, and physical environment planning combined (Wang Ying et al., 2009). However, the fundamental contents of age-friendly communities exclude economic development planning because of its differences in subjects, targets, planning features, public engagement, policy, etc., when compared with general comprehensive community planning, as shown in Figure 3.

With supplies of diversified, multi-lateral, and affordable housing, we can provide the elderly with living spaces fulfilling their physiological and psychological needs, and insure the disadvantaged elderly have access to housing. Health planning is such a concept of making use of medical treatment, health land, health housing, health transportation, etc., in order to safeguard a healthy living environment, while safety planning is designed to provide the elderly with secured living, leisure, entertainment, gymnastic, and transport environments. A safe community preventing harm will be more than important for the elderly when their physical functions degenerate, and elderly culture planning holds an objective of improving their well-being, enhancing their cultural accomplishments, enriching their spiritual life, and promoting positive ageing through various community culture constructions. Moreover, mixed-planning is a strategy, with tools of community engagement and mutual help, volunteers, and multi-generation planning, to help their integration and avoid separation among different ages. In addition, we can build a system serving the elderly to diversify community services, consisting of a hierarchy of free, low-cost, and stipendiary categories, in line with their well-being and financial conditions.

- Physical environment planning is composed of land utilization, residential units, friendly, open spaces, friendly transportation, and facilities designed to serve the elderly, etc.
- Land utilization planning, operating under the smart growth theory, is used for consultation in locating, land-use scale and model, and construction of age-friendly communities.
- Residential unit planning enables the elderly to live independently in an obstacle-free environment with optional living spaces, such as mixed residential units, independent residential units, and nursing homes.
- Friendly, open space planning, as a continuation of private space, creates an environment of communication space, rehabilitation green land, gymnastic areas, etc., in which relationships among different ages are cultivated, helping to eliminate loneliness and loss of identity in the elderly.
- Friendly transportation planning gives the elderly an accommodated traffic environment composed of obstacle-free infrastructures and a diverse, recognizable, and consecutive pedestrian network.
- Facilities and services planned with the elderly in mind.

4 INTEGRATION & COORDINATION OF DIFFERENT CONTENTS OF AGE-FRIENDLY COMMUNITY PLANNING

4.1 Integration and coordination of society and space

Coordinating work of different departments
The target of age-friendly communities cannot be achieved without the coordination of the housing department, the health department, the education department, and sports department, etc. Community planning is a comprehensive subject that combines not only physical environment planning, but also planning for health, safety, culture, etc. This quality of community planning requires both coordination among departments under the leadership of the Consultant and Coordinating body of the National Committee on Aging, and instruction by advisory groups with different expertise. For instance, the modifications of communities for the elderly lead to much construction and renovation of facilities. This process involves the planning dept., construction dept., land and resources department, the housing department, the civil affairs department, etc., in which they coordinate issues of land use transformation, cubage and density adjustment, change of property rights, fundraising, and voting rights of the elderly respectively.

Coordinating planning of different departments
The planning of age-friendly communities should be compatible with those of other departments. Relevant planning currently covers sections of national economy and social development, land utilization, ageing developments, urban developments, control detailed developments, site developments, developments of facilities for the elderly, etc.; as a non-statutory section of urban construction, the planning of age-friendly communities encounters difficulties in terms of housing, facilities, rehabilitation scenery, outdoor activity space, fitness facilities, etc. Therefore, in pursuing more space for the various needs of the elderly, we should include the planning of age-friendly communities in the drafting, implementation, and management of urban construction, control detailed planning, and residency planning, etc.

Coordinating segments within the planning of age-friendly communities
Separating the analysis of social planning and physical environment planning does not mean to contradict these two segments, but to emphasize mutual effects and correspondence between the social and physical environment, and to explore further how to synchronize effectively these two segments (Liu Jiayan, 2008). For example, planning of age-friendly communities essentially aims to meet the elders' social needs. However, it will be idealistic to rely solely on society. In order to effectively integrate the social and physical environment and to further realize the developments of the physical environment and society, social planning asks for different carriers for social activities, and

sound allocation of space resources among community land use, residential units, friendly open spaces, friendly transportation, public facilities, etc. In addition, in achieving fitness of the elderly, we have to coordinate soundly the planning and construction of healthy land use, healthy housing, healthy outdoor environments, medical care facilities, etc. Other goals of social planning need corresponding planning also.

4.2 Coordinating interests of different groups

Creating an elderly-centred coordination model among different interest groups
The planning and construction of age-friendly communities, with the support from government, businesses, public service units, non-governmental organizations, and citizens, should be devoted in creating a model of coordination among different interest groups that optimize the interests of the elderly. A necessary step to realize this goal is to transform and reposition sub-district offices and communities, making the change from management-oriented agencies to service-oriented agencies. Along with the transformation, autonomy shall be added to the functions of neighbourhood committees and community planning and management committees, and a new non-governmental agency that devotes its efforts in coordinating the formulation, managing the process, and supervising the implementation of the planning of age-friendly communities, shall be created. This should speed up the build-up of non-governmental organizations, encouraging their engagement in community planning and construction, and coordination with the government. During this process, non-profit organizations, such as foundations, philanthropy organizations, volunteer groups and associates for the elderly, should be prioritized to provide them with personalized services, which can both satisfy their wish of ageing in place, and ease the pressure on community nursing homes.

Encouraging elders' engagement in the planning
To construct a suitable plan for an age-friendly community from the bottom, up, engagements of different interest groups like the elderly themselves, the working-age adults, children, organizations, etc. are needed. Firstly, procedures need to be completed. Every procedure should contain a mechanism of public participation that, with its diverse and executable ways, enables the elderly, with different well-being needs and education, to take part in the planning process. For example, a forum is such a mechanism where elders can express their thinking freely, and for those disabled, a visit to the house is preferred. Secondly, lectures and multimedia should be made use of to improve the public's ability for participation, the elderly in particular. We can make community planning a compulsory course for education of the elderly, or encourage elders' participation in the planning when they when they attend lectures on on the theoretical knowledge. Thirdly, organizations' principle positions

should be established. Being led by community planning and a management committee, the opinions of the elders are collected by the owners' committee, expert groups, elders' associations, community organizations, etc., and transferred to the planning department, which should constantly optimize the planning of age-friendly communities.

Initiating a community architects programme

A messenger between different interests groups will be required during the planning and construction of age-friendly communities, and a community architect, as such a messenger, is influential in coordinating those interests' entities (Jiang Jinsong et al., 2004). Against a background where workers for affairs of the elderly lack professional knowledge, and elders' opinions are difficult to collect, the involvement of community architects appears to be paramount as a platform for governments, communities, planning departments, and the elders to communicate. Therefore, a community architect can be interpreted as a career attempting to build better communities that accommodate the elderly without infringing on the interests of other groups.

A community architect, aside from his professional expertise, shall be required to possess communication abilities with different departments and citizens, the elderly in particular. To instruct the elders' participation in community planning, a community architect should also be a good organizer. Accomplishing the work of the general community, community architects have to make more efforts in social welfare, medical care and healthcare, nursing agencies, and facilities for the elderly, and in promoting modifications that are more suitable for them. Coming from a different professional background, they will need relevant training in the sociology of the ageing process, the elderly and community life, and laws and regulations in order to carry out their jobs better.

5 CONCLUSION

As a relatively ideal residential model for the elderly in cities, ageing in place can only be realized through the planning and construction of age-friendly communities, which complement the lack of families' attendance and the shortage of available facilities, helps to avoid the elderly's detachment from society, which is the disadvantage of living in a nursing home, and to ease the pressure on care homes.

Accommodation for elderly people is a complex social problem that involves social sciences, the economy, psychology, medical science, social security, etc. This paper has analysed the planning of age-friendly communities, but some shortcomings and some segments for further research still exist. A complete community planning consists of the making, implementation, and evaluation of the plan. Future studies may intend to establish an evaluation system to assess the friendliness of a community. Research can be introduced further to investigate the planning of age-friendly communities in rural areas.

ACKNOWLEDGEMENTS

Sponsored by the Humanities and Social Science Project of Jiangxi Colleges and Universities (JC1434), "Twelfth Five-year Plan" of Jiangxi Provincial Social Sciences Planning Program (14SH05); Jiangxi Arts and Science Project (YG2014113).

REFERENCES

Alley, D., P. Liebig, J. Pynoos, T. Banerjee, I.H. Choi. Creating Age-friendly Communities [J]. Journal of Gerontological Social Work, 2007 49(1–2): 1–18.

Gordon, C., S.V.D. Pas, An Age-Friendly Community: Shaping the future for Waihi Beach. 2009, Waihi Beach Age-friendly Health Impact Assessment: Waihi Beach.

Jiang Jinsong, Lin Bingyao. The Thinking of the Theory, Method and Policy on Urban Community Planning and Constrution [J]. Urban Planning Forum, 2004(03): 57–59.

Kelley, M.L., R. Wilford, A. Gaudet, L. Speziale, J. McAnulty, Age-Friendly Thunder Bay. 2010, Centre for Research and Education on Aging and Health (CERAH): Thunder Bay.

Kloseck, M., R.G. Crilly, G.M. Gutman. Naturally Occurring Retirement Communities: Untapped Resources to Enable Optimal Aging at Home [J]. Journal of Housing for the Elderly, 2010 24(3–4): 392–412.

Lehning, A.J., A.E. Scharlach, T.S. Dal Santo. A Web-Based Approach for Helping Communities Become More "Aging Friendly" [J]. Journal of Applied Gerontology, 2010 29(4): 415–433.

Liu Jiayan. Toward Integration: Planning and Society [J]. Beijing Planning and Construction, 2008(01): 82–84.

Lui, C.-W., J.-A. Everingham, J. Warburton, M. Cuthill, H. Bartlett. What makes a community age-friendly: A review of international literature [J]. Australasian Journal on Ageing, 2009 28(3): 116–121.

Merrill, J., M.E. Hunt. Aging in place: a dilemma for retirement housing administrators [J]. Journal Of Applied Gerontology: The Official Journal Of The Southern Gerontological Society, 1990 9(1): 60–76.

Pynoos, J., C. Nishita, C. Cicero, R. Caraviello. Aging in Place, Housing, and the Law [J]. University of Illinois Elder Law 2008 16(1): 77–107.

Scharlach, A. Creating Aging-Friendly Communities in the United States [J]. Ageing International, 2012 37(1): 25–38.

Wang Ying, Yang Guiqing. The Construction of City Community in the Period of Social Transition[M]. Beijing: China Architecture & Building Press. 2009.

WHO. Global Age-Friendly Cities: A Guide[M]. Geneva, Switzerland: World Health Organization Press. 2007.

Advances in Civil Engineering and Building Materials IV – Chang et al (eds)
© 2015 Taylor & Francis Group, London, ISBN: 978-1-138-00088-9

Application of the interaction of design philosophy into landscape design

Dong Chu & Wenzhong Huo
Harbin University of Commerce, Harbin, Heilongjiang, China

ABSTRACT: The passive linear thinking of the industrial society has been replaced by the interactive cross-thinking resulting in interaction design, which is a new design method for observing the world. This design method is a synthesis of dynamics, virtualization, and experience, and emphasizes intersectionality, integrity, and scientificity. The thesis focuses on the ideological connotation and the according design method, taking landscape as a supporter, which expects to intervene in urban landscape design practice. Meanwhile, interdisciplinary research is also taken into consideration with conception integration to achieve the scientific design about the behaviour experience in the process of landscape interaction design. The goal is to solve the practical problems of urbanization in China with landscape design intervention.

Keywords: interaction design methodology, humanized experience, interactive virtuality, dynamic association.

With the change of the cognitive style, the invention of new technology, and the pluralism of aesthetic demands, one-dimensional linear thinking will certainly be broken. A new synthesis of organic design arises, which emphasizes the communication process between the subject and the object by the medium of digitalization, turning into lateral continuity of time and vertical expansion of space.

1 EXISTING CIRCUMSTANCE OF URBAN LANDSCAPE

In the process of urbanization in China, rapid removal and reconstruction lead to the absence of the comprehensive plan and ecological concept. Topographic and geological features, water, and vegetation are always neglected. Some of the projects are being mechanically applied and destroying the intrinsic natural relationship of the area. Because of the blindly copy from foreign countries, some of the cities in China are suffering from uniformity and they are facing similar problems. The European and the Roman styles of landscape are arising like mushrooms after the rain. The hard top, river diversion, and exotic tree species are appearing everywhere, which causes an ecological imbalance. This linear thinking neglects the secondary relationship between the physical forms and their inner core; the interaction relationship between humans and nature. The passive design method, which focuses on the forms, cannot bring the spirituality of humans in line with the texture of nature. This is the main reason that design forms deviating from our daily life and never playing their real role as the urban landscape.

2 WHAT IS INTERACTION DESIGN?

Every important invention does not achieved by disciplines of its own field, but to absorb the idea from other subjects. Modern design has gone through alternative changes between the form and the connotation. In the information age, human experience is being concerned, which means the interaction between complex system and human is taking into consideration. In brief, interaction design is about the behaviour of the artifacts, the environment, and its system by designing and defining the form elements. It is different from traditional design, which mainly focuses on the forms, but instead it pays much attention to the content and intention. Interaction design firstly aims at the planning and description of the manner of the behaviour of the objects, and then designs the most effective form to express the behaviour. As a subject of interactive experience, interaction design was proposed by Bill Moggridge, founder of the IDEO company in a meeting in 1984. Gillian Crampton Smith later applied this theory into design education and set up the Computer Related Design Department at the Royal College of Art. Now many research centers have set up similar departments in order to pursue quantitative and qualitative investigation from definition, methodology, and design criteria of different angles in the form of theory and practice.

3 TRANSFORMATION OF THE INTERACTION DESIGN METHODOLOGY

The concept of the interaction design was first applied into product design, focusing on the expectation of understanding the target user and the interactive behaviour between the product and the user. The effective interactive manners are designed by understanding human psychological and behavioural characteristics; meanwhile, the organic connection is built. In fact, design is not a verb, but practical process-seeking for the method. A comprehensive design method derived from interactive design gathers the dynamic, virtualization, and experience together and emphasizes the holistic, intersectionality, and scientificity aspects of these. It has removed the dependence of the individual subjectivity, and turns to the scientific method and tools, which pays much attention to human behaviour process design according to the holistic change.

With the development of modern landscape ecological theory, interaction design philosophy is applied to landscape design and emphasizes the quantitative analysis required to build the communication medium for human and nature. Landscape ecology research is about the space properties of the ecological system and their relationship, including the integration of human activities, structures, and functions of the ecological system. This design concept initially came from the book 'Design with Nature' written by Ian McHarg. The most basic principle is the overlay techniques of the layers. The importance of creating the interactive relationship is to obtain, analyse, and process all the geographic variables and data about space, resources, the environment, population, transportation, economy, education, culture, and finance.

Landscape design based on interaction design philosophy draws lessons from engineering theory and technology, which is a synthesis of the new method and practice, but not the overlay of the partial stack. Its design methodology has its own goal orientation to design the operation method for the users and redefine the relationship between the environmental behaviour and the environment, seeking for conversation among the material, culture, and history. This design methodology could also forecast the comprehension coming from the consumers, and show how to influence the relationship between the consumers and the environment. Because of the development of interaction design philosophy, the landscape is not only playing as a lens to observe the city, but also becoming a medium to reconstruct the city. The organic association will be built between humans and nature by the medium of landscape to achieve the interactive needs for people to experience nature.

3.1 Humanized experience

Landscape ecology, against the background of modern interaction design philosophy, aims at bringing all the related disciplines together with a thorough understanding. It takes energy, land, water, materials, and other natural resources simultaneously. After integrating all the elements, the best landscape plan will emerge based on the scientific quantitative analysis. By designing the natural resources, people could acquire entire experiences in the process of interactive

activity with nature, and enjoy all the aesthetic feelings coming from the experience.

3.2 Interactive virtuality

Roger Tomlinson created the Geographic Information System (GIS), a space analysis based on the technology of the data about the position of geographical objects and their morphological features. The GIS as a research tool was taken into the field of modern landscape design. The interactive connections between human and nature could be built by scientific technology based on geographical data, which applies a geographic model analysis to provide different space and geographic information. Because of the human-computer interaction of the digital period, design has changed the static forms. Designers can get a full review of the future landscape by dynamic and interactive means in the fictitious three-dimensional environment, considering the variability of the consumers and the field. Then they can create the process design of the interactive experience.

The GIS has been applied widely in developed countries. For example, Janet Silbernage, at Wisconsin University in the USA, has applied the GIS in the transition analysis of the landscape to the east of the Peninsulas of Michigan. Other designers have also used it to assess the landscape sensibility. Professor Kongjian Yu from College of Architecture and Landscape Design, Peking University, also applied GIS to analyse the Yellow Stone scenic area planning in Hubei Province. The traditional design is a static visual experience methodology using AutoCAD with Photoshop and 3DS MAX. To a certain extent, the old concept was built on the perceptual experience. The new landscape design based on the interaction design philosophy, analyses the space data created by GIS. It initially obtains the data by a remote sensing image, and then analyses its topography and space pattern by building a DEM (Digital Elevation Model) to overlay the three-dimensional remote image. Modern landscape design is built on the basis of scientific analysis after combining all the history features with all the space patterns.

3.3 Dynamic association

A traditional design method follows the nature of the problem itself, but modern interaction design emphasizes the connections with the problem. Design has become a dynamic process, and this design has changed into the relationship between the designers and the public. In the process of landscape design, advantages and useful information should be provided to guide the interactive behaviour, which will achieve the interactive connection between human and nature, but not forcibly imposed on people. Design is not a product with instructions to tell people how to use it. It can guide people to use it naturally with intuition.

4 CONCLUSION

What will the city look like in the twenty-first century? How can it be built with a more characteristic space experience and environmental quality? Designers' consciousness has to change first. Interaction in landscape design is not only an attitude, but also a way of thinking. The city should be interpreted as a life system process. The future structure and form of the space should be adaptable and flexible in order to meet future development. Based on the GIS as a medium to take a quantitative analysis, modern landscape design is not a hollow form. The key to interactive landscape design is not just to improve the environment quality; the shaping of humanity should also be taken. is also taken into consideration.

REFERENCES

Affordance and Design, [ID] Community, From the network: http://www.hi-id.com/?p=2732

Alan Cooper, Robert Reimann & Jianfeng Zhan translated. 2005. About Face2.0: The Essentials of interaction Design. Beijing: Electronics industry Press.

Danald, A. Norman & Qiufang Fu & Jinsan Cheng translated. 2005. Emotional Design: Why We Love (or Hate) Everyday Things. Beijing: Electronics industry Press.

Kongjian Yu. 2000. Lessons coming from the international city beautiful movement in China. Beijing: Chinese Garden.

Lan Lennox McHarg & Huang Jingwei translated. 2006. Design with Nature. Tianjin: Tianjin University Press.

Monica G. Turner, Robert H. Gardner & Robert V. O'Neill. 2001. Landscape Ecology in theory and practice: Pattern and Process. New York: Springer-Verlag New York Inc.

Qiang Niu. 2012. Application Guideline of GIS in the city planning. Beijing: China Architecture & Building press.

Advances in Civil Engineering and Building Materials IV – Chang et al (eds)
© 2015 Taylor & Francis Group, London, ISBN: 978-1-138-00088-9

A study on the public participation willingness of community development planning in Wuhan

Han Zou & Qi Kang

Institute of Civil Engineering and Architecture, Hubei University of Technology, Wuhan, China

ABSTRACT: Public participation is an important part of community development planning. Therefore, to be aware of the community members' willingness to participate is the foundation to guide the public in community-based planning participation. Four different typical communities in Wuhan were chosen in this paper. By analysing the survey data, the true willingness of the community members to participate in community development will be shown, and the future challenge of community planning will be forecasted. In this way, the approach and the feasibility of community development planning will be discussed. This paper also learns from the experience of several similar cases, to provide a reference for such communities updating the problem.

Keywords: Public participation, community planning, empirical research, Wuhan.

1 INTRODUCTION

In the early 1990s, with the transformation of social systems and the continuous progress of modernization, the concept of public participation in community planning began to be taken seriously by the urban planning professionals. As a method of social management, public participation aims to take 'bottom-up', the decision-making process so that more people can participate in the closely related development of policies and practices. Though, compared with the relatively mature and perfect system of public participation in developed countries, public participation of community development planning in China is still in its infancy. It needs to be further researched and advanced.

1.1 *Public participation situation in developed countries*

In terms of the concept and theory of public participation, concepts and theories were first proposed by the United States in the early part of the last century, and continued to be developed. In 1923, Clarence Perry created a form of 'neighbourhood unit' (Neighborhood Planning) for the United States as a solution for a series of social problems (social isolation and strangeness, the lack of public participation tools, etc.) (William. M. Rohe 2009); In 1965, Davidoff proposed advocacy planning theory which explained the appropriate policy should be reached through public discussion in a democratic institution (Paul Davidoff and Reiner Thomas 1962); In 1969, scholar SR Arnstein indicated that citizen participation is a categorical term for citizen power (Arnstein Sherry 1969).

In the aspect of public participation in planning practice, the citizens are free to take part in the organizations such as planning aid, neighbourhood planning, which are the foundation in the process of British city planning. The non-governmental organization is independent from the government, and some belongs to the volunteer or the spontaneous organizations, which played a positive role in the process of British city planning.

In terms of public participation and planning practice, Japanese law bestows the citizens with a high degree of planning participation rights. For some special areas, with the 'Urban planning Law', local residents even enjoy decision-making powers. Chapter II's 'decision of urban planning and change' section clearly states that the change in urban planning decisions and public participation. It takes the way of hearings, notice exhibitions, and notices applicable conditions and participants.

In addition, Canada, Australia, Germany, France, and other countries have established a system of public participation in urban planning with success.

1.2 *The development of community planning and public participation in China*

The achievement of the recent twenty years' development of public participation is primarily reflected in the gradual improvement of laws and regulations. The City Planning Law of the People's Republic of China, which was implemented in 1990, mentioned that all units and individuals should have the obligation to abide by the plan for a city, and shall have the right to report and bring charges against any action that runs counter to such a plan. The Measures for Formulating City Planning, which was implemented in 1991 by the Ministry of Construction, requires that the preparation of planning should solicit opinions from the relevant departments and local residents. The Urban and Rural Planning Law of the People's Republic of China, which was implemented in 2008, first established a mechanism for public participation in the planning, emphasizing the process which is organized by the government, led by the expert, and participated by the citizens.

In addition, domestic scholars have invested more in public participation and urban planning. Hong Kong scholar, Guo Yanhong, introduced the Western public participation in urban planning and development in the 1980s; Liu Qizhi published a master's thesis, 'Basic research of public participation in urban planning' in 1990; around 2000, there were dozens of papers about public participation in urban planning which were published in the 'Urban planning', 'Planner', 'Urban Planning Overseas', and other magazines; in 2004, Sun Shiwen defined 'Community Participation' as a process to develop a series of measures through formal or informal cooperation within the professionals, family members, community organizations, and other administration officials. Within a decade, public participation has attracted more attention in various disciplines.

However, compared with a more mature public system abroad, China's public participation and planning is still in its infancy period. Therefore, further research and development should be carried out.

2 RESPONDENTS AND RESEARCH METHOD

Four typical traditional communities, which are on the verge of alteration and rebuilding, were chosen based on the present situation of public participation, focusing on the willingness of community members participating in their community development. By taking different planning policies into consideration, this paper categorizes the four chosen communities into two types: the Demolition Type and the Preservation Type.

2.1 Demolition type community

2.1.1 Xidajie community in Hanyang district
The Xidajie Community, located in the centre of Jianqiao Street in Hanyang District, which is east of Beicheng Street, west of Yingwu Avenue, connects Xidajie Street to the south, and reaches north to Hanyang Avenue. This old and open urban district is $0.18 \, \text{km}^2$, including the Beicheng Marketplace, the Xidajie Business District, and the Dong Fang Hong Night Market. The majority of the residences in this community are low-rise residential buildings with ground-floor shops, brick-concrete in structure, and linear in layout. Older people consists the main part of residents in the community.

2.1.2 14th-Neighbour community of Honggangcheng in Qingshan district
The 14th Neighbour Community of Hong Gang Cheng was built between 1984 and 1990, and is located at 6th Jianshe Avenue in Qingshan District. This community is an employees' residential community of the Wuhan Iron and Steel (Group) Corporation (WISCO), thus most of the residents are employees of WISCO, and are mainly are older people. The streets are laid out in a grid system, and the buildings are mostly residential and multi-storied.

2.2 Preservation type community

2.2.1 Tanhualin community in Wuchang district
The Tanhualin Community is located to the north of the Old Town of Wuchang District, which is a result of Wuchang City's expansion from the Qing Dynasty. Since there are many modern buildings situated in the community, the Tanhualin Community is positioned as an historical and cultural community where partial renovation has taken place. The main street is of an east-west direction with the residents located radially from both sides of the branches. Large numbers of the residential buildings are brick-concrete structure in low-rise.

2.2.2 Sande Alley community in Jianghan district
The Sande Alley Community of Jianghan District is located from the north of Zhongshan Avenue to the east of Chenzhan Street in the Hankou Area. There are low-rise residential buildings with ground-floor shops along the street, together with townhouses behind the street, which all need to be repaired. The 1/3 of the residents are tenants in the community.

2.3 Research method

A questionnaire survey was performed among the residents of the four chosen communities during May 2014. In terms of the developing power source, category, planning process, and the needs of residents for a community, the questionnaire was composed of three parts including, the respondents' essential information part, the present situation of public participation in community development planning, and the willingness for community development planning. We set both Demolition Type and Preservation Type, two different types of questions among the topic of willingness for the community planning part. The Demolition type is emphasized in residents' willingness of demolition as well as their living experience, while the Preservation type is focused on the anticipation of the community development.

There were 30 questionnaires for each community. 120 questionnaires in total were sent, and 104 were replied to (24 from the Tanhualin Community, 28 from the Sande Alley Community, 22 from the Xidajie Community and 30 from the 14th Neighbour Community).

2.3.1 Respondents' essential information
The four chosen communities all are old communities, 82.7% of the respondents are permanent residents; middle-aged and over residents are in the majority, they consist of 37.4% of all the respondents; the primary level and senior high level makes up the great mass of the education level.

2.3.2 *The present situation of public participation in community planning*

The investigation result can be seen from the questionnaire survey of the present situation of public participation in community planning:

Firstly, referring to the question of opinions before a community rebuilding action 'is there any kind of activities held by neighbourhood committee to adapt the local residents' suggestion', there are 80.7% of all respondents said 'no.' While on the other hand, the respondents from the neighbourhood committee answered 'yes'.

82.7% of the respondents showed their aspiration concerning the question of how willing they were to participate in their community planning activities. Meanwhile, other people are unwilling to participate in the community planning apparently because they worry that their opinion will be ignored.

However, while residents pay strong interests taking part in their community planning, there is short of appropriate way to lead the residents to participate the community planning completely. Usually, the Neighbourhood Committees would like to announce the final planning result to the residents rather than organize the participant activities during the whole process of planning-programming. Seeing from the ladder theory of citizen participation, the present situation is still at the degree of Non-participation or Degree of tokenism, which forms a dramatic contrast.

When a further question about what way respondents would like to participate in the planning was asked, the majority of them took the 'provide advice' option. At the same time, a few of them preferred to 'participate in the process of planning-programming', 'participate in a proprietor committee', or 'participate fully.'

80.7% said 'yes' who replied to the question of 'whether respondents would like to take professional planning services', the overwhelming majority of them prefer to 'receive brochures regularly' as well as 'communicate with the planners who are designated to the community.' These were the following reasons: receiving brochures and communicating with planners are two effective ways to save time instead of taking professional planning courses; the planners can also improve the resident's suggestions which are more feasible to achieve.

3 WILLINGNESS FOR COMMUNITY PLANNING

3.1 *Demolition type community*

In the investigation of the Xidajie Community, there were 22 available responses, all of them are willing to stay as well as preferring reconstruction at the same place. The 'lifestyle habits', 'human relations', 'geographical advantage' are three main reasons made them stay. The 'aged accommodations' ranks the first toward all the options referring to the desire of conditions improvement.

Discussing about the preservation parts in the community, the residents regard the 'neighbourhood relationship' together with 'local resource' as the most important options of staying reason. To the extent that Xidajie Community is an old mixed-use commercial and residential urban district, the residents there own a common vision for their community to maintain the present situation. Meanwhile it is also an important response to the option of 'lifestyle habits'.

The findings of the 14th-Neighbour Community show that 22 in 30 available responses would like to stay. The rest 8 of them would like to choose the way of 'relocation' or 'economic compensation', and some of them 'reject to demolition'. Moreover, residents in the 14th-Neighbour Community also keep the similar options with those who living in the Xidajie Community referring to the questions about the preservation parts in the community and the hope of improving conditions. What makes the difference is from the point of 'lifestyle habits.' Residents in the 14th-Neighbour community prefer a residential community instead.

3.2 *Preservation type community*

There were 24 available responses from the investigation in the Tanhualin Community. Ten of them want to stay for the reasons of 'lifestyle habits' and 'geographical advantage.' The residents complain about the present situation of transportation problem, aged accommodations, and the outdated living service. 3/4 of the respondents agree with the 'preservation developing method'. They think it is necessary to repair the historic buildings with basic preservation method as well as the aged infrastructure. In addition, residents approve the current 'historical and cultural community' developing way, and hope that government-guidance should continue to be carried out.

In the investigation for the Sande Alley Community, 24 out of the 28 available responses show a desire to stay and live local. The 'lifestyle habits', and 'human relations', along with 'geographical advantage' are residents' reasons, with 'geographical advantage' being the major one. Accommodation needs to be updated. Concerning the issue of the community's future developing methods, 26 say they approve of the preservation way because the Wuhan Government has already taken steps to preserve the buildings. The Sande Alley Community has a long history in addition to its residential function, thus a renewed community with some tourism elements is welcomed by its residents.

4 CONCLUSIONS

The public participation possibility in community development planning in Wuhan can be judged by the ladder theory of citizen participation. A conclusion can be drawn from the data of the four chosen communities that public participation in community planning in Wuhan is still at an early stage. Usually, the Neighbourhood Committees would like to announce the final planning result to the residents rather than organize the participant activities during the whole process of planning-programming. Therefore, appropriate methods need to guide residents participating in the planning process and enhance their knowledge of urban planning field. In this case, one possible way is to utilize brochures which could make up for shortage of planning knowledge. Meanwhile, communication between the planners and the local residents by providing suggestions for the planning process should be carry out gradually.

In the Demolition Type Communities, 'lifestyle habits', 'human relations', and 'geographical advantages' are the top three reasons why residents prefer to stay; accommodation is the only part that residents want to update. It should be considered that the original type of community can be maintained for the community renewal.

In general, residents living in the Preservation Type Communities are in favour of this kind of developing method, furthermore looking forward to government guidance. For example, the Tanhualin Community has been renewed and developed into a historical and cultural tourism community for some time. However, the local residents are still in urgent need of further repairs of the aged buildings. On the other hand, alleys in the Hankou District, as a residential community with a long history, is to be expected to make the best use of its' historical value by the residents.

It shows that the relevant department of government and the urban planning professionals should pay more attention to the publics' appeals, and lead in expressing their idea of community development by providing professional services such as lectures, seminars, and professional service.

It can be concluded that the majority of community residents have a strong desire to participate in their community development planning. If different types of communities are taken into consideration, residents may place different emphasis related to their present living situation. Thus, the importance of public participation in community development planning should paid attention and reflection by the urban planning experts.

ACKNOWLEDGEMENT

The authors would like to thank the Doctor Scientific Research Start-up Found of Hubei University of Technology (No. BSQD12149), and the Building Technology Programme of Hubei Province (No. [2014] 54-66) for the support of this research.

REFERENCES

William. M. Rohe. One hundred years of neighborhood planning [J]. Journal of the American Planning Association, Vols: 75(2), 2009.

Paul Davidoff and Reiner Thomas. A choice theory of planning [J]. Journal of the American Institute of Planners, 1962.

Arnstein Sherry. A ladder of citizen participation [J]. Journal of the American Institute of Planners, 1969.

Luo Xiaolong, Zhang Jingxiang. Governance and Public Participation of Urban Planning in China[J]. Urban Planning Forum, Vols: 132(2), 2001. pp. 59–62.

Deng Lingyun. Study on the Public Participation Institution of Japan [J]. Urban Studies, 2011.

Wu Zuquan. Effect of Third-Party Participation in Urban Planning: a Case Study on Enning road in Guangzhou[J]. City Planning Review. Vols: 38(2), 2014. pp. 62–68.

Mo Wenjing, Xia Nankai. Planning Participation Approaches and Public Political Maturity[J]. Urban Planning Forum, 2012(4): 79–85.

Ni Meisheng, Chu Jinlong. Review Of Community Planning Research [J]. Planners, Vols: 29(9), 2013. pp. 104–108.

Jim Diers. Neighbor Power Building Community the Seattle Way [M]. US: the University of Washington, 2004.

Advances in Civil Engineering and Building Materials IV – Chang et al (eds)
© 2015 Taylor & Francis Group, London, ISBN: 978-1-138-00088-9

Relative analysis of factors of place attachment – Case study in the Tokushima urban area

S. Tsujioka
Shikoku University, Tokushima, Japan

A. Kondo & K. Watanabe
The University of Tokushima, Tokushima, Japan

ABSTRACT: This paper explains the factors of place attachment. Place attachment is an emotional bond between residents and their living places. We focused on the effect of facilities on place attachment. In this study, we proposed relationships among "satisfaction with facilities in a living area," "impression of living area," "participation in area / community activity," and "place attachment." We analyzed the relationships using structural equation modeling. A questionnaire survey was conducted to obtain data for the analysis in Tokushima urban area in Japan. The main results of the analysis are as follows: (1) the impression of a living area significantly influences place attachment and (2) satisfaction with the facilities in a living area affects the impression of the area. The results provide additional value to the establishment of facilities.

Keywords: place attachment, social identity theory, structural equation modeling, questionnaire survey

1 INTRODUCTION

1.1 *Objective of this study*

Recently, social vitality has decreased in rural areas in Japan, and place attachment is expected to improve the social vitality in these areas. In this study, place attachment is defined as the emotional bond between residents and their living places, as suggested by Hidalgo & Hernandez (2001). Strong place attachment of residents improves their participation in community activities, such as crime prevention and environmental beautification activities (Payton et al. 2005). Many of these activities involve cooperation among residents; thus, exchanges among participants become more active. Furthermore, their sense of belonging to the area increases (Brown et al. 2004) and their place attachments intensify. As described above, the correlation between place attachment and participation

in community activities is evident (Figure 1, part 1). Community activities improve security and convenience, which enhances the attractiveness of an area. Strong place attachment among residents enhances the attractiveness of an area (Figure 1, part 2). Such attractiveness improves the place attachment of the residents (Figure 1, part 3). In summary, there is a strong correlation between place attachment and attractiveness of an area.

Strong place attachment prevents the outflow of economic and resident to the other areas. High attractiveness of an area progresses the influx of economic and resident from the other areas (Figure 1, part 4). For these reasons, the Japan Tourism Agency (JTA) chants the slogan of " building a country good to live in, good to visit." The mission of the JTA is a regional construction by local residents (Japan Tourism Agency 2010).

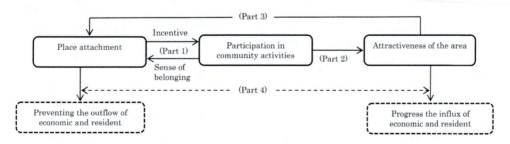

Figure 1. Relationship between place attachment and attractiveness of the area.

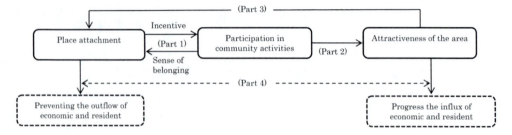

Figure 2. Relationship among our hypotheses.

The purpose of this study is to quantitatively clarify the factors of place attachment. The study focuses on satisfaction with an area environment (e.g. shopping, education, safety and medicine). We clarify the effect of the satisfaction on place attachment, which provides additional value to the establishment of facilities. Facilities have rarely been established to improve place attachment.

1.2 Previous studies

Numerous studies have investigated the factors related to place attachment. Brown et al. (2003) suggested that place attachment is influenced by individual attributes, area environment, and interaction with neighbors. They showed that the residence period extremely influence on place attachment. Hikichi et al. (2009) reported that a positive impression of neighbors has a greater influence than residence period on place attachment. A majority of studies of place attachment have focused on resident attributes and interaction with neighbors.

We could not locate any studies that quantitatively evaluate the influence of area environment on place attachment; in this study, this influence was analyzed. Area environment generates vitality, prosperity, and relaxation as well as improvements in convenience to the area. Furthermore, quantitatively clarifying the influence of the area environment on place attachment is useful for city planning and area administration.

2 DATA ANALYSIS, METHOD, AND RESULTS

2.1 Our hypothesis

In this study, the formation process of place attachment is based on social identity theory according to Hikichi et al. (2009). The theory demonstrates that a person's self-esteem includes the value recognition to his group (Tajfel & Turner 1986, Hogg 1992). Social identity is the emotional bond between a person and a group. Considering living area as belonging group, we can consider that place attachment is a part of social identity. Therefore, a high level of satisfaction with a living area environment produces high level of place attachment.

Four hypotheses are proposed in this study. These hypotheses are verified to clarify the formation process of place attachment. Our hypotheses are listed here. The relationships among these hypotheses are represented in Figure 2.

(Hypothesis 1) Impression of living area includes relaxation and prosperity. This impression significantly influences place attachment.
(Hypothesis 2) Satisfaction with a living area environment affects the impression of the living area.
(Hypothesis 3) Satisfaction with a living area environment affects place attachment.
(Hypothesis 4) Place attachment is evaluated by participation in area activities.

2.2 Questionnaire survey

A questionnaire survey was conducted to obtain data to verify our hypotheses in 2013. The questionnaire targeted the Tokushima urban area, which includes Tokushima city – the prefectural capital of Tokushima prefecture – and five nearby municipalities. The area contains the largest shopping facilities and public institutions in Tokushima prefecture (Figure 3). The population density of Tokushima prefecture was 187.1 person/km^2 in 2013. The population density is ranked as 33 of 47 prefectures in Japan. The gross annual product of the prefecture was 282 billion yen in 2013. This product is ranked as 44 among 47 prefectures. Based on these rankings, the Tokushima urban area is a typical rural area in Japan.

The questionnaire survey was mailed to 500 households, which were randomly sampled by each municipality to obtain standardized responses in all of target areas. A total of 3,000 questionnaires was distributed, and a total of 895 answers was obtained from the survey. The attributes of the questionnaire participants are shown in Figure 4. Many of the questionnaire participants were females over 44 years old. They account for more than 70% of the participants. Many of the questionnaire participants usually commute by car because suburban shopping centers have increased and downtown vitalities have decreased in the target area. Other useful facilities are located in places that are convenient and accessible by car. The questionnaire items are listed in Table 1.

2.3 Structural equation modeling

We analyzed the questionnaire data using structural equation modeling to verify our hypotheses as shown

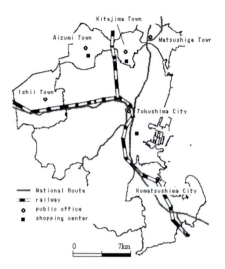

Figure 3. Tokushima urban area.

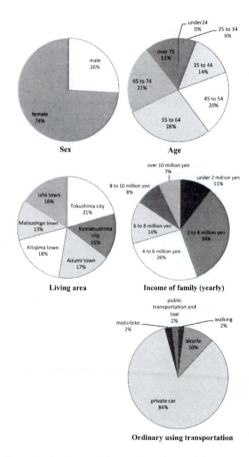

Sex

Age
- under24 0%
- 25 to 34 6%
- over 75 11%
- 35 to 44 14%
- 65 to 74 21%
- 45 to 54 20%
- 55 to 64 28%
- male 26%
- female 74%

Living area
- Ishii town 16%
- Tokushima city 21%
- Matsushige town 13%
- Komatsushima city 15%
- Kitajima town 18%
- Aizumi town 17%

Income of family (yearly)
- over 10 million yen 7%
- 8 to 10 million yen 8%
- under 2 million yen 11%
- 6 to 8 million yen 14%
- 2 to 4 million yen 34%
- 4 to 6 million yen 26%

Ordinary using transportation
- public transportation and taxi 2%
- walking 2%
- motorbike 2%
- bicycle 10%
- private car 84%

Figure 4. Attributes of the questionnaire participants.

Figure 2. The path diagram employed for our analysis is shown in Figure 5. The path coefficients in the path diagram were estimated with 895 questionnaire data points, which were used as indicator variables.

Table 1. Questionnaire items.

Category and Question (Answer candidates)	Variable name (*2)
(1) Place attachment How much attachment do you have to your living area? (1: Nothing -> 5: Much Attachments)	PA
(2) Impression of living area How relaxed do you feel in your living area? (1: Very Nervous -> 5: Very Relaxed)	Relax
How prosperous do you feel in your living area? (1: Very Deserted -> 5: Very Prosperous)	Prosperous
(3) Participation to area / community activity How often do you take a walk or use a bicycle in your living area? (1: Never -> 5: Daily)	Walk
How often do you talk with your neighbors? (1: Never -> 5: Daily)	Communicate
(4) Satisfaction to living area environment How satisfied are you with the medical environment in your living area? (1: Very Dissatisfied -> 5: Very Satisfied)	Medical
How satisfied are you with the educational environment in your living area? (1: Very Dissatisfied -> 5: Very Satisfied)	Education
How satisfied are you with the natural environment, such as park and public green, in your living area? (1: Very Dissatisfied -> 5: Very Satisfied)	Natural
How confortable do you feel with the security in your living area? (1: Very Uncomfortable -> 5: Very Comfortable)	Security
How satisfied are you with the shopping environment in your living area? (1: Very Dissatisfied -> 5: Very Satisfied)	Shopping

(*2 These are used in section 2.3.).

Let us discuss the validity of the model obtained by the analysis. The fitness of the model is represented by the goodness of fit index (GFI) and the comparative fit index (CFI). For this model, the GFI was estimated to be 0.948 and the CFI was estimated to be 0.914. It is suggested that GFI and CFI should exceed 0.9 (Muthen 1989). Therefore, the fitness of the model sufficiently describes the questionnaire data.

The path coefficients in the model were estimated to verify our hypotheses. The path diagram, with standardized path coefficients, is shown in Figure 6. In Figure 6, p values less than 0.05 are considered to be statistically significant, and the error terms and the covariance paths are omitted. The total effect of satisfaction with a living area environment on place attachment, as depicted by the path diagram, is shown in Table 2. The total effects are obtained by adding the direct and indirect effects. The direct effects are directly obtained as the path coefficients among the indicator variables. The indirect effect is an impact that

453

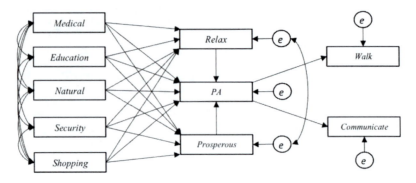

Figure 5. Path diagram to verify our hypotheses.
(In this figure, "*e*" represents the error term and the double-headed arrows represent the covariance paths.)

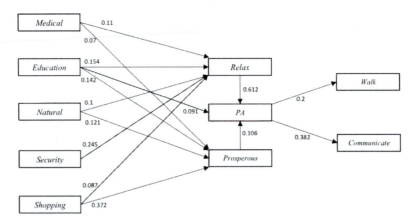

Figure 6. Path diagram with estimated values.

Table 2. Total effect of satisfaction with the living area environment on place attachment.

	Total effect
Medical	0.075
Education	0.2
Natural	0.074
Security	0.15
Shopping	0.093

An example of calculating a total effect (in *Medical*) : direct effect + indirect effect (*Medical* -> *Relax* -> *PA*) + indirect effect (Medical -> Prosperous -> PA) = 0 + 0.11 × 0.612 + 0.07 × 0.106 = 0.075.

is through at least one other variable (Muthen 1989). Our hypotheses were verified on the basis of the characteristics of the model obtained from Figure 6 and Table 2. The results of the verification are described as follows.

In the following sentences, italic words are the variables as described in Table 1.

Verification of hypothesis 1
(Impression of a living area includes relaxation and prosperity. This impression significantly affects place attachment.)

The impression of a living area, including relaxation and prosperity, significantly affects place attachment. *Relax* and *Prosperous* have significant path coefficients to *PA*, as shown in Figure 6. In particular, the path coefficient of *Relax* is more than fivefold better than *Prosperous* to *PA*. Therefore, *Education* and *Security*, which have large path coefficients to *Relax*, have a large total effect (Table 2). Relaxation is more important compared with prosperity for establishing facilities to improve place attachment.

Verification of hypothesis 2
(Satisfaction with a living area environment influences the impression of a living area.)

The impression of a living area is influenced by the satisfaction level for the living area environment. Specifically, *Relax* and *Prosperous* are both influenced by the satisfaction, with the exception of *Security*, as shown in Figure 6. *Shopping* significantly affects *Prosperous*, and *Security* significantly affects *Relax*. The estimation results are considered to be reasonable.

Verification of hypothesis 3
(Satisfaction with living area environment influences place attachment.)

Only *Education* directly affects *PA* (Figure 6). The questionnaire participants most likely do not include

454

students because of the age distribution of the participants. Therefore, convenience is not the reason because of which that *Education* influences *PA*. School attachment may be a factor of place attachment, such as social identity. *Shopping* was expected to directly affect *PA* because of its convenience prior to the analysis. However, it does not exhibit a significant path to *PA*. In the target area, public transportation is poor and many residents use private cars for transportation (Figure 4). Therefore, the convenience of facilities may not be a factor for improving their place attachment in the area.

Verification of hypothesis 4
(Place attachment is expressed by the participation in area activities.)

Place attachment is expressed by participation in area activities. *PA* has significant paths to *Walk* and *Communicate* (Figure 5). In particular, *Communicate* significantly affects *PA*.

3 CONCLUSION

In this study, factors of place attachment are analyzed using structural equation modeling. The analysis is based on a questionnaire. We conducted the questionnaire survey, which targeted the Tokushima urban area in Japan, in 2013. The results of the analysis verified our hypothesis: (1) the impression of a living area, which includes relaxation and prosperity, significantly influences place attachment, (2) the satisfaction with a living area environment affects the impression of a living area, (3) the satisfaction with a living area environment influences place attachment, and (4) place attachment is expressed by the participation in area activities.

This study is an important contribution because it presents quantitative relationships between facilities and place attachment. We clarify the effect of satisfaction with area environment on place attachment, which provides additional value to the establishment of facilities. Our future study will be focused on the comparative analysis of the relationship between place attachment and convenience.

REFERENCES

Brown, B., Perkins, DD. & Brown, G. 2003. Place attachment in a revitalizing neighborhood: Individual and block levels of analysis. *Journal of Environmental Psychology* 23(3); 259–271.

Brown, G., Brown, B. & Perkins, D. 2004. New housing as neighborhood revitalization – place attachment and confidence among residents. *Environmental and behavior* 36(6); 749–775.

Hidalgo, M. C. & Hernandez, B. 2001. Place Attachment: Conceptual and Empirical Questions. *Journal of Environmental Psychology* 21(3); 273–281.

Hikichi, H., Aoki, T. & Ohbuchi, K. 2009. Attachment to Residence: Affect of Physical Environment and Social Environment. *Journal of Japan Society of Civil Engineers D* 65(2); 101–110. (in Japanese)

Hogg, A. M. 1992. *The Social Psychology of Group Cohesiveness –From Attraction to Social Identity.* England. Harvester Wheatsheaf.

Japan Tourism Agency. 2010. Vision of JTA. (http://www.mlit.go.jp/kankocho/en/about/vision.html)

Muthen, B.O. 1989. Latent variable modeling in heterogeneous populations. *Psychometrika* 54(4); 557–585.

Payton, M., Fulton, D. & Anderson, D. 2005. Influence of Place Attachment and Trust on civic action; A study at Sherburne National Wild Refuge. *Society and Natural Resources* 18(6); 511–528.

Tajfel, H. & Turner, J. 1986. The Social Identity Theory of Intergroup Behavior. *Psychology of Intergroup Relations* 81: 7–24.

Advances in Civil Engineering and Building Materials IV – Chang et al (eds)
© 2015 Taylor & Francis Group, London, ISBN: 978-1-138-00088-9

A study of an integrated ICET system and its impact on an architectural planning and design

R. Hendarti
Department of Architecture, Faculty of Engineering, Bina Nusantara University, Jakarta, Indonesia

ABSTRACT: This paper presents the preliminary study of the application of the integrated Information, Communication and Energy Technology (ICET) system in order to manage the use of energy efficiently particularly for a building complex. A school complex was selected to simulate the implication of this integration on an architectural design. The concept itself is based on the principle of ICT, smart city and renewable energy and is constructed through literature review. The main deliverables of the analysis are the concept of the building form and the impact of the ICET integration system on the room organization of the school building design. The outcome of this study could be used as a basic consideration for architect in regards to the application of ICET system on architectural planning and design.

Keywords: Integrated ICET system, architecture, planning, design.

1 INTRODUCTION

Buildings in order to support human activities require energy, such as energy for its lighting and air conditioning. As the occupant increase the use of energy will definitely increase. On the other hand, currently, there is a crisis in energy production, particularly the one from fossil. Unfortunately, this type of energy is still being highly used in many sectors and buildings consume a big part of the production.

Figure 1 show that energy consumption in Indonesia is increasing by an average of 2% per year, while a significant increase of the energy consumption occurred in 2005 by around 5%. In contrast, the energy production was considered stagnant from 2003 to 2010 after decreasing of around 3% from 1999 to 2003. This fact shows that the need of conservation energy is essential to sustain the fuel procurement. Therefore a smart way is required to improve the efficiency and the effectiveness of energy usage.

Some studies on energy conservation have been conducted and one of them is the study of the use of Information and Communication Technology or ICT. The ICT technology is recognized enable to improve the economic growth and energy efficient (Kramers, 2014). Most leading cities in Europe and US are now actively adopting this technology in order to manage the energy use as well as to reduce the greenhouse emission.

Additionally, currently there is a further development of ICT, namely Information, Communication and Energy Technology or ICET. It merges the use of the two technologies and operates in one system, called "smart grid". The grid provides the flow of energy in

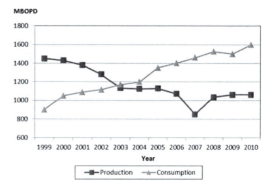

Figure 1. The energy consumption and production in Indonesia (Sutrisna and Rahardjo, 2009).

a network and Power Line Communication (PLC) is used to control the networking for "smart grid".

In relation to architectural design and energy, Stein (1985) stated that architecture will be on crisis when there is an energy crisis, because those two aspects are tangles. Therefore, the energy usage in a building should be planned for the next ten years. According to Gissen (2002) the effect of architectural design is also significant in terms of energy consumption. For example, a commercial building such as office building, 25% of the energy consumption is utilized by the electrical mechanical system, another 50% are influenced by the behavior of its inhabitants and the remaining balance (25%) is influenced by the design of the building.

This paper presents the preliminary study of the application of ICET on architectural design in a small scale; here is a building complex instead of city. The main objective of this study is to identify the impact of the use of ICT on architectural design such as the environment aspect, the human aspect and the building aspect. The environment aspect includes the site planning, the human aspect focuses on the list of activities and the room arrangement and the building aspect includes the building technology such as the application of alternative energy.

To address those objectives, a school complex is selected as a case study since the room organization and the variety of buildings are considered simple. The use of solar panel is proposed here as it is considered as the most potential microgeneration technology that can be implemented in Jakarta. This technology is suitable because of its climatic condition where Jakarta is rich in sunlight but low in wind flow. Moreover, the use of solar energy is planned to substitute the fossil fuel for about 2% for industry and 5% for household (Indonesia energy outlook 2010–2030).

2 FUNDAMENTAL THEORIES

This study is originally initiated from the thought that "a house is a machine for living in". It was defined by Le Corbusier around a century ago. His quote can be understood as "a built environment could shelter its habitant's activity as well as generate its occupant's productivity". Consequently, the need of energy is prominent in order to assist the inhabitant's activities. Nevertheless, it is necessary to wisely manage the use of energy. The use of Information and Communication Technology or ICT is now actively used to improve the energy efficiency as the way to increase economic growth (Ad-Hoc Advisory Group Report, 2008).

In general, ICT can be defined as a technology that can manage and access information through telecommunication. This technology is now being studied in many aspects including: its implementation on a concept of a smart city. Lee et al (2013) summarized that a smart city is designed to resolve various urban problems such as traffic and environment through ICT-based technology connected up as an urban structure. In relation to the energy conservation, Kamers et al (2014) reported some studies of the role of ICT on reducing the energy consumption and one of them is a study conducted by Mithcell (2000). It has been outlined that ICT is potential to reduce the use of energy by what so-called "intelligent operation" or a resource efficient operation of water, energy and transport system.

As such this study adopts the approach of "intelligent operation" which Kamers et al (2014) has also applied for their framework proposal for reducing energy usage in cities. As cited from REEB (2010), the "intelligent operation" could be applied to minimize the electricity usage particularly in household through some technologies. The management and operation of energy can be optimized through smart grid, advanced metering and microgeneration and sharing of energy between buildings (Hilty et al, 2004), and also the application of ICT-supported can facilitate the management and intelligent heating system.

A particular grid namely "smart grid' is developed to facilitate the "intelligent operation". It is a hybrid grid technology in which the use of traditional technologies and innovative digital solution is combined. This is to improve the management of the electricity network by advance monitoring, control and communication technologies. This technology also enables to integrate with other renewable energies such as solar panel and wind turbine with the electricity grids, where the outcome can reduce or decrease the load in real time, in other words, the energy is used based on the need of the system. The other benefits of smart grids are it can detect and mitigate faults, reducing power line losses and transmission and distribution networks and lowering energy consumption in individual building.

Another supporting device is "advance metering". It is an infrastructure that comprise of electronic or digital hardware and software in which interval data measurement is combined with continuously available remote communication. In short this technology allows two way communications between meter at the customer site and the central system (the service provider). The meter includes electric, gas and water utility. According to Electric Power Research Institute (EPRI) the use of this technology gives a good impact on three aspects: the system operation, the customer service and financial benefits. In regards to the system operation, advance metering can reduce meter reads, increase the accuracy of the meter reading, improve utility asset management, easier to detect the electrical theft, easier outage management and support the management and administration. As for the customer service, it helps to detect the meter failures and to improve the billing accuracy, service restoration and the flexibility of the billing cycles. The benefits of advance meter in financial are the reduction of equipment maintenance and support expenses as well as the improvement of inventory management.

Microgeneration technology is utilized to integrate the ICT technology with the alternative energy source. Microgeneration is an environmental friendly technology that produces energy on a small scale on site. This technology does not impact on the produce of CO2 into the atmosphere. The technology includes the power production from sun, wind and water. The advantage of microgeneration is to avoid the loss of power in transmittance and distribution.

To facilitate all those technologies, a Power Line Communication is constructed. This infrastructure is a technology that uses power lines for its data communication and it does not require any additional wiring (Morocco, 2011). This technology has been chosen because it can work at any place, like basement, or where the radio frequency (RF) cannot be operated (Power line communication, 2011).

Table 1. The main facilities which requires electricity.

Activity	Title of room	Physical requirement	Facility	Supporting facility
Studying	Class	Visual Comfort	LED light	Computer
		Thermal comfort	Air conditioning	
Reading	Library	Visual Comfort	LED light	Computer
		Thermal comfort	Air conditioning	
Experiment	Laboratories	Visual comfort	LED light	Computer
		Standard room temperature	Air conditioning	
Administration	Administration room	Visual comfort	LED light	Computer
		Standard room temperature	Air conditioning	

3 THE DEVELOPMENT OF THE CONCEPT OF THE INTEGRATED ICET SYSTEM AT A SCHOOL COMPLEX

3.1 School facilities

The word school, as described in oxford dictionary, is an institution for educating children or any institution at which instruction is given in a particular discipline. Based on that definition, the main activities are teaching and learning. In order to fulfill those activities, the main facilities can be defined as class room, library, administration room and laboratories. These activities can be allocated in a single building or a multiple buildings (complex). Each type of activity has some requirements such as the standard level of illumination and temperature (especially for laboratories). As such energy is needed to achieve that required standard.

Table 1 shows the main facilities in regards to the electricity requirement and energy. It can be analysed that there are two physical requirements that should be achieved. They are visual and thermal comfort, therefore the required facilities are lighting and air-conditioning. Air conditioning is provided to maintain the level of humidity, such as to avoid mold on laboratory equipment and books, as well as to prevent the excessive heat on computer.

3.2 Integrated ICET system

3.2.1 Concept of ICET system application

As mentioned previously that the fundamental approach for the integration of ICT technology for school building is the "intelligent operation", in consequence, this study tries to apply the concept of "smart grid" but in a smaller scale. This idea is also based on a report of the Ad-hoc Advisory Group of DG information society and media (2008). They stated that the management of power grids will be more efficient by the application of ICT. Furthermore, ICT enable to integrate any renewable energy source with grids. Furthermore, in the application, particularly for buildings, ICT can be in the form of smart metering and simulation. Simulation is applied to simulate the use of energy for buildings through optimizing the envelope measures and passive solar heating techniques.

As such, this project adopts that fundamental principles and modifying the ICT for a school area scale

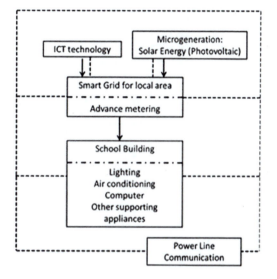

Figure 2. Diagram of Integrated ICET system for School.

and the ICT is modelled based on some following criteria:

- Affordable and simple in the application
- Enable to integrate with the local grid system (PLN network) and solar panel.

To facilitate those criteria, the ICET solution implements the following technology:

- The use of smart grid for local area
- The use of advance meter
- The use of Power Line Communication
- The use of microgeneration

The integration of the fourth technology is illustrated in Fig. 2.

3.2.2 Integration of ICET system

Figure 2 shows the diagram of ICT, Microgeneration and grid integration for a school complex. The schematic concept integration can be explained as follows:

- The traditional grid from Perusahaan listrik Negara (PLN) is still preserved. PLN is a local government company that provides electricity in Indonesia. To allow the integration with ICT and the microgeneration, an advance or smart metering is installed.

- The smart metering then is linked to the electrical system of the building, such as lighting, air conditioning, computer and other supporting appliances.
- Power line communication (PLC) is used as the medium for the data communication as well as the electricity connection. The infrastructure uses the existing wiring. The data communication will be carried over conductors where the conductors itself work as electric power transmission.
- Since the distribution of the electricity is short and inside the building, therefore the electrical power distribution deploys a "low-voltage" domain.

The smart grid, particularly for this project, is a local grid or an electrical connection between rooms, which has two functions: as the electrical connection and data communication.

Solar panels are selected as the energy source for the electricity. The capacity of each source is not determined in this study, since the study is only concerned on the concept of the integration.

4 THE IMPACT OF ICET SYSTEM ON ARCHITECTURAL PLANNING AND DESIGN

The deployment of the ICET integration, to some extent, will have some impacts on the planning and design of the building. As stated previously in section 1 that the energy system should be planned for the next 10 years, therefore, the planning and design of a building needs to accommodate the technology of renewable energy application.

Based on the concept of ICET mentioned above, the source of energy will use Solar panel as the additional energy source. The infrastructure for managing the ICET integration will deploy data communication; therefore, the implication in architectural planning and design is constructed based on the characteristics of the ICET systems.

The concept of architectural design is explained in Table 2. It can be seen from that table that there are some additional room, such as battery room and monitoring room. In terms of design type, an integrated building photovoltaic is applied to harmonize the use of solar panel and the building. Lastly, a particular structured wiring is applied to accommodate the power line communication.

5 CONCLUDING REMARKS

A literature review on ICET system as well as other supporting technologies has been discussed and is then used as the fundamental theory of the integration of ICET system for a small area, here: a school complex. The essential points of this study are concluded as follows:

Table 2. The implication of integrating ICET on the modification of planning and design of school building.

ICET system	Definition	Equipment	Physical requirement	Design Aspect		
				Environment Site plan	Human activities Room arrangement	Building technology
Solar energy	An energy source from sunlight	Solar panels, inverter, battery	A space for installing the solar panels, inverter and battery	Integrated with the building	Battery room	Building integrated Photovoltaic (BIPV)
ICT system	Managing data information through communication technology	Communications network devices such as PABX, gateways and network infrastructure such as cabling and patch panels	A designated space on the floor of a building used to house or process large quantities of data	Inside the building	Monitoring room	The use of air conditioning to maintain the performance of the equipment. The use of thermal sensor
Power line communication	Lines for data communication	The infrastructure used to carry information between workstations and servers or other network devices	Cabling, junction boxes, patch panels, fibre distribution panels	Inside the building	–	Structured wiring enclosures with a inflammable materials.

- The integration of ICET system for managing the use of energy for a building can use the principle of "intelligent operation". This case study proposes the use of advance metering and power line communication (PLC).
- This study proposes the use of solar energy as the alternative energy source which based on the local condition whereas it is rich with sunlight but less wind flow.
- The impact of the ICET system on architectural planning and design is also analysed. The results are: (1) the form of the building is the building integrated photovoltaic (BIPV), and (2) some additional rooms are provided to address the need of the ICET system such as monitoring room and battery room.
- The construction and the building material for covering the wires should be inflammable materials and the room is equipped with thermal sensors.

ACKNOWLEDGMENT

The authors would like to thank the Department of Architecture, Faculty of Engineering, Bina Nusantara University, Bina Nusantara Grant 2014, for the financial support for this project.

REFERENCES

Ad-Hoc Advisory Group Report, (2008, October 24), ICT for Energy Efficiency, DG-Information Society and Media, Brussels.

Kramers, A, et al., Smart sustainable cities – Exploring ICT solutions for reduced energy use in cities, Environmental Modelling and Software (2014), http://dx.doi.org/10.1016/j.envsoft.2013.12.019

Lee, J.H., et al., Towards an effective framework for building smart cities: Lessons from Seoul and San Fransisco, Technol. Forecast. Soc. Change (2013), http://dx.doi.org/10.1016/j.techfor.2013.08.033

Marocco, G, et al., Fundamentals of Power Line Communication Systems. FTW-TR-2011-004, 2011. Smart Grid. Retrieved 15 January 2014 from Bonneville Power Administration. Web: http://www.bpa.gov/energy/n/Smart_Grid-Demand_Response/Smart_Grid/

Power line communication. Retrieved 15 January 2014 from Echelon. Web: https://www.echelon.com/technology/power-line/Smart Grid. Retrieved 15 January 2014 from Bonneville Power Administration. Web: www.bpa.gov

Advances in Civil Engineering and Building Materials IV – Chang et al (eds)
© 2015 Taylor & Francis Group, London, ISBN: 978-1-138-00088-9

The creative movement of 'Park(ing) Day' in Jakarta

Albertus Prawata
Bina Nusantara University, Jakarta, Indonesia

ABSTRACT: The government has a strong role in making plans and shaping the city. The planning establishment of the government is based on capital-intensive strategic actions, so they can shape the urban spaces according to a certain set of values. These values are clearly made in the patterns of resource consumption. However, they often create a hierarchical gap between the people and the communities. In the city, the economy becomes the basis of how the urban spaces are shaped and created. New economic activities often have an impact of degrading the quality of the spaces. As a result, the city will lose its attractions, they become aesthetically unpleasant, and people feel alienated. The purpose of this paper is to discuss the interaction and appreciation of creative users and the citizens of the city towards urban spaces, and how this will encourage endless collaboration amongst local citizens to create thoughtful and meaningful designs for the public. The discussion and arguments will be based on some creative movements such as 'Park(ing) Day.' It has the importance to be a city icon and become an urban fabric that supports the city, becoming a part of the sustainable city concept. The engagement and ideas from the creative user can be a strong foundation for creative urbanism that has the power to re-shape the city spaces to be more liveable. Therefore, it can also bring a new identity and vibrant atmosphere to a certain area as well as to the city.

Keywords: Creative, creative user, urban spaces, Park(ing) Day.

1 INTRODUCTION

The quality and quantity of a city's creative aspects have a direct impact on creating a better and vibrant city, including its economic development. It will bring benefits to the city; it will attract a skilled workforce, more entrepreneurship, cultural and local tourism, and be a strong social connection. These will make the city a great place to live. By providing opportunities to attract, retain, and develop the best and brightest local culture and creative aspects and atmosphere, a city can create an outstanding and culturally creative experience for its residents. The world's population continues to grow at an alarming rate. At the same time, urbanization will also increase. Forecasts from the UN suggest that by the year 2050, it is estimated that there will be 9.3 billion people on earth, and almost 7 billion will live in the cities (UN, 2011). With their high concentration of people, infrastructure, business, and investment, cities are the engine rooms of the new global economy. A new consumption economy has emerged, and it is driven by talented creative innovators. The best and brightest creative talents have the power to drive the new economy; they create unique products and services that are highly profitable. When they have access to the right conditions, environments and opportunities, they will thrive, generate new products, services, and boost innovation. Other cities around the world have employed lifestyle-development strategies to capture global attention and create an economic environment in which these talented people can prosper. To date, others have found success in exploring options such as city branding, employing successful place making and event strategies, and enriching creative settings.

Jakarta, Indonesia, is a mega city with a population of more than 9.7 million people, and is one of the most congested cities in the world. As the capital city of Indonesia, Jakarta is not only a centre of government, but is also becoming the centre of information, trade, and the economic sector in Indonesia. Jakarta is also fast becoming the place for creative talents and communities to thrive. Many events such as exhibitions and performances are being held in the capital city of Indonesia. There are also a lot of communities and creative talents exploring new creative engagement in public spaces during certain days, such as the car-free day.

This paper will share the interaction of creative users and the citizens of Jakarta towards the urban spaces through Park(ing) Day, which would be a great design addition to the public space in Jakarta.

2 CREATIVE DESIGN AND THE MAKING OF PLACE

Solving urban challenges has become the key to addressing global challenges. It is important for cities to be ready for the challenges ahead, transforming city

Table 1. Potential metrics for projects goals.

Goal	Potential Metrics
Safety	• Crashes and injuries for motorists, pedestrians, and cyclists • Traffic speeds
Access/ Mobility	• Volume of vehicles, bus passengers, bicycle riders and users of public space • Efficiency in parking/loading • Traffic speeds
Economic Vitality	• Number of businesses; employment • Retail sales; visitor spending
Public Health	• Minutes of physical activity per day • Rates of obesity, asthma, diabetes, etc.
Environmental Quality	• Air quality; water quality • Urban heat island; energy use
Liveability/ Quality of Life	• User satisfaction • Public space usage

Source: The New York City Department of Transportation (2012).

spaces into more efficient and welcoming spaces that better accommodate all users. A city that has been a leader in creating new models for sustainable urban development in recent years is New York. The New York City Department of Transportation (2012) has already developed a robust set of metrics to evaluate the outcomes of its projects with respect to the agency's policy goals, both in the service of continually improving project designs, and because the public increasingly expects such data-driven decision-making from government.

The table shows that the goals of the project design in New York simply to create a better and safer environment for its residents. Ted Mann (2012) stated that the changes of the New York City (NYC) streetscape are making pedestrians, cyclists, and drivers safer. It shows that bike lanes and pedestrian plazas have been good for business too. Using data from the city's Department of Finance, the Department of Transportation found an increase of as much as 49% in retail sales on 9th Avenue from 23rd to 31st Streets since the bike lane & plaza was initiated in the fall of 2007. The New York City Department of Transportation (2010) also stated that a large body of experience and research from around the world indicates that attractive streetscapes and urban public spaces can enhance local business performance in retail and real estate sectors. It shows that a good and creative design for pedestrians and local communities has a strong impact on economic activity. The data below by the New York City Department of Transportation (2010) shows that a good and attractive public space will draw people:

- 84% more people are staying (e.g., reading, eating, taking photographs) in Times and Herald Squares.
- 42% of NYC residents surveyed in Times Square say they shop in the neighbourhood more often since the changes.
- 70% of theatregoers say the plazas have had a positive impact on their experience.

- 26% of Times Square employees report leaving their offices for lunch more frequently.

In an effort to reclaim street space for pedestrians, the New York City Department of Transportation has temporarily created nine pedestrian plazas in four of the city's five boroughs, in many cases simply marking the street with paint and orange cones and adding movable chairs and tables. Now the city aims to refashion the spaces with more permanent materials (Baker, 2010). There is also a more formal, city-sponsored capital program, the NYC Plaza Program, which reviews proposals for pedestrian plazas from local community organizations. In other cities, San Francisco launched a similar programme called Pavement to Parks. It has carved three makeshift pedestrian spaces out of existing streets. Another initiative is unfolding in Seattle, where the city aims to reclaim downtown alleys for pedestrians. Collectively, these initiatives point to a new generation of car-free spaces, a movement that emphasizes simple, flexible, and inexpensively creative designed spaces, managed by local organizations and integrated into the surrounding neighbourhood. The projects also reimagine street spaces as places to build the community.

These examples show that creative and initiative designs from communities and local organizations can lead to positive changes and impact towards a better urban space and place. Creative and initiative movements or activities certainly can play a major role in starting to build a better community. Park(ing) Day is one of the initiatives that can be easily created as a movement to push the city or local authorities to create a better public space.

3 PARK(ING) DAY

Park(ing) Day is an annual event where on-street parking spaces are converted into park-like public spaces. The initiative is intended to draw attention to the sheer amount of space devoted to the storage of private automobiles. The movement first occurred in 2005 when the interdisciplinary design group, Rebar, converted a single San Francisco parking space into a mini-park. The group simply laid down sod, added a bench and a tree, and fed the meter with quarters. Instantly garnering national attention, Park(ing) Day has spread rapidly amongst liveable city advocates and is thought to be the pre-cursor to New York and San Francisco's Parklet and Pavement to Parks programs (Lydon, 2012).

Lydon (2012) also stated that Park(ing) Day encourages collaboration amongst local citizens to create thoughtful, but temporary additions to the public realm. Parking spaces are programmed in any number of ways; many focus on local, national, or international advocacy issues, while others adopt specific themes or activities. The possibilities and designs are as endless as they are fun. The movement has created a worldwide phenomenon, and it is becoming a worldwide campaign every year. Every year Park(ing) Day movement is held on Friday of the third week of September. The action to convert parking spaces into mini-parks has delivered a strong message to city councils and governments that public space is important for a city. Public spaces will act as activity generators for communities, and therefore, a creative environment will emerge, and it will benefit the particular area and also the whole city itself.

The Park(ing) Day movement in Jakarta was started by the Institute for Transportation and Development Policy (ITDP) in Indonesia in 2011. As the main objective of Park(ing) Day is prioritizing public space for human interaction rather than giving it to private motor vehicles, ITDP Indonesia has urged for a reformation of the parking management system in Jakarta. ITDP Indonesia (2013) has proposed suggestions for parking draft regulations in Jakarta. These entries are:

1. Determination of a tariff through parking zoning to restrict the entry of vehicles into the city centre.
2. The use of advanced technology (smart cards, parking machines, etc.) to manage the parking (especially on street).
3. A progressive tariff system for denser areas.
4. A change of regulations required for the provision of a minimum set of parking spaces (SRP) in the building/activity centre in order to limit the number of SRPs in the building/activity centre.
5. The provision of park and ride facilities for commuters, especially from the Jakarta suburbs.
6. An improvement in the means of mass transportation, sidewalks, and bike lanes as an alternative provision for public transport.
7. To convert the function of the parking lot (especially illegal parking lots) into sidewalks, bike paths, parks, or open spaces for community interaction.
8. Planning for an integrated parking area along with building and pavements.
9. Transparency regarding charges derived from the parking lot and ensuring it goes to community development.
10. Law enforcement for any violations of parking regulations.

Although the draft that was adopted from 2011 to date has not been approved by the city council, the Park(ing) Day movement could be the catalyst for a better city, through passing a new law for greater purpose.

In the past couple of years, the Park(ing) Day movement has also been celebrated by communities and creative enterprises in Jakarta. In 2012, several communities and creative enterprises (Volume Factory, Indonesia Berkebun, Kopi Keliling, and Kemang89) held a Park(ing) Day venue in the parking lot of Kemang89 building, in Kemang, South Jakarta. The event attracted creative communities to participate. Kopi Keliling made a public gallery, displaying graphic artworks by its members and collaborators. They also set up a pop-up café that served coffee and tea. Indonesia Berkebun set up their organic

Figure 1. Park(ing) Day ONX Idea Studio, Menteng Square parking lot.
Source: Albertus Prawata.

products, and there were lectures from creative enterprises regarding public spaces and how important they are to the liveability of a city and community. The event lasted from 3 p.m. to 8 p.m., and it was a pleasant experience concerning the exchange of ideas, stories, and creative engagement between communities and the public. In 2013, and recently on 19th September 2014, the ONX Idea Studio and the Volume Factory hosted Park(ing) Day in the ONX Idea Studio's parking lot. The event brought together creative individuals and enterprises to share ideas and stories for their creative projects. The event also sparked an immediate impact on its surroundings. Many people started to gather around the mini-park, and enjoyed refreshments.

4 CONCLUSION

Jakarta, as the capital city of Indonesia, has great potential to be a creative city. Many creative events have been held in Jakarta and in the future, these events could add to the further development of the city. The development of the creative economy in recent years could lead to multiple events based on arts, culture, and education. With all of the resources of a particular site, an historical site, a building, or even a region, Jakarta offers a great variety of strategies to trigger creative acts and talents in order to boost its economy.

The government, urban planners, and developers who have the role of planning and shaping the city could use creative strategies to increase the liveability of a city. In the future, they should engage the broader public and communities as the generators of creative strategies. They could create a new urban design guideline for creative strategies that would define the physical development of an urban area. The guidelines would involve a strong legal aspect, special regulations, and incentives. Creative movements such as Park(ing) Day is a great design addition to the public space in Jakarta, contributing towards better and lively urban spaces. It is only one of several examples of creative movements that could act as an urban fabric that would bring variety and liveability to an area.

For the future, the Park(ing) Day movement could lead to a permanent park in a busy area or street, and permanent guidelines and regulations for public spaces in certain areas. It could create liveable communities; it could be an urban icon, a generator of creative talents that would create creative movements based on arts and culture.

REFERENCES

Baker, Linda (2010) *Walking Wins Out. Pedestrian streets are in style again.* Planning, the magazine of the American Planning Association.

ITDP Indonesia (2013) *Reconsider The Value of Space (60 Minutes for Public Space).* http://www.itdpindonesia.org//index.php?option=com_content&task=view&id=1642&Itemid=98. 28 March 2014.

Lyndon, Mike (2012) *Tactical Urbanism Vol. 2.* Miami, New York.

New York City Department of Transportation (2010) *Green Light for Midtown Evaluation Report.* New York City DOT, New York.

New York City Department of Transportation (2012) *The Economics Benefit of Sustainable Streets.* New York City DOT, New York.

United Nations (2011) *World Population to reach 10 billion by 2100 if Fertility in all Countries Converges to Replacement Level,* United Nation Press Release, New York.

Advances in Civil Engineering and Building Materials IV – Chang et al (eds)
© *2015 Taylor & Francis Group, London, ISBN: 978-1-138-00088-9*

Parametric urban design: Shaping the neighbourhood by optimizing the urban density and considering Road Service Level

Michael Isnaeni Djimantoro & Firza Utama Sjarifudin
Bina Nusantara University, Jakarta, Indonesia

ABSTRACT: The shaping of a city is always dynamic, adjusting what is in it. One of the factors that affect the formation of a city is the increasing population of the region. When the horizontal development has not been done for some reason, an alternative that is often used is vertical development, which will increase the density in the region. However, to be considered, that the increase of density in a region should also lead to the increase of life supporting infrastructure, such as transport infrastructure and so on. More importantly, everyone needs to meet their daily needs, including Indonesia. The development of transport infrastructure in Indonesia is still focused on the development of the road network. Therefore, there is a need to increase the density with careful consideration of the existing road network to avoid negative impacts such as the traffic jam. This simulation aims to seek the shape of a neighbourhood through optimizing the density with consideration of the roads' level of service.

Keywords: Level of Service, optimum density, shape the urban neighbourhood

1 INTRODUCTION

The development of a city occurs with the growth of the population, from low density into a higher density. The better health level decreased the mortality number and increased the birth rate. In addition, unequal regional development has driven the increases in urbanization. People are looking for a better life than in the village. Thus, it is important for urban designers to anticipate the city's total population growth.

Increasing the population number will increase the density of population in an area, which indirectly changes the form of the city. It is marked by the emergence of high-rise buildings in the city, though not all areas in a city can be developed with high-rise buildings, because the increase in population density must be balanced with the development of supporting infrastructure. Without good infrastructure development, there will be a negative impact for the city region.

One of the necessary infrastructures in the development of a city is the transportation infrastructure. Because everyone in the city has their daily activities, they want to go to the office, to school, to the hospital, shopping, and so on, which will increase trip generation in a certain region.

In developing countries such as Indonesia, the development of a transport infrastructure is still based on the 'old style' transportation system – the road network. This is reflected in the many major cities in Indonesia, which still do not have a good mass transit system. Therefore, trips made by residents of the city will be more reliant on the ability of their networks.

However, other data shows that urban density development is so rapid compared to the development of road networks, which eventually leads to road congestion.

So it becomes a question for the urban designers, how much does the development of road infrastructure needed to anticipate the increasing density in the region? Or, put another way, how much is the maximum density that can be accommodated by the existing road conditions?

2 RELATION BETWEEN DENSITY CAPACITY WITH ROAD LEVEL OF SERVICE

Level of Service is a measurement used to determine the quality of a particular road segment in serving the flow of traffic through it. Road Service Level (Level Of Service/LOS) is a description of the operational condition of traffic flow and motorists perception in terms of speed, travel time, comfort, freedom of movement, security, and safety. The Level of Service consists of service levels (depending on flow) and service levels (depending on the facility). (Tamin, 2000). The Level of Service (LOS) in road planning is expressed by the letters A to F where the stated Level of Service is from the best to the worst (Equation 1).

$$LOS = \frac{V}{C} \tag{1}$$

Table 1. Generated traffic of land use.

Housing	Apartment	5.7 trips per unit
	Condominium	5.1 trips per unit
	Elderly house	3.3 trips per unit
Institution	Higher Education	2.2 trips per Student
	High School	1.3 trips per Student
	Elementary School	1.0 trips per Student
	Hospital	9.4 trips per beds
	Library	58.4 trips per employee
	Government Buildings	64.6 trips per square feet
Commercial	Regional Shopping Centre	315 trips per nett area
	Local Shopping Centre	949 trips per nett area
	Office	15 trips per 100 square feet
	Bank	43 trips per employee
	Car Repair	57 trips per employee
Industries	Industry	79 trips per nett area
	Industry area	64 trips per nett area
	Warehouses	81 trips per nett area

LOS: Level of Service
V: Traffic Volume (passenger car unit/hour)
C: Road Capacity (passenger car unit/hour)

The capacity of the road is defined as the maximum current through a point on the road which can be maintained per unit hour in certain conditions. For a two-lane roads with two-way, capacity is determined by the current two-way (two-way combinations), but for roads with many lanes, separated flow is determined by the direction and the capacity per lane. The basic equation for determining the capacity is as follows (Equation 2):

$$C = C0 \times FCW \; FCsf \times x \times FCsp \; FCCS \; .... \quad (2)$$

C = Road Capacity (passenger car unit / hour)
C0 = Basic road capacity (passenger car unit / hour)
FCW = Road width adjustment factor
FCSP = Direction separation adjustment factor
FCSF = Shoulder and side barrier adjustment factor
FCCS = Size of the city adjustment factor

The volume of vehicles consists of the volume of vehicles passing through the area and also the volume of vehicles into the area. All vehicle volume is calculated in passenger cars unit where there are different coefficients between cars, motorcycles, buses, and trucks.

Any development in a city would increase traffic generating in that region of the city. Generating massive traffic resulting from the development of the city is described in Table 1 (Stover and Koepke, 1983).

On the other hand, the population density in the city can be controlled by adjusting the covered area, because more peoples need more space to live in. It is governed by a city government with a set number FAR. FAR is the amount of space that is calculated from a comparison of figures for the entire floor area

Figure 1. Aggregation of building code parameters; building lot, maximum height, rear-front yard provision.

of the building to the ground area. (Capital Regional Regulation No. 6 of 1999, on Regional Spatial Plan of Jakarta). The determination of FAR rate eeds to pay attention to sustainability, harmony, and the carrying capacity of the environment and urban development strategies (Jakarta Governor Decree No. 678 of 1994, on Improving Building intensity in the Jakarta region).

There are several factors that determine the shape of an urban neighbourhood, but the main thing is the relationship between the area of the ground floor, the total area that can be built, and the height of the building. A balanced composition between these three components can shape a better urban neighbourhood.

3 METHODOLOGY

Information regarding the LOS and building regulations are mostly presented in the form of text, although some are also found in the form of graphics; it is rarely made in the form of a volumetric representation. Municipal government has made the information clear enough so that it can be used to create a parametric model. Parametric modelling tools allow for the translation of the building regulations and LOS information from text into dynamic visual forms of design variations.

The study was initially conducted by creating zoning regulations of a region in Jakarta in terms of the type of site and building, envelopingshapes using parametric tools to be able to represent it visually (Figure 1).

This study aims to offer a flexible way to design the framework of zoning regulations by presenting the design possibilities that can be generated. To do this, the specific site conditions and zoning regulations should be integrated into the model to be made. The next approach is to develop a method for integrating the site data into an LOS-based parametric model, and applying the model to the zoning regulations.

The LOS data needs to be combined with the building regulations and the impact of information on the site and the schematic design. There is a large number of building regulation data that can be obtained, such as capacity, average speed, travel time, delay, and queue that can be used to analyse the building volume and its envelope.

This workflow is implemented using a parametric editor *Grasshopper* that is a plug-in from a 3D CAD program *Mc Neel-Rhinoceros*. An important goal of this study is to be able to read data from a separate

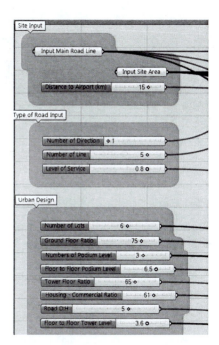

Figure 2. Input and read data in Grasshopper.

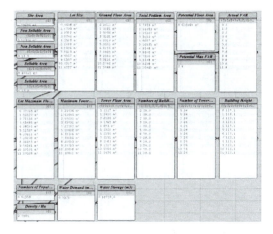

Figure 3. Output data previews.

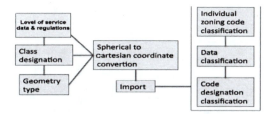

Figure 4. Workflow of data reading interface.

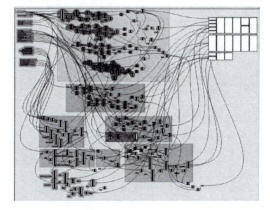

Figure 5. Grasshopper algorithm model interface.

3.1 Data visualization

The developed parametric models can describe the neighbouring placement including street width and building setback zone specifically for each region. Figures 4 and 5 show the data workflow and algorithm model in Grasshopper.

3.2 Setback

Ground level setback can be divided into three conditions, namely: building corresponding lines, the boundary between the building, and the back yard. Each site has at least one of these conditions, The second variable determines whether the neighbouring site has an existing building and the setback value. In the proposed algorithm, the setback condition was made with minimum offset value.

3.3 Floor area ratio and building height

This characteristic is calculated simultaneously for the reason that the FAR value is lower than the maximum value allowed in the site that can be easily obtained on the type of large buildings. By the limit of the height in calculation, the visualization is created automatically by the height of the building at the end of the completed floor level. The proposed algorithm allows for adjusting the height from floor to floor to show the maximum FAR. In addition, it is also possible to set the thickness of the floor.

file and write the variation in the Grasshopper components without having to import/export from other software packages. Using this can abridge the system and make it easier for architects who have limited time in designing a project (Figure 2).

By using the method developed by Nicholas Monchaux et al. (2010) to manage the imported files in Grasshopper, the LOS data is then inserted into the Grasshopper component using VB.NET programming language contained in the current build of Grasshopper. LOS output data from the component reads directly from the spreadsheet file so it is not necessary to first import the data into Rhinoceros (Figure 3).

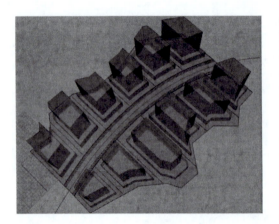

Figure 6. Output 3D preview of density simulation.

3.4 *Density*

The determination of the building height and permissible FAR of zoning in an area is the population density of the area. In general, the higher the FAR value, the higher the possibilities to increase density, especially in residential buildings.

As an alternative to the planning scheme, this study used the potential population density as the main controller for the development of the city. The proposed algorithm covers a maximum height factor and the percentage of building/land use (Figure 6).

4 CONCLUSION

When the site and zoning information are represented in graphical form through dynamic tools, architects can get a chance to explore variation with the placement of the building design, building organization, and building envelope. The result is a parametric model that covers the site and zoning information as inputs that can generate dynamic representations of various possible solutions. In addition, the parametric model is also able to explore variations of the design that go beyond the general rules that apply, especially for architects who want to experiment with new possibilities. Although zoning regulations differ from one city to another, zoning and the LOS terminology used for this study were consistent enough to be used with this parametric model.

REFERENCES

Directorate General of Spatial Planning Jakarta (1999), Regional Spatial Plan of Jakarta, Capital Regional Regulation No. 6.

Jakarta Governor Decree No. 678 of 1994, on Improving Building intensity in the Jakarta region.

Monchaux, Nicholas, Shivang Patwa, Benjamin Golder, Sara Jensen, David Lung (2010). Local Code: The Critical Use of Geographic Information Systems in Parametric Urban Design. Proceedings of ACADIA Life In: Formation, NY, NY.

PT. Bina Karya (1997), Indonesian Highway Capacity Manual, Bina Karya Press.

Stover, Koepke (1983), *Traffic access and impact studies for site development: a recommended practice*, Institute of Transportation Engineers, Transportation Planners Council.

Tamin, Ofyar Z. (2000), Perencanaan dan pemodelan transportasi, ITB Publishing, Bandung.

Advances in Civil Engineering and Building Materials IV – Chang et al (eds)
© 2015 Taylor & Francis Group, London, ISBN: 978-1-138-00088-9

Experimental studies of the outlet tube of a pump power plant model with a horizontal capsular unit

S.V. Artsiamchuk
Belorussian State Agrarian Technical University, Minsk, Belarus

G.I. Sidorenko
St. Petersburg State Polytechnical University, St. Petersburg, Russia

ABSTRACT: This article describes the results of experimental studies of the outlet tube of a pump power plant model with a horizontal capsular unit. Straight and bent outlet tubes were investigated. The energy performance of the pump power plant unit with the outlet tube of straight and bent types at the diffuser transition to round, square, and rectangular cross-sections were obtained.

Keywords: pump power plant, horizontal capsular unit, outlet tube, experimental studies, energy performance

1   INTRODUCTION

In connection with the increased capacity of hydropower plants, increased demands are made on the effectiveness of individual elements of feeder tracts and, in particular, on the outlet tube of pump power plants with horizontal capsular units. For low-pressure hydropower settings with horizontal units, the cost of the outlet tube construction is up to 30% of the construction works cost (Clifford 1981, Kline, Abbott & Fox 1959).

The outlet tube, being one of the main elements of the feeder tract, largely determines the technical and economic readings of the pump power plant (Artsiamchuk 1985).

It is designed for the following technological functions:

1. The conversion of kinetic flow energy into the pressure energy with minimal hydraulic losses in the outlet tube.
2. The formation of the flow pattern at the exit of the pump impeller.
3. Overflow and lock-up device of the pump power plant unit.

To perform these functions the most effective elements of the outlet tube were determined: variable guide vanes, the output stator, the outlet tube diffuser, and the transition tube of the diffuser with the channel. Based on analysis of the energy–loss distribution in the flow path of hydraulic machines, it is known that there is great power waste in the outlet tubes (Zubarev 1977, Gubin 1970, Varlamov 1976).

On the other hand, the level of the hydraulic losses in the elements of the feeder tract largely determines the energy performance of a low-pressure hydroelectric generator block of a pump power plant with horizontal capsular unit.

Analysis of the ways to improve axial flow pumps shows that the possibility of increasing efficiency by improving the system blade impeller and straightening machine is relatively remote (Zubareb 1977, Varlamov & Yablonskiy 1977).

At the same time, the outlet tubes, which are associated with intense flow separation irreversible losses of energy, have considerable reserves for increasing the efficiency of aggregation block aggregates horizontal capsular units (Zubarev 1977).

The relatively large velocity head of the impeller axial pump and the expansion of the flow in the outlet tube, cause the largest hydraulic energy loss. The influence of export on the position of the discharge point of maximum efficiency of the performance of the axial pump was also established (Bogdanovskiy 1958).

Size and shape of the outlet tube exercise a significant influence on the energy characteristics of a pumping station unit. In some cases, the increase in the overall dimensions of the outlet tube is accompanied by an increase in capital investments for construction and assembly work on the pump power plant building with a horizontal capsule unit. Hence, there is an optimal variant of the outlet tube, in which the estimated expenditure of the pump power plant unit will be minimal.

Layout solutions with a straight location of the feeder tract received the greatest distribution in the Tidal Power Plant (TPP) and the Hydro Power Plant (HPP) with capsular units, as this reduced the flow resistance in the hydroelectric generator block. Reduction of hydraulic losses can be achieved by providing

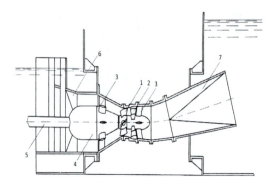

Figure 1. A pump power plant model with a horizontal capsular unit (1–impeller, 2–straightening machine, 3–stator, 4–streamlined capsule, 5–shaft housing, 6–closed water-supply chamber, 7–outlet tube diffuser).

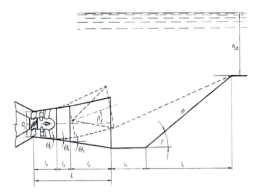

Figure 2. Geometric parameters of the outlet unit.

optimum flow conditions at the inlet of the diffuser (Gubin 1970, Purdy 1979, Neve & Wirasinghe 1978, Klein 1981, McDonald & Fox 1966, Zuykov 2010, Kharkov 2010). In some cases, outlet tubes with small angles of axle lift of the deflection diffuser can be used; this will reduce the amount of excavation and concrete work on the underground part of the pump power plant building (Rebernik 1974). In this connection, two types of outlet tubes were investigated – straight and bent. Figure 1 shows a pump power plant block model with a horizontal capsular unit.

2 STATEMENT OF THE PROBLEM

2.1 Optimization task

The study was conducted in order to construct a computer model to optimize the parameters of the outlet tubes of large pumping stations with horizontal capsular units. This model is a part of the general method of optimizing the parameters of pumping stations with water supply systems.

To construct a model for optimizing the shape of the outlet tube, the main parameters were identified; their optimization should be carried out in conjunction (Clifford 1981). These are the geometrical parameters and the type of the outlet tube. Geometrical parameters: the length of the outlet tube elements (l_0, l_K, l_D, l_V) and the opening angle of the outlet tube elements (θ_0, θ_K, θ_D). The geometric parameters are shown in Figure 2.

At selected geometric parameters and with a certain type of outlet tube, the dependence of the change of the cross-sectional area along the length F(l) is uniquely determined. These parameters enable us to give a complete mathematical description of the outlet tube.

2.2 Experimental set-up and models of outlet tubes

Studies of the energy and hydraulic characteristics of the pump station block model with an impeller diameter, $D_1 = 0.35$ m, were performed in the laboratory of the Department of Hydropower use of

Figure 3. Photo of the experimental stand.

water power on a special stand (Vissarionov, Belyaev & Elistratov 1984) (Figure 3).

Simulation is carried out in compliance with the conditions, Fr = idem, in the area of self-similarity at numbers, Re > $5 \cdot 10^5$. Energy and hydraulic studies of the outlet tube of the pump station block with the capsule unit were made in accordance with international guidelines for model tests and evaluations of the results (Mezdunarodniy kod ... 1974). The construction of the working energy performance of the pump power plant unit was made for fixed angles of blade setting of the impeller and straightening machine.

Two types of outlet tubes were investigated – straight and bent models are given in Figure 4.

The following models of outlet tubes were studied: straight divergent cone of circular section; with the transition to a square section; with the transition to a rectangular section. We have also investigated the models of outlet units with corners lift diffuser $\beta = 0°$, 7°, 15°, 25°. Experimental studies were conducted at a constant shape and size of the flow part.

3 RESULTS OF ENERGY RESEARCH

As a result of experimental studies, the dependence of the forms, (1) and (2), has been constructed:

$$\eta = f_\eta(T, \bar{l}, \beta, \theta, K_Q), \qquad (1)$$

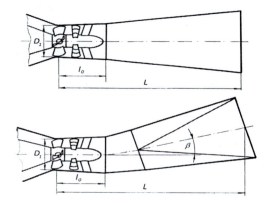

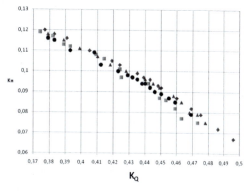

Figure 4. Investigated types of outlet tubes: straight; bent.

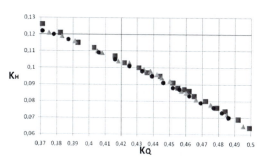

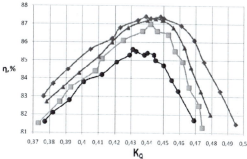

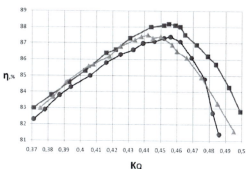

Figure 5. The performance of the pump power plant unit with the outlet tubes of straight type $(L = 4.5D_1,$ $F_{outlet}/F_{inlet} = 4)$. Outlet tubes with: —■— diffusers of circular section; —▲— diffuser transitions to square; —•— rectangular cross section formed on the length 2.5 D_1.

$$K_H = f_H(T, \bar{l}, \beta, \theta, K_\varrho), \qquad (2)$$

Model (1) and (2) are based on the optimization of the outlet tubes of pumping stations with horizontal capsular units. Some simplified dependences (1) and (2) are shown in Figure 5 and Figure 6. These are the results of experimental studies of the outlet tubes, straight and crank types, which have been constructed in dimensionless coordinates of pressure K_H and K_Q:

$$K_H = \frac{H}{n^2 \cdot D_1^2}, \qquad (3)$$

Figure 6. The performance of the unit of the pump power plant with the crank type outlet tubes $(L = 4.5D_1,$ $F_{outlet}/F_{inlet} = 4)$, with the transition to a square section at the output at different angles of climb β: —◆— $\beta = 0°$, —▲— $\beta = 7°30$, —■— $\beta = 15°$, —•— $\beta = 25°$.

$$K_\varrho = \frac{Q}{n \cdot D_1^3}. \qquad (4)$$

4 CONCLUSIONS

1. It is an established fact that the size and shape of the flow part of the outlet tube in the feeding tract of the low-pressure pump power plant have a profound influence on the energy characteristics of the pumping station unit. Therefore, the effectiveness of the pump power plant with the capsular units is intimately associated with the determination of the optimum shape and dimensions of the outlet tube.

2. It was established experimentally that for the outlet tubes of a straight type with a diffuser of a circular section, maxima of efficiency occurs when the length of the outlet tube is equal to 5D_1 and ratio $F_{outlet}/F_{inlet} = 4$. The outlet tubes with diffuser transitions to square and rectangular cross-sections formed on the length 2.5D_1 of the output section with a total length $(L = 4.5D_1)$, reduce the maxima of the efficiency of the pump power plant unit, respectively to 0.7 and 0.9 % in comparison with the diffusers of circular section.

3. The use of the outlet tube of a cranked type with an angle of ascent of the diffuser axis at 15°, starting at a distance 2D_1 of the impeller axis of the pump,

473

reduces the value of efficiency to 0.5% in the zone of the optimum characteristics in comparison with the outlet tube of the straight type (for $L = 4.5D_1$, $F_{outlet}/F_{inlet} = 4$).

REFERENCES

Artsiamchuk, S.V. 1985. Obosnovanie parametrov otvodyashchikh ustroystv krupnykh nasosnykh stantsyy s gorizontalnymi kapsulnymi agregatami. Avtoreferat dissertatsii na soiskanie uchenoy stepeni kandidata tekhnicheskikh nauk, SPb, LPI, 1985. (rus).

Bogdanovskiy, V.I., 1958, Issledovanie form podvodov i otvodov osevykh nasosov. Trudy BNIIGidromash, Vypusk 22, s. 92–113. (rus).

Clifford, A. 1981. Intakes and Outlets for Low-Head Hydropower, Journal of the Hydraulics Division, ASCE, Vol. 107, No. HY9, Sept., pp. 1029–1045.

Gubin, M. F., 1970. Otsasyvayushchie truby gidroelektrostantsyy. M.: Eneriya, 270 s. (rus).

Kharkov, N. S. 2010. Poteri napora po dline v vintovom tsirkulyatsionnom potoke (oblast nizkikh zakrutok). Avtoreferat dissertatsii na soiskanie uchenoy stepeni kandidata tekhnicheskikh nauk, SPb, SPU. (rus).

Kline, S.J., Abbott, D.E., Fox, R.W. 1959. Optimum Design of Straight-Walled Diffusers. – Journal of Basic Engineering ASME 81, Sept., pp. 321–331.

Klein, A, 1981. Review Effects of Inlet Conditions on conical diffuser performance, ASME, Journal of Fluids Engineering, Vol. 103, June, pp. 250–257.

McDonald, A.T., Fox, R.W. 1966. An experimental investigation of incompressible flow in conical diffusers, International Journal of Mechanical Sciences, Vol. 8, No. 2, pp. 125–139.

Mezhdunarodnyy kod modelnykh priemno-sdatochnykh ispytaniy gidroturbin. Publikatsiya MEK No. 193. Perevod s angl. L: ONTI TSKTI, 1967. Publikatsiya MEK No. 193A. Perevod s angl. L: ONTI TSKTI, 1974. (rus).

Neve, R.S., Wirasinghe, N.E.A, 1978. Changes in conical diffusers performance by swirl addition, The Aeronautical Quarterly, August, pp. 131–143.

Purdy, C.C. 1979, Energy Losses at draft tube exists and in penstoks, Water Power and Dam Construction. October.

Rebernik, B. 1974. Hydraulic Investigation on Water Intakes For Pump Storage Plants, International Association For Hydraulic Research. 7th. Symp., Vienna, Trans. Part 1, s.1., s.a. 14/1–14/12.

Varlamov, A.A., 1976. Vybor optimalnykh razmerov diffuzornykh kanalov gidromashyn. M. Energomashinostroenie, No. 5, s. 11–13. (rus).

Varlamov, A.A., Yablonskiy, G.A., 1977. O razrabotke nasosov vysokoy proizvoditelnosti dlya trass perebroski stoka severnykh rek v yuzhnye rayony strany. M. Energomashinostroenie, No. 2, pp. 3–6. (rus).

Vissarionov, V.I., Belyaev S.G., Elistratov V.V., 1984. Laboratornye ustanovki dlya kompleksnykh issledovaniy nasosnykh stantsiy, prednaznachennykh dlya raboty v rezhime gidroakkumulirovaniya, Trudy Leningradskogo politekhnicheskogo instituta, No. 401, s. 60–65. (rus).

Zubarev, N.I., 1977. O gidravlicheskikh poteryakh v râzlichnykh komponovkakh bloka krupnogo osevogo nasosa. M. Energomashinostroenie, No. 5, s. 17–19. (rus).

Zuykov, A.V. 2010. Dinamika vyazkikh tsirkulyatsionnykh techeniy v trubakh i poverkhnostnykh voronkakh. Avtoreferat dissertatsii na soiskanie uchenoy stepeni doktora tekhnicheskikh nauk, M (rus).

Author index